A MODERN ILLUSTRATED MILITARY HISTORY

AIR
POWER

# A MODERN ILLUSTRATED MILITARY HISTORY

# AIR POWER

 Phoebus

Material previously published in *Fighters 1914–1939*
© 1978 Phoebus Publishing Co/BPC Publishing Ltd;
*Fighters 1939–1945* © 1978 Phoebus Publishing Co/
BPC Publishing Ltd; *Jet Fighters* © 1975 Phoebus
Publishing Co/BPC Publishing Ltd; *Bombers 1914–1939*
© 1974 Phoebus Publishing Co/BPC Publishing Ltd;
*Bombers 1939–1945* © 1974 Phoebus Publishing Co/
BPC Publishing Ltd and *Jet Bombers* © 1976 Phoebus
Publishing Co/BPC Publishing Ltd

Made and printed in Great Britain by
Redwood Burn Limited

ISBN 0 7026 0047 4

**JOHN BATCHELOR**, after serving in the RAF, worked
in the technical publications departments of several
British aircraft firms, and went on to contribute on a free-
lance basis to many technical magazines. Since then, his
work for Purnell's Histories of the World Wars, and
subsequently the Purnell's World War Specials, has established him as one of the most outstanding artists in his field.

**BILL GUNSTON** is an aviation historian who has spent
many years flying and writing about all types of aircraft.
With experience as a flying instructor behind him, he joined
the editorial staff of *Flight International* in 1951, and was
appointed Technical Editor in April 1955. In 1964 he
moved on to become the Technology Editor of *Science
Journal* until in December 1970 he became a freelance
writer and editor. As well as countless articles he has written
many books for aerospace publishers throughout the English-
speaking world. He is Assistant Compiler of *Jane's All the
World's Aircraft*, and European Editor of the Australian
journal *Aircraft*.

**DAVID A. ANDERTON** is a writer and photographer
who specialises in aviation and space subjects, and who
has written extensively in both areas for more than 25 years.
He graduated as an aeronautical engineer, worked on a
variety of aircraft and projects at Grumman, and on
missile design and development at General Electric. He
joined *Aviation Week* as an Associate Editor, was the first
European Editor of the magazine and was then appointed
Technical Editor before leaving in 1963 to freelance. He
has followed technical developments in aircraft design for
many years, and is well known for his published work in a
large number of aviation journals and in many specialised
educational publications for the National Aeronautics and
Space Administration.

**BRYAN COOPER** started his career as a journalist and,
during service with the RAF, as a feature writer on *Flying
Review*. He is the author of a number of books on military
subjects, including *The Ironclads of Cambria*, *Battle of the
Torpedo Boats*, *The Buccaneers*, *Tank Battles of the Great
War* and *Fighter*.

# INTRODUCTION

From the frail biplanes spotting for rival guns in the First World War to the most modern jet aircraft, this book tells the story of Air Power. John Batchelor, who takes every opportunity to fly, shoot with, or climb over and photograph any piece of military equipment he can find, has illustrated the book with hundreds of dazzling full colour drawings.

The first section describes the fighter, whose role demands that it should fly higher than anything else. This principle remains constant, from 1915 to today, only the technology has changed. Here we present a fascinating description of the development of fighter aircraft from the first experiments in the earliest days of the First World War, through the trials of

four years of warfare and the period of development between the Wars, to the Second World War when fighter performance, engines and armament were tested in combat. We also cover the development of the jet fighter from the early days of Heinkel and Whittle, through the Second World War and all the wars since then, up to the present day and beyond.

The second section traces the development of the bomber from its earliest beginnings during the First World War, when pilots dropped steel darts, artillery shells and cans of kerosene, through the heavy bombers of the interwar years, to the production of jet bombers, both subsonic and supersonic, during and after the Second World War.

*A Fokker E.III Eindekker (monoplane), the first production fighter to be armed with synchronized machine-guns and perpetrator of the 'Fokker scourge' of 1915*

A Gotha G.V, one of the main German heavy bombers of the First World War

# CONTENTS

*Royal Flying Corps pilot in the cockpit of his Nieuport 17. The Lewis gun mounted above the top wing to clear the propeller arc was fired by cable from within the cockpit and the whole gun swung down for ammunition drums to be changed*

# FIGHTERS 1914–1939
## *Bill Gunston*

When Europe mobilised for war in August 1914, powered flight was only eleven years old. In a matter of months the Western Front turned to stalemate and the pattern of fighting in the air above the trench-lines was set — unarmed two-seat reconnaissance aircraft spotting for the guns, the massed firepower directed by their reports.

With offensive armament of duck-guns or rifles, mechanical failure was more of a hazard for the first warplanes than enemy action. When the first synchronised machine-gun was taken aloft, however, a new breed of aircraft was born. It was the single-seat fighter designed to deny air-space to the enemy.

# CONTENTS

*US naval fighters of the 1930s wore some of the most dazzling colour schemes of all time. This Boeing F4B has been immaculately restored at Chino, its paintwork shimmering in the Californian sun*

John Batchelor

# A CONCEPT EMERGES

On 5 October, 1914, a Voisin LA, called a Type III by the French *Aviation Militaire,* was droning lustily over a Western Front that had yet to degenerate into the bloody stalemate of trench warfare. The Voisin was a biplane pusher, with a strong steel framework but not possessed of a sparkling performance. The engine was put at the back of the nacelle so that the crew, comprising a pilot and an observer in the extreme nose, should have a better view. The Voisin's role was reconnaissance, though by October 1914 it had become the custom to carry small bombs in the front cockpit and throw them over the side in the vague hope of hitting suitable targets. But on the whole the business of reconnaissance was routine, and it was also practised by the hated Boche.

Suddenly the observer Louis Quénault spotted an enemy aircraft, an Aviatik with the distinctive black crosses. His pilot, Joseph Franz, opened the throttle and they closed the range. It was Quénault's big moment. Pivoted to the edge of his cockpit was a Hotchkiss machine gun, and as the Aviatik came within range he opened up with clip after clip of ammunition. Soon the Aviatik was spinning down out of control, to crash in a pyre of smoke. Eventually the smoke had dispersed and the little incident was but a memory to those who had witnessed it. But its ripples spread and multiplied, and reverberated in a thousand offices, parliaments, factories and front-line mess-tents. It was precisely what a very few people – virtually all of them not 'qualified' to express an opinion – had been predicting for four or five years. Air warfare was a concept so radically new that the official military mind had found it singularly hard to grasp.

By 1909 the flying machine was no longer part of mythology, laughed to scorn by newspaper editors and the general public alike. In the summer of that year vast crowds had flocked to Reims to watch the world's first great aviation meeting. Blériot boldly flew across the Channel to England. Nobody could scornfully claim that these things were delusions or fakes. The aeroplane had unquestionably arrived, and – though a few claimed that, like the horseless carriage, it was a mere flash in the pan, doomed soon to fade from the scene – most people thought it more likely that each new year would bring more and better aeroplanes capable of doing useful jobs. Visionaries published stories and sketches of future aerial travel – and future aerial warfare.

*German Army officers experiment with an Otto, 1913. In the earliest days of military flying the observer was the officer and the pilot or 'chauffeur' was the NCO*

Nor was the concept of air warfare confined to paper. The US Army was widely chastised for its sluggish reaction to the achievements of the Wright Brothers, but in fact it was years ahead of all other military services in forming an embryo 'air force' (the Aeronautical Section of the Signal Corps in 1907) and issuing a detailed specification for a military aircraft, in the same year. The US Army was not concerned with particular missions; it rightly judged that the ability to fly was at that time a worthwhile end in itself. But the enthusiasm of its young officers ran somewhat ahead of the old staff officers in Washington – whom the young pilots scathingly called 'mossbacks' – and, without any authority from above, experiments began in 1910 to see what warlike things could be done in the air. In August of that year Lt Jake Fickel, of the 29th Infantry at Governor's

*Maurice Farman Longhorn, the almost ubiquitous training and reconnaissance machine of the first Allied airmen*

*Albatros L.1, B-type reconnaissance biplane – like the Farman an easy prey for an armed aircraft*

*Albatros-built* Taube *– the 'dove'-like monoplane built by several German and Austrian manufacturers*

Island, New York, went to the public Air Meet at Sheepshead Bay. There he persuaded two famous aviators, Glenn Curtiss and Charles Willard, to let him try air-to-ground firing with a rifle. The civilian pilots had been apprehensive about the recoil, and also pointed out that, as there was no proper seat, their previous passengers had needed both hands to hold on with. But Fickel set up a ground target and, on many runs, scored many hits.

This was nothing more than a rather dangerous escapade by an enthusiast. The only official report, apart from word-of-mouth, was in a letter by Fickel to brother-officer Henry Arnold, who later became chief of the whole US Army Air Corps. Fickel commented that 'you have to allow for the motion of the plane'. It was easiest, he said, when you were in a direct line with the target so that you passed overhead. Shooting to either side was difficult. Fickel had tried in practice what would later be called deflection shooting. How to aim a gun was to be central to almost every part of air combat right up to the present day. There was just one time, in the 1950s, when it seemed likely that the ability to shoot straight would no longer be important to the fighter pilot. A combination of radar, autopilot and rockets appeared to do the job automatically, and later the guided missile seemed to render the gun obsolete. Today we know better, and to a considerable degree the top-scoring fighter pilots of any future conflicts will still be the good shots.

In August 1910 two other US Army officers did something which rather put poor Fickel in the shade. Lt Paul Beck and 2nd Lt Myron T Crissy went to a shop in San Francisco, bought some 2½-in pipe and made the world's first aerial bomb. They also put fins on a live 3-in artillery shell. Then they flew over a big Air Meet at San Francisco's Tanforan racecourse and startled – if not physically endangered – the crowds with the world's first live aerial bombing raid. The bombs worked, exploding with very satisfying bangs.

By this time developments were taking place in aerial communications. The US Army experimented with an air-to-ground Morse system, with a wing-mounted device rather like a modernised Red Indian smoke-signal generator. Compressed air was released by a pilot-actuated valve to blow past a can of lamp-black, emitting short blobs of black soot for dots and long ones for dashes – provided the turbulent slipstream did not disperse the signal before they could be read. Far more significant was the pioneer experiment of Canadian J A D McCurdy, at the 1910 Sheepshead Bay meet: he had successfully transmitted messages by air-to-ground radio.

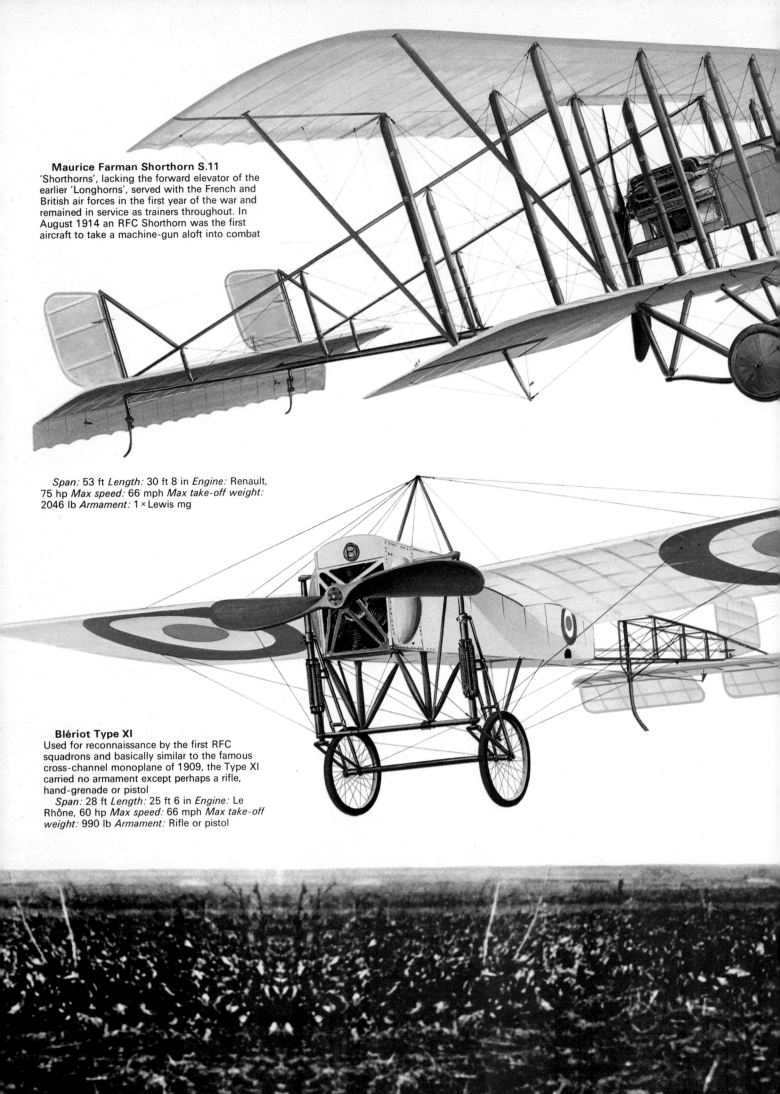

**Maurice Farman Shorthorn S.11**
'Shorthorns', lacking the forward elevator of the
earlier 'Longhorns', served with the French and
British air forces in the first year of the war and
remained in service as trainers throughout. In
August 1914 an RFC Shorthorn was the first
aircraft to take a machine-gun aloft into combat

*Span:* 53 ft *Length:* 30 ft 8 in *Engine:* Renault,
75 hp *Max speed:* 66 mph *Max take-off weight:*
2046 lb *Armament:* 1 × Lewis mg

**Blériot Type XI**
Used for reconnaissance by the first RFC
squadrons and basically similar to the famous
cross-channel monoplane of 1909, the Type XI
carried no armament except perhaps a rifle,
hand-grenade or pistol
    *Span:* 28 ft *Length:* 25 ft 6 in *Engine:* Le
Rhône, 60 hp *Max speed:* 66 mph *Max take-off
weight:* 990 lb *Armament:* Rifle or pistol

Imperial War Museum

*Early arrivals to organised air warfare – men of No 1 Squadron RNAS pose around their airfield klaxon, late 1914*

*Unarmed Albatros takes off from a Flanders airfield. Any aircraft that might shoot it down, even with a 12-bore duck-gun, would be amongst the first fighters*

In 1911 a former US Army officer, Riley E Scott, arrived at the world's first military air base at College Park, Maryland, with the world's first bombing system. It comprised a frame designed to be readily attached to a Wright or Curtiss, on which could be hung two aerial bombs. The frame carried an inclined telescope, with a graticule marked on one of the lenses, and a fixed sheet of numbers which related the passage of the target down the telescope scale to the speed and altitude of the aircraft. Probably the chief problem was that there were no reliable airspeed indicators or altimeters, and estimates were inaccurate. But the Riley Scott bombsight was at least a workable device which, once it had been calibrated by trial and error by a pilot used to judging speed and height, gave

encouraging results. Riley Scott took it to Europe, and at Villacoublay won the Michelin Prize of $5000 by extremely accurate bombing of marked-out ground targets from heights of 200 and 800 metres. Later the Riley Scott was the basis for the first production bombsight.

In 1911 war flared up between Turkey and Italy in Tripolitania. It was the first opportunity for aircraft to demonstrate whether or not they were of military value. The Italians had several aircraft in North Africa, and on 22 October, Capt Piazza took off in a Blériot on the world's first combat mission. He landed after having spent a full hour studying the Turkish positions, and his report was extremely detailed. Unlike the observer in a balloon he was able to fly over and around the

enemy, and the picture he gathered was as full as one could wish. Further reconnaissance missions followed, and on 1 November, Lt Gavotti flew over the Turkish camp at 2500 ft and dropped four bombs by hand. Though he had no Riley Scott sight, he caused casualties and consternation, and provoked an angry reply alleging a criminal act. In 1912 there was plenty of air activity in the Balkan War. Turkish-held Adrianople was bombed several times, and the air-minded Bulgars not only had pilots of their own but hired mercenaries with their own aircraft. Sakoff, a Russian, appears to have been the first bomber pilot to return to base with bullet holes in his machine. One mercenary, the American Bert Hall, knew of the Riley Scott sight and went into action with a locally made copy. He did much better than the Russian aviators who merely hung their 22-lb bombs from their feet with a slip knot and released them with a sharp kick! Later one of the Russians, Kolchin, was shot down and killed by Turkish rifle fire. He was the first aviator to be killed in air warfare.

Britain, France and Germany all had embryonic air forces by 1910, and in 1911 officers in all three countries had practised firing rifles in the air. The Briton, Capt Brooke-Popham of the Air Battalion, Royal Engineers, was promptly ordered to cease such a practice, but about 18 months later official sanction was given for him, as a major commanding No 3 Sqn of the new Royal Flying Corps, to practice shooting at kites from a Henry Farman. The hazards to people on the ground are obvious, and instead of being a serious matter of training and research the firing of guns in the air seems to have been regarded as sport for senior officers. It required that the marksman should take both hands off the control column for several seconds at a time, so in rough air it was a tricky business. Pilots kept personal score by putting a dab of coloured paint on the bullets taken aloft, which left a ring round any hole in the kite. Such a practice never became official.

### Active Combat
In the United States Col Isaac N Lewis was finding no success in marketing the outstanding machine gun, designed around the turn of the century by Samuel McClean, which he had improved and which now bore his name. The Lewis gun was in fact the best in the world in 1912 for aerial use, because it was light and compact, had a high rate of fire, was reliable, and was fed by a handy drum of 47 or 97 rounds. In June 1912 Lewis visited College Park and got one of his guns mounted on a US Army Wright. Lt Tom Milling did the flying, and the CO, Capt Charles de Forest Chandler, blasted away with the new gun at a ground target made of cheesecloth. The results were excitingly good, and the gun made a most favourable impression. But the 'mossbacks' in the War Department insisted that there was absolutely no question of the aeroplane ever carrying guns and engaging in active combat. In any case, even if it ever did, the standard Army machine gun was the Benet-Mercier, not the Lewis. Unfortunately when the Benet-Mercier was mounted on a Wright it got in the way of the pilot's control column, because it had an 18-in ammunition feed sticking out on one side and an equally long ejection chute on the other.

Other experiments were going on in many countries to see how useful aeroplanes, and

airships, might be as carriers of guns and droppers of bombs. One of the longest-established air units was the airship (*Luftschiff*) branch of the Imperial German Navy. This used the well-tried Zeppelin airships, which, compared with early aeroplanes, carried much heavier loads and could fly further. Though still frail and extremely vulnerable in severe weather, the Navy Zeppelins were manned by courageous and efficient crews and directed by officers who never doubted that in wartime their mission would be an active bomb-dropping one, as well as ocean reconnaissance. By 1912 they were possibly the first air service in the world to indulge in bomb-dropping practice as a routine matter of training, and their effectiveness became considerable because they dropped heavy bombs with increasing accuracy. The German Army, which used Zeppelins and Schütte-Lanz ships, saw its role mainly as front-line reconnaissance and possibly bombing in support of its ground troops, and learned the hard way that airships do not survive long over battlefields.

When The First World War began on 4 August, 1914, all these things were public knowledge. Altogether it added up to a substantial body of experience of air warfare, yet to most of the general staffs and members of governments air warfare was still something written about by semi-lunatics. Almost universally the official

### Morane-Saulnier Type L
The Type L is probably the most famous parasol monoplane of its time, and was the first to carry a machine-gun firing forward through the propeller disc. Steel wedges deflected bullets that hit the blades, and this innovation was discovered by the Germans when a Type L flown by Roland Garros was shot down at Courtrai in April 1915. Within weeks the Germans had deployed synchronised machine-guns as a response

*Span:* 36 ft 9 in *Length:* 22 ft 6¾ in *Engine:* Gnome seven-cylinder or Le Rhône 9C nine-cylinder air-cooled rotary, 80 hp *Max speed:* 72 mph at 6500 ft *Max take-off weight:* 1440 lb *Armament:* 1 × 8-mm Hotchkiss or 0.303-in Lewis mg

view was that the aeroplane might have some use as a reconnaissance vehicle, because it might be able to have a closer or more varied look at the enemy than would be possible with a tethered balloon. Even this was a reluctant admission, and many officers – who, of course, were all officers in an established army or navy, because no

*The Royal Flying Corps, established in 1912, began work with a series of Royal Aircraft Factory products including the B.E.8*

*An Albatros B-type was still enough of a novelty to attract the curious gaze of German officers*

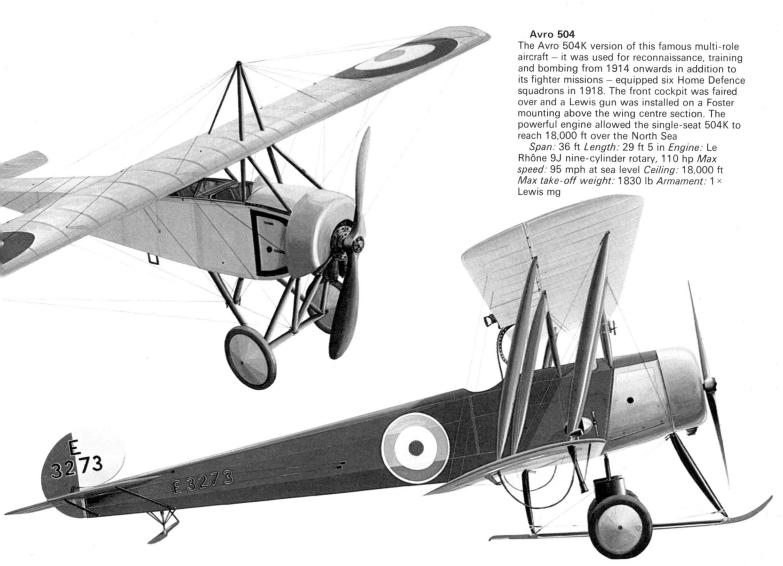

**Avro 504**
The Avro 504K version of this famous multi-role aircraft – it was used for reconnaissance, training and bombing from 1914 onwards in addition to its fighter missions – equipped six Home Defence squadrons in 1918. The front cockpit was faired over and a Lewis gun was installed on a Foster mounting above the wing centre section. The powerful engine allowed the single-seat 504K to reach 18,000 ft over the North Sea
*Span:* 36 ft *Length:* 29 ft 5 in *Engine:* Le Rhône 9J nine-cylinder rotary, 110 hp *Max speed:* 95 mph at sea level *Ceiling:* 18,000 ft *Max take-off weight:* 1830 lb *Armament:* 1 × Lewis mg

such thing as an air force existed – regarded the aeroplane as being similar to the submarine as something that no gentleman would wish to be associated with. This was especially the case in Britain, where the Air Battalion, and from 1912 the Royal Flying Corps, had had to get along on budgets so meagre that its handful of enthusiastic officers had often met official bills out of their own pockets. Even in Germany, where military aviation funding was at least twelve times greater, there was no official grasp of the concept of air warfare whatever.

Even when one makes allowance for the fact that in the years before 1914 elderly men had not yet got used to a world of rapid technological change, the failure to plan for an aerial war is difficult to comprehend. Today the subject of forecasting – not just the weather but the future of almost every human activity – is a highly developed art, science and big business. People of all kinds, and certainly people in military uniform, need to know what is probably, or even possibly, going to happen in the years ahead. Had there been any forecasters before the First World War, other than the writers of fiction, air warfare would have immediately been identified in stark reality.

In the absence of knowledge of their motives it would be wrong to pour scorn on the important people of the pre-1914 era who dismissed the whole notion of air warfare as ridiculous. It may be that their minds could not accept so new an idea. It may be that they had watched the flying of the early aviators and, rightly judging that some could only just get off the ground,

were short-sighted enough to conclude that the aeroplane would never be able to carry a gun or a bomb. Perhaps they considered the aeroplane would always be a rare and exotic species, and that machines belonging to opposing sides would be unlikely to meet. Yet, no matter how one tries to view things through pre-1914 eyes, the absence of any official action to plan for aerial warfare remains an enigma.

It was especially surprising in Britain. The British had for generations felt totally secure in their island, guarded by the unchallenged might of the Royal Navy.

When Blériot landed his frail flying machine in the grounds of Dover Castle in 1909 it was a profound shock. 'No Longer an Island' proclaimed the newspaper headlines, and more than any other event this caused the British to look at their own non-performance in aviation. Three years later an ominous throbbing along the Thames heralded the visit of a large airship, whose lights were seen by many Britons. Later the Admiralty admitted it was 'not one of our ships'. It could only be a Zeppelin, and Zeppelins could do a lot of damage. Yet not one anti-aircraft gun was ordered, and not

*Mechanical failure was a greater danger than enemy action in the earliest days, as this crashed Farman testifies*

*A B.E.2c displays one of the earliest forms of British marking, a simple red disc on a white background on the upper wing with a Union Jack on the tail*

one aeroplane was bought with a gun. No action whatsoever was taken to provide any form of air defence.

Nine months earlier, when the Royal Flying Corps was formed, it was announced that its strength was to be '131 aeroplanes', but in December 1912, two months after the Zeppelin's nocturnal visit, the Secretary of State for War admitted the RFC possessed 14 aeroplanes, of which three were under repair. In August of that year the War Office had organised a Military Aircraft Trials competition to decide upon the best aeroplane for the RFC. Far and away the best was the BE.2 designed and built at the government's own Royal Aircraft Factory at Farnborough; but the RAF (factory) had no authority to build aircraft and its superintendent was one of the judges, so the winner had to be an obsolete and un-suitable machine (Cody's large biplane nicknamed 'The Cathedral'). Even in 1912 this was outclassed and had no chance of flying useful military missions; but the very idea of a 'military mission' was thought not to exist, despite all the shooting and bomb-ing that had already been done by a few enthusiasts.

When war did finally break out the puny Royal Flying Corps was hastily ordered to despatch aircraft to France. What was to happen if they should meet a Zeppelin coming the other way? This was a fair enough question, and a prospect which could have been discussed and armed against at any time in the previous two years. The answer was that, as there was no other method, they should attempt to destroy the Zeppelin by ramming it. Admit-tedly at this time RFC machines were carrying out the first operational patrols over the Thames estuary with observers carrying rifles, and canvas bandoliers con-taining 50 rounds of ammunition. But as 'fighter' aircraft such machines were un-impressive; one could have done better in 1911, and a quick answer was needed.

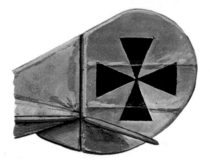

**Germany**

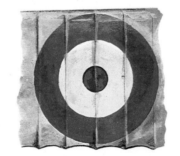

**France**

**Britain**

**National Markings: 1914**

Up to this point nobody had apparently thought about the problem of aircraft recognition until inability to tell one machine from another had already caused trouble. The first aeroplane buffs, many of whom were schoolboys, could tell at a distance not only what type an aeroplane was but which particular machine it was, and probably who owned it. Such capability was not possessed by those in positions of authority, nor by the ordinary private soldier with an itchy trigger finger. One of the first RFC machines to reach France had to land at Senlis, short of its destination. As the aircraft bore no national markings the pilot was arrested, and though he protested in what must have been recog-nised as English he was locked up in the town gaol – according to published reports, for a week. It was mainly because of this comic-opera situation that the RFC and Royal Naval Air Service (RNAS) were required to display some form of national marking. France had adopted a tricolour roundel and Germany the Croix Paté – often called the Maltese cross. As a temp-orary measure the RFC sewed small Union Jacks on the fuselage or nacelle, but before long much larger 6-ft Union Jacks were painted on the wings and smaller flags on the tail. Later, in 1915, the British adopted the French roundel and rudder strips, and then reversed the colours so that the British and French machines, though obviously Allied, could be distinguished from each other. The Union Jack had, it was said, too much similarity to the German cross.

In France the problem of aircraft recog-nition was judged so important that it resulted in an across-the-board regula-tion regarding aircraft design. German aeroplanes had tractor (front-mounted) pro-pellers, so French aircraft, especially bombers, were by General Staff decree designed as pushers. By this brutish but effective rule the French, at least, hoped to avoid mistaken identity.

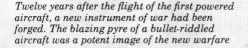

*Twelve years after the flight of the first powered aircraft, a new instrument of war had been forged. The blazing pyre of a bullet-riddled aircraft was a potent image of the new warfare*

# THE OPENING ROUNDS

Before the First World War it had been difficult enough just to build a reliable and safe flying machine. But in 1914–15 this aim was no longer enough. First, aircraft had not only to fly but to carry armament which could be used effectively. Secondly, the armament had to be developed. Thirdly, tactics had to be worked out so that the new species of combat aircraft could be used in the most efficient way.

In only one respect was the designer's task made easier. By 1914 several families of aero-engine had sufficient background of flight experience for the propulsion of the aircraft to be no longer a problem. The Central Powers – Germany and Austria-Hungary – almost completely standardized on a range of water-cooled engines with four or six cylinders in line, developed directly from pre-war car engines. The Benz, the Mercedes and the Austro-Daimler were the chief members of a closely related and highly refined family of engines which had been tempered in the fire of Grand Prix racing. They were robust and reliable, but massive. They needed heavy water cooling systems, and the radiator caused extra drag when it was simply bolted on the side of the fuselage. By late 1915 most German and Austro-Hungarian fighting scouts had the radiator recessed inside the upper wing, with a cooling-water pipe sloping up to it from the front cylinder. This reduced drag, but the water circuit was vulnerable, and if a bullet hit the radiator the pilot was drenched in near-boiling water.

The 1914 Mercedes Grand Prix engine was also used in Britain. The winner of the British Grand Prix was towed up to Derby and given to Rolls-Royce, with instructions that the famous car company should build an aero-engine, using the proven Mercedes

*The perils of actually fighting from the first military aircraft are amply illustrated in this contemporary French print of a Hotchkiss-armed Deperdussin*

technology where they thought it would help. The result was a six-in-line called the Hawk, used for blimps (non-rigid airships) and a few aircraft, and a much better engine called the Falcon with two banks of Hawk cylinders arranged in vee formation. It was to become a famed fighter engine, and the bigger V12 Eagle was one of the most powerful Allied engines used in large numbers in the war, rated at up to 375 hp. There were many satisfactory water-cooled in-line and vee engines used by the Allies, most of them of French design.

Throughout the war a very large role was also played by the rotary engines derived from the Gnome of 1907. These had broken away entirely from established automobile practice and offered aviators an engine which, for its power, was much lighter and in some ways simpler. A drawback of the Gnome was that it was machined from high-strength steel forgings, and though beautifully made it was time-consuming and expensive to construct. It also behaved as a powerful gyroscope, because the fact that the whole engine (apart from the crankshaft) spun round with the propeller was a mixed blessing. It kept the cylinders cool, but made rapid flight manoeuvres difficult, especially in small aircraft with a single large engine. As will be explained, several rotary-engined aircraft became notorious for being able to turn like lightning in one direction, while being most reluctant to turn the opposite way. This powerfully influenced the way a pilot handled his aircraft in combat, and sometimes made his actions predictable. The rotaries were also liable to fling their castor-oil lubricant from every small crack in their rotating parts, particularly from the valve gear (which differed fundamentally between types, some having a flap-valve in the piston itself). Even today one can whiff burned castor-oil near any surviving rotary, and it was an odour familiar to most fighter pilots of the First World War.

At first it did not make very much difference whether the engine was put in front of the crew, in the so-called Blériot arrangement, or behind a short nacelle, in the so-called Farman arrangement (this nomenclature was responsible for the British B.E. designation for early tractor aircraft and F.E. for early pushers). In the pre-1914 era few aircraft could exceed 70 mph and neither configuration seemed to have a marked performance advantage. The Central Powers displayed a Teutonic love of standardization and throughout the war made virtually every one of their aircraft of the tractor type. The French initially had a preponderance of pushers, because this gave the crew a good view and, by late 1914, because it made it easier to mount a machine gun. The British went into production with both tractors and pushers, but most of the earliest fighting scouts were pushers. Important examples included the Airco D.H.1 and 2, the F.E.2 and the Vickers 'Gunbus', the last two of which were designed long before the war.

The tragedy with these machines is that so much time was wasted in getting them into service. A R Low and G H Challenger designed the Vickers Type 18 'Destroyer' in early 1913, and it was exhibited at the 1913 Aero Show at Olympia with a Maxim machine-gun in the nose. The design was then refined in a succession of further prototypes until the E.F.B.5 (Experimental Fighting Biplane) was ordered for the RFC. These had a much handier gun, the new Lewis, but were not impressive performers, being just able to work up to a maximum of 70 mph and taking half an hour to climb to a ceiling of about 9000 feet. The very first 'fighter squadron' in history was RFC No 11, which arrived in France with the F.B.5 Gunbus in July 1915. The original F.E.2a was also designed in early 1913, in this case at the Royal Aircraft Factory at Farnborough. Like the Vickers it was a two-seater, the gunner (called the observer, because the intention had been to build an armed reconnaissance machine) occupying the front cockpit where his Lewis could sweep the whole sky in the forward hemisphere. No urgency appears to have been put behind development or production, and it was not until a year later, in August 1914 (after the start of the war), that 12 were ordered. These F.E.2a aircraft had a poor British engine, the 100-hp water-cooled Green, and an unimpressive performance. By 1915 the decision had been taken to fit the 120-hp Beardmore, and this resulted in a very useful machine – though still no great performer, with a maximum speed of 80 mph and ceiling of 9000 feet. Many 'Fees' had a second Lewis on a tall pillar behind the observer to cover the upper rear. Later these tough machines became night bombers and even night fighters. The original F.E. designer, Geoffrey de Havilland, produced a rather smaller fighter/reconnaissance aircraft in early 1915, the D.H.1. Powered by either the 70-hp Renault or (D.H.1a) 120-hp Beardmore, it was a very manoeuvrable machine and with the Beardmore was a much better performer than the 'Fee'; only 73 were delivered, however. The D.H.2 was a small single-seater with quite

**Vickers F.B.5 Gunbus**
The F.B.5 achieved distinction by being the aeroplane operated by the world's first single-type squadron formed specifically for air combat: No 11 Sqn, Royal Flying Corps, in July 1915. The Gunbus did not have an outstanding performance, and its engine was unreliable, but the type achieved a considerable reputation with the Germans. By the spring of 1916 the F.B.5

was outclassed by Fokker's monoplanes and began to be retired
 *Span:* 36 ft 6 in *Length:* 27 ft 2 in *Engine:* Gnome Monosoupape, 100 hp, or Clerget 9Z, 110 hp *Max speed:* 70 mph at 5000 ft *Ceiling:* 9000 ft *Max take-off weight:* 2050 lb *Armament:* 1×0.303-in Lewis or Vickers mg, occasionally with supplementary Lewis gun or rifle

**Mercedes In-line Six-cylinder Engine**
German designers concentrated on in-line engines rather than rotary or radial layouts. The Mercedes powered such aircraft as the Fokker D. VII and Albatros D.V
 *Power:* 180 hp at 1400 rpm

good performance on the 100 hp of a Gnome Monosoupape (Monosoupape = single valve, the other valve being in the crown of the piston).

Flying a D.H.2 in RFC No 5 Sqn during early trials on the Western front in July 1915 was a full-time job. Though the aircraft itself was fine, the pilot also had to aim his forward-pointing Lewis gun. Pilots argued over whether the dominant hand (in most people the right) should fly the aircraft or aim and fire the gun. In either event it left no hand free to work the throttle or do anything else, and changing ammunition drums was even more tricky. In most early D.H.2s the gun was pivoted on the left side, but many pilots (not necessarily the left-handed ones) found it better on the right. But the best answer was staring everyone in the face, and some time in February 1916 the first unit fully equipped with the production D.H.2, RFC No 24 Sqn, hit on the idea of simply fixing the gun to fire straight ahead. This for the first time gave the Allies an effective dogfighter, and by 1916 such a thing was desperately overdue.

One of the greatest puzzles of the genesis of fighter aircraft is why, when so many people had given thought to the matter of fighter armament and come up with an answer, nobody in any official position in any country took any action whatsoever. By far the biggest single advance was the machine-gun fixed to fire ahead past the blades of a tractor propeller. This could be done in several ways. The crudest was merely to fix steel deflectors on each blade at the correct radius, but this had several drawbacks. It imposed numerous severe shocks on the wooden blade, which might eventually lead to failure; the deflectors and attachment holes themselves weakened the blade and impaired its efficiency; and the ricocheting bullets posed a significant hazard. A much better solution was to link the mechanism of the gun with the engine so that it could fire only if there was no propeller blade about to pass across the line of fire. In general, blades passed ahead of the gun at a faster rate than the gun could fire bullets. A typical 1914 machine gun fired 550 rounds per minute, but a 1914 aero-engine ran at 1000–1500 rpm (which had to be multiplied by two or four depending on the number of blades on the propeller). As early as 1913 Franz Schneider of the German LVG company and Lt Poplavko of the Imperial Russian Air Service had devised and published rather crude interrupter gears and actually tested them. Schneider, at least, took out a patent. In

*Drum-fed Hotchkiss arming a Voisin*

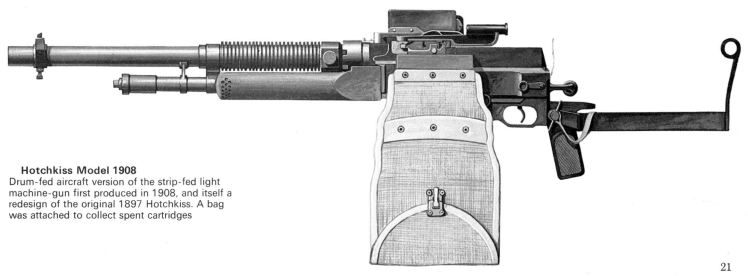

**Hotchkiss Model 1908**
Drum-fed aircraft version of the strip-fed light machine-gun first produced in 1908, and itself a redesign of the original 1897 Hotchkiss. A bag was attached to collect spent cartridges

1914 the British Edwards brothers took out a patent for a better system, and demonstrated a working model to the War Office, while in France Raymond Saulnier, of the famed Morane-Saulnier aircraft company, linked the gun trigger with a special two-lobed cam on the engine (on the crankshaft, although it was a rotary) and demonstrated it actually working.

Despite all this, there was no support whatever for any such scheme from a single official in any country. Aircraft designers, who may not have known what others had achieved, thus had to fall back on a grotesque array of aircraft configurations to try to build a fighter. At Farnborough the mass-produced B.E.2c – virtually the standard equipment of the 1914–15 RFC, but in no sense a fighter – was turned into the B.E.9 by swapping the observer and the engine. The result was a pathetic travesty of a fighter, with the observer sitting in a separate cockpit carried on struts and wires in front of the great four-blade screw. Dubbed 'The Pulpit' by RFC 16 Sqn, it was potentially lethal to its observer, who could not communicate with the pilot and would have been cut to pieces or crushed in any bad nose-over landing, such as then were frequent. An even stranger answer was found by Louis Becherau, technical director of the French Spad company, in the Spad A2. This had a front cockpit that could be bodily hinged down to permit access to the engine. When clipped in place the front gunner's nacelle was fastened by two pins engaging in trunnions on the upper wing and by a ball-race fixture on a front extension of the propeller shaft! Small numbers proved most unpopular with the French and Russians, and later there was even a more powerful series designated A3, A4 and A5, some of which had flight controls and a gun in both cockpits so that either man could fly the aircraft or fire at the Hun.

## F.E.8

Designed, like the D.H.2, to counter the Fokker monoplanes on the Western Front, the first F.E.8 flew in October 1916. It had some success but was soon outclassed by new tractor types and it was the last of the pusher scouts
*Span:* 31 ft 6 in *Length:* 23 ft 8 in *Engine:* Gnome, 100 hp *Max speed:* 94 mph *Max take-off weight:* 1346 lb *Armament:* 1 × Lewis mg

## de Havilland D.H.2

Geoffrey de Havilland's pusher fighting scout equipped the Royal Flying Corps' No 24 Sqn, which became the RFC's first single-seat fighter unit to see combat (in February 1916). Once the D.H.2's unforgiving handling characteristics had been mastered it became a useful addition to the fighter force, but the superiority of the new German Albatros and Halberstadt tractor biplanes was emphasised when the 24 Sqn commander, Maj Hawker, fell to von Richthofen's guns in November 1916

*Span:* 28 ft 3 in *Length:* 25 ft 2½ in *Engine:* Gnome Monosoupape, 100 hp, or Le Rhône 9J, 110 hp *Max speed:* 93 mph at sea level *Ceiling:* 14,000 ft *Max take-off weight:* 1440 lb *Armament:* 0.303-in Lewis mg

*One solution to the problem of firing a machine-gun in the line of flight was the pusher layout. The Lewis-armed D.H.2 could just hold its own against the Fokker E.III with its synchronised machine-gun*

The need to develop aircraft to shoot down Zeppelins led to equally strange arrangements. In Lincoln the firm of Robey & Co, which made large numbers of aircraft to others' designs, teamed with J A Peters to build two rather different prototypes of a three-seat anti-Zeppelin fighter. Basically they were conventional biplanes, of quite good design and performance; the odd part was that the armament was carried in large streamlined nacelles, each housing a gun and gunner, on the left and right upper mainplanes. Of course, thousands of combat aircraft were built with guns pivoted or fixed to the upper wing, even after methods had been found to fire safely past the propeller. Sometimes the pilot had to grip the rear of the gun and aim it up at the belly of the enemy, while he flew (probably in violent manoeuvres) with the other hand. In other cases the gun was fixed to the upper wing at an inclination that would clear the propeller. In many aircraft, such as the Sopwith Dolphin and the Parnall Scout, one or more machine guns were fixed firing up at an oblique angle from the fuselage, presaging the *Schräge Musik* type armament used by Luftwaffe night fighters from 1943. One of the best schemes was the British Foster mount, comprising a curved rail arching in a quadrant from the upper wing to the cockpit of a single seater. A Lewis gun could be pulled back down this rail until it was almost pointing straight up, with the breech right in front of the pilot. The pilot could fire it in this position, but the reason for the rail was to facilitate changing ammunition drums. The reloaded gun could then be pushed back up the rail to its normal position firing ahead, just above the tips of the propeller blades.

There were many other schemes, all intended to enable a machine-gun to be fired accurately from a fast aircraft. Dufaux in France and Gallaudet in the United States put the propeller on a hub rotating around the main structural member of the fuselage, just behind the wing, driving through gears. Thus the advantages of a pusher, with clear field of fire ahead, could be combined with a high-speed streamlined form. Of course with two engines there was no problem, and one could have machine-gunners from nose to tail. There were not many twin-engined fighters in the First World War, but the Caudron R.11 was very successful and used in large numbers, while there were numerous prototypes by other companies. A cunning arrangement by Mann & Grimmer in Britain was the single-engined twin-pusher. A 100- or 125-hp Anzani radial engine in the fuselage nose was installed back to front, driving a shaft on which were two gearwheels. These in turn drove chains turning pusher propellers carried on struts between the wings. Thus the nose was left free for a manually aimed Lewis; had the scheme been taken further, one might in a more powerful machine have had a regular battery of fixed guns firing ahead. Another scheme was to accept the tractor propeller on the nose and put the gunners on each side of it, as was done in the Armstrong Whitworth F.K.12. This strange-looking fighter had a Lewis gunner in left and right nacelles projecting well ahead of the propeller – where they could have shot at each other!

Many companies, including Nieuport in France, Lloyd and various Brandenburg C.I types in Austria-Hungary and Sage in

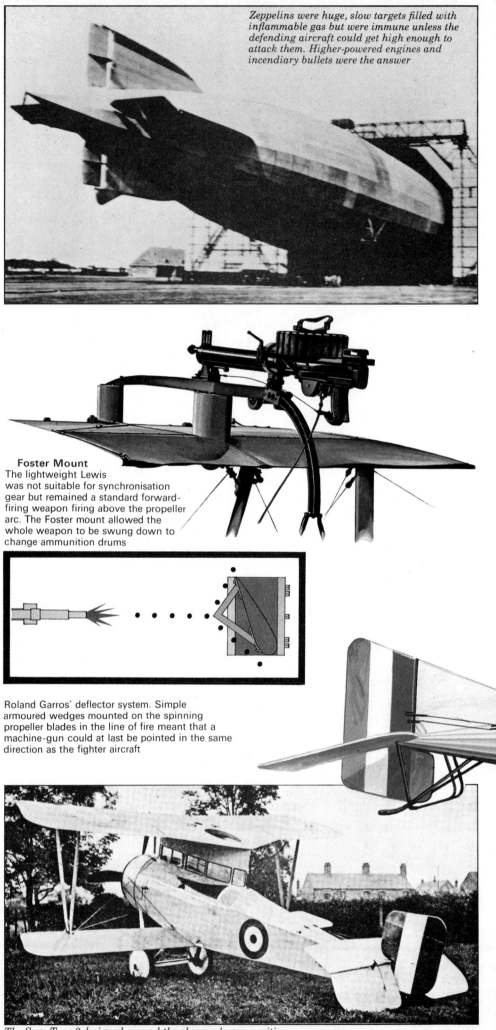

*Zeppelins were huge, slow targets filled with inflammable gas but were immune unless the defending aircraft could get high enough to attack them. Higher-powered engines and incendiary bullets were the answer*

**Foster Mount**
The lightweight Lewis was not suitable for synchronisation gear but remained a standard forward-firing weapon firing above the propeller arc. The Foster mount allowed the whole weapon to be swung down to change ammunition drums

Roland Garros' deflector system. Simple armoured wedges mounted on the spinning propeller blades in the line of fire meant that a machine-gun could at last be pointed in the same direction as the fighter aircraft

*The Sage Type 2 designed around the observer's gun position.*

*Business end of an RFC Morane, the Lewis machine-gun and 'Garros-wedges' clearly showing*

Britain, all arranged for a gunner to stand upright in a deep fairing (sometimes it was virtually a cabin) to fire a gun mounted above the upper wing, usually with the mounting having limited depression to avoid hitting the propeller. There was yet another possibility. From as early as 1912 engineers and inventors had investigated the possibility of arranging a gun to fire through the hub of the propeller. Many early fighters – including the first scheme for one of the best fighters of the entire war, the S.E.5 – were planned to have a machine-gun or even a large-calibre cannon mounted close beside or between the cylinder blocks of a geared engine, and with the barrel passing through the gearbox and hollow propeller shaft. This became common only in the years after 1918, but there was one really remarkable fighter of early 1917 that solved the problem in a novel way. The ace Charles Nungesser asked Armand Dufaux if he could build a small high-speed fighter capable of mounting a heavy-calibre cannon firing ahead. Dufaux's answer was to use two rotary engines mounted sideways in the nose driving the hollow propeller shaft through bevel gears, with the big gun down the centreline.

But all these clever and not-so-clever schemes were destined to be mere side issues. The mainstream of fighter development was to centre on the gun arranged to fire safely past the blades of a propeller, and like so many other human achievements it was ignored until it was adopted by the enemy. By the outbreak of war there cannot have been anyone interested in the subject who did not know at least some of the patents and firing trials that had related to interrupter gears or synchronising gears, but no action was taken. The way such armament finally came about was circuitous. In his pre-war trials on the ground Saulnier had got a good mechanism working but had been completely thwarted by

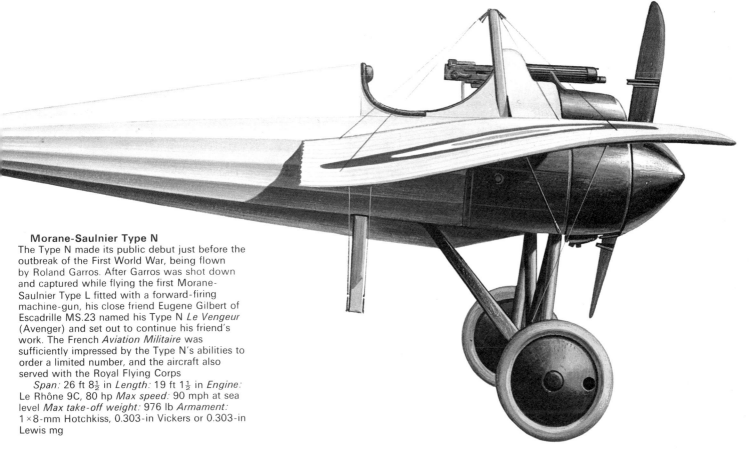

**Morane-Saulnier Type N**
The Type N made its public debut just before the outbreak of the First World War, being flown by Roland Garros. After Garros was shot down and captured while flying the first Morane-Saulnier Type L fitted with a forward-firing machine-gun, his close friend Eugene Gilbert of Escadrille MS.23 named his Type N *Le Vengeur* (Avenger) and set out to continue his friend's work. The French *Aviation Militaire* was sufficiently impressed by the Type N's abilities to order a limited number, and the aircraft also served with the Royal Flying Corps
*Span:* 26 ft 8½ in *Length:* 19 ft 1½ in *Engine:* Le Rhône 9C, 80 hp *Max speed:* 90 mph at sea level *Max take-off weight:* 976 lb *Armament:* 1 ×8-mm Hotchkiss, 0.303-in Vickers or 0.303-in Lewis mg

the Hotchkiss ammunition, which was prone to 'hang fire' – a moment's delay before the round fired – and this was enough to hit the propeller blade with the scheme Saulnier had devised. After trying to find an answer Saulnier gave up and resorted to 'brute force and ignorance' by simply fitting the propeller with deflectors.

Morane-Saulnier's brilliant pilot Roland Garros, who had done more than anyone to spread the fame of the company from 1912 onward, joined the Aviation Militaire and obtained permission to try out the bullet-deflector idea in flight. His main contribution was to reshape the propeller with a narrow portion in line with the gun, so that the deflector could be smaller. He experimented for months, sometimes with catastrophic results, until after patient work by his mechanic Jules Hué he had a system that worked reliably. This was in February 1915. Features of the definitive deflectors included a channel shape, to guide the bullets, and strong braces linking them to the propeller shaft; it was also important to use copper or thick-jacketed bullets that did not shatter when they hit the wedge-shaped deflector. Garros returned to his combat unit in March 1915 eager to try out his new forward-firing gun. The French authorities still showed no interest, and even cancelled an order to convert more Morane-Saulnier Type L parasol monoplanes to carry such a gun. But Garros flew as often as he could, and on 1 April, 1915, chanced on a formation of four Albatros two-seaters. The German crews doubtless saw the little Morane but gave it little thought as it dived at them. Then, as if by magic, machine-gun fire spurted out in front of it. After three clips from his Hotchkiss the first Albatross was spinning down in flames. The others beat it for home, and reported what had happened. On 15 April Garros got another Hun and on the 18th a third (many reports insist that he scored five confirmed victories between 1 and 18 April, 1915). But on 18 April, he foolishly made a low-level bombing mission on Courtrai railway station; hit by ground fire, he force-landed and only partly succeeded in burning his aircraft.

### Deadly Fokker

The Germans were intensely interested in Garros' aircraft. The gun and engine/propeller installation were removed and set up as a display exhibit to study how it worked. The Dutch aircraft designer Anthony Fokker was shown the Garros gear and ordered to fit a copy on his new monoplane. But Fokker did better than this. He had tried to design aircraft for Britain, but been turned down. In Germany he flew a copy of the Morane-Saulnier Type H in 1913. His copy, called M.5, was developed with short-span wings (M.5K) and long-span (M.5L). By 1915 both were in small-scale service, but as they could not carry armament they were not especially useful. Fokker at once saw how ideal his nimble monoplane would be as a carrier of a forward-firing gun. In his autobiography he claimed to have invented synchronising gear, but what actually happened is that his team of engineers – who knew of Schneider's pre-war scheme – decided that a proper synchronising gear was better than mere deflectors. Within a week they had designed, built, tested, improved and fully developed a simple interrupter gear that was to bring about a dramatic change in

fortunes in the air war.

It so happened that Fokker's mechanism, which normally fired one round for every two revolutions of the propeller (ie, four blades), arrived just as the Imperial Aviation Service was being completely overhauled and turned into a much more dynamic and aggressive force by the new *Feldflugchef*, Major Hermann von der Leith-Thomsen. The Fokker with the forward-firing gun was put into urgent mass-production (numbers were still modest, but they were measured in hundreds, where before there had not even been tens). It was called the E.I (E=Eindecker=monoplane), and the first to reach the Western front was given to an exceptional pilot, Oswald Boelcke, in June 1915. By July the E.II was in service, with an Oberursel rotary (similar to the French Le Rhône) of 100 hp, instead of only 80 hp. Though still light and seemingly flimsy aircraft, they were deadly, especially in the hands of such pilots as Boelcke, Max Immelmann, Ernst Udet and most of the future aces of the Imperial Aviation Service.

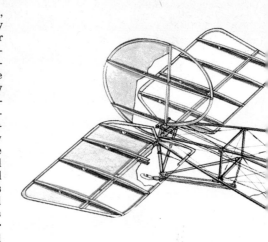

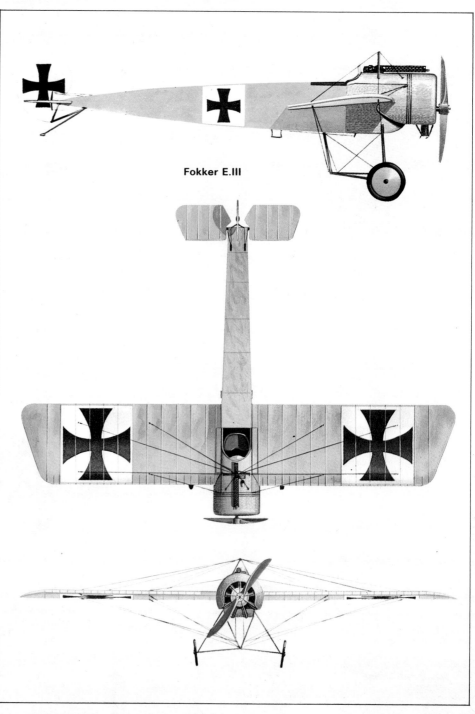

Fokker E.III

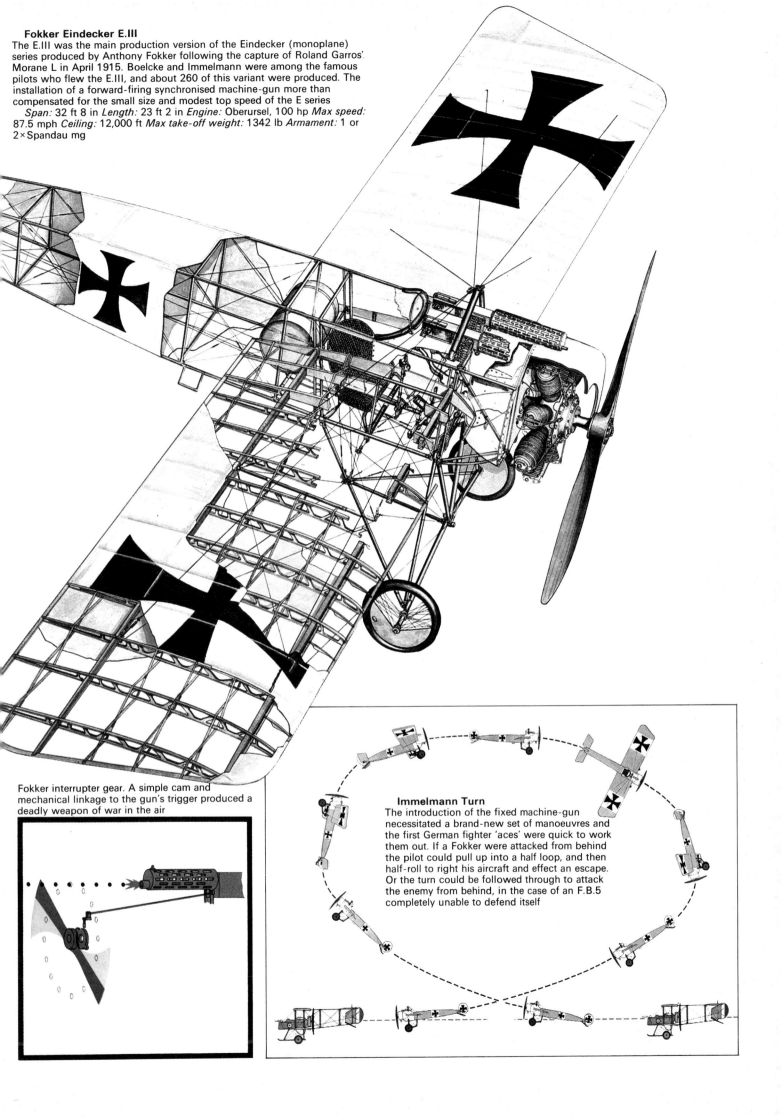

**Fokker Eindecker E.III**
The E.III was the main production version of the Eindecker (monoplane) series produced by Anthony Fokker following the capture of Roland Garros' Morane L in April 1915. Boelcke and Immelmann were among the famous pilots who flew the E.III, and about 260 of this variant were produced. The installation of a forward-firing synchronised machine-gun more than compensated for the small size and modest top speed of the E series
*Span:* 32 ft 8 in *Length:* 23 ft 2 in *Engine:* Oberursel, 100 hp *Max speed:* 87.5 mph *Ceiling:* 12,000 ft *Max take-off weight:* 1342 lb *Armament:* 1 or 2×Spandau mg

Fokker interrupter gear. A simple cam and mechanical linkage to the gun's trigger produced a deadly weapon of war in the air

**Immelmann Turn**
The introduction of the fixed machine-gun necessitated a brand-new set of manoeuvres and the first German fighter 'aces' were quick to work them out. If a Fokker were attacked from behind the pilot could pull up into a half loop, and then half-roll to right his aircraft and effect an escape. Or the turn could be followed through to attack the enemy from behind, in the case of an F.B.5 completely unable to defend itself

# FORGING THE WEAPONS

On 1 August, 1915, the RFC raided the home base of *Fliegerabteilung* 62 at Douai. This happened to be the E.II squadron that had led all others in receiving new equipment, and its pilots included Boelcke and Immelmann. The two outstanding Germans took off, chased the raiders and brought one down (it would have been two, but Boelcke's gun jammed). This can fairly be described as the day the fighter came into existence. From then onwards a quite small number of Fokker monoplanes completely turned the tables on the Allies, who were bringing to the front thousands of hastily trained young aircrew and thousands of almost defenceless aircraft. The most numerous RFC type was the B.E.2c, which not only could not defend itself but was so strongly stable that it could not be manoeuvred out of the line of fire. By October the life-expectancy of an Allied pilot on the Western front had been reduced to a week or two – unless his aircraft was kept on the ground by unserviceability or bad weather. Newspapers wrote of the 'Fokker scourge', while the British Parliament described the enormous numbers of replacement RFC pilots as 'Fokker fodder'. What was not explained was why the Allies had consistently refused to show any interest in fighter armament.

The Eindecker itself was merely a pleasant little platform on which to mount a machine gun. With a speed of 83 mph it was only just fast enough to catch most Allied machines, and its success came increasingly from carefully planned tactics, bold flying and the skill of its pilots. 'Beware the Hun in the Sun' was a vital piece of advice as early as January 1916, and though the German pilots (and the Austro-Hungarians, who also used the E-types) had yet to work out how to operate as a team, they individually learned techniques that would save time, maximise the time their guns could be brought to bear on a target, and increase their own chances of survival. At this time no pilots had parachutes, very few aircraft

had armour or any other protection, and everything depended on the pilot's own alertness (in all directions), vision from the cockpit, aircraft performance and manoeuvrability, and the reliability of his guns. On the whole the last factor was poor. Especially in the air, guns frequently jammed, as explained in a later chapter. The pilots who could handle all these factors, and never forget any of them for a moment, were those who racked up scores of victories. Until the Fokker E-type there had never been any real 'fighter pilots'. Suddenly Germany realized that the public and front-line troops wanted heroes, and in Boelcke and Immelmann they found them. The two aces, the first in the world, later split and worked in different sectors in friendly rivalry. Eventually Immelmann, who gave his name to the climbing 180° turn formed by a half-loop followed by a half-roll, plunged to his death on 28 June, 1916, with 15 victories to his credit. The great Boelcke was killed after a mid-air collision on 28, October, 1916, with 40 victories.

The little monoplane that struck terror into the Allied flyers, a mere copy of a 1913 French machine, eventually drove home to the Allied leaders that fighting aircraft were important. Without a synchronised gun, the only available aircraft were the Gunbus, the D.H.2, the 'Fee' and, probably best of all, the French Nieuport XI. Called the *Bébé* (Baby) when it entered Aviation Militaire and RNAS service in mid-1915, the trim Nieuport was one of a long and very important series of combat aircraft designed by Gustave Delage. The XI was possibly the smallest major warplane in history, with very compact dimensions and a gross weight of only 1058 lb, compared with about 1400 lb for the Fokker monoplanes. Thus, though it had only an 80-hp Le Rhône

rotary engine, it could reach almost 100 mph, and it was as nimble as anything in the sky. The only drawback was that the Lewis machine gun had to be mounted above the upper wing to clear the propeller, and the pilot had to fly with one hand, sight while looking straight ahead and reach up with one arm in the full blast of the slipstream to grasp the pistol grip and fire. Changing drums was not normally tried.

### Hanriot HD.1
The HD.1 appeared slightly later than the Spad 7.C1 and failed to impress the *Aviation Militaire* sufficiently to be selected as a replacement for the earlier type. Despite this setback, however, the HD.1 was adopted by the Italians and was built in substantial numbers by Nieuport Macchi
*Span:* 28 ft 6½ in *Length:* 29 ft 2 in *Engine:* Le Rhône 9Ja, 110 hp; Le Rhône 9Jb, 120 hp; Le Rhône 9Jby, 130 hp; or Clerget 9B, 130 hp *Max speed:* 115 mph at sea level with Le Rhône 9Jb *Ceiling:* 20,000 ft *Max take-off weight:* 1330 lb *Armament:* 1 × Vickers 0.303-in mg

*The redoubtable Cdr Samson RNAS and his Nieuport XI*

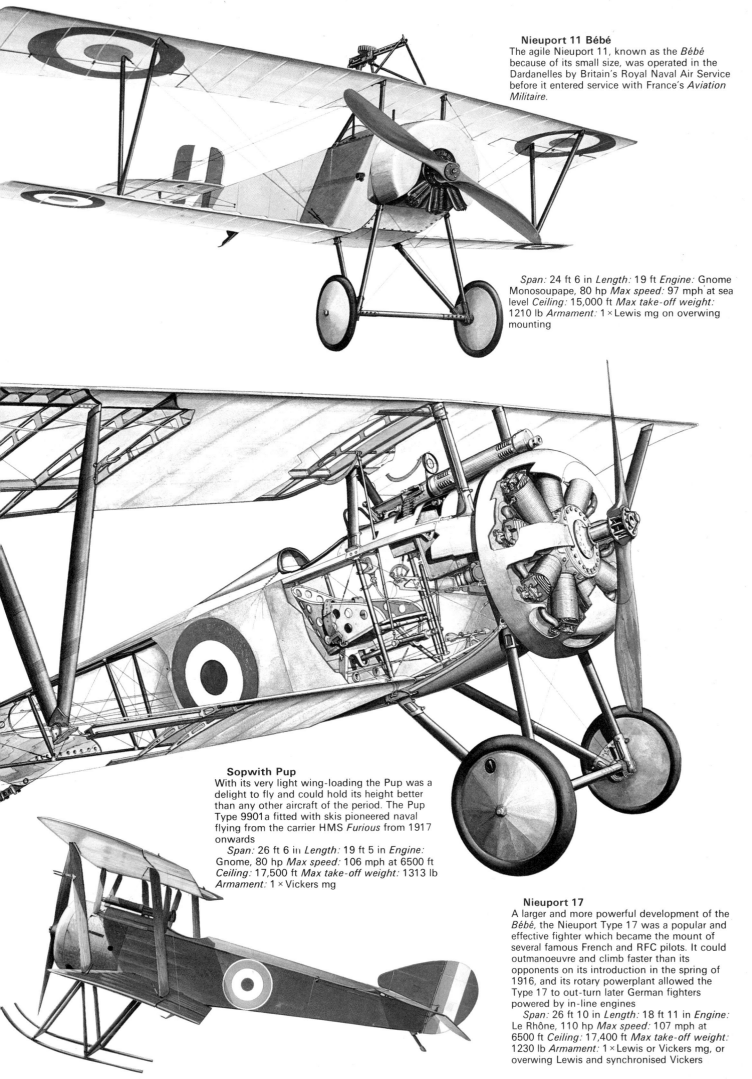

**Nieuport 11 Bébé**
The agile Nieuport 11, known as the *Bébé* because of its small size, was operated in the Dardanelles by Britain's Royal Naval Air Service before it entered service with France's *Aviation Militaire*.

*Span:* 24 ft 6 in *Length:* 19 ft *Engine:* Gnome Monosoupape, 80 hp *Max speed:* 97 mph at sea level *Ceiling:* 15,000 ft *Max take-off weight:* 1210 lb *Armament:* 1 × Lewis mg on overwing mounting

**Sopwith Pup**
With its very light wing-loading the Pup was a delight to fly and could hold its height better than any other aircraft of the period. The Pup Type 9901a fitted with skis pioneered naval flying from the carrier HMS *Furious* from 1917 onwards
*Span:* 26 ft 6 in *Length:* 19 ft 5 in *Engine:* Gnome, 80 hp *Max speed:* 106 mph at 6500 ft *Ceiling:* 17,500 ft *Max take-off weight:* 1313 lb *Armament:* 1 × Vickers mg

**Nieuport 17**
A larger and more powerful development of the *Bébé*, the Nieuport Type 17 was a popular and effective fighter which became the mount of several famous French and RFC pilots. It could outmanoeuvre and climb faster than its opponents on its introduction in the spring of 1916, and its rotary powerplant allowed the Type 17 to out-turn later German fighters powered by in-line engines
*Span:* 26 ft 10 in *Length:* 18 ft 11 in *Engine:* Le Rhône, 110 hp *Max speed:* 107 mph at 6500 ft *Ceiling:* 17,400 ft *Max take-off weight:* 1230 lb *Armament:* 1 × Lewis or Vickers mg, or overwing Lewis and synchronised Vickers

At this time fighters were still lightly built and powered by small engines. Though most were reasonably safe in normal flight, the stresses imposed in the new art of dogfighting were still partly unknown. Even local damage from a single bullet could cause total collapse of primary structure, or the fabric to peel from a wing, while the basic flying characteristics of many aircraft were highly dangerous. Some had engines prone to catch fire without help from the enemy, while others were almost impossible to recover from a spin. About three-quarters of the primary structural members – more in some aircraft, fewer in others – were of wood, and in many aircraft the wood itself was below nominal strength, faulty or even too short for the part and thus connected by joints of untried strength. Gradually the strength of airframes and materials became better understood in precise numerical terms, but this was counterbalanced by the fact that, as the war continued, the best structural hardwoods were used up and poorer materials took their place.

With the earliest fighting scouts it made quite a difference whether the pilot was a big man or a small one. A single machine-gun and ammunition represented as much as one-tenth of the total laden weight, and the recoil of a Lewis on the upper wing could tilt the nose up sufficiently to cause the bullets to pass harmlessly above the enemy. Two guns was often too much, and after Immelmann had burdened his E.III with three Spandaus he thought better of it and removed one. The obvious next stage in fighter development was to increase engine power, to carry heavier loads, climb faster and more steeply and reach higher speeds.

It was here that the traditional rotary engines began to show their limitations. Though at the 80–100-hp level they were neat and light, as power was increased they began to become more complex and relatively heavier, and their spinning mass exerted such a powerful gyroscopic effect as to dominate the aircraft in which they were fitted. Even the 130-hp Clerget 9B made the otherwise excellent Sopwith F.1 Camel an exceedingly tricky machine to fly, and with the inexorable demand for more power the future for rotaries was not promising. The original engines were mostly of nine cylinders, and it was found possible to squeeze 11 into a single row and push power up to 200 hp. Some engine manufacturers developed two-row rotary engines, while others worked hard to wring more power from each cylinder. One of the latter was Lt W O Bentley (later to be famed for his cars) who in 1917 produced a nine-cylinder rotary rated at 230 hp yet posing less of a gyroscopic problem than the Clerget. Such engines remained in mass production to the end of the war, because they combined the essentials needed.

### Albatros D.III
The D.III, with a narrow-chord lower wing and V-shaped interplane struts *à la Nieuport* to improve visibility, enjoyed a brief period of ascendancy in the spring of 1917 before it came up against the new British fighters
*Span:* 29 ft 6¼ in *Length:* 24 ft *Engine:* Mercedes, 160 hp *Max speed:* 109 mph at 3200 ft *Ceiling:* 18,000 ft *Max take-off weight:* 1950 lb *Armament:* 2×Spandau mg

### Roland C.II
The dumpy Roland C.II, nicknamed *Walfisch* (Whale), had a substantial endurance – thanks partly to its low drag, achieved by keeping interplane bracing to a minimum – and was employed as a long-range escort in addition to its primary role of reconnaissance. The type entered service in early 1916, and nearly 300 were built in all
*Span:* 33 ft 10¾ in *Length:* 24 ft 8 in *Engine:* Mercedes, 160 hp *Max speed:* 103 mph *Max take-off weight:* 2885 lb *Armament:* 1 × Parabellum mg plus, in some aircraft, Spandau mg

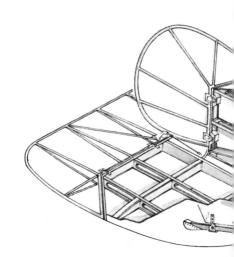

Imperial War Museum

*Lt Danhuber in his Albatros D.Va. Accessibility of the guns was an important factor if they jammed*

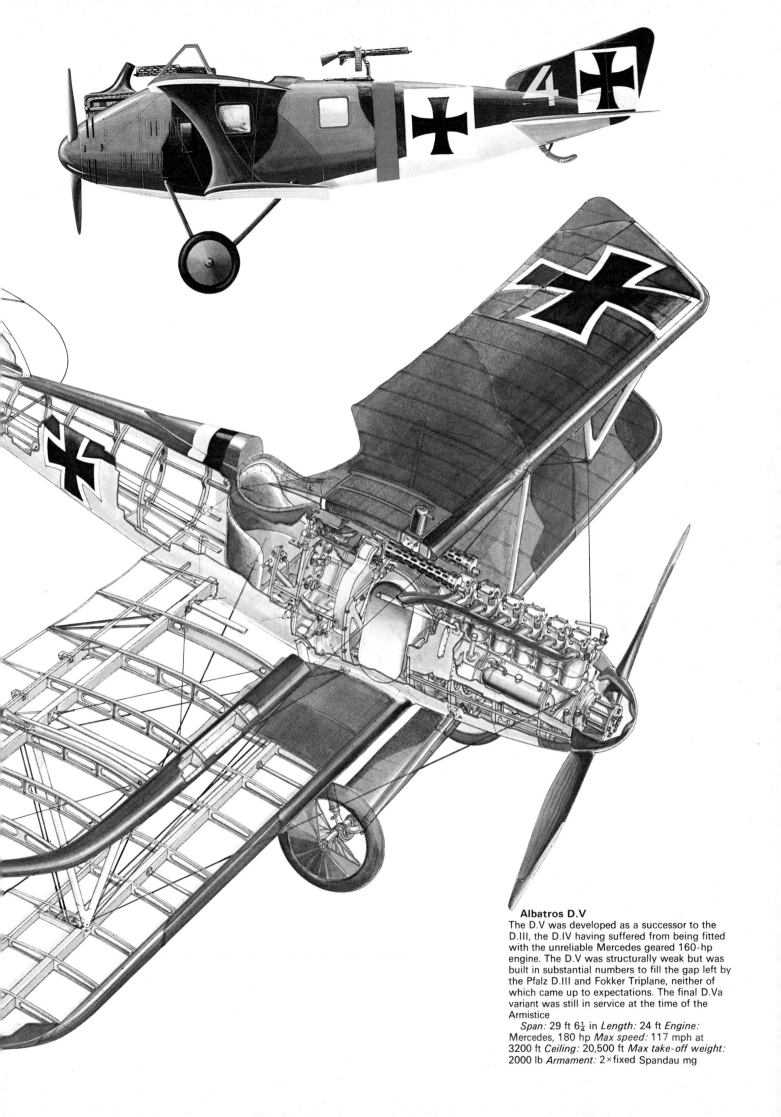

**Albatros D.V**
The D.V was developed as a successor to the
D.III, the D.IV having suffered from being fitted
with the unreliable Mercedes geared 160-hp
engine. The D.V was structurally weak but was
built in substantial numbers to fill the gap left by
the Pfalz D.III and Fokker Triplane, neither of
which came up to expectations. The final D.Va
variant was still in service at the time of the
Armistice
  *Span:* 29 ft 6¼ in *Length:* 24 ft *Engine:*
Mercedes, 180 hp *Max speed:* 117 mph at
3200 ft *Ceiling:* 20,500 ft *Max take-off weight:*
2000 lb *Armament:* 2×fixed Spandau mg

Engine designers knew in their hearts that the future belonged to the in-line and the static radial. At first there had been few static radials, one of the types to get into production being the Salmson (Canton-Unné) which had water-cooled cylinders. Nearly all the non-rotary engines had cylinders arranged in line, the Central Powers never departing from their generous and reliable four- and six-cylinder engines and the Allies building large numbers with six, eight or 12 cylinders in vee formation. Three of the most important early builders, the Royal Aircraft Factory (RAF), Beardmore and Renault, were gradually eclipsed by the Hispano-Suiza and Rolls-Royce companies which succeeded brilliantly in pushing up the power of a single engine to 300 hp and beyond. In 1917 the United States entered the war and a consortium of car manufacturers, assisted by Allied engineers, very quickly designed an advanced V12 which was soon in production on an unprecedented scale at a power of 400 hp. This engine, the Liberty, was no short-term lash-up and many were in service in the 1930s.

With ample power available, fighters became larger, faster and more heavily armed. As described in the next chapter, numerous schemes were perfected for synchronising fixed guns, and by 1916 the usual forward-

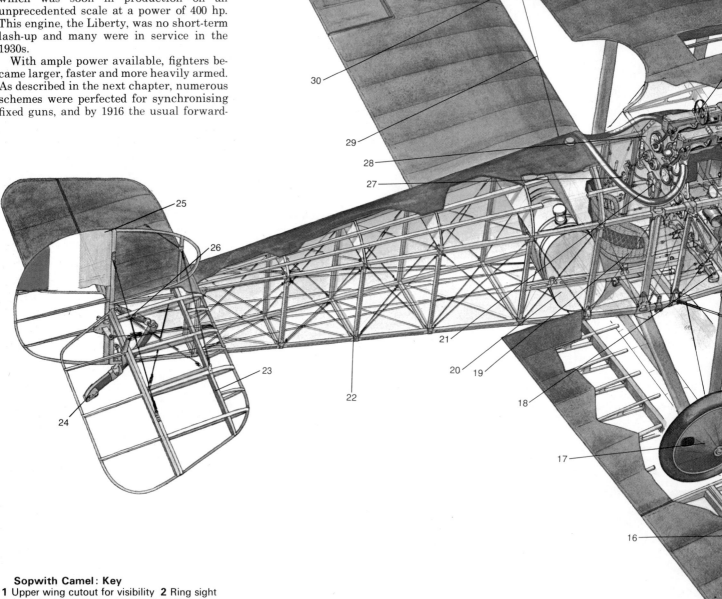

**Sopwith Camel: Key**
1 Upper wing cutout for visibility 2 Ring sight
3 Vickers gun 4 Ammunition tank 5 Wooden
propeller 6 Aluminium cowling 7 Rotary engine
8 Oil tank 9 Wing rib 10 Aileron control wire
11 Compression rib 12 and 13 Wing bracing
wires 14 Main spars 15 Aileron operating horn
16 Aileron connecting wire 17 Bungee-sprung
wheel 18 Rudder bar 19 Wicker seat 20 Fuel
tank 21 Control column 22 Wire-braced wooden
fuselage 23 Tailplane structure 24 Steel shod
skid 25 Fin and rudder 26 Bungee skid spring
27 Throttle and mixture controls 28 Instrument
panel 29 Flying wire 30 Landing wire
31 Incidence bracing wires

**Sopwith Triplane**
Designed for good visibility and manoeuvrability, the 'Tripehound' could out-climb its German contemporaries and gave rise to a whole family of German and Austrian triplanes. Deliveries began in late 1916 and the Black Flight of five Canadians, flying Triplanes, shot down 87 enemy aircraft between May and July 1917
*Span:* 26 ft 6 in *Length:* 18 ft 10 in *Engine:* Clerget, 110 or 130 hp *Max speed:* 117 mph *Ceiling:* 20,500 ft *Endurance:* 2¾ hr *Armament:* Vickers mg

**Sopwith 5F.1 Dolphin**
The Dolphin has been described as the most seriously underrated British fighter of the First World War. It was more manoeuvrable than the S.E.5a and had an excellent high-altitude performance, although the Hispano-Suiza engine was unreliable and in short supply. Pilot reports of the Dolphin's flying qualities were enthusiastic, although the negative-stagger wings were reminiscent of the D.H.5's layout (thought of as dangerous) and increased the risk of injury in landing accidents

*Span:* 32 ft 6 in *Length:* 22 ft 3 in *Engine:* Hispano-Suiza of 200, 220 or 300 hp *Max speed:* 128 mph at 10,000 ft *Ceiling:* 21,000 ft *Max take-off weight:* 2000 lb *Armament:* 2×Vickers mg, or 1 or 2×Lewis mg

**Sopwith F.1 Camel**
The Camel was the first British fighter fitted with twin synchronised Vickers guns, and the hump enclosing the breeches gave the aircraft its name. The Camel needed careful handling but was a formidable weapon, accounting for more kills than any other type in the First World War. Nearly 5500 were built, some being operated from ships as the 2F.1
*Span:* 28 ft *Length:* 18 ft 9 in *Engine:* Clerget, 130 hp *Max speed:* 115 mph at 6500 ft *Ceiling:* 19,000 ft *Max take-off weight:* 1453 lb *Armament:* two Vickers mg

firing armament was not one gun but two. The Nieuport scouts became large and even more lethal, and in the middle war years had an unsurpassed reputation for dog-fight manoeuvrability. The Spad company ran Nieuport close, and by 1917 were in front with some of the finest fighters of the entire war, helped by the fact that from the start they had been matched with powerful Hispano-Suiza engines. The British Sopwith company seldom put a foot badly wrong, following the 80-hp Tabloid and Baby with the similarly powered Pup (often held to be the most perfect aerobatic machine of 1914–18), the multi-role 1½-Strutter (often described as the first British aircraft with a proper scheme of armament) with 110–130 hp, the 130-hp single-seat Camel with twin Vickers firing ahead (the tricky hump-backed Camel set a record in destroying at least 1294 hostile aircraft) and the 230-hp Snipe, which represented the pinnacle of rotary-engined aircraft. From the Royal Aircraft Factory came the S.E.5 series, one of the best and most popular fighters, fitted with a French Hispano water-cooled V8 engine. Early S.E.5s had an engine of 150 hp, but the mass-produced 5a had the geared 200-hp version which, though it improved performance, also caused endless trouble. The geared drive failed repeatedly, so that completed aircraft littered British factories waiting for engines. It was offici-ally decided to fit them with faulty engines, this being judged better than no engines at all; eventually the British Wolseley com-pany developed a high-compression direct-drive version of the French engine (it was named the Viper) which solved all the problems, and the aircraft was a winner.

During the first quarter of 1916 the Sopwith design team, while delighted at the outstanding combat manoeuvrability of the Pup, considered how this might be improved further. The ability to roll or turn quickly had never before been very important in aircraft design, but by 1916 this had been recognised as central to the very concept of a fighter, especially one whose armament was fixed to fire ahead. Sopwith's designers, led by Herbert Smith, thought it worth trying a triplane version of the Pup, and the prototype Triplane flew in May 1916 and, with a Vickers gun added, was tested on the Western front the following month. Its unusual appearance emphasised its excellent manoeuvrability and rate of climb, and though only 266 were built – all used by the RNAS – the nimble Triplane had a fantastic impact on the enemy. General von Hoeppner, commander of the Imperial Air Service, went into raptures over it, and within days almost every fighter builder in the Central Powers was trying to build an 'answer'. Eventually 14 German and Austro-Hungarian fighter triplanes were developed, but the only one to see extensive service was Fokker's Dr.I (Dr.=Dreidecker=three-winger). Fokker had been so desperate to find out about Sopwith's machine that he

had – so the story goes – improperly arranged for his factory to receive the wreck of the first RNAS Triplane to be shot down. In fact Fokker's brilliant designer, Reinhold Platz, was not sold on the triplane at all, and in any case produced a totally different machine which in its original form did not have interplane struts (later a single, streamlined strut was added to link the three wings). Like the Sopwith, the Dr.I was extremely manoeuvrable, and though it was smaller and lighter it carried two guns. Yet, like the Sopwith, it was made in only modest numbers. It was never completely outclassed, and the greatest of all First World War aces, Manfred von Richthofen, was flying a Dr.I when he met his death on 21 April, 1918. But the triplane and the numerous quadruplanes played only a minor part in the overall conflict.

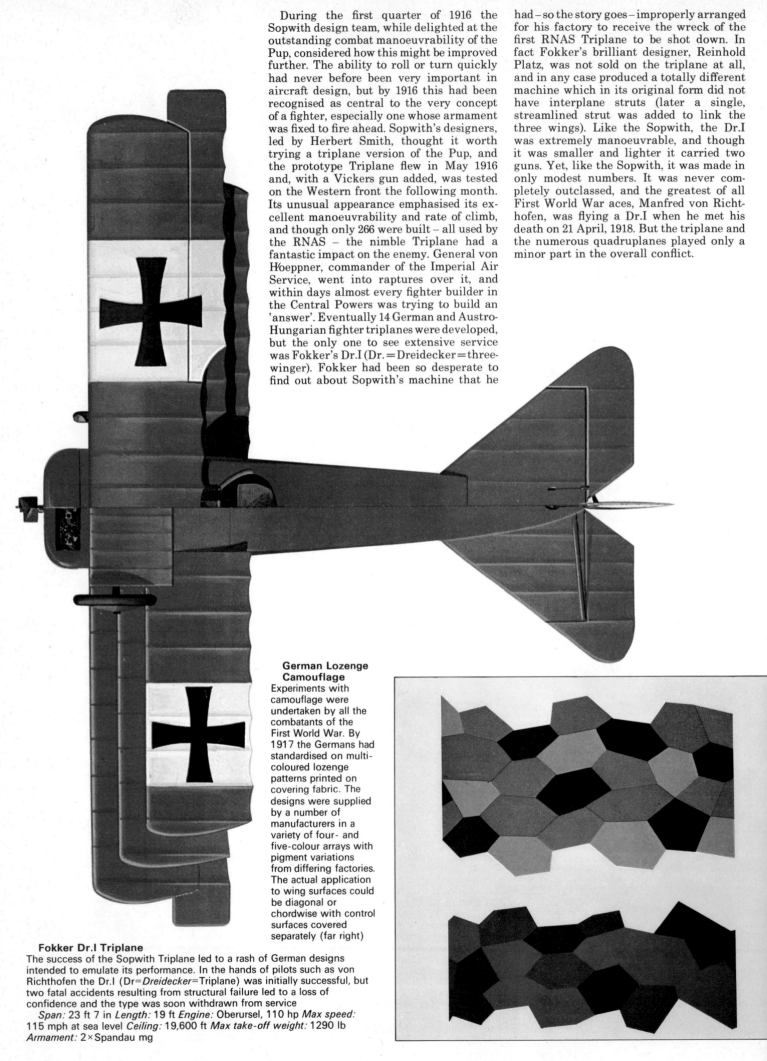

**German Lozenge Camouflage**
Experiments with camouflage were undertaken by all the combatants of the First World War. By 1917 the Germans had standardised on multi-coloured lozenge patterns printed on covering fabric. The designs were supplied by a number of manufacturers in a variety of four- and five-colour arrays with pigment variations from differing factories. The actual application to wing surfaces could be diagonal or chordwise with control surfaces covered separately (far right)

**Fokker Dr.I Triplane**
The success of the Sopwith Triplane led to a rash of German designs intended to emulate its performance. In the hands of pilots such as von Richthofen the Dr.I (Dr=Dreidecker=Triplane) was initially successful, but two fatal accidents resulting from structural failure led to a loss of confidence and the type was soon withdrawn from service
    *Span:* 23 ft 7 in *Length:* 19 ft *Engine:* Oberursel, 110 hp *Max speed:* 115 mph at sea level *Ceiling:* 19,600 ft *Max take-off weight:* 1290 lb
*Armament:* 2×Spandau mg

*The moments before take-off at a Fokker Dr.I-equipped* Jasta

**Halberstadt D.II**
The D.II came into service in 1916 just as the Fokker monoplanes were losing their ascendancy. For a time the D.II outclassed any Allied opposition, combining the climb and handling of a biplane with a synchronised machine-gun. The Allies soon caught up, however, and the frail tail assembly was a weakness. The D.II had disappeared from the Western Front by May 1917
*Span:* 28 ft 11½ in *Length:* 21 ft 5 in *Engine:* Argus As II, 120 hp *Max speed:* 90 mph
*Armament:* 2×Spandau mg

*The great von Richthofen discusses the qualities of the Fokker Dr.I with his fellow officers. The aircraft's climb and manouevrability made it a favourite mount of the most skilful pilots*

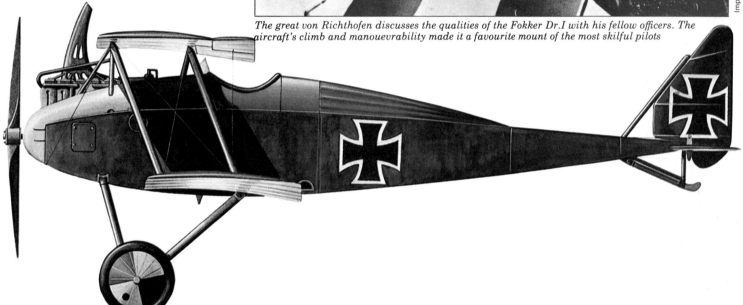

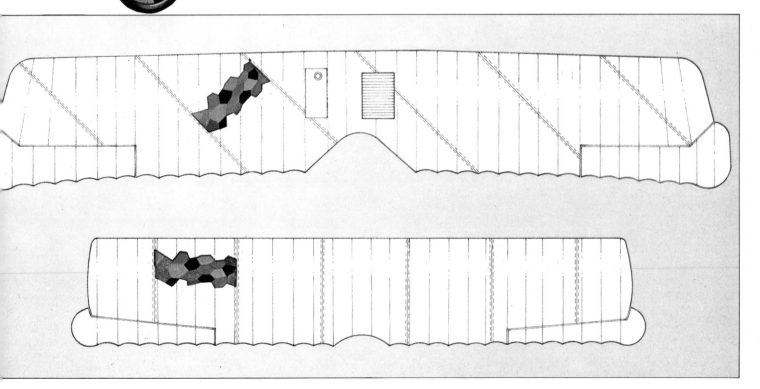

## Ground Attack Fighters: Keeping the enemy's head down

By 1917 the pattern of aerial warfare was set – Reconnaissance aircraft spotting for the guns while ever more potent fighters duelled for air superiority. Then the Germans led the way with a new concept – ground-attack fighters spearheading offensives by actually engaging battle on the front line – and British designers were quick to follow

Far more important were the improved two-seat fighters which emerged in 1916–17. The Germans introduced the CL category of two-seat fighters in early 1917 which led to an excellent series of tough and versatile machines. Though naturally bigger and more ponderous than the single-seaters, the 'Hannoveranas', the Halberstadt CL series and the all-metal Junkers monoplanes were quite manoeuvrable, difficult to shoot down and worthy opponents in battle. Carrying various bombs and grenades they also operated in the dangerous close-support role low over the battlefield and at times succeeded in inflicting severe casualties on Allied troops and lowering their morale. Sopwith tried to develop a good aircraft in the same class, but in the Hippo and Bulldog only scored near-misses. The company did achieve great success with a similar design on a slightly smaller scale, the single-seat Dolphin which often had as many as four

machine-guns, and also with heavily armoured derivatives of the small single-seaters designed specifically for 'trench fighting' (ie, close support), the Salamander being the most notable. But by far the most important of all First World War two-seat fighters was the Bristol Fighter.

*A Halberstadt CL.II is bombed up with bundles of fragmentation grenades, ready for a close-support mission*

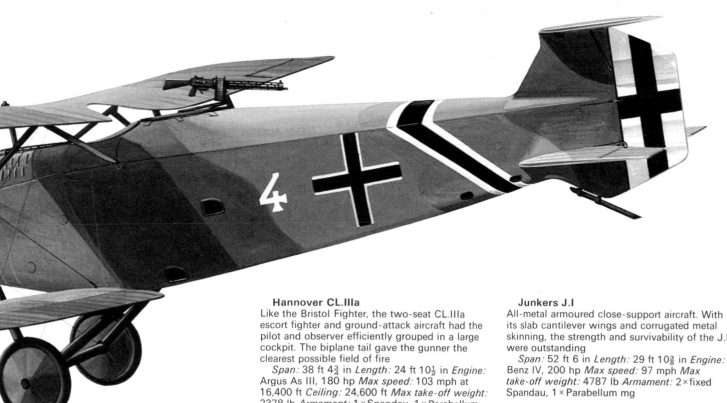

### Hannover CL.IIIa

Like the Bristol Fighter, the two-seat CL.IIIa escort fighter and ground-attack aircraft had the pilot and observer efficiently grouped in a large cockpit. The biplane tail gave the gunner the clearest possible field of fire

*Span:* 38 ft 4¾ in *Length:* 24 ft 10½ in *Engine:* Argus As III, 180 hp *Max speed:* 103 mph at 16,400 ft *Ceiling:* 24,600 ft *Max take-off weight:* 2378 lb *Armament:* 1 × Spandau, 1 × Parabellum mg

### Junkers J.I

All-metal armoured close-support aircraft. With its slab cantilever wings and corrugated metal skinning, the strength and survivability of the J.I were outstanding

*Span:* 52 ft 6 in *Length:* 29 ft 10⅜ in *Engine:* Benz IV, 200 hp *Max speed:* 97 mph *Max take-off weight:* 4787 lb *Armament:* 2 × fixed Spandau, 1 × Parabellum mg

*A Hannover CL.IIIa (foreground) on the airfield of a Schlasta (Schlachtstaffel – Battle-Squadron), as the German ground-attack units were known*

*Looking like a curious throwback when it first appeared in 1917, the Vickers FB.26 Vampire might have been a formidable ground-attack aircraft if it had been developed. The pusher layout made several further comebacks during the 1920s*

### de Havilland D.H.5

The marked back-stagger of the D.H.5 was designed to give the pilot the best possible view, but the loss of aerodynamic efficiency made it difficult to fly. The first batches reached France in 1917 but they were soon relegated to escort work and ground-attack

*Span:* 25 ft 8 in *Length:* 21 ft 9 in *Engine:* Le Rhône, 110 hp *Max speed:* 102 mph at 10,000 ft *Ceiling:* 16,000 ft *Max take-off weight:* 1492 lb *Armament:* 1 × Vickers mg

*With the Ludendorff offensive of March 1918, the Schlastas came into their own. Here a CL.III flies over the troops it is supporting*

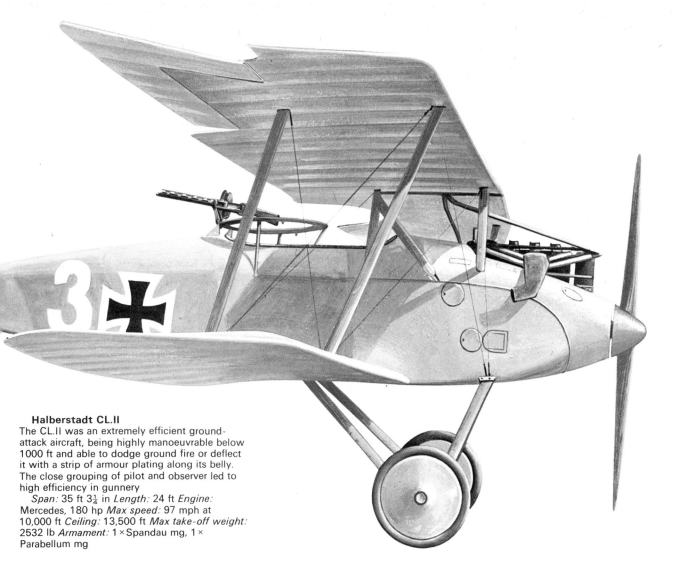

**Halberstadt CL.II**
The CL.II was an extremely efficient ground-
attack aircraft, being highly manoeuvrable below
1000 ft and able to dodge ground fire or deflect
it with a strip of armour plating along its belly.
The close grouping of pilot and observer led to
high efficiency in gunnery
    *Span:* 35 ft 3$\frac{1}{4}$ in *Length:* 24 ft *Engine:*
Mercedes, 180 hp *Max speed:* 97 mph at
10,000 ft *Ceiling:* 13,500 ft *Max take-off weight:*
2532 lb *Armament:* 1 × Spandau mg, 1 ×
Parabellum mg

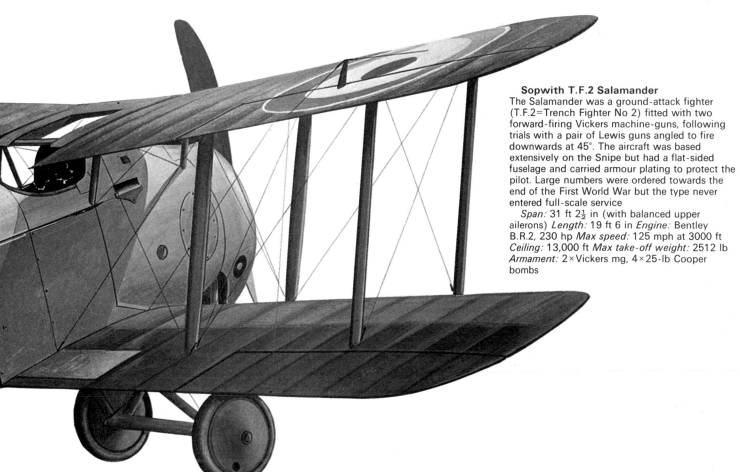

**Sopwith T.F.2 Salamander**
The Salamander was a ground-attack fighter
(T.F.2=Trench Fighter No 2) fitted with two
forward-firing Vickers machine-guns, following
trials with a pair of Lewis guns angled to fire
downwards at 45°. The aircraft was based
extensively on the Snipe but had a flat-sided
fuselage and carried armour plating to protect the
pilot. Large numbers were ordered towards the
end of the First World War but the type never
entered full-scale service
    *Span:* 31 ft 2$\frac{1}{2}$ in (with balanced upper
ailerons) *Length:* 19 ft 6 in *Engine:* Bentley
B.R.2, 230 hp *Max speed:* 125 mph at 3000 ft
*Ceiling:* 13,000 ft *Max take-off weight:* 2512 lb
*Armament:* 2 × Vickers mg, 4 × 25-lb Cooper
bombs

# The Bristol Fighter

Perhaps the best general-purpose aircraft of the First World War, the Bristol Fighter was outstandingly strong, manoeuvrable and efficient in air combat

This was originally planned as an improved reconnaissance type, but entered production with the 190-hp Rolls-Royce Falcon engine as a fighter, with a fixed Vickers and free Lewis. At first it proved easy meat, because pilots flew straight and level and relied on the observer's gun. Later it was discovered that the tough Bristol could be flung about in a dogfight, and that the observer could still get in shots even under these conditions. Fitted with various engines of 190 to 400 hp, the splendid F.2B version soon gained such a reputation that German fighters would never attack a formation of three or more – so the Bristols deliberately flew in pairs or singly to try and bring the Hun to battle. Squadron after squadron in 1917–18 established complete mastery of local airspace with this quite large two-seater, which succeeded so well simply by having no faults. It was unbreak-

able in combat manoeuvres, the crew were close and could talk, the armament was wholly satisfactory and performance was very good. To show what a Bristol could do, one Canadian crew, Lt (later Maj) A E McKeever and Sgt (later Lt) L F Powell, shot down 30 enemy aircraft in the second half of 1917.

Eventually there were more than 3000 Bristols in action, but by 1918 the main dogfighters were the Camel, S.E.5a, Spad VII and XIII, and the formidable German Albatros D.III and D.V and Fokker D.VII. The German scouts were not exceptional, but merely good conventional machines boldly flown by experienced and aggressive pilots. It says much for the skill of the German designers and pilots that these aircraft were not outclassed even by the time of the Armistice, despite the fact that the Germans lacked engines suitable for fighters giving more than 185 hp. There were small numbers of more powerful engines, including advanced rotaries, but the best fighter of all, the D.VII, usually had only 160 hp. Yet this was the type singled out for special attention by the Allied disarmament control commission in 1918–19. Every D.VII had to be handed over to the Allies; but with a bright young man like Fokker behind it the next stage is not hard to guess.

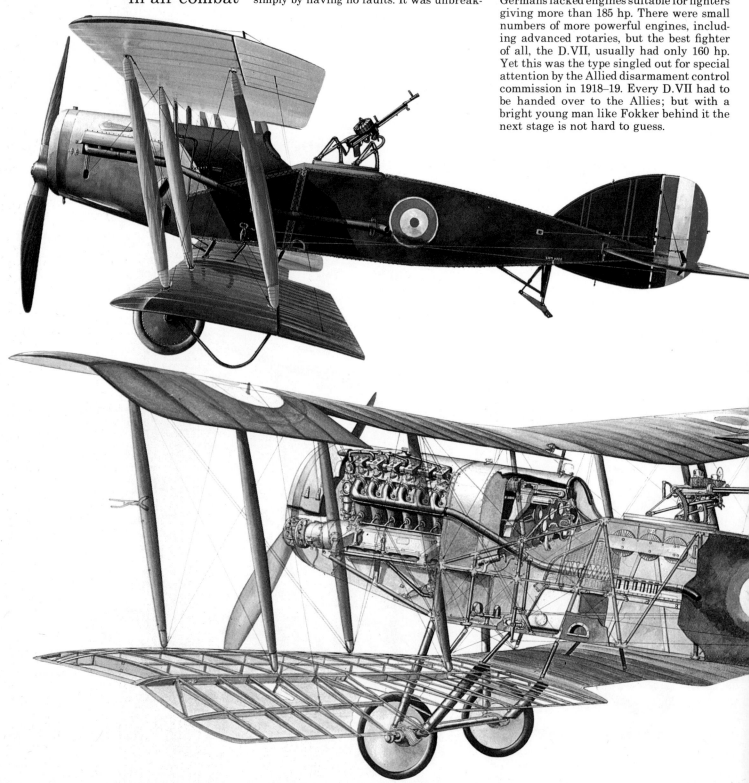

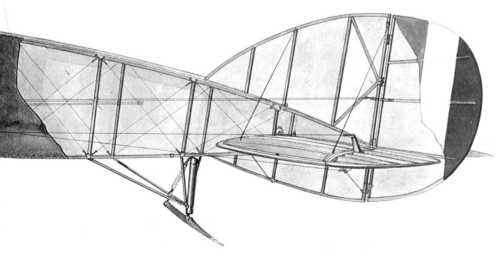

### Bristol F.2B Fighter

The F.2B version of the ''Brisfit'' was fitted with a Rolls-Royce Falcon of up to 275 hp, conferring a performance superior to that of many single-seat contemporaries. The Bristol Fighter was designed to be operated as a conventional fighter with the addition of a sting in the tail in the form of the free-mounted weapons. The type remained in RAF service until 1932

*Span:* 39 ft 3 in *Length:* 26 ft 2 in *Engine:* Rolls-Royce Falcon of 190, 220 or 275 hp *Max speed:* 125 mph at 3000 ft *Ceiling:* 22,000 ft *Max take-off weight:* 2590 lb *Armament:* 1 × Vickers mg, 1 or 2 × Lewis mg

# BIRDS OF PREY

*The trenchbound armies of the First World War saw the duels of the airmen at a distance – unless they crashed earthwards, usually in flames. This Camel has crashed almost intact, however, emphasising the structural strength of this machine*

Most of the armament carried by aircraft in the First World War was to some degree an improvisation; indeed, even the supposed properly designed armament schemes often failed to work properly (or at all) until they had been completely changed by pilots and fitters in the front-line squadrons. At the same time, the major powers did set up capable research establishments where such questions as aircraft armament were debated and improved. The whole technology of military aviation, and the flying machine itself, was advancing at an unnatural pace, and in this chapter it is necessary to stop for a moment and look at some of the problems.

We are concerned mainly with armament, but what had to be done was to turn the aeroplane from a dangerous and temperamental toy, that with luck could usually be flown in good weather, into a reliable weapon-platform or reconnaissance vehicle that could be flown without fail at a given time no matter what the weather. In fact the true night and all-weather machine could not be achieved until many years later, but advances were nevertheless enormous. Flying by night had been attempted as early as 1910, and the equipment later included an electric accumulator (in large aircraft it was recharged by a small windmill-driven generator) and cockpit lighting, some form of airfield lighting such as flares made from petrol-soaked asbestos carried above the ground in small containers, and searchlights whose crews were skilled in using the lights to convey information. In some

*In 1918, 297 Nieuport 28s were purchased from France as equipment for the first US fighter squadrons. This beautifully preserved example is flown at Chino, California*

*The Canadian ace Capt W Bishop demonstrates how to change a drum on a Lewis-equipped Nieuport 17*

**Lewis Gun (left)**
Designed by Col I N Lewis
(US Army rtd) in 1911, the
lightweight gas-operated
Lewis was an ideal aircraft
weapon and was
manufactured in Britain by
BSA. The Lewis was taken
aloft in 1912 and first used
in combat in August 1914.
The 'Scarff' ring for observers
was standard until well into
the Second World War and
the Foster mount made it a
standard forward-firing
weapon without the
synchronising gear of a
fuselage-mounted Vickers
*Weight:* 27 lb *Rate of fire:*
600 rpm

*An RFC squadron lines up with its S.E.5a fighters - one of the best British fighters of the war. The Lewis guns on their Foster mounts are clearly visible but the squadron codes have been obscured by the censor*

## Vickers F.B.12

The pusher layout effectively expired with the D.H.2 and F.E.8 but Vickers produced a rotary-engine powered pusher fighter prototype, the F.B.12, in late 1916. Thirty were ordered in November but poor performance and improved tractor types curtailed any further development

*Span:* 29 ft 7 in *Length:* 21 ft 10 in *Engine:* Le Rhône, 80 hp *Max speed:* 86 mph at 6500 ft *Ceiling:* 14,500 ft *Max take-off weight:* 1447 lb *Armament:* 1 × Lewis mg

cases searchlights could illuminate hostile aeroplanes or airships, for interception by fighters, while in others the lights could give position information or even send simple prearranged messages. But everyone had to learn the hard way. RFC Lt Slessor, who 40 years later became Chief of the Air Staff, tried to land after a primitive night anti-Zeppelin mission in 1915 but found the airfield shrouded in dense white mist. The airfield searchlight crew decided to help, but instead of aiming at where he was going they pointed the beam right at him. He was, of course, blinded and was no longer able to see even brief glimpses of the ground.

A year later bold pilots of RFC 44 Sqn flew the Camel on night patrol, and from 1916 onwards this very demanding little aircraft became an important night fighter. Why was the Camel difficult to fly? The reasons centred on the engine, which imparted such a strong gyroscopic force that the Camel had to be 'flown' constantly. It could turn like lightning to the right, but in turning to the left the pilot had to fight the effect of the engine (so Camel pilots sometimes turned left by making a 270° turn to the right). Enemy pilots were not slow to do what they could to take advantage of this lopsided characteristic, though the Camel and other rotary-engined machines were never to be taken lightly. The tricky handling of light scouts powered by large spinning engines was, of course, fundamental and incapable of much improvement.

Other basic areas of technology where very considerable progress was made by 1918 concerned flight control and stability, instruments, navigation and pilot equipment. In 1914 the fact that aircraft often appeared to get out of control and spiral straight down to the ground was the subject of apprehensive discussion among pilots. Not until British test pilots deliberately investigated spinning was the problem solved. Lt Wilfred Parke, a Naval pilot, had recovered from a spin in August 1912 but the deadly locked-in situation was not mastered until deep study, rounded off by courageous trials by F A Lindemann (later Lord Cherwell, Britain's Chief Scientist in the Second World War). This removed the terror of spinning, so that by winter 1917–18 it could even be taught to pupils along with a foolproof method of recovery. At the same time the instruments in the cockpit were made more accurate, giving better and more immediate indications of what the aircraft and engine were doing – but still with severe limitations. By 1918 the pilot could safely fly in clear weather, and navigate with map and magnetic compass (and in large aircraft with a mariner's type sextant and plotting charts), but he was in peril if he entered cloud or met fog. Inexplicable

crashes gradually rammed home the lesson that pilots could either lose control in a cloud or lose their orientation. Once they could no longer tell which way was up, they became totally disoriented and could come out of cloud in a screaming dive and discover the ground above them. Unless there was plenty of room to recover, they would just hit the ground. As for pilot equipment, this had from the earliest days included goggles and warm windproof clothing, and during the First World War Sir Geoffrey Taylor solved the basic design problems for a safe parachute (though air forces did not in general adopt this life-saving device until long after 1918).

Almost every other facet of the design of combat aircraft could also be described as a lash-up, but the central factor where fighters are concerned has always been armament. At first there were only infantry weapons, small grenades, bombs and mortars, and a few larger quick-firing guns mainly of naval origin. The standard service rifle was unwieldy, being rather long and difficult to aim in an airstream going past at 60 to 100 mph, yet it was the only authorised RFC weapon at the outbreak of war. Most handguns, such as pistols and revolvers, were so inaccurate and difficult to aim in

air combat as to be of very limited value, though the long-barrelled German pistols (especially the Mauser) could be aimed much more accurately when clipped to their wooden holsters, turning them almost into low-power rifles.

Until 1914 little thought had been given to armament actually attached to the aircraft. In any case the most likely weapon, the machine-gun, was still considered rather new and ungentlemanly, and the machine-guns adopted as standard army weapons were almost all large and cumbersome, and needed water to cool the barrel. One of the very few reasonably light guns was the Benet-Mercier, the standard machine-gun of the US Army Signal Corps, and was thus the world's first 'fighter armament', but simply did not fit. The Wright biplanes sat the pilot and machine-gunner side-by-side, and the Benet-Mercier had a long rigid ammunition feed projecting on one side and an equally long ejector chute on the other. No matter which way round the pilot and gunner sat the gun fouled the control columns and made flight impossible. Later the neat drum-fed Lewis was admitted, and this compact air-cooled weapon was made in larger numbers than any other for air use in the First World War.

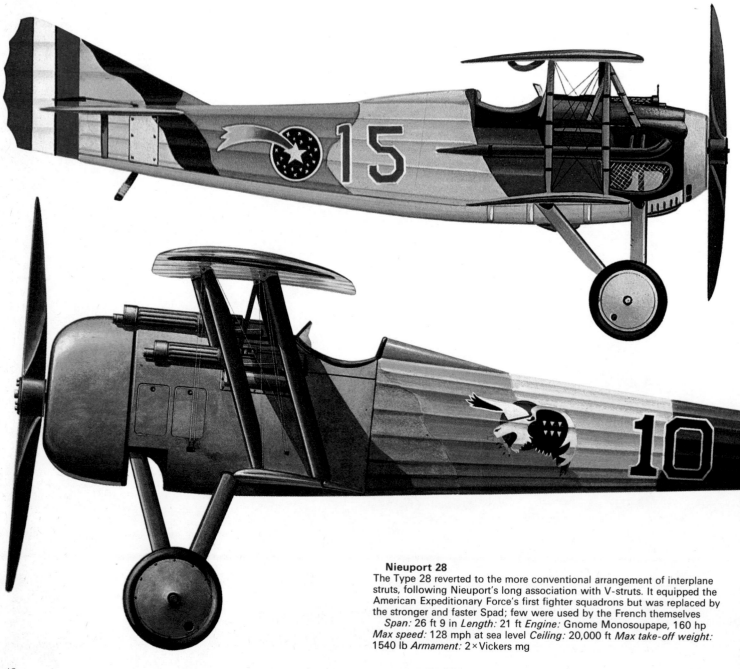

**Nieuport 28**
The Type 28 reverted to the more conventional arrangement of interplane struts, following Nieuport's long association with V-struts. It equipped the American Expeditionary Force's first fighter squadrons but was replaced by the stronger and faster Spad; few were used by the French themselves
*Span:* 26 ft 9 in *Length:* 21 ft *Engine:* Gnome Monosoupape, 160 hp
*Max speed:* 128 mph at sea level *Ceiling:* 20,000 ft *Max take-off weight:*
1540 lb *Armament:* 2 × Vickers mg

*Italy entered the war in 1915 with some experience of air fighting – over Tripoli in 1911. Macchi built under licence the Nieuport XI seen equipping this Squadriglia in 1917*

Imperial War Museum

## Ansaldo SVA.5

The SVA series of fighters (Savoia-Verduzio-Ansaldo) represented the best of Italy's indigenous wartime fighter aircraft. If not quite up to other Allied types, it served successfully as a bomber and scout and some remained in service until the mid-1930s

*Span:* 29 ft 10¼ in *Length:* 26 ft 7 in *Engine:* SPA 6A, 220 hp *Max speed:* 143 mph *Max take-off weight:* 2315 lb *Armament:* 2×Vickers mg

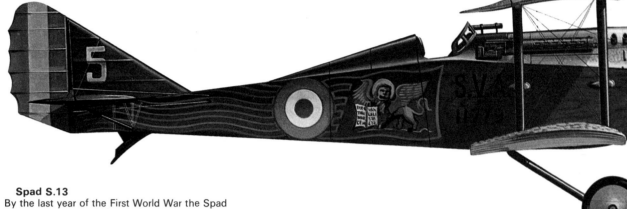

## Spad S.13

By the last year of the First World War the Spad 13 was the standard French fighter, having succeeded the smaller and less powerful S.7. The S.13 was strong and fast, although less manoeuvrable at height than its German contemporaries. Nearly 8500 were built, and the type remained in service with the *Aviation Militaire* until 1923. The S.13 was also operated by Italy and Belgium; it would have been built in large numbers in the United States had the war continued

*Span:* 26 ft 11 in *Length:* 20 ft 8 in *Engine:* Hispano-Suiza, 200 hp *Max speed:* 130 mph at 6500 ft *Ceiling:* 22,300 ft *Max take-off weight:* 1815 lb *Armament:* 2×Vickers mg

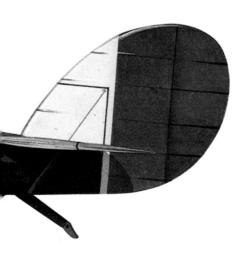

*Right: This searchlight-equipped Spad IX was an unsuccessful attempt to produce an anti-Zeppelin night fighter. Bottom: The Spad SVII and its derivatives were perhaps the best Allied fighters of the war*

Fortunately the RFC and RNAS were both permitted to use the Lewis at an early date, before August 1914. Two years earlier the first trials were held at Farnborough with a B.E., and later with a prototype F.E.2, carrying a weighty Maxim machine gun, but fortunately such a weapon was never issued to the RFC. Instead the light gas-operated Lewis was standardised, and it was used both with and without the fat air-cooling casing that was necessary on non-flying Lewises. The latter invariably had a 47-round drum magazine, but in aerial use a deeper 97-round drum was often used. This enabled longer bursts, or more bursts, to be fired without changing drums, but the actual task of changing the heavy drums was no joke in a single-seater. It was no joke to the observer either, if the aircraft was wildly manoeuvring in a dogfight, but the single-seat fighter pilot had to change drums with the Lewis pulled down from its

*Imperial War Museum*

*Above: The observer on a German two-seater services his Parabellum machine-gun. Below: An RFC wall chart on just how to attack a two-seater and avoid giving the rear gunner his chance*

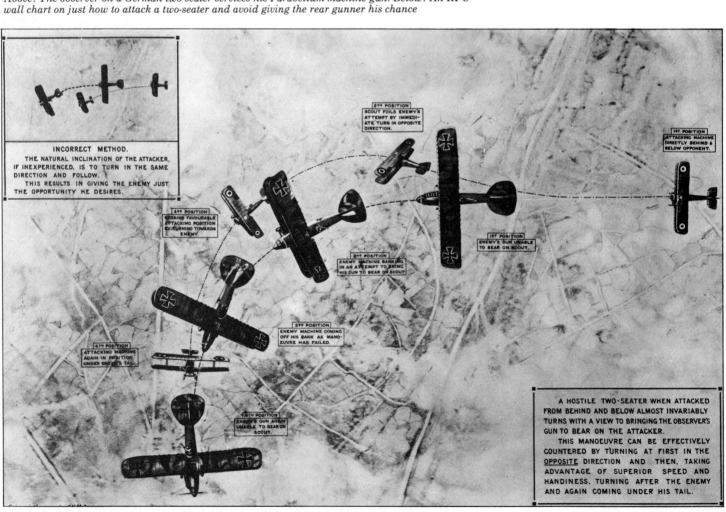

INCORRECT METHOD.
THE NATURAL INCLINATION OF THE ATTACKER, IF INEXPERIENCED, IS TO TURN IN THE SAME DIRECTION AND FOLLOW.
THIS RESULTS IN GIVING THE ENEMY JUST THE OPPORTUNITY HE DESIRES.

2ND POSITION
SCOUT FOILS ENEMY'S ATTEMPT BY IMMEDIATE TURN IN OPPOSITE DIRECTION.

1ST POSITION
ATTACKING MACHINE DIRECTLY BEHIND & BELOW OPPONENT.

3RD POSITION
REGAINS FAVOURABLE ATTACKING POSITION BY TURNING TOWARDS ENEMY.

1ST POSITION
ENEMY'S GUN UNABLE TO BEAR ON SCOUT.

2ND POSITION
ENEMY MACHINE BANKING IN AN ATTEMPT TO BRING HIS GUN TO BEAR ON SCOUT

3RD POSITION
ENEMY MACHINE COMING OFF HIS BANK AS MANOEUVRE HAS FAILED.

4TH POSITION
ATTACKING MACHINE AGAIN IN POSITION UNDER ENEMY'S TAIL.

4TH POSITION
ENEMY'S GUN AGAIN UNABLE TO BEAR ON SCOUT.

A HOSTILE TWO-SEATER WHEN ATTACKED FROM BEHIND AND BELOW ALMOST INVARIABLY TURNS WITH A VIEW TO BRINGING THE OBSERVER'S GUN TO BEAR ON THE ATTACKER.
THIS MANOEUVRE CAN BE EFFECTIVELY COUNTERED BY TURNING AT FIRST IN THE OPPOSITE DIRECTION AND THEN, TAKING ADVANTAGE OF SUPERIOR SPEED AND HANDINESS, TURNING AFTER THE ENEMY AND AGAIN COMING UNDER HIS TAIL.

**Pfalz D.XII**
The D.XII saw only a few weeks' action in the autumn of 1918 as a substitute for the Fokker D.VII, which was in short supply. It could withstand considerable punishment but was sluggish, had a poor rate of climb and was difficult to maintain
*Span:* 29 ft 6 in *Length:* 20 ft 11 in *Engine:* Mercedes, 160 hp *Max speed:* 120 mph at sea level *Ceiling:* 18,500 ft *Max take-off weight:* 1960 lb *Armament:* 2 × Spandau mg

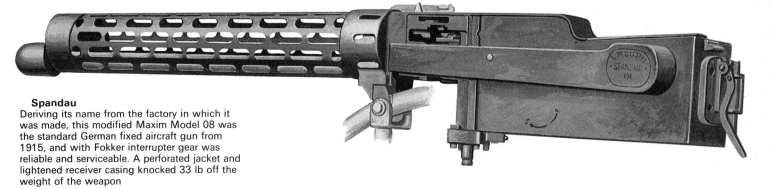

**Spandau**
Deriving its name from the factory in which it was made, this modified Maxim Model 08 was the standard German fixed aircraft gun from 1915, and with Fokker interrupter gear was reliable and serviceable. A perforated jacket and lightened receiver casing knocked 33 lb off the weight of the weapon

**Parabellum**
An early modification of the Maxim to provide a lightweight aircraft gun for flexible mountings, the 7.92-mm gun was standard on nearly every German two-seater
*Weight:* 22 lb *Rate of fire:* 700 rpm

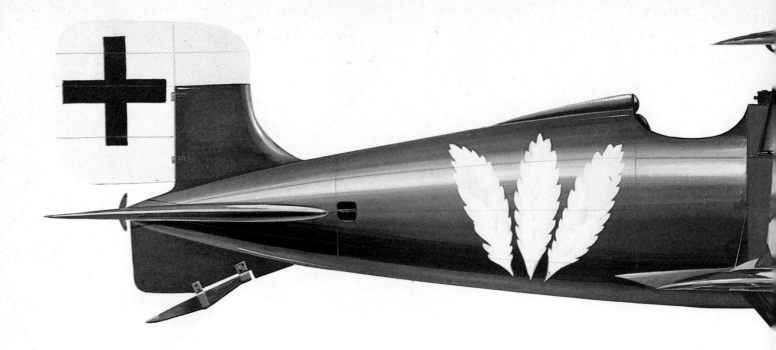

Foster mount into a high-speed airflow while he also tried to fly the aircraft (indeed, he was likely to be engaged in a fierce dogfight). Sometimes the sheer mass of the 97-round drum, combined with 'g' and wind loads, would almost break the pilot's wrist.

Many Lewis were fixed obliquely, firing outwards past the tractor propeller, or upwards. Observers began with plain spigot mounts, some aircraft having as many as four spigot sockets around the cockpit into which the gun could be placed. RFC Capt L A Strange devised a series of mounts, one of which fired obliquely ahead, so that the pilot had to formate on the enemy's quarter (diagonally behind) or else crab along with full rudder from astern; another was a cranked pillar for a Lewis fired by an observer. Many aces, such as Albert Ball, tried various armament schemes and often seemed unable to decide what was best. For observers there was soon no doubt; two-seat fighters, and other aircraft with manually aimed guns, universally adopted the ring mount devised by Warrant Officer F W Scarff of the Admiralty Air Department. The Scarff No 2 ring mount was developed with the help of Sopwith and first used on the 1½-Strutter. It at last allowed the observer to aim freely throughout his entire arc of fire and was a standard fitment in many air forces well beyond 1939. The Bristol F.2B and many other types, including French and American fighters, often had Scarff mounts for twin Lewis, or for the American 0.30-in Marlin or Browning.

Scarff was also one of the inventors of a synchronising gear for fixed guns. Once Fokker's engineers had perfected the mechanical interrupter gear used in the E-series monoplanes, there was no longer any question but that the myopic Allied top brass would swiftly follow suit. One of the earliest Allied synchronising gears was developed by Vickers (Challenger). It was purely mechanical, and so were those developed by Ross, by Sopwith in partnership with Kauper, and Scarff in partnership with the Russian Lt-Cdr Dibovski. In France several schemes, notably the Alkan, followed similar arrangements, answering the urgent need for linkages that would work reliably. But a much better scheme was developed

later in 1916 by a Romanian mining engineer, Georges Constantinesco, in partnership with an Artillery officer Maj G C Colley. The Constantinesco CC gear was not mechanical but hydraulic, linking a motor on the gun trigger by sealed oil pipes with plungers driven by the engine. The pilot had to pump up a suitable pressure in the piping, after which the CC gear linked the engine and gun smoothly and (after some months of trouble in the S.E.5 and 5a, the first installation in service) reliably. One advantage was that the same basic CC gear could be used for almost any kind of engine and any kind of gun, or any number of guns. With modifications, this gear remained standard with many air forces until the Second World War.

Most of the fixed machine guns of the Allies were Vickers, based on the Maxim. This famous name encompassed a great variety of sub-types and calibres from 6.5- to 11-mm, but all were fed by a long belt of ammunition usually joined by metal links. The dominant free gun was the Lewis, designed by an American and made chiefly in Belgium, but another extremely important gun was the Hotchkiss. This was virtually the same as the Benet-Mercier, and thus had the same drawback of being fed by short clips of, usually, 25 rounds. Revelli made a family of guns in Italy, but these were little used in aircraft until the 1920s. By far the most important family of guns of the Central Powers were those based on the Maxim, and they thus had an action identical to that of the Vickers. The basic fixed fighter gun was the MG 08/15, commonly called the Spandau from the armoury near Berlin where it was first made. On board Zeppelins it was used as a free gun with the same water-cooling jacket as the ground guns, but for combat aeroplanes a simple slotted casing admitted cooling air. Ammunition was the standard German 7.92-mm (0.311-in) rimless type. The German observer gun was a development commonly called the Parabellum, after the code-name for the Deutsche Waffen und Munitionsfabriken where it was developed. It was not easy to make the basic Maxim action any better, but the air-cooled Parabellum was notably lighter, fired at up to 700 rds/min and was

**Siemens-Schuckert D.IV**
Tough, manoeuvrable and with a very high power-to-weight ratio, the D.IV had an incredible rate of climb but arrived too late and in too few numbers to restore Germany's declining fortunes on the Western Front
*Span:* 27 ft 4¾ in *Length:* 18 ft 8½ in *Engine:* Siemens-Halske Sh IIIa, 200 hp *Max speed*: 119 mph *Ceiling:* 26,240 ft *Max take-off weight:* 1620 lb *Armament:* 2 × Spandau mg

*German pilots confer moments before take-off. By mid-1918 the Allies had wrested almost complete mastery in the air over the Western Front*

**Fokker D.VII**
The D.VII won a design competition for a single-seat fighter in January 1918 and reached the front only four months later. Advanced features included a thick-section cantilever wing, fuselage of welded steel tubing and car-type radiator. These combined to give good high-altitude manoeuvrability and resistance to battle damage combined with docile handling qualities, although the top speed was not impressive
*Span:* 29 ft 3½ in *Length:* 22 ft 9 in *Engine:* Mercedes, 160 hp *Max speed:* 117 mph at 3200 ft *Ceiling:* 20,000 ft *Max take-off weight:* 1936 lb *Armament:* 2 × Spandau mg

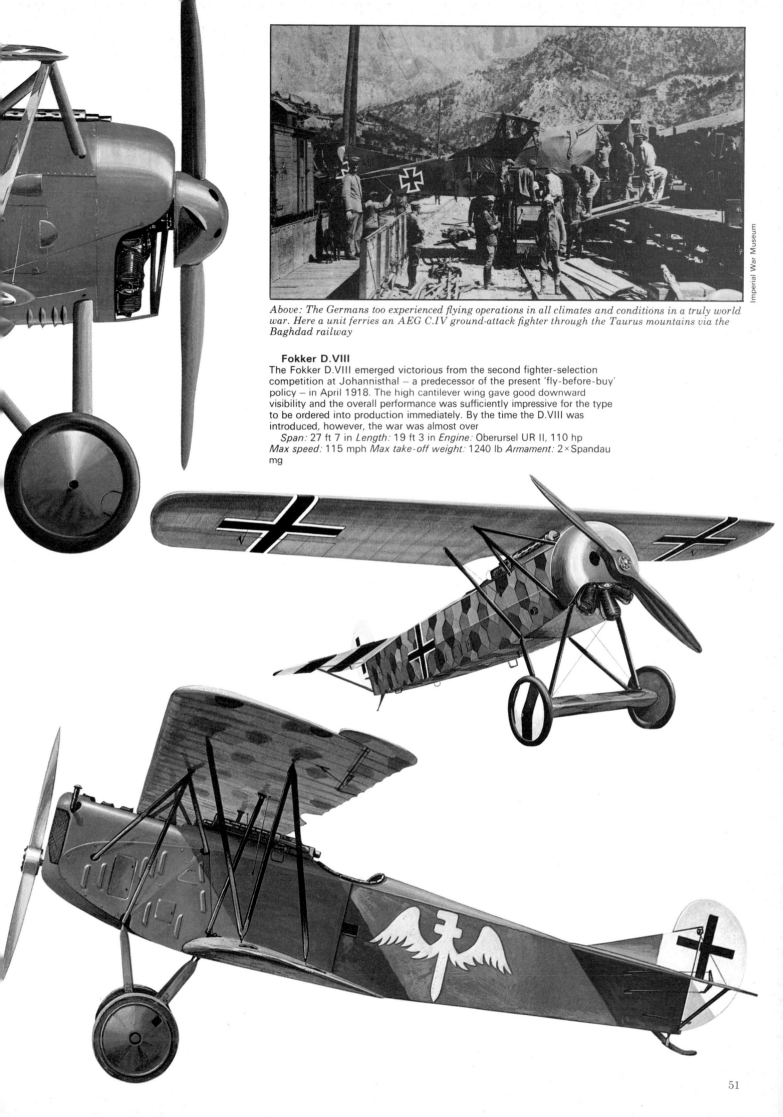

*Above: The Germans too experienced flying operations in all climates and conditions in a truly world war. Here a unit ferries an AEG C.IV ground-attack fighter through the Taurus mountains via the Baghdad railway*

**Fokker D.VIII**

The Fokker D.VIII emerged victorious from the second fighter-selection competition at Johannisthal — a predecessor of the present 'fly-before-buy' policy — in April 1918. The high cantilever wing gave good downward visibility and the overall performance was sufficiently impressive for the type to be ordered into production immediately. By the time the D.VIII was introduced, however, the war was almost over

*Span:* 27 ft 7 in *Length:* 19 ft 3 in *Engine:* Oberursel UR II, 110 hp *Max speed:* 115 mph *Max take-off weight:* 1240 lb *Armament:* 2×Spandau mg

# Austria-Hungary: Fighters of the Dual Monarchy

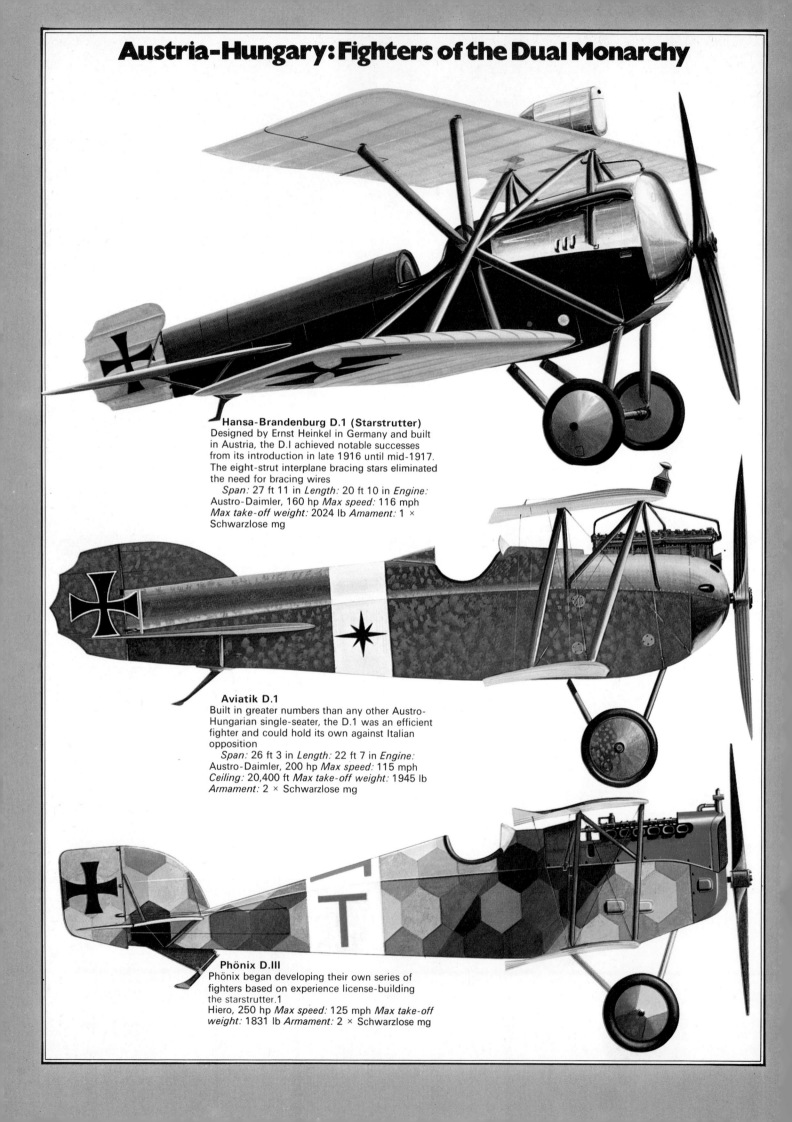

**Hansa-Brandenburg D.1 (Starstrutter)**
Designed by Ernst Heinkel in Germany and built
in Austria, the D.I achieved notable successes
from its introduction in late 1916 until mid-1917.
The eight-strut interplane bracing stars eliminated
the need for bracing wires
   *Span:* 27 ft 11 in *Length:* 20 ft 10 in *Engine:*
Austro-Daimler, 160 hp *Max speed:* 116 mph
*Max take-off weight:* 2024 lb *Amament:* 1 ×
Schwarzlose mg

**Aviatik D.1**
Built in greater numbers than any other Austro-
Hungarian single-seater, the D.1 was an efficient
fighter and could hold its own against Italian
opposition
   *Span:* 26 ft 3 in *Length:* 22 ft 7 in *Engine:*
Austro-Daimler, 200 hp *Max speed:* 115 mph
*Ceiling:* 20,400 ft *Max take-off weight:* 1945 lb
*Armament:* 2 × Schwarzlose mg

**Phönix D.III**
Phönix began developing their own series of
fighters based on experience license-building
the starstrutter.1
Hiero, 250 hp *Max speed:* 125 mph *Max take-off
weight:* 1831 lb *Armament:* 2 × Schwarzlose mg

fed by a long fabric belt neatly wrapped in a large spool on one side. The only other aircraft machine gun of the pre-1918 era that deserves mention is the outstanding Gast, based on the British Bethel Burton of 1886 but re-patented by Carl Gast in Germany in 1916. Its action served left and right recoiling barrels, each of which locked and fired the other. Fed by twin drums clipped on from above as one unit, the Gast fired at the unprecedented rate of 1600 rds/min (with spring-loaded buffer, 1800). It was in very large-scale production in mid-1918, but it did not see action and the existence of the Gast was not even known to the Allies until 1921!

With a gun such as the Gast even a single second of well-aimed fire could be decisive. But from before 1914 a few far-seeing people had begun to debate the possible advantage of having a gun which, while firing more slowly, could destroy its enemy with one shot. This argument raged until long after the Second World War and has not been stilled even today. Certainly, in 1913–14 the French official view was that, should air combat ever come to pass, it would be better to have a large-calibre gun, and throughout the First World War the French were the leading exponents of the *Avion Canon*. The 37-mm Hotchkiss cannon, firing a shell weighing about 700 grams ($1\frac{1}{2}$ lb), was mounted for the first time in a pusher Voisin at the Paris air show in 1911, but apparently did not fly until 1913. The gun was first used in the primitive-looking but tough and quite popular Breguet de Chasse (Breguet fighter) types BUC and BLC in late 1915. This slow pusher did not have the performance to catch targets, though the 37-mm was effective against surface targets. Later Breguets were faster, and in the formidable Type 11 Corsaire there were twin gun nacelles, one having one or two machine guns and the other the cannon. The same gun was used in many other aircraft, and by 1918 had even been fired in the air in a number of small single-seaters (mainly advanced Spads) with the gun firing through the hub of the propeller. This was to remain a favoured type of cannon installation until the end of the piston era.

Britain made little use of guns of calibre larger than the traditional 0.303-in, and confined such use to trials with the Vickers

*The Supermarine Nighthawk, an anti-Zeppelin fighter designed to fly standing patrols and intercept raiders*

Maxim in the 0.45-in calibre, the Vickers 1-pdr 'Pom Pom' and a small range of unadopted guns, plus the only one to see operational service, the Coventry Ordnance Works (C.O.W.) $1\frac{1}{2}$-pdr. This big and weighty gun was much longer and harder-hitting than the low-velocity Hotchkiss of similar 37-mm calibre, and was mainly confined to large aircraft, though it did form the main armament of the three-seat F.E.4 fighter. Likewise the Germans restricted their excellent Becker 20-mm cannon to bombers and Zeppelins, and would probably have done the same with their very advanced later cannon such as the Semag, Rhein-metall (Ehrhart) and Szakats.

There was yet another type of gun developed specifically for fighters: the re-coilless type. Britain developed two important members of this family, the Davis and the Vickers-Crayford. The Davis fired normal shells (the most common model fired 900-gram, 2-lb projectiles) and simul-taneously expelled propellant gases and the spent case to the rear to approximately cancel out the recoil. The Robey-Peters

prototypes were designed to carry at least one Davis 2-pdr, aimed by a gunner over whose shoulder passed the rear of the barrel with its venturi nozzle. Other Davis installations were fixed, two of them (AD Scout and Blackburn Triplane) having a pusher rotary engine just behind the gun! How these guns worked in such an instal-lation has become a mystery. The gun by the Vickers works at Crayford was totally different, for it fired rocket-propelled shells. Two of the aircraft built to carry this fearsome weapon were the Royal Aircraft Factory N.E.1 and Vickers' own F.B.25, both of which were anti-Zeppelin night fighters. Another aircraft in this class, built to carry the Davis in the $1\frac{1}{2}$-pdr size as well as two Lewis, was the Pemberton-Billing P.B.31E. This incredible aircraft, flown in February 1917, was intended to loiter for hours waiting for Zeppelins. A veritable flying fortress, it had two engines, four wings, an electric generating set and searchlight, a battery of gunners and a pilot who could hardly see out. Strange to think that the P.B. company was renamed Supermarine, so that this weird 'fighter' was an ancestor of the Spitfire.

### Sopwith 7F.1 Snipe
The Snipe, intended as a replacement for the Camel, became operational only in September 1918 and therefore saw most of its service after the First World War. Electrical heating and oxygen equipment were standard, and the Snipe would have replaced Camels in home-defence squadrons and on shipboard duties if the war had continued. The type was finally withdrawn in 1927
*Span:* 31 ft 1 in *Length:* 19 ft 10 in *Engine:* Bentley B.R.2, 230 hp; Clerget 11Eb, 200 hp; or ABC Dragonfly, 320 hp *Max speed:* 121 mph at 10,000 ft *Ceiling:* 19,500 ft *Max take-off weight:* 2020 lb *Armament:* 2×Vickers mg, 4×25-lb Cooper bombs

# THE TWENTIES

While those who survived the First World War danced the Charleston or bootlegged whisky into the United States, fighter technology progressed in different countries at paces that ranged from mediocre to very slow indeed. Paradoxically, the victorious Allies were those who led the race to be slowest, and their young air services (in the case of Britain, an independent Royal Air Force, which had to fight for its very existence) had to shrink to the stature of an exclusive flying club, where everyone knew everyone else. Airfields lay derelict, littered with hundreds of rotting airframes and rusting engines. Only relatively small numbers of equipment items were needed for the small peacetime air forces, and every penny and cent for defence was argued by governments before being grudgingly allocated. Yet in such countries as Japan and the Soviet Union nothing rated higher than fighter technology, and budgets grew annually. Even the defeated Germans found ways of advancing the art of fighter design, mainly under cover or in friendly states such as the Soviet Union, Sweden, Switzerland and Spain.

To young Fokker the Armistice meant only the slightest of hiccups. Under the noses of the Allied Control Commission, he packed up tons of aeronautical products, including more than 400 of the latest engines, about 120 Fokker D.VII aircraft and 20 of the new D.VIII monoplanes, and smuggled them across the frontier into his native Holland. There he established a prosperous and rapidly growing aircraft works building everything from fighters to airline transport aircraft. Others tried to do the same, but orders were few and competition fierce. For the engine-builders times were slightly better, because in the years following 1918 engines developed very greatly. First World War engines had been produced in a great hurry, to state-of-the-art designs that generally went back in concept to 1907–13. Reliability was exceedingly poor, and total life seldom greater than 100 hours and often only half as much. But in the new peacetime air forces engines had to work reliably for many years, often in extremes of climate. So new engines were developed, better planned and designed, and made of improved steels and other alloys. The concept of fatigue, which in 1918 had scarcely been identified, was reduced to precise numerical terms so that highly stressed parts could last for 1000 operating hours, or longer, without a hint of failure. The whole business of aircraft design and manufacture was put on a sound footing, with precise inspection of the designs, the raw material and the finished products. Structural breakup in flight, common in the First World War, became almost unheard-of, despite growing engine power, growing weights and progressive increases in speed, climb and ceiling.

One of the few wartime engines to remain important through the 1920s was the American Liberty. This had from the start shown that hasty design and rapid mass-production could still yield a superior product, and though it had failed to improve certain European combat types (notably the Bristol Fighter) the Liberty gave exemplary service in many others (such as the D.H.9a). In the American Packard Le Père LUSAC-II fighter a 425-hp Liberty gave sparkling performance despite large size, and had the war continued these machines would for the first time since the Wrights have shown the quality of US aeronautical products. Such encouragement was needed, because in 1917 the US government had grandly decided on central Federal control of aircraft design and production and got the potentially vast programme into a sad mess. In 1918 it turned the job back to industry, the LUSAC being a good result. Another, too late for the war, was the Thomas-Morse MB-3, designed to beat the Spad XIII which was the pride of the American Expeditionary Force which established a fine reputation on the Western front in 1918.

Though the MB-3 suffered from many faults, and had the same Hispano-Suiza engine as the Spad but made by Wright, it had an important influence on a little-known company in the far northwest of the United States. In 1920, a year after the prototype MB-3 had flown, the US Army decided to buy 200 (a remarkable number for the lean year 1920). The policy in the USA at that time was that the Army could put any production job out to the lowest bidder, no matter who had designed the item, and in this case the low bidder was a company called Boeing. Building the 200 improved MB-3A scouts established Boeing as a major force in aviation, and began one of the greatest families of fighters of the interwar era. The US Army Air Service called fighters 'pursuits', and gave them numbered designations prefaced by 'PW', for 'pursuit, water-cooled engine'.

*First flown in April 1918, the Westland Wagtail was so hampered by problems with its ABC engine that it missed the war and the prototypes served as flying testbeds*

**Packard-Le Père LUSAC II**
Two-seat fighter which would have been the US equivalent of the Bristol Fighter had it appeared before the armistice. Only 30 were built
*Span:* 41 ft 7 in *Length:* 25 ft 6 in *Engine:* Liberty 12A, 400 hp *Max speed:* 132 mph at 2000 ft *Ceiling:* 20,000 ft *Max take-off weight:* 3746 lb *Armament:* 2× Marlin mg, 2× Lewis mg

**Thomas Morse S-4C**
Equipped almost entirely with French types, the American Expeditionary Forces received only a trickle of home-built designs before the armistice. One of these was the S-4C, a US-built fighter trainer of 1917. Experience with the S-4C gave the company a lead in designing the MB-3, produced in large numbers by Boeing
*Span:* 26 ft 6 in *Length:* 19 ft 10 in *Engine:* Le Rhône, 80 hp *Max speed:* 95 mph *Ceiling:* 15,000 ft *Max take-off weight:* 1374 lb

*The Nieuport 29, the superlative fighter design of Gustave Delage built around an in-line Hispano-Suiza engine. It arrived too late to see combat but figured prominently in several post-war air races*

Boeing's first original fighter design was the XPW-9 (X for experimental) of 1923. Boeing had no order for this new pursuit, but recognised that, in the tough aviation climate of the early 1920s, it would have to work hard to survive. The thing to do was demonstrate what the US Army did not fully appreciate: that the MB-3A represented a wartime technology that could be improved upon significantly. Like the Spads on which it was based, the Thomas-Morse fighters were made entirely of wood, with fabric covering. Boeing did not object to wooden wings, made efficiently by glueing and pinning, but considered the fuselage a poor structural concept. Traditional wartime fighters had a fuselage made from four strong longerons of hardwood, held together by numerous vertical and cross-members. The parts could not just be glued; they had to be linked by complicated fittings, usually of steel, which were attached by multiple bolts. The fittings also anchored steel bracing wires which had to be tightened after assembly. Not only was the construction complex and expensive but it resulted in a fuselage whose shape depended on the tightness of each wire. Throughout the service life of the aircraft, maintenance engineers – usually called airframe fitters, or riggers – had to crawl up and down the fuselage adjusting the tightness of the wires, to make the aircraft fly in a straight line and with proper loads on all the parts. But Fokker had perfected a way of making the fuselage of his wartime fighters from welded steel tubing. This was not new; the Wright brothers, who were bicycle makers, used the same method. What Fokker, and his designer Platz, did was make gas-welded joints in a complete fuselage while all the parts were held in their exact positions in a fixture. It was much cheaper, weighed about the same, and meant that the structure needed no maintenance because its shape never changed. But Boeing could see how even this could be improved.

When Frank Tyndall took off at Seattle in the Boeing Model 15 on 29 April, 1923, he was flying the newest fighter, technically, in the world. The fuselage was made of steel tubes joined by an improved method of welding with an electric arc. The engine was the new Curtiss D-12, rated at 435 hp. The water radiators were moved from the sides of the fuselage to a new position under the engine, with the air ducted through a kind of tunnel. The two guns, one of 0.30-in and the other of the hard-hitting 0.50-in calibre, were neatly fixed above the engine under the cowling. The wooden wings had more span than most earlier fighters, giving increased area for tighter turns. Boeing again copied Platz's methods in using a relatively thick high-lift section, with a wooden spar built up into a slender box. Though considerably larger than the Thomas-Morse, the Model 15 flew at 163 mph, compared with 140 (itself an excellent speed), and after testing by the Army at McCook Field it was accepted as the XPW-9. There followed a long series of production PW-9 fighters, and the similar FB-1 to FB-5 for the Navy and Marine Corps (FB= Fighter, Boeing).

Competition for Boeing came from Curtiss, who used their D-12 engine in the first of the famous Hawk series, the PW-8 of 1923. This was if anything even faster, partly because it had ply-covered wings with smooth and accurate profile, the skin of the upper plane incorporating completely flush cooling radiators. The latter enabled the streamlined pointed-nose installation of a water-cooled in-line engine to be achieved without the drag of a radiator (which generally made the speeds of water-cooled and air-cooled radial-engined fighters about the same, despite the fact that radials seemed to be blunt and badly streamlined). Curtiss had perfected the skin radiator in racers of 1921–22, but in military service it gave too much trouble, besides being extremely vulnerable to gunfire. But on 23 June, 1924, one of the speedy PW-8 Hawks achieved fame by flying across America in a day. Lt Russell Maughan took off from Long Island and flew in stages, travelling with the Sun, until he reached San Francisco 21 hr 48 min later, an average of 117 mph including all the refuelling stops. From the PW-8 series came the long succession of Army Hawks (P-1 onwards) and the Navy F6C series.

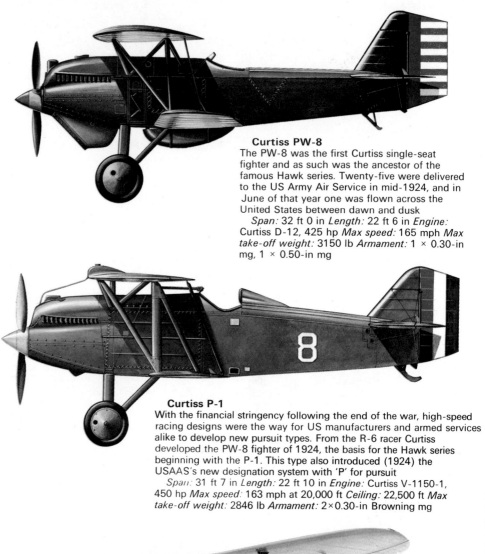

**Curtiss PW-8**
The PW-8 was the first Curtiss single-seat fighter and as such was the ancestor of the famous Hawk series. Twenty-five were delivered to the US Army Air Service in mid-1924, and in June of that year one was flown across the United States between dawn and dusk
*Span:* 32 ft 0 in *Length:* 22 ft 6 in *Engine:* Curtiss D-12, 425 hp *Max speed:* 165 mph *Max take-off weight:* 3150 lb *Armament:* 1 × 0.30-in mg, 1 × 0.50-in mg

**Curtiss P-1**
With the financial stringency following the end of the war, high-speed racing designs were the way for US manufacturers and armed services alike to develop new pursuit types. From the R-6 racer Curtiss developed the PW-8 fighter of 1924, the basis for the Hawk series beginning with the P-1. This type also introduced (1924) the USAAS's new designation system with 'P' for pursuit
*Span:* 31 ft 7 in *Length:* 22 ft 10 in *Engine:* Curtiss V-1150-1, 450 hp *Max speed:* 163 mph at 20,000 ft *Ceiling:* 22,500 ft *Max take-off weight:* 2846 lb *Armament:* 2×0.30-in Browning mg

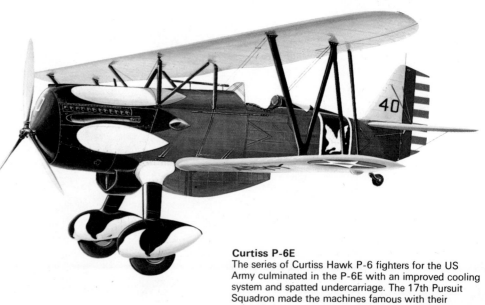

**Curtiss P-6E**
The series of Curtiss Hawk P-6 fighters for the US Army culminated in the P-6E with an improved cooling system and spatted undercarriage. The 17th Pursuit Squadron made the machines famous with their painted claws on the spats and diving eagle insignia
*Span:* 31 ft 6 in *Length:* 22 ft 7 in *Engine:* Curtiss V-1570-23 Conqueror C, 600 hp *Max speed:* 198 mph *Ceiling:* 24,700 ft *Max take-off weight:* 3392 lb *Armament:* 2×0.30-in mg

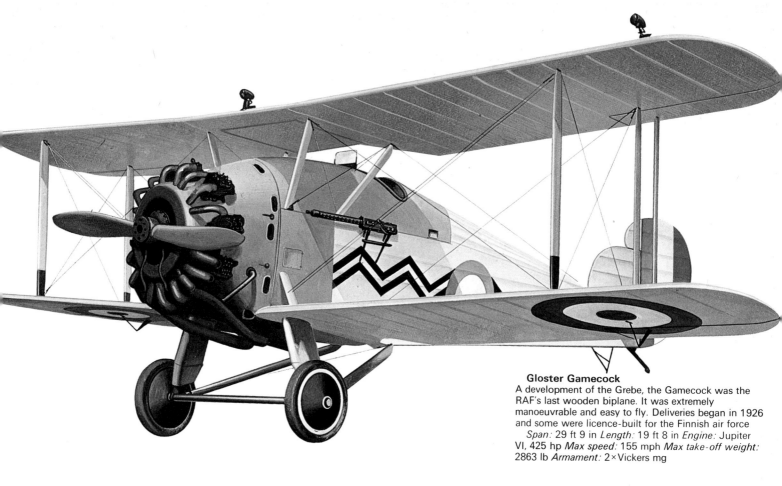

**Gloster Gamecock**
A development of the Grebe, the Gamecock was the RAF's last wooden biplane. It was extremely manoeuvrable and easy to fly. Deliveries began in 1926 and some were licence-built for the Finnish air force
*Span:* 29 ft 9 in *Length:* 19 ft 8 in *Engine:* Jupiter VI, 425 hp *Max speed:* 155 mph *Max take-off weight:* 2863 lb *Armament:* 2 × Vickers mg

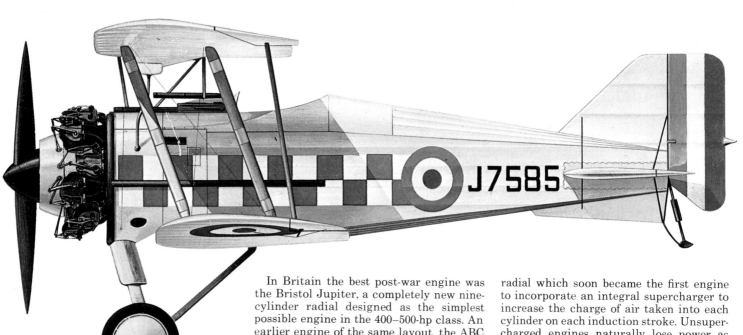

**Gloster Grebe**
The Grebe was, together with the Armstrong Whitworth Siskin, the first new fighter to be ordered into quantity production for the Royal Air Force after the First World War. The high-lift upper wing gave good take-off and climb performance, while the lower wing was designed for the minimum drag at cruising speeds. The total order was for 130 Grebes, some of which took part in parasite-fighter experiments with the R.33 airship
*Span:* 29 ft 4 in *Length:* 20 ft 3 in *Engine:* Armstrong Siddeley Jaguar, IVA, 425 hp *Max speed:* 151 mph at 5000 ft *Ceiling:* 23,000 ft *Max take-off weight:* 2622 lb *Armament:* 2 × Vickers mg

In Britain the best post-war engine was the Bristol Jupiter, a completely new nine-cylinder radial designed as the simplest possible engine in the 400–500-hp class. An earlier engine of the same layout, the ABC Dragonfly, had been selected in 1918 for almost all the new crop of fighters that were coming into mass-production at the Armistice. Had the war continued there would have been a crisis in British fighters, because the Dragonfly was not only a failure but its faults could not be cured, and over 1000 had been built when this was discovered. A few aircraft, such as the Badger and Nighthawk, were re-engined with the Jupiter, but it was not until 1923 that the RAF wanted any new post-war fighters and the choice fell on the Jupiter-engined Hawker Woodcock. Another type bought in quantity was the Armstrong Whitworth Siskin, developed from the Siddeley Siskin first flown with a Dragonfly in 1919. In this case the engine was the Siddeley company's own Jaguar, a neat 14-cylinder two-row

radial which soon became the first engine to incorporate an integral supercharger to increase the charge of air taken into each cylinder on each induction stroke. Unsupercharged engines naturally lose power as the aircraft climbs into thinner air, and this was especially serious for fighters, which need the highest possible performance at all heights. Supercharging enabled the full flight performance to be maintained up to high altitudes, giving such aircraft a pronounced edge in combat. In 1924 the RAF received the first Siskin III fighters, with a completely new structure made entirely from thin steel tubing and strip (but used in a direct replacement for wood, and with fabric covering). The Mk IIIA Siskin, with the supercharged engine, followed in 1927.

Two other RAF fighters of this period were the Gloucestershire (Gloster) Grebe, with Jaguar engine, and Gamecock, with Jupiter. Suddenly attracted by the 'streamlined' Curtiss D-12 engine, Fairey built a

**Bristol Bulldog**
The basis of the Bulldog's design was an immensely strong all-steel frame with a forward-placed cockpit giving the pilot an excellent field of view. Many examples were exported, and some were still in front-line service with several air forces ten years after entering RAF service in 1929
*Span:* 33 ft 10 in *Length:* 24 ft 9 in *Engine:* Bristol Jupiter VII F, 530 hp *Max speed:* 174 mph at 10,000 ft
*Ceiling:* 27,000 ft *Max take-off weight:* 3490 lb
*Armament:* 2×Vickers mg

fast bomber called the Fox that was faster than most fighters, and followed with the Firefly fighter later used by Belgium. But in 1927 the Jupiter radial was again chosen for the RAF's next standard fighter, the Bristol Bulldog. The water-cooled Rolls-Royce Kestrel was selected for RAF Hawker Fury single-seaters, and two-seat Demon fighters that succeeded the Bristol Fighter of 1917, but overseas customers chose the air-cooled radial engines. The RAF's final biplane fighters were the Gloster Gauntlet of 1935 and Gladiator of 1937, both powered by a development of the Jupiter called the Mercury. Like the Sopwith Snark of 1918 the Gloster company had fitted six machine-guns into the S.S.19 of 1932, but when the S.S.19 became the Gauntlet the armament reverted to the standard two Vickers guns that had been on almost every RAF fighter since the Camel. But there had been many interesting variations.

For example, in 1924 the Air Ministry issued Specification 4/24 for a heavy twin-engined fighter carrying two cannon. Between 1918 and 1935 no British aircraft was designed to carry any cannon other than the ponderous 37-mm C.O.W. gun, though plenty of smaller, faster-firing cannon were available from other countries. The two aircraft built to the 4/24 specification were poor designs, the Westland Westbury being a biplane and the Bristol Bagshot a braced monoplane, both having two Jupiters, and the cannon aimed by hand from the nose and a mid-upper cockpit. Soon the officials forgot about such machines, but with Specification F.29/27 in 1927 they suddenly asked for a cannon-armed single-seater. The result was the Westland C.O.W. Gun Fighter, a trim low-wing monoplane, and the archaic-looking Vickers 161 biplane pusher, both of which carried the heavy gun firing obliquely upwards at 55°. Again, once

built, these aircraft were forgotten.

Other countries had a wider choice of armament, and many used calibres of 0.5-in (12.7-mm), 20-mm and 23-mm, though the rifle-calibre gun remained by far the most common. The standard French fighters of the 1920s were the Loire-Gourdou-Leseurre LGL 32 and a family of Nieuport-Delage machines of which the main type was the ND 62 and its derivative the 622. The LGL was a tough high-wing monoplane with Gnome-Rhône-built Jupiter, while the ND 62 was a sesquiplane (biplane with the lower plane extremely small) with water-cooled Hispano-Suiza 12M engine. A small number of Wibault scouts also entered service, notable for their patented form of all-metal construction with light-alloy skin. Though the skin did not bear flight loads it was much stronger than fabric or ply, and was widely copied or built under licence (for example, by Vickers for fighters in Britain). In Japan all military aircraft before the 1930s were designed by Western engineers, almost all from Britain, while the Soviet Union was so engrossed in settling the protracted civil war and re-building the economy it had little time for breaking new ground in aviation (though Polikarpov's I-3 single-seat biplane fighter of 1927 reached the excellent speed of 186 mph).

Fokker continued to build fighters in Holland, but competition was severe. In newly created Czechoslovakia several companies built outstanding fighters. Fiat in Italy began a long series of C.R. (*Caccia Rosatelli*, ie fighter designed by Rosatelli) models, Emile Dewoitine produced an equally long series of outstanding high-wing monoplane fighters in France and Switzerland, and a brilliant Polish designer, Zygmunt Pulawski, chose the same layout for a series of PZL fighters that were even

**PZL P.7**
This high-wing all-metal monoplane led the world when it entered service with the Polish air force in 1931
*Span:* 33 ft 10 in *Length:* 23 ft 6 in *Engine:* Skoda-Jupiter F. VII, 485 hp *Max speed:* 203 mph *Max take-off weight:* 3047 lb *Armament:* 2×Vickers mg

better (beginning with the 195-mph P.1 of 1929). Meanwhile, in the United States in 1925 a team of gifted engineers broke away from the powerful Wright company and founded Pratt & Whitney Aircraft, building not aircraft but engines. Their first product was an outstanding 400-hp radial, named the Wasp. It transformed the fighter scene, sweeping away the pointed noses of the water-cooled engines and making the air-cooled radial dominant throughout military aviation. Boeing's PW-9 and FB fighters turned into the F2B, F3B and F4B for the Navy and Marines and P-12 family for the Army, while Curtiss developed a diverse family of Hawks of different sizes and shapes and eventually retractable landing gear in the 1930s.

*Wibault W.11 experimental French parasol-winged fighter of 1928*

### Hawker Demon

A fighter version of the highly successful 1928 Hawker Hart day bomber, the Demon two-seat escort fighter went into service from April 1933. The Osprey was a navalised reconnaissance version for service with the fleet

*Span:* 37 ft 3 in *Length:* 29 ft 1 in *Engine:* Rolls-Royce Kestrel II, 581 hp *Max speed:* 181 mph at 15,000 ft *Ceiling:* 24,500 ft *Max take-off weight:* 4464 lb *Armament:* 2×Vickers mg, 1×Lewis mg

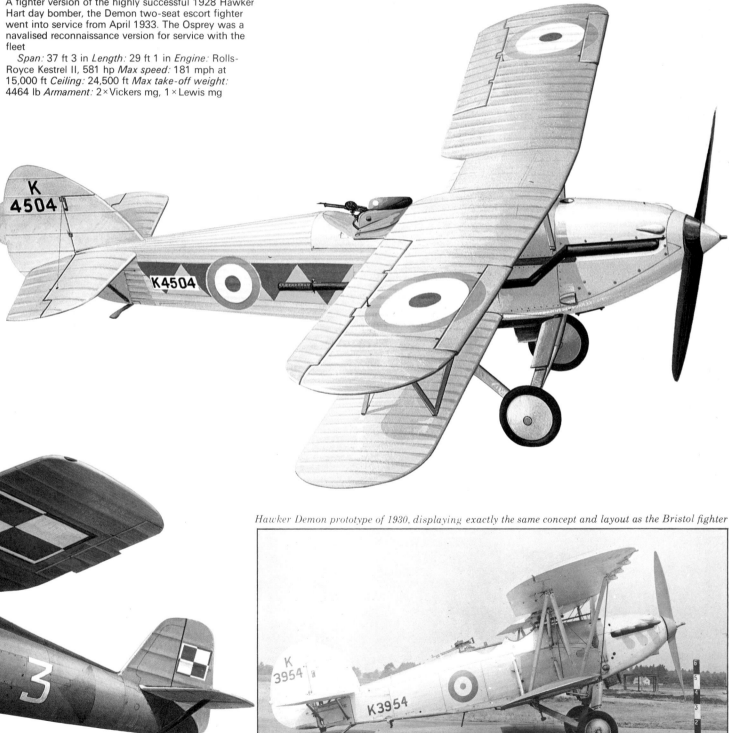

*Hawker Demon prototype of 1930, displaying exactly the same concept and layout as the Bristol fighter*

# THE THIRTIES: TIME OF TESTING

The decade that led up to the Second World War was as crucial and uncertain as any in the whole of fighter history. None of the old problems had been resolved. Senior officers, procurement officials, chief designers and combat-ready pilots, all had firm views on what was best, often failing to listen to those whose views were in violent disagreement.

Sydney Camm, of Hawker Aircraft, liked water-cooled in-line engines because the aircraft built around them looked streamlined. Almost everyone in the United States ignored such engines, until the last Curtiss V-1570 Conqueror came off the line in 1935 and pointed-nose Americans were temporarily extinct. Italian pilots passionately clung to the traditional fabric-covered biplane, sacrificing speed, firepower and everything else for the best possible manoeuvrability and pilot view. Japan's rapidly growing army and navy air forces thought along similar lines, but had no objection to the new stressed-skin monoplanes. In the vast air fleets of the Soviet Union, fighters were sound and conventional, but at the test establishments were unconventional experiments that explored more radical ideas than in any other country. In Britain immense talent was seldom permitted to depart from tradition – the fabric-covered biplane with two Vickers machine-guns – though in industry and research airfields many new ideas were being tried. And in some countries, not including Britain, bigger fighters were continually appearing, either as long-range escorts for bombers or as vaguely conceived aerial battleships with guns pointing in all directions.

France was especially drawn towards the latter, despite the poor showing of the Blériot 127/2 of 1929 which in the early 1930s saw service with the *Aviation Militaire* (*Armée de l'Air* from 1933). Like the types which followed, the 127/2 was a lumbering monoplane with the appearance of a bomber, an unusual feature being the location of gunners in the tails of the long nacelles. Various *Multiplace de Combat* designs followed, but the next to serve in quantity was the Potez 540, first delivered in 1934. This again looked like a bomber, and could also fly bombing and reconnaissance missions. In 1934 its speed of 193 mph was excellent for so large an aircraft, but like the Amiot 143 which followed it the ungainly Potez was really a bomber, and quite incapable of holding its own in air combat. There were numerous attempts in other countries to build large multi-seat fighters relying for their effectiveness not on manoeuvrability but on firepower, and the concept was resurrected in 1943 with special escort versions of the US Army Air Force heavy bombers. It was the exact opposite of the concept followed by the Italians and Japanese, in which everything possible was done to make the fighter lighter, even at the expense of leaving out all but the lightest armament.

Indeed, at intervals from 1915 onwards there have appeared 'light fighters' of one sort or another, intended either to have better manoeuvrability, or to be cheaper (and thus purchasable in greater quantity), or for various other reasons. In the jet era in the early 1950s there was a sudden crop of small lightweight fighters largely because the ordinary kind seemed to be getting impossibly large, complex and expensive, with the prospect that they could never be kept serviceable. But between the wars there were other reasons. France named its small fighter prototypes *Le Type Jockey*, but though many were built none was adopted as a standard type. The nearest miss was the beautiful little Caudron-Renault C.714, which was just getting into production at the very end of the 1930s after the start of the Second World War. Powered by a slender 450-hp Renault air-cooled V12 engine, this stemmed from a long line of extremely efficient little racers. Built cheaply of wood, the C.714 had four machine-guns in fairings under the wings, and might have been effective if it had been ordered earlier. Its speed and climb were comparable with those of the standard M.S.406 of the *Armée de l'Air*, which had twice the power.

*On the flight line at Boeing's Seattle plant is this beautiful F4B-4 for the US Navy.*

**Arado Ar 68E**
After the *Luftwaffe* came into the open in 1935, its chief fighters such as the Ar 68 and the He 51 stuck to the proven biplane formula despite the imminent appearance of the *Luftwaffe's* first monoplane fighters. Developed later than the He 51, the Ar 68 was little better and by 1940 they were relegated to training
*Span:* 36 ft 1 in *Length:* 31 ft 2 in *Engine:* Junkers Jumo 210Ea, 680 hp *Max speed:* 202 mph at 13,125 ft *Max take-off weight:* 4435 lb *Ceiling:* 26,575 ft *Armament:* 2 × 7.9-mm MG 17 mg

Smithsonian Institution

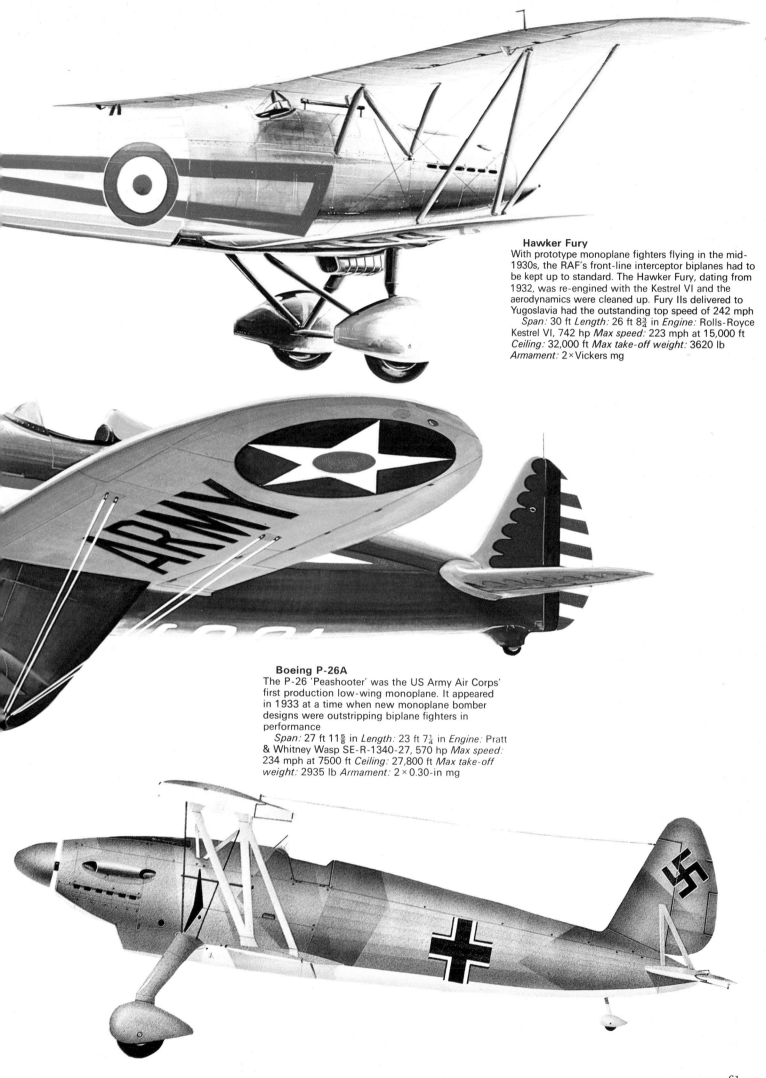

**Hawker Fury**
With prototype monoplane fighters flying in the mid-1930s, the RAF's front-line interceptor biplanes had to be kept up to standard. The Hawker Fury, dating from 1932, was re-engined with the Kestrel VI and the aerodynamics were cleaned up. Fury IIs delivered to Yugoslavia had the outstanding top speed of 242 mph
*Span:* 30 ft *Length:* 26 ft $8\frac{3}{4}$ in *Engine:* Rolls-Royce Kestrel VI, 742 hp *Max speed:* 223 mph at 15,000 ft *Ceiling:* 32,000 ft *Max take-off weight:* 3620 lb *Armament:* 2×Vickers mg

**Boeing P-26A**
The P-26 'Peashooter' was the US Army Air Corps' first production low-wing monoplane. It appeared in 1933 at a time when new monoplane bomber designs were outstripping biplane fighters in performance
*Span:* 27 ft $11\frac{5}{8}$ in *Length:* 23 ft $7\frac{1}{4}$ in *Engine:* Pratt & Whitney Wasp SE-R-1340-27, 570 hp *Max speed:* 234 mph at 7500 ft *Ceiling:* 27,800 ft *Max take-off weight:* 2935 lb *Armament:* 2 × 0.30-in mg

# Fighters over Spain: Dress rehearsal for a second round

**Polikarpov I-15**
*Span:* 29 ft 11½ in *Length:* 20 ft 7½ in *Engine:* M-25, 700 hp *Max speed:* 224 mph *Ceiling:* 32,800 ft *Max take-off weight:* 3135 lb *Armament:* 4×7.62-mm ShKAS mg

**Polikarpov I-16**
*Span:* 29 ft 6½ in *Length:* 20 ft 1¼ in *Engine:* Shvetstov M-62, 1000 hp *Max speed:* 362 mph at sea level *Max take-off weight:* 4564 lb *Ceiling:* 29,530 ft *Armament:* 4×7.62-mm ShKAS mg

**Messerschmitt Bf 109B-2**
*Span:* 32 ft 6 in *Length:* 28 ft 7 in *Engine:* Jumo 210Da, 650 hp *Max speed:* 280 mph at 10,826 ft *Ceiling:* 29,530 ft *Max take-off weight:* 4680 lb *Armament:* 5×7.62-mm MG 17 mg

### Polikarpov I-16

When the I-16 entered service in 1934 it was of extremely advanced conception, being the first low-wing interceptor monoplane with a retractable undercarriage to enter service in any air force. It served with the Republicans in Spain, where it held its own with the first Bf 109s, and fought over China and Manchuria. The I-16 was a sitting target for the first Luftwaffe attacks of 1941, however, when thousands were destroyed on the ground and in the air

### Polikarpov I-15

First appearing in 1933, the I-15 went to Spain in large numbers where it was dubbed *Chato* (The Flat-nosed One). The type fought in Finland and a few survived only to be shot down in droves during the opening weeks of *Barbarossa*

### Messerschmitt Bf 109B

Spain was the laboratory in which the new monoplane fighters of the late 1930s could be tested in combat. In late 1937 the Bf 109 went to Spain, where its speed and handling were shown to be excellent. The poor armament soon became apparent, though, and this was increased on the subsequent C-series

### Fiat C.R.32

Typical of the Italian fighter designers' preference for manoeuvrability, the C.R.32 fought in some numbers over Spain with the *Aviación Legionara* in the hands of Italian pilots and was supplied to the Nationalist Government itself. The type, which was widely exported, was still Italy's most important front-line fighter in the Autumn of 1939

*Span:* 31 ft 2 in *Length:* 24 ft 5⅜ in *Engine:* Fiat A30, 600 hp *Max speed:* 221 mph at 9480 ft *Ceiling:* 25,750 ft *Max take-off weight:* 4200 lb *Armament:* 2×12-mm Breda-SAFAT mg

Other countries generally avoided being sidetracked into the 'light fighter' concept, though many examples were built. British prototypes included the Avro Avocet and Blackburn Lincock, both powered by Lynx engines of well under 250 hp. In Germany the careful plans for rebuilding the Luftwaffe in the early 1930s led to an RLM (Air Ministry) specification in 1933 for a 'home-defence fighter' which could also serve as an advanced trainer. In this case the concept was a good one. Germany lacked sufficient supplies of engines suitable for full-blooded fighters, but could build aircraft that could train fighter pilots to become proficient in air combat. Several aircraft were built to this requirement, all powered by the 240-hp Argus As 10C air-cooled inverted-V8 engine, and the winner was a parasol monoplane, the Focke-Wulf Fw 56 *Stösser* (Falcon). About 1000 were built, armed with one or two synchronised machine guns and racks for light bombs. Virtually unbreakable, the Fw 56 could be used for vertical dive-bombing, and its general combat capability was close to the limit attainable from 240 hp. One has only to compare its speed of 173 mph with that of more powerful machines in the First World War to appreciate its efficiency, and in countries such as Britain and the United States where defence funding was niggardly in the extreme it is remarkable that more was not done to explore the limits attainable with low-powered fighters.

Of course, everything hinged around the engine and armament, and these aspects are discussed in the relevant chapters. But there began in the 1930s a process of fundamental change. At the start of the 1930s the normal fighter was a fabric-covered biplane with two machine-guns and with a gross weight of about 3500 lb, while ten years later it was a stressed-skin monoplane weighing almost twice as much and with much heavier armament. This alone represented a complete revolution, and it ran parallel with the deeper revolution in all aircraft as the forces of civil or military competition made designers adopt the advanced ideas and techniques that others had demonstrated many years earlier. It is important now to list what these changes were.

Structurally, designers had had a series of choices since 1910. By that time pioneers in Switzerland and France had devised ways of making monocoque structures, a word which means that the strength is in the outer shell instead of in a large framework braced by wires both within and without. During the final years before 1914 this technique had been used to make beautifully streamlined fuselages for racing aircraft, but the structural methods at that time involved wrapping multiple layers of tulip-wood or mahogany, fastened by glue and hundreds of brass pins. Rather similar methods were later used in production of seaplane floats and flying-boat hulls, but only because there was no simple alternative. Fighters were made more cheaply by cruder methods that gave a tough structure that could easily be repaired after a combat or a minor crash. In any case, wings could not yet be made by such a method, though the Fokker fighters of Reinhold Platz had thick wings that were strong enough to need no external bracing. Fokker added struts to his triplane and D.VII mainly to please the officials (and perhaps the pilots).

During the 1920s the change to metal structures had been accompanied by the emergence of all-metal stressed-skin aircraft in which the skin carried a major part of the loads. All-metal fighters had seen service in the First World War, notable examples being the Junkers D.I and CL.I. Likewise the mass-produced French Wibault 7 series of the 1920s excited widespread interest, yet curiously the modern type of stressed-skin airframe was introduced not on fighters but in slow flying boats and fast American airliners. It was Emile Dewoitine who built the first truly all-stressed-skin fighter, in 1931. Though French, he had become disenchanted with his own country in the 1920s because it would not buy his aircraft, and most of his fighters of that decade were parasol monoplanes built in Switzerland. But in 1930 Dewoitine began work on a bold low-wing cantilever (unbraced) fighter to meet the future need of the French air force and this, the D.500, flew on 19 June, 1932. It was notable in that the all-metal stressed-skin philosophy was applied to the ultimate, even the ailerons and tail control surfaces being skinned in light alloy. It was of generous size, so that despite having a long and bulky water-cooled engine it could manoeuvre as well as the best of the biplanes. In spite of having a prominent spatted landing gear, it was very fast; the D.500 and 501 reached 224 mph and the more powerful 510 about 250 mph. By no means least it could carry a Hispano 20-mm cannon and up to six machine-guns. This showed the way things were going.

In the United States Boeing made the switch in pursuit design from biplane to monoplane with the Model 248, flown in March 1932. But they knew their customer, and deliberately made the 248 old-fashioned, with plenty of wire bracing (though the B-9 bomber of 1931 was a perfect stressed-skin cantilever). In January 1934 the delightful little Boeing was delivered to the Army Air Corps as the P-26A, becoming famed as 'the Peashooter'. Compared with the Dewoitine it was much smaller; and though it could not easily carry more than two rifle-calibre guns, its small wing limited radius of turn and made the landing rather fast for the small, rough grass airfields of the mid-1930s. Undoubtedly Boeing should have made a bigger P-26 with stressed-skin cantilever wing and retractable landing gear, able to take the much more powerful engines that appeared in the mid-1930s.

In the late 1930s the whole question of just how a fighter should be conceived was far from settled. Most of the squadron pilots, all over the world, were certain it ought to be a biplane, with open cockpit, and their view of future development was simply that more powerful engines would allow them to increase performance and armament. They looked askance at the sleek new monoplanes, with highly loaded wings, fast landing speeds, complicated systems to work landing gear and flaps (which would need armies of ground crew and would probably prove unreliable and vulnerable in combat) and claustrophobic enclosed cockpits which would interfere with the vital requirement of all-round vision. The only point of the new monoplanes was that they could go faster, but that was of no value if biplanes could out-manoeuvre them and shoot them down.

In the Soviet Union a little fighter even more compact than the Boeing Peashooter appeared at the same time, but in those days Russian designers were hardly taken

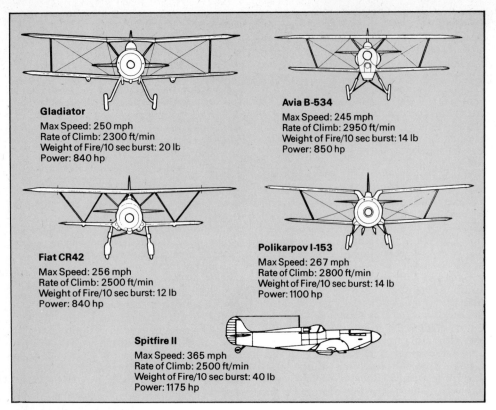

**Gladiator**
Max Speed: 250 mph
Rate of Climb: 2300 ft/min
Weight of Fire/10 sec burst: 20 lb
Power: 840 hp

**Avia B-534**
Max Speed: 245 mph
Rate of Climb: 2950 ft/min
Weight of Fire/10 sec burst: 14 lb
Power: 850 hp

**Fiat CR42**
Max Speed: 256 mph
Rate of Climb: 2500 ft/min
Weight of Fire/10 sec burst: 12 lb
Power: 840 hp

**Polikarpov I-153**
Max Speed: 267 mph
Rate of Climb: 2800 ft/min
Weight of Fire/10 sec burst: 14 lb
Power: 1100 hp

**Spitfire II**
Max Speed: 365 mph
Rate of Climb: 2500 ft/min
Weight of Fire/10 sec burst: 40 lb
Power: 1175 hp

seriously. The designer concerned was Nikolai N. Polikarpov, by far the most experienced Soviet designer and already responsible for several important programmes of combat aircraft. In 1927 he had written a classic report on fighter design, pointing out how one could achieve the best compromise between high power, low weight, a low wing loading (for manoeuvrability) and high speed. One of his conclusions was that the air-cooled radial was clearly superior as an engine, as discussed in the next chapter. His basic conclusions led naturally to a new fighter that was unlike anything seen before. His guidelines were: make the whole aircraft as small as possible, but make the wing area large; use a large radial engine, with variable-pitch propeller; use some form of trailing-edge flap; and make the landing gear retract. The result was the TsKB-12, later adopted as a standard fighter as the I-16. This was one of the great warplanes of history, and at least 7000 were built despite the fact that it was obsolescent by the time Germany invaded the Soviet Union in June 1941. It was developed through at least 24 major versions with increasing power and speed. In the Spanish Civil War the I-16 Model 10 showed itself generally superior to the later and much more expensive Bf 109B. Many I-16s operated on skis in winter, and were among the first aircraft to fire the RS-82 rocket against both air and surface targets.

Soviet philosophy tried to match the speedy I-16 with the even more manoeuvrable I-15 biplane from the same design bureau. The idea was that enemy formations would be caught and brought to battle by the monoplane fighter and then, some minutes later, outmanoeuvred and defeated by the slow biplanes that had had time to catch up. In practice this never worked; indeed in the large-scale air fighting over Mongolia in the first half of 1939 the rigid Soviet political doctrine negated the performance of the I-16 and the manoeuvrability of the I-15, letting the Japanese establish almost complete superiority.

Nevertheless the I-15 biplane ran parallel in timing with the I-16, and was succeeded by the I-15bis in 1937. Today we regard this era as one in which the biplane was fast becoming outmoded; the Gladiator, first delivered to the RAF in 1937, had little chance in 1940 even of surviving. Yet the Soviet Union decided to cut back the speedy I-16 and boost output of the biplanes, and in the I-15bis introduced a completely new design. Moreover, in 1938 yet a third generation of biplanes appeared, the I-153 with retractable landing gear.

It is important not to fall into the trap of concluding that the Russians were backward in clinging to biplanes, or (as one Western 'expert' writes) that in June 1941 they were 'even driven to throwing into the battle obsolete biplanes', and no nation conducted more exhaustive tests on the contrasting kinds of aircraft than the Soviet Union. Polikarpov's bureau was by far the most important in the field of fighters, and throughout the first half of the 1930s concentrated on monoplanes, backed up by the extremely competitive I-15 biplane. It looked as if the future lay with the monoplane, provided it had adequate manoeuvrability. On 1 September, 1934, Polikarpov's team watched the first flight of the TsKB-15, the first fighter in the world to have completely stressed-skin construction, completely retracting landing gear, and all the other modern attributes. It had a liquid-cooled engine, but this long and heavy unit made the aircraft less manoeuvrable than the existing I-16. Though development continued, and small numbers of an improved version saw service as the I-17-2, the emphasis swung increasingly in favour of the biplane. The Spanish Civil War and the intermittent fighting with the Japanese on the Mongolian and Chinese frontiers tended to confirm the view that dogfight manoeuvrability counted for more than speed. Even as late as 1941 the biplane I-153 was still in full production. Every possible effort was made to push this battle-proven type of aircraft to the ultimate. The engine power reached 1000 hp, and variable-

pitch propellers were introduced. The landing gear, which could be wheels or skis, was made to retract, and the armament was increased to four of the outstanding ShKAS machine-guns plus bombs, rockets or drop tanks (a new idea pioneered in the Soviet Union).

At the end of the decade new Soviet designers – Mikoyan and Gurevich, Lavochkin (assisted by Gorbunov and Gudkov) and the more experienced Yakovlev – all produced completely new fighters to the cantilever monoplane formula that by then was sweeping the board everywhere. These were to prove the salvation of their country in the grim days after June 1941, partly because the low-powered I-16 and the biplanes were simply outclassed and partly because they were destroyed in their hundreds and thousands in the catastrophic first weeks of Operation *Barbarossa*, as the Germans had planned. But despite the promise of the new and much faster monoplanes the Soviet engineers and air staff were extremely reluctant to give up the apparent dogfighting advantage of the biplane. An important secondary factor was that the highly loaded monoplanes needed long, smooth runways and were almost incompatible with the short-rough fields that served as bases for most Soviet fighter units. The last attempt to get the best of both worlds was one of the most remarkable fighters in history. In 1939 V. V. Nikitin and V. Sevchenko took an I-153 and rebuilt it with a retractable lower wing! There had previously been various research aircraft with wings that could be extended in span or modified in other ways, but nothing like this. There were no interplane struts, and when the pilot took off he first selected 'wheels up' (the main gears folding into the inboard lower wings) and then selected 'lower wing up'. The lower wing hinged at the root and almost half-way to the tip, the inboard section folding into a recess in the side of the fuselage and the outer panel fitting into a recess under the upper wing. The problems are obvious. For example, any recess in the upper wing would cut into the depth of the spars and weaken the whole wing, and the entire scheme was complex, heavy and fraught with mechanical difficulties. Yet, surprisingly, the prototype flew in late 1940 quite successfully. It is doubtful that anyone seriously expected the Nikitin-Sevchenko IS-1 to go into production, but it shows the boldness of Soviet design thinking. Slightly later Britain experimented with a simpler idea: the 'slip-wing' fighter in which an otherwise conventional monoplane could have an extra upper wing for short take-off, the biplane wing then being jettisoned. This was a good idea in that it left the fighter ready for battle unencumbered by any added devices, but denied the fighter any added dogfight manoeuvrability or slow landing capability.

**Gloster Gladiator**
One of the last fighter biplanes to see operational service, the Gladiator was obsolete even before it flew. It nevertheless fought over Finland, Norway, during the Battle of Britain (over the Orkneys), and distinguished itself over North Africa and the Mediterranean

*Span:* 32 ft 3 in *Length:* 27 ft 5 in *Engine:* Bristol Mercury IX, 840 hp *Max speed:* 253 mph *Ceiling:* 33,000 ft *Max take-off weight:* 5420 lb *Armament:* 4×0.303-in mg

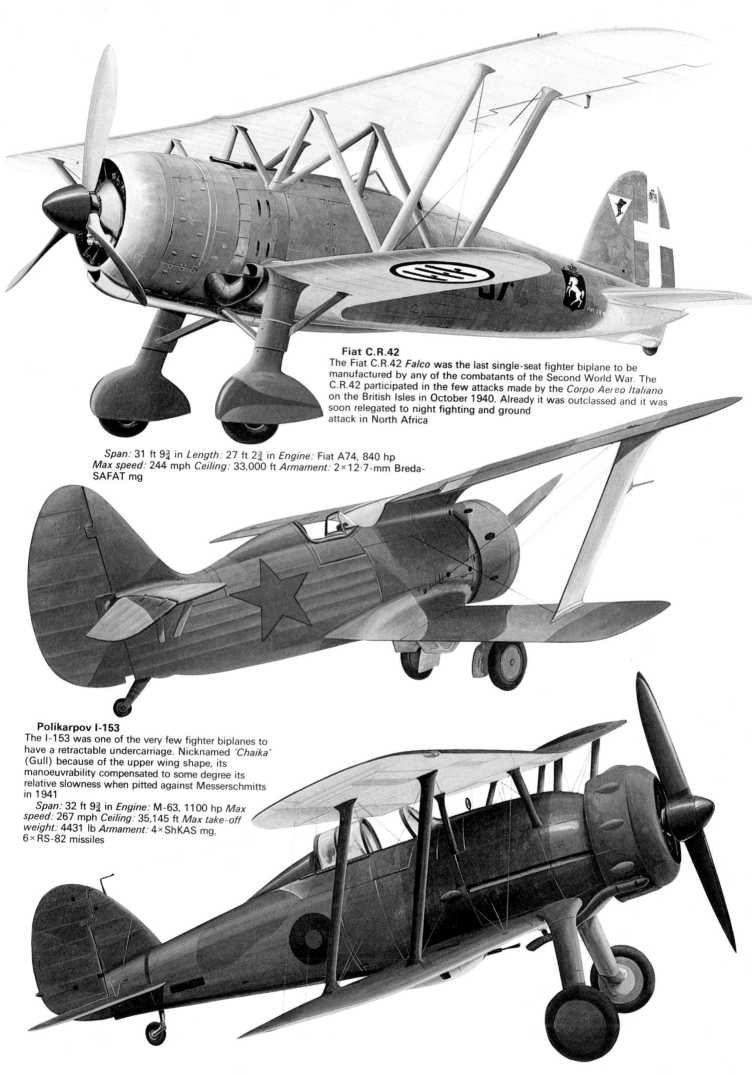

**Fiat C.R.42**

The Fiat C.R.42 *Falco* was the last single-seat fighter biplane to be manufactured by any of the combatants of the Second World War. The C.R.42 participated in the few attacks made by the *Corpo Aereo Italiano* on the British Isles in October 1940. Already it was outclassed and it was soon relegated to night fighting and ground attack in North Africa

*Span:* 31 ft 9¾ in *Length:* 27 ft 2¾ in *Engine:* Fiat A74, 840 hp *Max speed:* 244 mph *Ceiling:* 33,000 ft *Armament:* 2×12·7-mm Breda-SAFAT mg

**Polikarpov I-153**

The I-153 was one of the very few fighter biplanes to have a retractable undercarriage. Nicknamed *'Chaika'* (Gull) because of the upper wing shape, its manoeuvrability compensated to some degree its relative slowness when pitted against Messerschmitts in 1941

*Span:* 32 ft 9¾ in *Engine:* M-63, 1100 hp *Max speed:* 267 mph *Ceiling:* 35,145 ft *Max take-off weight:* 4431 lb *Armament:* 4×ShKAS mg, 6×RS-82 missiles

# THE MONOPLANE SUPREME

In Britain, the vital decision had been taken by 1934 to use not just two machine-guns but a whole battery. Inevitably this demanded a much more powerful fighter, and the first of the new eight-gun monoplanes to go into production was Sydney Camm's Hurricane, by Hawker Aircraft. Partly because of its early start, and derivation from an earlier 'Interceptor Monoplane' with spatted landing year, this was a far from bold design – nothing like as advanced as Messerschmitt's new monoplane that ran parallel in timing in Germany. First flown on 6 November, 1935, the Hurricane was of considerable size, with a span of 40 feet, and it had a thick wing and rather deep and lumpy fuselage. Structure was of traditional Hawker form, with a fuselage based on a girder of tubes held together by multiple bolted fittings and with fabric covering everywhere except the nose and the wing leading-edges. The main advantage of the Hurricane was that it was put into production early and at a high rate, and enough were available to halt the Luftwaffe in its tracks in 1940. As a fighter it was outstanding in being a good steady gun platform, strong and reliable, and easy to repair; and if the pilot flew really boldly it was a fair dogfighter. But by 1940 it was severely handicapped in confrontations with the Bf 109E. which was smaller, faster and had cannon armament.

Partner to the Hurricane, Reginald Mitchell's Spitfire was smaller and of later concept, with stressed-skin construction. It was about as different as it could be from the Supermarine company's seaplanes, also designed by Mitchell, that had won the Schneider Trophy three times in succession. It was a fortunate thing for humanity that this was the case, because the fact that the Spitfire was of modern conception enabled it not only to beat all other fighters in direct combat in 1940 but also made it suitable for progressive development throughout the war. By 1945 the same basic design had double the power, much heavier armament and much higher performance. Its structural basis was a wing with a single spar at 25% chord (one-quarter of the way back from the leading edge) which, together with a thick skin wrapped round the leading edge, formed a D-section 'torsion box' that gave the wing great strength. As everyone knows, the wing was elliptical in plan-form, partly because this is an efficient shape and partly because it gave adequate depth well outboard for the outermost of the eight guns. But for a fighter to be built with minimum effort in huge numbers it was a curious choice.

Another series of fighters with elliptical wings came from the Seversky Aircraft Corporation in the United States. Designer Alexander Kartveli built an experimental two-seat fighter (the SEV-2XP) in early 1935 that established this form of wing as an efficient shape for an all-metal stressed-skin machine, and it was to last right through three generations of fighters of which the last was the famed P-47 Thunderbolt of the Second World War. From the original prototype emerged a rebuild called SEV-1XP with the rear cockpit removed, which led to a succession of attractive fighters called P-35 and 35A by the US Army Air Corps and EP-106 by Sweden. Fitted with landing gear that folded to the rear, projecting under the wing in stream-

lined fairings, these shapely machines led to the XP-41 of early 1939 and the turbocharged AP-4 ordered into production in 1939 as the P-43 Lancer, the company having by this time been renamed Republic Aviation.

Chief rival to Kartveli's fighters were the new monoplane pursuits from the drawing board of Don R Berlin at Curtiss-Wright. His Design 75 was ready in May 1935, and was thus one of the first stressed-skin retractable-gear fighters to fly. Though beaten for Army orders by the Seversky P-35, Curtiss kept improving the aircraft, replacing the experimental XR-1670 by a familiar R-1820 Cyclone, and eventually (as did Seversky) finding success with the Pratt & Whitney Twin Wasp. In July 1937 the Army ordered 200 as the P-36, thus launching one of the biggest and most diverse families of fighters in all history. Fractionally larger than the P-35, the P-36 was a better basis for development, and while Republic had to suffer two total redesigns to increase the size, Curtiss kept in production with the same size of fighter until December 1944. A feature of virtually all the 15,000-odd monoplane 'Hawk fighters was that the landing legs rotated as they folded backwards so that the wheels could lie flat inside the wing. An exception was the Hawk 75 export model of 1937 which, following a policy of simplification, had fixed gear. This was sold to China and Thailand and mass-produced in Argentina.

In the second half of the 1930s the French industry, racked by the problems of an enforced nationalisation which still left a vast profusion of separate organisations as well as numerous private firms (which were deliberately deprived of priorities and subjected to many unnecessary delays and pinpricks), contrived to develop no fewer than 31 types of fighter. Of these the only one that came off the assembly line in large numbers before 1940 was the Morane-

Saulnier M.S.406, an heir to a famous fighter name but a far from outstanding aircraft. Having only 860 hp and an armament of one 20-mm cannon and two machine guns it was described by one of its pilots as 'too slow to catch the enemy and too poorly armed and armoured to avoid being shot down'. Marcel Bloch, who later changed his name to Dassault (as today's world knows), produced one of the few fighters in history that actually refused to get off the ground. After a year of redesign it took to the air in October 1937 and eventually led to a family of radial-engined fighters greatly superior to the Morane. The late 1930s were marked by chaos and confusion in French defence production, with a liberal sprinkling of sabotage (caused mainly by the political left, despite the fact that the arms were needed for defence against Fascist Germany), and this crippled the Bloch fighters more than most. Not until 1940 were gunsights and propellers available so that pilot training could begin. Another French fighter that was even better, but available in quantity only after February 1940, was the Dewoitine D.520, a beautiful little machine that ought to have been started several years earlier as a natural follow-on to the D.510.

Italy and Japan doggedly clung to the old tradition of supreme manoeuvrability at all costs. Celestino Rosatelli of Fiat created a famous series of biplane fighters which had this quality in full, and they were especially notable for the Warren bracing of their wings (the diagonal struts having a W shape seen from the front). The only concession made to firepower was that the machine guns changed during the 1930s from 7.7-mm (rifle calibre) to 12.7-mm (the same as the US 0.50), and two 12.7-mm remained the armament of the new monoplanes produced by Fiat's other designer, Gabrielli, and the rival companies Macchi, Reggiane and Caproni-Vizzola. The basic

*Even with a trousered undercarriage, the Martin-Baker MB.1 of 1934 looks elegant. The size of the tiny fin was increased after the first flight*

*Prototype Hurricane. The first flight was in November 1935*

*XP-40, a re-engined P-36 with an Allison in-line*

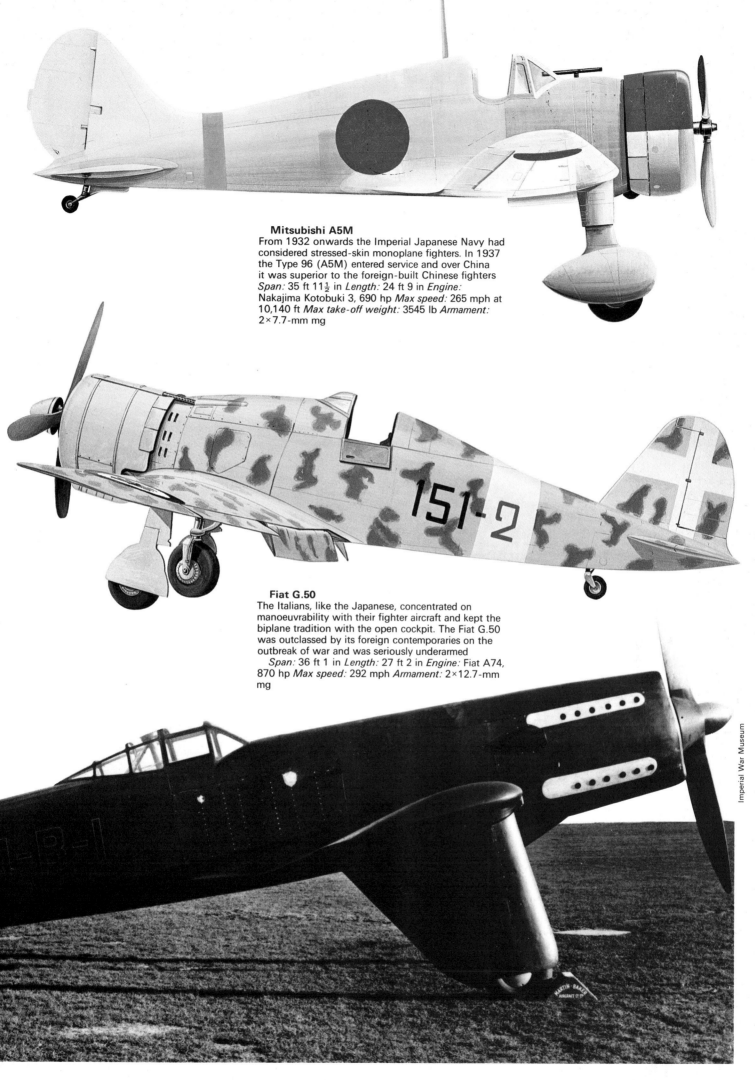

**Mitsubishi A5M**
From 1932 onwards the Imperial Japanese Navy had
considered stressed-skin monoplane fighters. In 1937
the Type 96 (A5M) entered service and over China
it was superior to the foreign-built Chinese fighters
*Span:* 35 ft 11½ in *Length:* 24 ft 9 in *Engine:*
Nakajima Kotobuki 3, 690 hp *Max speed:* 265 mph at
10,140 ft *Max take-off weight:* 3545 lb *Armament:*
2×7.7-mm mg

**Fiat G.50**
The Italians, like the Japanese, concentrated on
manoeuvrability with their fighter aircraft and kept the
biplane tradition with the open cockpit. The Fiat G.50
was outclassed by its foreign contemporaries on the
outbreak of war and was seriously underarmed
   *Span:* 36 ft 1 in *Length:* 27 ft 2 in *Engine:* Fiat A74,
870 hp *Max speed:* 292 mph *Armament:* 2×12.7-mm
mg

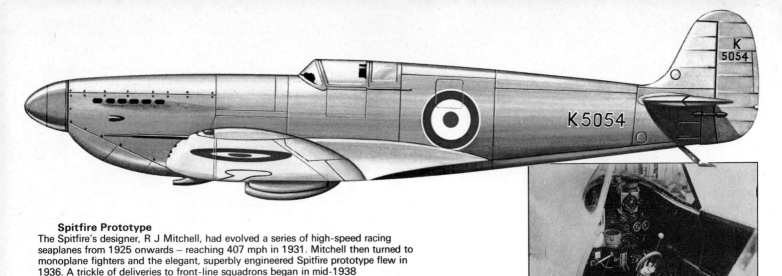

**Spitfire Prototype**
The Spitfire's designer, R J Mitchell, had evolved a series of high-speed racing seaplanes from 1925 onwards – reaching 407 mph in 1931. Mitchell then turned to monoplane fighters and the elegant, superbly engineered Spitfire prototype flew in 1936. A trickle of deliveries to front-line squadrons began in mid-1938

*Inside the prototype Spitfire, a long way from the open cockpits of its fixed-undercarriage biplane contemporaries*

fault lay in a shortage of high-power engines, which condemned Italy to mediocre performance despite the light armament. In Japan the engine position was slightly better, though the chief fighters of the late 1930s were designed around developments of the Bristol Jupiter rated at a mere 600–700 hp. The Imperial Navy's main fighter was the Mitsubishi A5M, first flown in February 1935; that of the Army was the Nakajima Ki-27, flown in October 1936. Both were low-wing monoplanes with spatted wheels and open cockpits, and armed with a pair of Vickers guns just like a Sopwith Camel. But they were among the most nimble fighters ever made, and the prototype A5M reached 279 mph on only 585 hp, and climbed to 16,400 ft in less than six minutes. The loaded weight of the fully equipped service models was more than 1000 lb less than that of a Gladiator and not much more than half the weight of a Curtiss P-36 or Spitfire I.

These little monoplanes 'danced like a butterfly and stung like a bee', and in countless bitter battles that were hardly even mentioned in Western newspapers achieved dominance over the carefully designed monoplanes and biplanes of the Soviet Union. In the mid-1930s nobody could say with certainty how a fighter ought to be planned. The variables all fight each other. Less armament means less chance of downing the enemy, but more means a sluggish fighter with poor performance and reduced manoeuvrability. There is a fair element of luck in hitting the best basic design, and this certainly attended German Willi Messerschmitt when he bent over his drawing board at the Bayerische Flugzeugwerke in the winter 1934–35 and sketched his company's first military aircraft. Back in 1929 Messerschmitt had had a flaming row with Erhard Milch, the future Nazi Air Minister, in a dispute which bankrupted the company. Back in business in 1933, he got no work from the Nazis, and it was only when he was criticised for building aircraft for export (replying 'then give me a contract') that he was invited to build a prototype to meet the Luftwaffe's requirement for a new fighter. The aircraft flew, with a British Rolls-Royce engine, in September 1935.

Nobody devoted much attention to the Bf 109 at first, and it was regarded as such a rank outsider as to be almost a joke. Even Ernst Udet, the World War I ace who became head of the technical procurement section at the RLM (air ministry), scornfully said, 'That thing will never make a fighter'. It was small, long and rakish, with a remarkably small wing and extremely severe lines. The cockpit seemed cramped, and had a hinged canopy that enclosed it completely. It looked tricky, unpleasant and utterly unlike the nimble, open-cockpit kind of machine Udet and other pilots hoped for. Trials at Travemünde began in October 1935, and it was very soon evident that the Bf 109 and its closest counterpart the Heinkel He 112 were way ahead of the other prototypes. It was expected that the Heinkel would win, but prolonged comparative testing kept bringing out the fact that the severe 109 could outfly the pretty 112. The narrow-track landing gear did not seem to bother pilots, and the fact the wheels were far forward made it possible to clamp the brakes on without nosing over. Full-span slats on the wings allowed extremely tight turns to be made, and altogether the 109 seemed to have the best chance of being developed with more powerful engines and heavier armament.

**Fighter-Bomber**
To the amazement of Heinkel and many others, an order for ten Bf 109s was placed in January 1936, and by May 1937 the Bf 109B was in large-scale production with the 635-hp Jumo 210 engine and armament of three machine guns. This sub-type proved extremely formidable in Spain, and the 109C, with four guns (two of them in the wings) was followed by the 109D with the 1000-hp DB 601 engine. Many D models had a 20-mm cannon firing through the propeller hub, and this was also common in the 1100-hp DB 601-engined Bf 109E, the first mass-produced series, which was the standard Luftwaffe fighter from 1938 until 1941. Some E models had two machine guns and no fewer than three cannon, and many also carried bombs weighing up to 551 lb, becoming the first land-based fighter-bombers in history.

The qualification 'land-based' must be added because from the late 1920s the US Navy had devoted great attention to hitting ships and other surface targets with relatively small carrier-based aircraft. While Britain's Fleet Air Arm made do with biplane fighters carrying two Vickers guns, the US Navy bought such advanced monoplanes as the Brewster F2A and Grumman G-36, both of which could carry two 100-lb bombs as well as four 0.50-in guns. Even the previous US Navy generation, exemplified by the tubby Grumman F3F biplanes, had been fast and powerful, and able to carry the same bomb load. It was sheer bad luck that in Japan a designer named Jiro Horikoshi at the Mitsubishi company was

creating a monoplane that would out-manoeuvre even the land-based fighters, outrange all other fighters, and carry two 20-mm cannon as well as two machine guns.

There remains one other category of fighter that came strongly to the fore after 1935. Whereas the *Multiplace de Combat* was virtually a non-starter, there seemed to be a need for a modern high-speed monoplane with two seats and/or two engines to serve as a bomber escort and to fly other long-range missions. Most countries bought such machines in numbers. France began with the Potez 63 family, the first of which flew in April 1936, so enough were on hand to make a real contribution to defence in May 1940 (and versatile

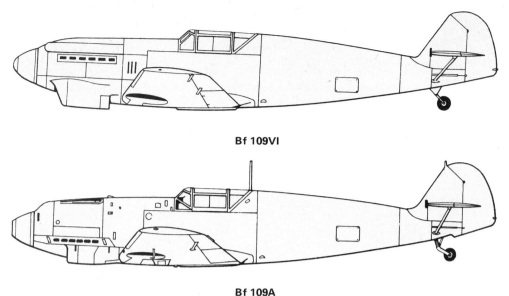

**Bf 109VI**

**Bf 109A**

### Heinkel He 100
Extremely fast but with very poor handling, the 12 production He 100D-1s were really high-speed research aircraft. They made excellent propaganda when repeatedly repainted and photographed, baffling British intelligence into beleiving that large numbers of these 'fighters' existed
*Span:* 30 ft 10¾ in *Length:* 26 ft 10¼ in *Engine:* DB 601M, 1020 hp *Max speed:* 416 mph at 13,120 ft *Max take-off weight:* 5512 lb *Armament:* 2×7.92-mm MG 17 mg, 1×20-mm MG FF/M cannon

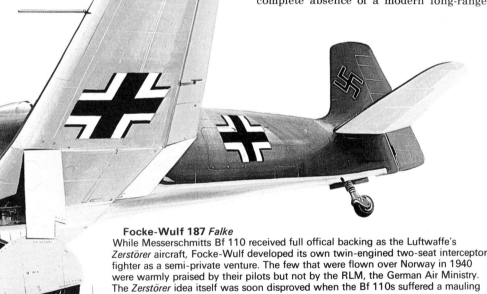

### Focke-Wulf 187 *Falke*
While Messerschmitts Bf 110 received full offical backing as the Luftwaffe's *Zerstörer* aircraft, Focke-Wulf developed its own twin-engined two-seat interceptor fighter as a semi-private venture. The few that were flown over Norway in 1940 were warmly praised by their pilots but not by the RLM, the German Air Ministry. The *Zerstörer* idea itself was soon disproved when the Bf 110s suffered a mauling during the Battle of Britain

*Span:* 50 ft 2⅜ in *Length:* 36 ft 5 in *Engine:* 2×Jumo 210G, 700 hp *Max speed:* 326 mph at 13,800 ft *Ceiling:* 32,800 ft *Armament:* 4×7.92-mm MG 17 mg, 2×20-mm MG FF cannon

enough to be developed for bombing and reconnaissance also). The Breguet 690 family were potentially even better, but suffered from faulty operational techniques that (among other things) made them sitting ducks for the German flak, in their role of attack bomber. The British Whirlwind was simply ill-conceived. Planned as a fighter armed with 20-mm cannon, it had a small wing and useless low-power engines of poor reliability and made only a minor contribution to the war. Likewise the Blenheim IF was a pathetic lash-up, converted from a bomber by fitting four puny machine guns under the belly to try to fill the yawning chasm in the RAF caused by complete absence of a modern long-range

fighter. The Blenheim was also the only RAF night fighter, fitted with the newly invented AI radar, until the arrival of the Beaufighter (first flown in July 1939) which the RAF had never asked for.

Possibly the most successful of the early long-range fighters was the Messerschmitt Bf 110, which proved outstandingly useful in numerous duties until the summer of 1940 when it encountered the formidable and determined single-seaters of the RAF and suddenly proved vulnerable. Earlier the Luftwaffe had tried to buy an all-purpose aircraft with a *'Kampfzerstörer'* specification, but this proved a mistake (the Fw 57, Hs 127 and 124, Bf 162 and Ju 85/88 competed). Even the 110 would have proved a long-term error had it not been just right much later as a basis for a formidable radar-equipped night fighter. In the United States the Bell company chose as its first prototype in 1937 an extraordinary throwback to the concepts of 1916 but translated into modern stressed-skin and near-300-mph speed. The XFM-1 Airacuda had two pusher Allison engines in enormous overwing nacelles, in the front of which were gunners manning 37-mm cannon. But this was just one of many strange fighters. Britain's Hotspur and Defiant ought to have been regarded as freaks, because their entire armament was four machine guns in a power-driven turret behind the pilot; but the Defiant was actually mass-produced (it eventually made a good target tug).

Following a principle first explored by Dunne in 1907, the Westland Pterodactyl V of 1932 was a tailless fighter-bomber with pilot and rear gunner in a stubby nacelle hung under a swept wing with fins on the tips (it flew beautifully). In Holland Koolhoven built the F.K.55 with contra-rotating propellers driven by an engine behind the pilot, while Fokker built the G.1 with twin tail booms and a battery of eight guns in the nose, and the push-pull D.23 with an engine at each end of the central nacelle. To show the Dutch really could not make their minds up, De Schelde built the S.21 with a central nacelle housing one pilot, four machine-guns, a forward-firing 23-mm Madsen cannon and a second big Madsen firing aft through the hub of the pusher propeller driven by a DB 601 engine. This 370-mph machine could have been really formidable, but for some reason that will possibly never be explained was thought in Britain to be a German fighter called 'Focke-Wulf Fw 198'. Needless to say, several RAF pilots reported seeing the 'Fw 198', just as an even larger number reported the 'He 113' which was dreamed up by Goebbels using pictures of the prototypes of the defunct He 100. All fighters, good and bad, were at the mercy of faulty aircraft recognition until long into the era when electronics was supposed to take care of such matters.

# POWER FOR COMBAT

During the First World War the basic choice open to the designer of a fighter was to use either a water-cooled in-line or vee engine, closely related in concept to the engines of automobiles, or a totally different air-cooled rotary. As explained earlier, the rotary was subject to inherent shortcomings which made it gradually fade from the scene, but its place was taken by another engine that owed little to road vehicles. This was the static radial, with its cylinders arranged in the same way as those of the rotary, like spokes of a wheel, all the connecting rods acting on a single crankpin. The first radial in military aircraft was the Swiss-designed Canton-Unné, mass-produced by the Salmson company, and this was cooled by water. Virtually all subsequent radials were cooled by air flowing past the finned cylinders.

Throughout the inter-war period (1918–39) the question of which kind of engine was best for a fighter was the subject of prolonged and heated argument. Until the mid-1920s the water-cooled engines were dominant, partly because they were highly developed whereas the radials were a relatively new species dating from after 1917. But the proven advantages of such engines as the Bristol Jupiter, perfected in 1920–21, and the Pratt & Whitney Wasp, of 1925–26, gradually gave them the dominant share of the market. Really the only factor in favour of the water-cooled in-line was that it looked slimmer and more streamlined than the bluff, flat-fronted radial. This was not too apparent at first, but by 1922 the designers of racers were installing water-cooled engines in a way that seemed almost perfectly streamlined, with a pointed propeller spinner faired into a beautifully smooth engine cowl. Of course, the water cooling radiator spoilt things, but eventually ways were found to use flush cooling radiators in the wings, as had sometimes been done in the First World War.

Those who preferred the radial pointed out that the flush radiator systems of the racers were useless for fighters because they were troublesome, vulnerable to bullets and incapable of working in harsh service conditions. They demonstrated that the simple air-cooled radial, with a short crankshaft with all cylinders working on one crankpin, was lighter, cheaper to make and much easier to maintain. Its finned cylinders, much hotter in relation to the surrounding air than a water radiator, got rid of excess heat perfectly even in the hottest desert climate. Unlike the water-cooled engine, there was nothing to freeze in Arctic winter conditions, and no water circuit to need warming up. Absence of the water circuit and radiator saved weight and drag, while the shorter and more compact engine was especially suited to a fighter because it improved manoeuvrability and handling.

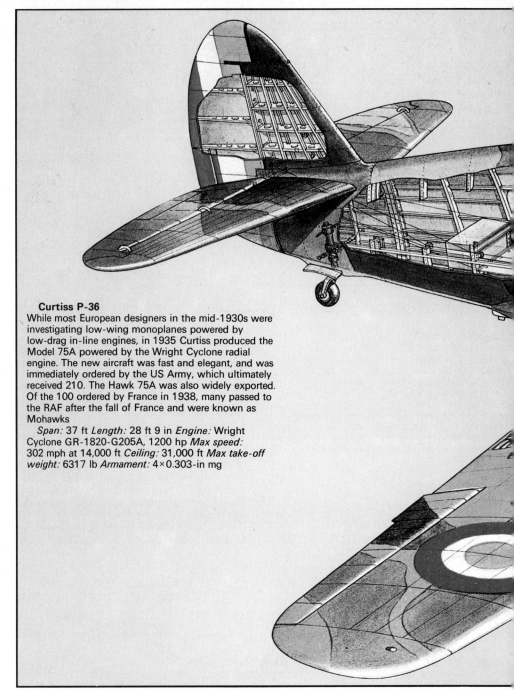

**Curtiss P-36**
While most European designers in the mid-1930s were investigating low-wing monoplanes powered by low-drag in-line engines, in 1935 Curtiss produced the Model 75A powered by the Wright Cyclone radial engine. The new aircraft was fast and elegant, and was immediately ordered by the US Army, which ultimately received 210. The Hawk 75A was also widely exported. Of the 100 ordered by France in 1938, many passed to the RAF after the fall of France and were known as Mohawks
*Span:* 37 ft *Length:* 28 ft 9 in *Engine:* Wright Cyclone GR-1820-G205A, 1200 hp *Max speed:* 302 mph at 14,000 ft *Ceiling:* 31,000 ft *Max take-off weight:* 6317 lb *Armament:* 4×0.303-in mg

After extensive comparative trials the US Navy declared in 1928 it would buy no more water-cooled aircraft, and even discarded or re-engined almost new machines with such engines. The US Army kept using Liberty, Packard and, above all, Curtiss water-cooled engines until 1935, though by that time the radial had come to the fore in the Army also. Most air forces used either the radial or a mixture. By 1929 more than 70 per cent of the horsepower delivered for the French air force and navy was comprised of one radial, the Jupiter (built under licence by Gnome-Rhône), and as late as 1939 Bristol radial engines accounted for 52 per cent of the horsepower in the RAF. But behind the squadrons the influence of racing was distorting the picture.

During the 1920s the water-cooled V-engine was especially well suited to out-and-out racing. It could develop higher output from a given cubic capacity (engine cylinder size), was more readily installed in a streamlined way, and so tended to be used almost exclusively for the seaplanes that battled to win the coveted Schneider Trophy. In 1927, 1929 and 1931 British Supermarine seaplanes powered by special Rolls-Royce racing engines won the trophy three times consecutively, thus bringing the contests to an end with the trophy in British hands for ever. This had the effect of putting a giant spurt behind high-power water-cooled engines and of planting in people's minds the belief that the radial could not compete. This was the situation when, in 1934–35, the future monoplane fighters were designed for the RAF (Hurricane and Spitfire) and Luftwaffe (Bf 109). By 1940 the resulting aircraft were locked in combat, powered by the Merlin and DB 601. The only fighters with radial engines were poor second-class citizens, such as the Curtiss Mohawk, the Polish PZL family and the Italians. The Russian and Japanese fighters were hardly taken seriously. It was a cosy feeling, but unfortunately totally erroneous.

Even today the powerful emotional link between the Schneider Trophy success, the Merlin engine and the Hurricane and Spitfire makes it difficult to accept the true facts. The superiority of such engines as the Merlin lay partly in the intense effort applied to their ancestors in the years of

racing, partly in the fact that they were simply bigger than most radials (and thus more powerful) and, above all, in the fact that it took a long time for designers to learn how to instal a radial in a high-speed aircraft. At first they simply stuck the cylinders out in the airstream. Then they tried to fair the cylinders in with individual 'helmets', or with tapering fairings behind each 'pot'. By 1929 the Townend ring was added, behaving like a slender circular wing helping to pull the aircraft along. In the early 1930s the NACA (US National Advisory Committee for Aeronautics) had developed long-chord cowls that enclosed the whole engine. Bristol then refined this by ducting the exhaust into the leading edge of the cowl and adding controllable flaps around the rear edge to adjust the cooling airflow. Each month, in 1929–39, the power of radial engines went up, the cooling was improved and the drag reduced.

Liquid-cooled engines fought back by switching from plain water to ethylene glycol, which prevented freezing, and allowed the transfer of heat to be greatly speeded up by using a high-pressure system

cooled by a smaller radiator offering less drag to the aircraft. Many fighters flew with no ordinary radiator at all, with steam cooling dissipating heat through surface condensers. The steam-cooled Rolls-Royce Goshawk was much in evidence in fighters in 1931–34 but eventually the steam-cooling philosophy was abandoned. Extremely attractive to the British Air Ministry, it resulted in an engine installation even more complicated and troublesome than ordinary water-cooling, and Rolls-Royce were glad to let it drop.

One of the fighters with a steam-cooled Goshawk was the Westland P.V.4 (F.7/30) of 1934. This reverted to an idea tried in one or two fighters in the First World War, but for different reasons. The heavy engine was placed in the mid-fuselage, between the biplane wings, while the pilot's cockpit and four Vickers guns were put in the nose. This seemed a sensible arrangement in that it gave the pilot a perfect view with an excellent armament installation, and having the engine on the centre of gravity improved manoeuvrability. Others thought so too. Among a crop of mid-engined fighters

emerged the Koolhoven FK.55 with an 860-hp Lorraine Petrel driving two contra-rotating propellers via a long shaft passing through the cockpit, the Italian Piaggio P.119 with a 1700 hp air-cooled radial buried in the centre-fuselage behind the cockpit, and the American Bell XP-39 Airacobra and carrier-based Airabonita. The latter differed in that the Airacobra was even more radical in having a nosewheel-type landing gear, yet it alone was built in large numbers and gave good service in the Second World War. Yet another way of putting the engine on the centre of gravity was to use a pusher propeller and carry the tail on booms, as was done by the De Schelde company with the S.21. An even more radical arrangement was to have no conventional tail at all, as in the Pterodactyl V and the Italian SAI S.S.4, first flown in May 1939, which had such revolutionary features as a steerable nosewheel, canard horizontal controls on the nose, and armament of two 20-mm and one 30-mm cannon. The tail-first idea was to be pursued by several fighter builders during the Second World War.

On balance, posterity must conclude that

the best place for the engine in a single-piston-engined fighter was in the nose, but there remained problems. As engines increased in power, from the 450 hp of the 1920s to 750 hp by 1935, 1000 hp by 1938 and 2000 hp in the prototypes of 1940, so did the area of the propeller blades to absorb the power have to increase. This end could be attained by increasing the number of blades, increasing the chord (width) of each blade or increasing the propeller diameter, but there were drawbacks to each of these choices. Later, all these answers were to be seen in abundant measure, together with the increasing use of the heavy, complex and possibly troublesome contraprop. A great advantage of the contraprop was that it cancelled out the reaction from a single large propeller which tends to rotate the whole aircraft in the opposite direction, whilst surrounding the tail in a powerful slipstream of air rotating in a vicious spiral. During the 1930s fighter pilots for the first time learned (often 'the hard way') to live with powerful engines in small fighters. Opening the throttle on take off caused a violent pull to left or right, depending on the direction of propeller rotation, and the torque (twist) imparted to the fuselage dug one mainwheel into the ground and tended to lift the other off, which in turn caused offset wheel drag during the take off run. The answer was 'a bootful of rudder', possibly accompanied by aileron, which had to be eased off as speed increased and the fighter left the ground. Of course, another and usually unrelated problem was that many fighters tended to 'swing' badly on takeoff, landing or both. A fighter would 'swing' by trying to pirouette round in a sudden half-circle on the ground and proceed tail-first, or sit on its belly amidst portions of smashed landing gear. This became an everyday occurrence in the

Second World War, and the only aircraft usually free from the hazard were the new breed with nosewheels.

More important than mere handling problems – which, authority decreed, competent pilots should learn to cope with – were those associated with maintaining power at high altitude and in turning shaft power into thrust. As outlined earlier, the answer to the need for power at height was the supercharger. At first this was a centrifugal blower, geared up from the crankshaft to rotate at very high speed to pump air into the cylinders. Special clutches were soon found to be necessary to remove shocks from the drive gears, eliminate vibration and, by the late 1930s, allow for a gearbox much like that in a road vehicle to drive the blower at different speeds. Later more complicated superchargers were developed with two blowers in series, the air being rammed in at a forward-facing inlet, compressed in the first blower, cooled in an intercooler to increase its density, and then pumped into the cylinders by a second blower. But in the United States the technically difficult turbo-supercharger had been under development since before the First World War, and during the late 1920s began to come into US Army service in advanced high-altitude biplane fighters. In 1935 a turbocharged Conqueror engine powered the Consolidated P-30 (PB-2A) two-seat monoplane fighter, giving it a speed of almost 250 mph at 25,000 ft and a ceiling of almost 29,000 ft. Turbochargers followed on many American fighters, the advantage being that, whereas the mechanically driven blower took power out of the engine via the gearwheels, the turbocharger merely used some of the power otherwise wasted in the hot exhaust gas.

Turning the power of the engine into thrust had traditionally been done by a

propeller carved from laminations of mahogany or other hardwood. By 1930 metal propellers were becoming common, especially for high-speed aircraft, because they were no longer significantly heavier and in many cases not only lasted longer but could be 'bent straight' after hitting the ground and put back into use. By this time the variable-pitch propeller had been more than ten years under development, and the two-position Hamilton was about to come into use in the USA. This relatively simple variable-pitch propeller could operate either in fine pitch, with the blades set at a small angle to the air, for maximum engine speed and power on takeoff or landing, or in coarse pitch, set at a large angle, for high-speed flight. During the 1930s these propellers came into general use, mainly on airliners and bombers rather than fighters until after 1936, when the absolute need for improved propellers could be seen even by the British (who were slowest off the mark, having been the first to work on the variable pitch propeller in the First World War).

By 1939 virtually every fighter in production in every country had an advanced variable-pitch or constant-speed propeller, the constant-speed type having a governor to adjust pitch automatically to any desired setting for best engine power or efficiency, without input from the pilot. The main French propellers were the Chauvière and Ratier, the Germans used the extremely good VDM and the Americans the pioneer Hamilton and the newer Curtiss electric. The only exceptions were the otherwise excellent Hurricane and Spitfire, which poured off the assembly lines fitted with solid wooden propellers just like those used on a Sopwith Camel. This disgraceful state of affairs was eventually put right by frantic efforts in 1940.

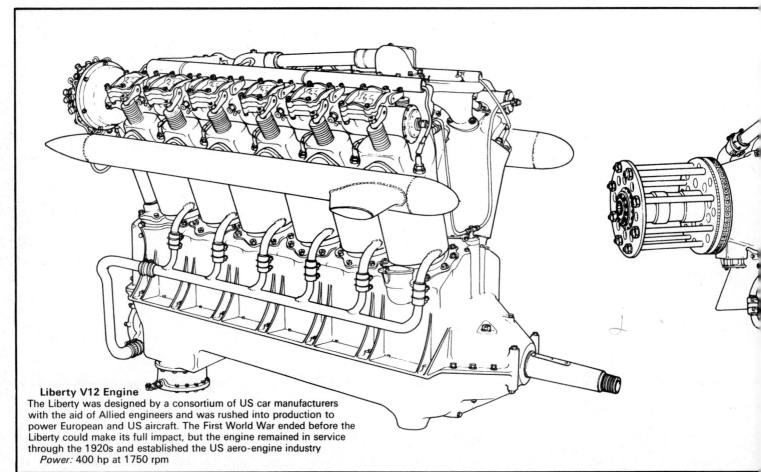

**Liberty V12 Engine**
The Liberty was designed by a consortium of US car manufacturers with the aid of Allied engineers and was rushed into production to power European and US aircraft. The First World War ended before the Liberty could make its full impact, but the engine remained in service through the 1920s and established the US aero-engine industry
*Power:* 400 hp at 1750 rpm

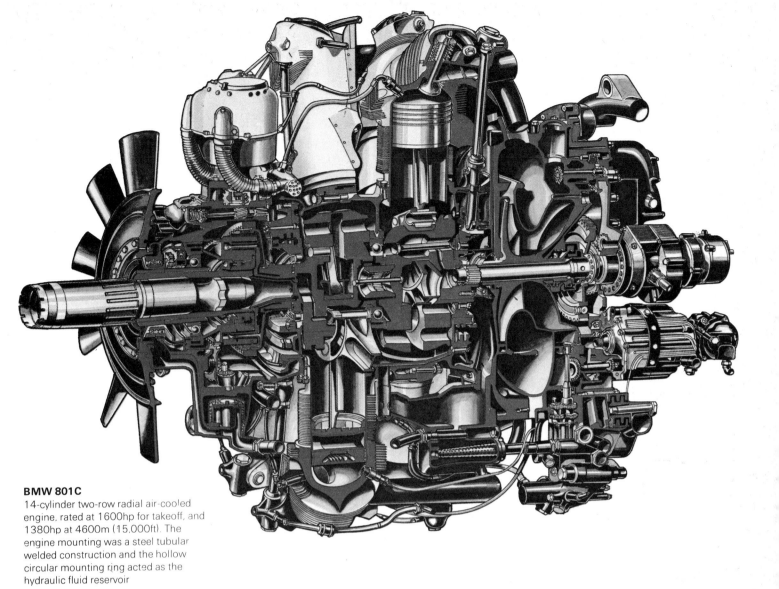

**BMW 801C**
14-cylinder two-row radial air-cooled engine, rated at 1600hp for takeoff, and 1380hp at 4600m (15,000ft). The engine mounting was a steel tubular welded construction and the hollow circular mounting ring acted as the hydraulic fluid reservoir

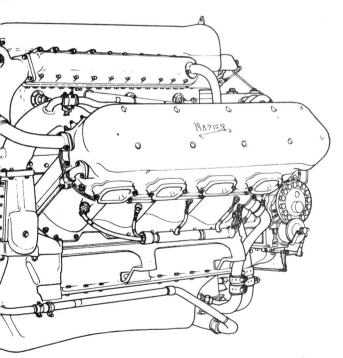

**Napier Lion**
When the British government issued a specification in 1916 for a 'high-power altitude engine', the Napier Lion with its 'broad arrow' W12 cylinder formation was a brilliant response and the engine was one of the most important of the 1920s undergoing continuous development
*Power:* (1927 VIIB) 875 hp at 2600 rpm

## Aero-Engine Development 1909–1945

| Date | Engine | Type | Power |
|------|--------|------|-------|
| 1909 | Anzani | 3 cyl semi-radial | 25 hp |
| 1909 | Gnome | rotary | 50 hp |
| 1912 | Salmson/Canton-Unné | radial | 110-130 hp |
| 1913 | Gnome Monosoupape | rotary | 100 hp |
| 1915 | Oberursel | rotary | 100 hp |
| 1915 | Rolls-Royce Eagle | V12 | 225 hp |
| 1915 | Benz Bz IV | 6 cyl in-line | 200 hp |
| 1916 | Le Rhone 9J | 9 cyl rotary | 110 hp |
| 1917 | Bentley B.R.2 | 9 cyl rotary | 235 hp |
| 1917 | Siemens-Halske Shl | 9 cyl geared rotary | 110 hp |
| 1917 | Liberty | V12 | 400 hp |
| 1920 | Bristol Jupiter | 9 cyl radial | 480 hp |
| 1923 | Curtiss D-12 | V12 | 435 hp |
| 1924 | Napier Lion | W12 | 450 hp |
| 1930 | Rolls-Royce Kestrel | V12 | 575 hp |
| 1934 | Pratt & Whitney R-1830 Twin Wasp | 9 cyl radial | 1200 hp |
| 1935 | Rolls-Royce Merlin | V12 | 990 hp |
| 1936 | Nakajima Kotobuki 2-kai-1 | 9 cyl radial | 585 hp |
| 1939 | Daimler Benz DB 601A | V12 | 1100 hp |
| 1940 | Hispano-Suiza 12Y-45 | V12 | 910 hp |
| 1940 | BMW 801C | 14 cyl radial | 1600 hp |
| 1940 | Nakajima NK1C Sakae 12 | 14 cyl radial | 925 hp |
| 1940 | Fiat A-74 | 14 cyl radial | 840 hp |
| 1941 | Napier Sabre IIA | H24 | 2180 hp |
| 1941 | Mikulin AM-35A | V12 | 1350 hp |
| 1941 | Daimler Benz DB 605A | V12 | 1475 hp |
| 1943 | Shvetsov M-82FN | 14 cyl radial | 1700 hp |
| 1944 | Junkers Jumo 004B | turbojet | 1980 lb |
| 1944 | Walter HWK 509 | liquid rocket | 3750 lb |

# FIGHTERS 1939–1945
## *Bill Gunston*

The new generation of monoplane fighters that emerged in the mid thirties had sent a great curve arching across the graph of military science. When the test came in 1939–40 actions were fought at altitudes and speeds and with firepower that would have been unthinkable ten years earlier. Yet it was the War itself that gave the greatest impetus to fighter development.

Some types like the Spitfire and Bf 109 were 'stretched' successfully and were still fighting at the end of the war. Other new designs like the P-47 Thunderbolt and Hawker Tempest relied on brute power for high performance, yet the true classics of the wartime years were the superbly balanced designs such as the Fw 190 and P-51 Mustang which pushed piston-engined aircraft design to its limit.

New roles produced new aircraft – like the stalking night-fighters and the many successful adaptations of fighter aircraft for ground-attack.

*The Flying Tigers fly again – superbly restored Curtiss P-40s of the Confederate Air Force*

# CONTENTS

John Batchelor

# THE EAGLE ASCENDANT

The opening of the Second World War was carefully stage-managed by Hitler after a faked 'incident' on the frontier between Germany and Poland. The start of operations, code-named *Ostmarkflug* in the Luftwaffe's secret orders, was timed for 04.45 on Friday 1 September, 1939, but the Ju 87B dive bombers of *Stuka Geschwader 1* were over-eager, and three of them, led by *Oberst* B Dilley, swooped on the bridge over the Vistula at Dirschau fifteen minutes too early. But this was of small consequence. The Poles had virtually no chance of stemming the German invasion, and the campaign that opened the greatest of all wars was completed in twenty-seven days of fierce action.

Hitler's Generals and Air Fleet Commanders liked to plan their campaigns in detail. Though strategic planning was almost totally absent – and in the long term this was to make a very large contribution to final defeat – in the short term everything possible was done to catch the enemy off-guard, hit him hard and, especially, knock out his fighting power in the shortest time. So on 1 September most of the initial strikes by the Luftwaffe were flown against Polish airfields, and within hours all the known military bases had been reduced to a shambles. More than 70% of the Polish front-line air strength was destroyed or incapacitated on the ground, though some aircraft escaped the initial attack, having been secretly dispersed a few days earlier.

By 3 September fewer than 30 Polish fighters were operating, out of the inventory total of 148, all of the P.11c type. The record of combat losses is confused. One frequently quoted report states that the P.11c squadrons shot down 126 German aircraft for the loss of 114 of their own number. This is erroneous. About 126 Luftwaffe aircraft were almost certainly shot down in air combat in the Polish campaign, but 114 is more than the total of Polish fighters available to be shot down. Almost the whole force of 148 was lost, but the best estimate is that more than 100 of these were written off in attacks on airfields, or in accidents in trying to operate from bombed airstrips.

On the whole the Polish pilots were experienced, and they lacked nothing in courage; but by 1939 the P.11c was obsolescent. The first P.11 had flown in August 1931, and though at that time it was one of the best fighters in the world, eight years later it was not in the same class as the opposition. The P.11c's structure was mixed. The landing gear was fixed, the gull wing externally braced, and the 645-hp Skoda-built Bristol Mercury radial was housed in a simple Townend ring cowl, driving a fixed-pitch wooden screw. Though the later P.11c, first flown in 1933, was able to carry the above-average armament of four 7.7-mm KMWz.33 machine guns, most of the 175 aircraft delivered to the Polish Air Force had only two, and they still had only two in 1939.

The gallant Polish Air Force set the pattern for most of the air forces that were to confront the Luftwaffe in the first year of war. Its strength was inadequate, and its attempts to remedy this had been hampered by late starts, muddles, delays and bad luck.

Compared with the Bf 109E, the P.11c had only one advantage: its radius of turn at speeds around 200 mph was marginally better. In the relatively few air combats that took place over Poland, the well-trained and confident Germans soon learned that even the early models of Bf 109 were excellent dogfighters, while the new E-1 sub-types, delivered in the six weeks prior to the Polish campaign, had two MG FF cannon and could pick off their targets at a distance. Earlier E-1s, and the E-0 and D variants, all of which participated in the Polish invasion, were armed with only four MG 17 machine guns.

The basic Messerschmitt design was an inspired conception by a design team with no previous experience of military aircraft, or even high-speed aircraft. Though at the start of its career it was widely derided, principally by pilots suspicious of any fighter with an enclosed cockpit and highly loaded monoplane wing, the Bf 109 gradually emerged as the most important combat aircraft of the war. It was built in much larger numbers than any other aircraft in history, save the Russian I1-2 armoured ground-attacker, and held its own to the bitter end despite the emergence of seemingly much superior designs and even jet aircraft.

During the Polish campaign the Luftwaffe was still using the Bf 109B, C and D in front-line fighter wings, but all were being replaced as fast as possible by the E. The B and C had modest performance on the Junkers Jumo 210 engine of 720 or 730 hp, the maximum speed falling well short of 300 mph. Armament of the B, which had equipped the *Legion Kondor* in Spain from 1937, was only three 7.92-mm MG17 machine guns, the middle gun lying between the engine cylinder-blocks and giving prolonged trouble from overheating. The C had four MG 17s, two of them in the wings. The D was the first model with the 1000-hp Daimler-Benz DB 600 engine, and the usual armament was one 20-mm MG FF cannon firing through the propeller hub, and two wing-mounted MG 17s. It was an adequate fighter, but the E (*Emil*) was even better, and supplanted the D when only about 200 of the latter had been built. The most important change was the DB 601A engine, which suffered development snags that held back the E until 1939, but in return gave 1175 hp (with promise of a lot more to come) and featured direct injection of fuel into the cylinders. This enabled the *Emil*, and many

**PZL P.11**
The P.11, based on the P.7, was in its P.11c variant the most numerous Polish fighter to engage the invading German forces in September 1939. The licence-built P.11f version was operated by the Romanian Air Force as late as 1941, but the fall of Poland prevented the introduction of the improved P.11g *Kobuz* variant

*Span:* 35 ft 2 in *Length:* 24 ft 9½ in *Engine:* PZL-built Bristol Mercury VI S.2, 645 hp *Max speed:* 242 mph at 18,000 ft *Ceiling:* 36,000 ft *Max take-off weight:* 3960 lb *Armament:* 2 or 4×7.7-mm KM Wz.33 mg, 2×27-lb bombs

other Luftwaffe aircraft, to ignore cold weather and outmanoeuvre their opponents. Not least of the *Emil's* assets was its powerful armament of two 20-mm MG FF cannon in the wings and two synchronised MG 17s.

Admittedly the MG FF was a relatively puny 20-mm weapon, but it was reliable and could destroy opposing fighters with a single good hit. Moreover, it was effective from ranges more than double the effective limit for rifle-calibre weapons, and in some of the Polish fighting this advantage was exploited to the full (it had not been possible in the limited, close dogfighting in Spain to come to any conclusions about the few E-1s that fought there, often with only four MG 17s).

Combat over Poland invariably took place between isolated aircraft, and the fact that it was one-sided prevented the Luftwaffe from learning useful lessons. There was only one aeroplane in Poland that approached the 109 in performance, and even that fell far short. The PZL P.50 *Jastrzab* was started too late and was

## Messerschmitt Bf 109E-3

The Bf 109E series were the first true mass-production models, and the E-1 was standard equipment when Germany went to war in 1939. The E-3 with a 20-mm cannon mounted to fire through the spinner entered production later that year and by mid-1940 was the main type

*Span:* 32 ft 4½ in *Length:* 28 ft 4 in *Engine:* Daimler-Benz DB 601A, 1100 hp *Max speed:* 354 mph at 12,300 ft *Ceiling:* 37,500 ft *Max take-off weight:* 5523 lb *Armament:* 2×7.92-mm MG 17 mg, 3×20-mm MG FF cannon

*Business end of a Bf 109E-3 with a 20-mm cannon firing through the airscrew hub. (Inset) Professor Willy Messerschmitt, designer of some of Germany's most important wartime aircraft*

John Batchelor

grossly underpowered (840-hp Mercury) and generally undistinguished; the solitary prototype was mistakenly shot down by Polish anti-aircraft fire.

In the Norwegian campaign the Luftwaffe came up against the RAF. The first RAF unit in Norway was 263 Sqn, equipped with Gladiators, which had flown from HMS *Glorious*, lying off the Norwegian coast, on 24 April 1940. Thereafter they operated from the frozen Lake Lesja.

The Gladiator was the ultimate answer to the RAF's F.7/30 requirement, and in 1930, the year it was drawn up, it would have been outstanding. Although it could have been in service (with a slightly earlier Mercury engine) in 1932, the Gladiator was so delayed by official indecision that it did not fly until 1934, and did not reach the RAF until well into 1937. At first armed with two synchronised Vickers machine guns and two Lewis under the lower wing, the Gladiator was totally obsolete by 1939, even though by then the armament had been changed to four rifle-calibre Brownings. It was a pleasant and manoeuvrable machine, but with a fixed-pitch wooden screw was not in the same class as the 109E; moreover, it lacked any form of armour protection. So, too, did most Bf 109s at this time, but the Gladiator was much more in need of it. The Norwegian Air Force had bought six Gladiator Is and six Mk IIs with three-blade fixed-pitch metal propeller, and these were still on their winter skis when the Luftwaffe arrived.

Nevertheless the Gladiators did extremely well. Nine Norwegian Gladiators had remained in April 1940, and these destroyed considerably more than nine of the enemy. Although all were eventually destroyed, only one was shot down. The same pattern characterised 263 Sqn. The RAF machines had wheels instead of skis, and on the ice were barely controllable. Subjected to daily bombing and strafing, it is a marvel that any survived to the second day, but in fact five were still serviceable three days later. In that time 13 had been destroyed on the ground, but only one in the air. The embattled 263 Sqn claimed 15 aerial victories, including BF 110s, He 111s and He 115s, Ju 52/3m transports, and at least one Bf 109. When fuel ran out on 27 April, the five survivors were burned, and it took three weeks to steam back to Britain and return with fresh aircraft. This time 16 Gladiators operated for two weeks in May–June 1940 from Bardufoss airfield, where there were at least hangars and huts. They claimed a further 26 confirmed victories for the loss of only five in the air and ten on the ground. Sadly, the rest were sunk whilst homeward bound in *Glorious*.

These remarkable results run counter to the assertion that the Gladiator was an obsolete aircraft. The fact that it was so judged is inescapable, and the type played virtually no part in the desperate battles over France or England. Gladiators, and the very similar Sea Gladiator, saw much action in the Mediterranean theatre in 1940–41, but this was simply because they were better than nothing. Although it possessed an adequate turn radius, by 1940 standards the Gladiator was deficient in every aspect of performance and lacked firepower and protection. The Gladiator was an exact parallel to the Italian C.R.42, but its scoreboard was dramatically better, in Norway at least.

In Finland, Gladiators saw fierce action in the 1939–40 war against the Soviet invader. The Finnish Air Force had 30 Gladiators, and these were soon condemned as virtually useless. According to the Finns, the I-16 could always get the better of a Gladiator unless the latter had an exceptional pilot; against the I-15 *bis* and I-153 biplanes a Gladiator had no chance, and 13 were destroyed in air combat. Against Soviet bombers the weak firepower of the British fighter was compounded by its inability to catch either an SB-2 or an Il-4. Yet the volunteer Swedish squadron of J8A Gladiators sent to help the Finns downed six fighters and six bombers for the loss of two Gladiators in the air and one on the ground. The reason lay in the small engine, of only 645 or 840 hp, which by 1940 was unable to carry adequate firepower, armour and equipment and reach a competitive performance.

However, there are always exceptions to most rules. In the early 1930s the French Caudron-Renault company designed a series of outstanding baby racers, chiefly for the Coupe Deutsch de la Meurthe trophy, which were truly remarkable for the performance they obtained on 450-hp Renault engines. These inverted V-12 engines had a frontal area smaller than that of the seated pilot, and unlike other in-line engines were air-cooled and thus needed no plumbing or radiator.

### Polikarpov I-16
The I-16 remained the Russians' most important front-line fighter well into 1941 and thousands were destroyed on the ground and in the air during the opening stages of the German attack. During the Winter War skis were standard equipment
*Span:* 29 ft 6½ in *Length:* 20 ft 1¾ in *Engine:* M-62 radial, 700 hp *Max speed:* 300 mph *Ceiling:* 31,500 ft *Max take-off weight* 4546 lb *Armament:* 4 × 7.6-mm mg

### Gloster Sea Gladiator
The Sea Gladiator, the Fleet Air Arm's last single-seat biplane fighter, differed from its land-based counterpart in having an arrester hook, catapult attachment points and provision for a dinghy beneath the fuselage
*Span:* 32 ft 3 in *Length:* 27 ft 5 in *Engine:* Bristol Mercury VIIIA, 840 hp *Max speed:* 245 mph at 15,000 ft *Ceiling:* 32,000 ft *Max take-off weight:* 5420 lb *Armament:* 4 × 0.303-in Browning mg

By 1939 the *Armée de l'Air* had extensively tested several prototypes or racers and ordered 100 C.714 fighters. One Caudron had carried two cannon, but the 714 had just four machine guns. Later the French cancelled the contract, claiming that rate of climb was inadequate (though it was only fractionally worse than for an M.S. 406 or Hurricane I). Eventually, between 50 and 100 were built, a few reaching Finland and the rest equipping GC I/145, an *Armée de l'Air* fighter unit manned by escaped Poles. The slim Caudrons saw brief but violent combat over northern France, and certainly gained a much better than 1:1 kill ratio. In Finland, however, they proved too 'hot' for the short, boggy airstrips.

This neat and significant machine was built of wood, needed only 5000 man-hours to manufacture, and had a gross weight of 3858 lb. Moreover, in January 1940 testing began of the C.R.760, with more powerful engine, metal fuselage, and six guns with the exceptional ammunition capacity of 500 rounds per gun. It proved to be a splendid performer, reaching 334 mph, having outstanding manoeuvrability and being very simple to maintain. In May 1940 the C.R.770, powered by an 800-hp Renault, reached 367 mph. France, and possibly all the Allies, may well have done better to have concentrated on aircraft of this class in the

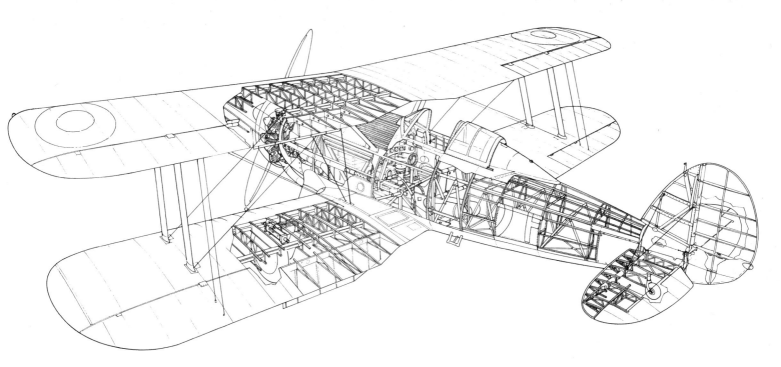

late 1930s rather than on a profusion of much larger and more complex machines which could not be made available in quantity in time.

In almost every respect the French aircraft programme of the late 1930s was a disaster. To some degree the situation resulted from an excess of planning, with the plans continually being changed. The industry had been thrown into chaos by nationalisation under a law of 1936, which grouped most manufacturing capacity into geographical units quite unrelated to their products and left other companies in private hands hamstrung by their inability to get delivery of vital parts (which were reserved for the priority use of the nationalised plants). But the worst problem of all was the multiplicity of different types, and when this was combined with the activities of what must have been a substantial number of saboteurs and troublemakers, the result was that few aircraft reached the *Armée de l'Air*.

The only fighter to reach the *Groupes de Chasse* in reasonable numbers was the Morane-Saulnier M.S.406, and this was a second-rate machine that reflected its early conception. Although developed at the same time as the Bf 109 and Hurricane, the root cause of the aircraft's inadequacy was the mere 860 hp of its Hispano-Suiza 12Y engine. Production began at Bouguenais in June 1938, and by the outbreak of war had reached eleven per day, backed up by a second line at Puteaux. By the Armistice of 25 June, 1940, some 1080 Moranes had been delivered, but according to their pilots they were, 'though free from vices and very manoeuvrable, too slow to catch the German bombers and too badly armed to shoot them down. Ill-armoured, our losses were high.'

Second in importance among French fighters was the Bloch 150 series. This had started with a 1936 prototype which refused to fly, but by 1939 the Bloch 151 was in production and could have given the *Armée de l'Air* some much-needed muscle. Bloch had been nationalised as SNCASO, but though this organisation enjoyed priority over such private companies as Breguet, the prevailing muddle and sabotage prevented delivery of more than a trickle of propellers and then halted output of gunsights completely. Eventually, 140 moderately useful 151s, with 920-hp Gnome-Rhône radial and four machine guns, were followed by 488 152s with an improved engine of just over 1000 hp and two 20-mm cannon and two machine guns. Unfortunately they arrived too late, and in many cases the Bloch pilot's first combat mission was also his first flight on the type. But the record of 188 victories for the loss of 86 Bloch pilots dead, wounded or taken prisoner is testimony to a sound combat aircraft, which was extremely strong, excellent to fly, and possessed adequate performance and more than adequate hitting power.

### Fokker D.XXI
First flown in 1936, the D.XXI was licence-built in Finland from 1939 and was immediately involved in the fighting with Russia, becoming responsible for the first Finnish air victory in the Winter War
*Span:* 36 ft 1 in *Length:* 26 ft 3 in *Engine:* Pratt & Whitney Twin Wasp Junior, 825 hp *Max speed:* 272 mph at 9000 ft *Ceiling:* 32,000 ft *Max take-off weight:* 4820 lb *Armament:* 4 × 7.9-mm mg

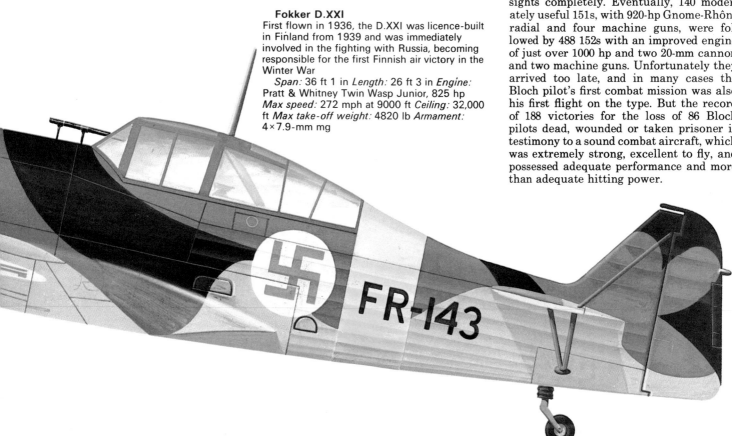

France's third most numerous fighter was the Hawk 75A, bought from Curtiss in the United States. Though only 291 actually reached the *Armée de l'Air* before the Armistice, they did much more fighting than any other type except the Morane for the simple reason that Curtiss (unlike the French industry) delivered ahead of time, and there was plenty of time to get pilots and ground crews trained and squadrons operational. The standard Hawk 75A had a Pratt & Whitney Twin Wasp of 1050 or 1200 hp and armament of four or six FN-built Browning machine guns. Speed was comfortably over 300 mph, most reaching 311-323 mph; structural strength, manoeuvrability, handling qualities and ease of maintenance were all good. The Curtiss equipped five groups – GC I/4, II/4, I/5, II/5 and III/2 – and these were credited with a remarkable 311 victories. In the 1er *Escadrille* (squadron) of I/5 were *Lt* Marin-la-Meslée (20 victories), *Capts* Accart (15) and Vasatko (15), *Sous-Lt* Perina (13) and *Sgt-chef* Morel (12); virtually all these victories were gained between 10 May and 23 June 1940. The Curtiss units had the advantage of proper equipment and thorough training, which in turn resulted in high morale even in the adversity and chaos of the great retreat.

France did possess superior fighters, but these made only a small impact on the battle. The Dewoitine D.520 is generally judged the best of the fighters that saw action, but though the nationalised SNCA du Midi worked production up to an excellent 299 aircraft during June 1–25, no aircraft reached a squadron until March 1940, when GC I/3 began to convert. A neat and extremely manoeuvrable machine, the 520 reached an adequate 329 mph on only 910 hp and carried one Hispano cannon and four machine guns. Though most of the pilots who flew it into action had no experience on the type, they were credited with 147 victories for the loss of 85 aircraft and 44 pilots. Beyond the 520 were many excellent prototypes and projects, such as the Merlin-engined 521, the even more powerful 520Z, and the redesigned 551. Likewise the Bloch 155 was in production in June 1940, followed by the outstanding Bloch 157 which, with a 1700-hp Gnome-Rhône 14R radial, reached 441 mph and easily out-performed all other fighters of the first half of the war.

A list of the remaining French fighters of 1940 seems endless. By far the most important of the twin-engined long-range escort and night fighters was the Potez 631, first flown in 1936 and in production with Hispano-Suiza 640-hp radials the following year. The main production variant had 670-hp Gnome-Rhône engines, reached approximately 280 mph and carried two 20-mm cannon under the slim two-seat fuselage. In February 1940, when just over 200 were in use, six machine guns were added in trays under the outer wings, making the Potez the most heavily armed machine in service. It was an attractive aircraft, and pleasant to fly, but as a combat aircraft simply lacked power. It took nine minutes to climb to 4000 m (13,120 ft) and was much slower even than a Ju 88, let alone a Bf 109. The Potez plant turned increasingly to attack and reconnaissance versions of the 630 family, a novel single-engined fighter, the Potez 230, remaining a one-off.

Breguet's neat 690 family were used mainly in the attack role, and to speed

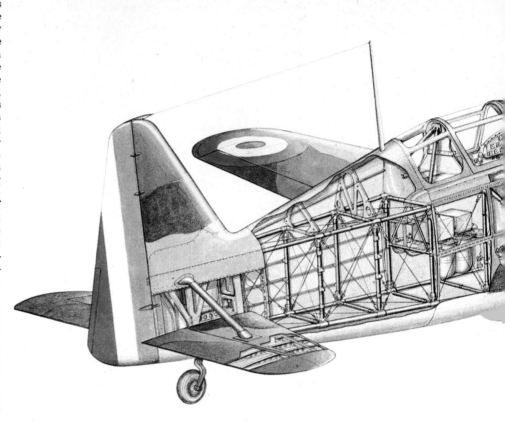

output of these machines the SNCAC Bourges plant dropped the Hanriot NC 600 twin-engined fighter despite the latter's excellent performance (337 mph), heavy armament (three Hispano cannon and two machine guns) and generally good handling.

An even more impressive twin-engined fighter was the S.E.100, first flown on 2 April, 1939. This was a remarkable aircraft, with ailerons forming the wing-tips, a stumpy body, a single nosewheel and three small wheels on the rear tip of the body and under the twin fins. However, there was nothing wrong with the S.E.100's speed (360 mph), handling or armament of four forward-firing Hispano cannon and a fifth firing from the rear cockpit. In 1940 a second S.E.100 was built with six forward-firing cannon, two in a rear dorsal turret and one in a ventral tunnel. To mount nine 20-mm cannon in a 360-mph fighter in the spring of 1940 was an extraordinary accomplishment.

Probably the best of all the French fighters were the Galtier/Vernisse designs at the Arsenal de l'Aéronautique, which reached mass production by the spring of 1940 as the VG 33. This aircraft did the very best it could with an Hispano-Suiza of only 860 hp, and with one cannon and four machine guns reached the excellent speed of 347 mph. It was a fine-looking aircraft, and 160 had taken shape on the assembly line by the Armistice, though only seven reached the squadrons (beginning with GC I/2) and pilot conversion had not started. Prototypes had flown of the 910-hp VG 34, 1100-hp VG 35 and 1280-hp VG 39, which reached 388 mph with one cannon and six machine guns.

The strangest Arsenal fighter was the Delanne 10, which achieved the excellent speed of 342 mph on only 860 hp. Delanne was the pioneer of the tandem-wing aircraft, for which he claimed many advantages in centre of gravity travel, manoeuvrability and safety. The Delanne 10

had slats on the forward wing and a rear wing which could be considered as a large tailplane/elevator combination. At the rear sat the pilot and rear-facing gunner with two machine guns to discourage tail-chasers. Forward-firing armament comprised one cannon and two machine guns, but this had not been fitted when the extraordinary fighter first flew, under German supervision, in October 1941.

There were many other French fighters that never reached the starting line, including the CAO.200 (culmination of the Loire Nieuport 160 and 161 families) and even seaplanes. But one fighter that did get into action was the Dutch Koolhoven F.K.58, ordered for the *Armée de l'Air* in the winter of 1938 to help make up for the lateness of nearly all the French programmes. Fritz Koolhoven's team at Waalhaven, near Rotterdam, had produced many interesting aircraft, including the F.K.55 fighter with contra-props driven by an engine behind the pilot. But the F.K.58 was a straightforward machine, constructed in a mere two months in August–September 1938, which reached 313 mph on its first flight, powered by a 1080-hp Hispano-Suiza radial. Armed with four FN-Browning machine guns, it was merely an adequate aircraft; only 18, delivered to France in the late summer of 1939, reached squadrons (one reason was that later French batches specified Gnome-Rhône engines which were never delivered). But the 18 that reached France were formed into local-defence *Patrouilles de Protection*, manned by escaped Polish pilots.

No F.K.58 reached the LVA (Dutch Air Force), whose principal fighter in May 1940 was the outdated Fokker D.XXI. First flown in 1936, this was a serviceable and workmanlike machine, but too advanced in concept. It had mixed wood/metal construction, with covering of fabric or metal panels, just as in aircraft of the 1920s; though it was a cantilever monoplane with small split flaps, it

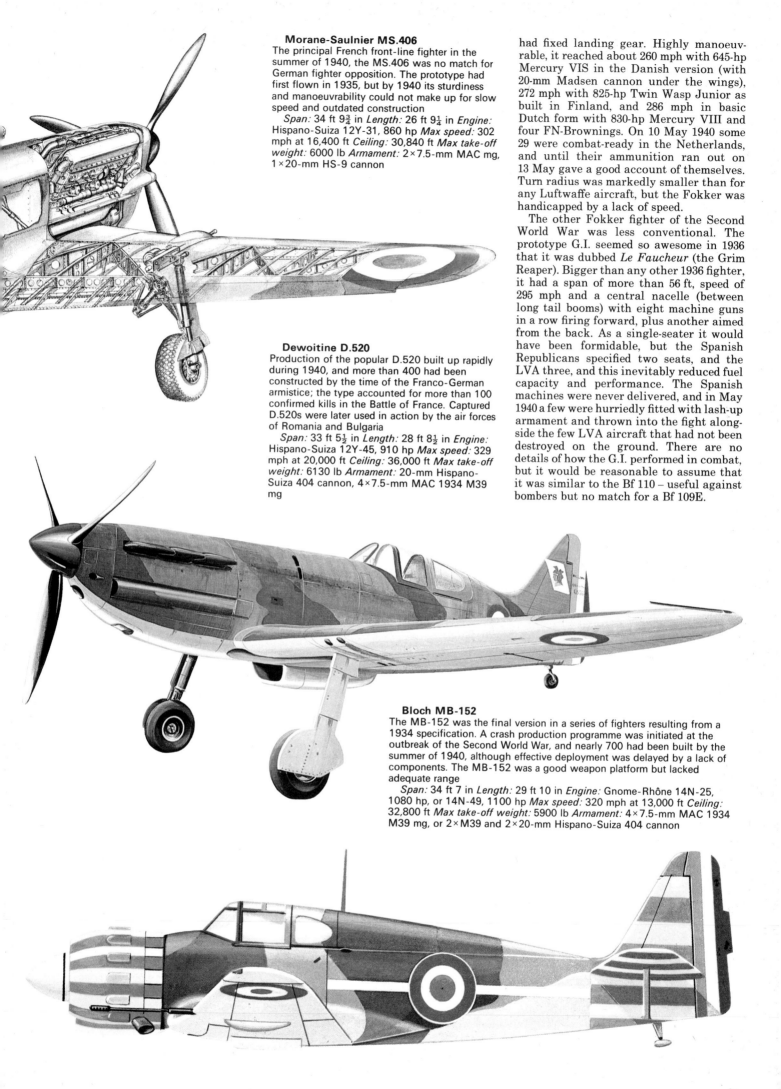

## Morane-Saulnier MS.406
The principal French front-line fighter in the summer of 1940, the MS.406 was no match for German fighter opposition. The prototype had first flown in 1935, but by 1940 its sturdiness and manoeuvrability could not make up for slow speed and outdated construction
  *Span:* 34 ft 9¾ in *Length:* 26 ft 9¼ in *Engine:* Hispano-Suiza 12Y-31, 860 hp *Max speed:* 302 mph at 16,400 ft *Ceiling:* 30,840 ft *Max take-off weight:* 6000 lb *Armament:* 2 × 7.5-mm MAC mg, 1 × 20-mm HS-9 cannon

## Dewoitine D.520
Production of the popular D.520 built up rapidly during 1940, and more than 400 had been constructed by the time of the Franco-German armistice; the type accounted for more than 100 confirmed kills in the Battle of France. Captured D.520s were later used in action by the air forces of Romania and Bulgaria
  *Span:* 33 ft 5½ in *Length:* 28 ft 8½ in *Engine:* Hispano-Suiza 12Y-45, 910 hp *Max speed:* 329 mph at 20,000 ft *Ceiling:* 36,000 ft *Max take-off weight:* 6130 lb *Armament:* 20-mm Hispano-Suiza 404 cannon, 4 × 7.5-mm MAC 1934 M39 mg

## Bloch MB-152
The MB-152 was the final version in a series of fighters resulting from a 1934 specification. A crash production programme was initiated at the outbreak of the Second World War, and nearly 700 had been built by the summer of 1940, although effective deployment was delayed by a lack of components. The MB-152 was a good weapon platform but lacked adequate range
  *Span:* 34 ft 7 in *Length:* 29 ft 10 in *Engine:* Gnome-Rhône 14N-25, 1080 hp, or 14N-49, 1100 hp *Max speed:* 320 mph at 13,000 ft *Ceiling:* 32,800 ft *Max take-off weight:* 5900 lb *Armament:* 4 × 7.5-mm MAC 1934 M39 mg, or 2 × M39 and 2 × 20-mm Hispano-Suiza 404 cannon

had fixed landing gear. Highly manoeuvrable, it reached about 260 mph with 645-hp Mercury VIS in the Danish version (with 20-mm Madsen cannon under the wings), 272 mph with 825-hp Twin Wasp Junior as built in Finland, and 286 mph in basic Dutch form with 830-hp Mercury VIII and four FN-Brownings. On 10 May 1940 some 29 were combat-ready in the Netherlands, and until their ammunition ran out on 13 May gave a good account of themselves. Turn radius was markedly smaller than for any Luftwaffe aircraft, but the Fokker was handicapped by a lack of speed.

The other Fokker fighter of the Second World War was less conventional. The prototype G.I. seemed so awesome in 1936 that it was dubbed *Le Faucheur* (the Grim Reaper). Bigger than any other 1936 fighter, it had a span of more than 56 ft, speed of 295 mph and a central nacelle (between long tail booms) with eight machine guns in a row firing forward, plus another aimed from the back. As a single-seater it would have been formidable, but the Spanish Republicans specified two seats, and the LVA three, and this inevitably reduced fuel capacity and performance. The Spanish machines were never delivered, and in May 1940 a few were hurriedly fitted with lash-up armament and thrown into the fight alongside the few LVA aircraft that had not been destroyed on the ground. There are no details of how the G.I. performed in combat, but it would be reasonable to assume that it was similar to the Bf 110 – useful against bombers but no match for a Bf 109E.

# THE BATTLE OF BRITAIN

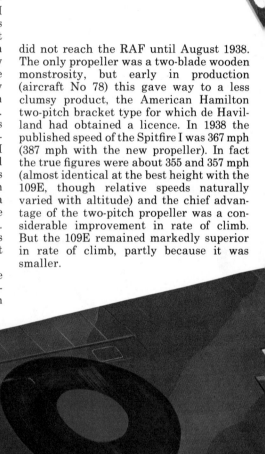

The Battle of Britain was one of the truly decisive battles in the history of warfare, and, like Waterloo, was 'a damned close-run thing'. It was not entirely a case of David besting Goliath; aircraft and men on both sides were outstanding, and often the fighting was on equal terms, or even in the favour of the Luftwaffe. Nor was either side 'defeated' in any accepted sense of the word; both air forces were approximately as strong at the end of the daylight battle as at the beginning. But imperfect top-level planning and direction by Germany allowed the desperately hard-pressed RAF a vital breathing-space; and geography restricted the Bf 109 to penetrations hardly extending beyond Kent and Sussex. Moreover, Sir Robert Watson-Watt's early-warning radar chain (combined with the inventions of IFF – Identification Friend or Foe – and VHF voice links with a ground-based fighter controller) made an incalculable difference to the effectiveness of every British fighter.

Over Europe the Bf 109 dominated the skies, though the D.520 had run it close, and sheer pilot skill and courage had often narrowed the gap with the Curtiss and Bloch. From the earliest days of the war the 109E had met the Hurricane I over the so-called Western Front in sporadic and usually individual combats. Ultimately no fewer than twelve Hurricane squadrons had fought in France, and, as few returned, the RAF lost about one-quarter of its front-line strength in trying to delay the German advance. Nearly all these Hurricanes had been of the Mk I type with a 1030-hp Merlin I driving a Watts fixed-pitch wooden propeller; armament comprised eight 0.303-in Browning machine guns in a fabric-covered wing. Their performance was markedly inferior to the Messerschmitt in every respect other than turn radius; and even this advantage was seldom exploited because of pilots' inexperience and timidity. To survive, Hurricane pilots needed a wide-open throttle and a stick boldly placed in the far corners of the cockpit. Even then, the direct-injection Daimler-Benz engine, with its complete indifference to negative-g manoeuvres, gave the 109E a major advantage; but again it was pilot inexperience that often left this advantage unused. In any case, if a 109 stuffed its nose down and dived, it made little difference that the Merlin would cough and stop running if the British pilot tried to follow, because – surprisingly – both the Hurricane and Spitfire could catch a diving 109.

From its inception in 1934, at the same time as the 109, the Hurricane had possessed a great asset in the Merlin engine, which was significantly more powerful than the engines of the Luftwaffe's Continental opponents and made a considerable difference to aircraft performance. At the same time the Hurricane was handicapped by being large and cumbersome, having a primitive structure, a crude propeller and rifle-calibre guns. At first it also lacked protection, as did its adversary, but in 1938–40 the Hurricane progressively received an armoured bulkhead forward of the cockpit, bullet-proof windscreen, constant-speed three-blade propeller, metal stressed-skin wings, and finally, from February 1940, seal-sealing tanks and rear armour. Cannon were also fitted, but these did not enter service until 1941.

The heavy attrition in France, amounting to 386 Hurricanes between 10 May and 20 June, 1940, resulted in the urgent re-issue to Fighter Command of many old Mk I Hurricanes with fabric wings and Watts propeller, and with these a dogfight against an experienced 109E pilot was a tough proposition. Maximum speed of a new Hurricane was around 316 mph, but the much-repaired veterans, burdened now by armour and extra equipment, did not, in Lord Dowding's experience, exceed 305 mph. Yet this did not quite put them in the class of such also-rans as the Morane. Determinedly flown, even a patched-up Mk I could out-manoeuvre a 109 sufficiently well to avoid being shot down. The Mk I was also the best and steadiest gun-platform, an extremely forgiving aircraft (so that a badly wounded pilot could get back on the grass in one piece) and easy to repair. Though ability to take punishment does not of itself win wars, it helps. By contrast the 109 did not absorb punishment so well.

The Spitfire also lacked the ability to take heavy punishment. Though much less important numerically than the Hurricane in the first year of war, and not used over France (except on occasion over Dunkirk), the Spitfire was the only fighter that could claim to be marginally superior to the 109E – though this remains a matter for argument. It was slightly later in conception, and from the start the structure was unnecessarily complex and difficult to make and repair, remaining so throughout the war until the development of the completely new airframe of the Spiteful in 1944. These manufacturing problems were responsible for delaying production, and the first Spitfire did not reach the RAF until August 1938. The only propeller was a two-blade wooden monstrosity, but early in production (aircraft No 78) this gave way to a less clumsy product, the American Hamilton two-pitch bracket type for which de Havilland had obtained a licence. In 1938 the published speed of the Spitfire I was 367 mph (387 mph with the new propeller). In fact the true figures were about 355 and 357 mph (almost identical at the best height with the 109E, though relative speeds naturally varied with altitude) and the chief advantage of the two-pitch propeller was a considerable improvement in rate of climb. But the 109E remained markedly superior in rate of climb, partly because it was smaller.

**Hawker Hurricane Mk I**
The Hurricane was numerically the most important RAF fighter during the Battle of Britain, the first squadron having become operational in January 1938; more than 500 were in service on the outbreak of war. The decision to pull the Hurricane squadrons back from the debacle in France was crucial when the battle over southern England began
*Span:* 40 ft *Length:* 31 ft 4 in *Engine:* Rolls-Royce Merlin II, 979 hp *Max speed:* 316 mph at 17, 750 ft *Ceiling:* 33,750 ft *Max take-off weight:* 6040 lb *Armament:* 8 × .303-in Browning mg

Hurricane squadron scramble. The wooden two-blade propellers were progressively replaced by constant-speed airscrews during the heat of battle. The faster Spitfires had priority, but the Hurricane I remained the most important RAF fighter numerically throughout the summer and autumn of 1940

*Bf 109 pilots relax on a Luftwaffe Channel airfield*

Possibly Messerschmitt's team placed too much emphasis on compactness. Reginald Mitchell made the Spitfire's cockpit larger and more comfortable, and in planning for the expected ultimate armament of eight machine guns (only four were fitted to begin with) was forced to give the wing generous area, with considerable chord well outboard to accommodate the Nos 1 and 8 guns. This resulted in the characteristic elliptical plan-form. Gross wing area was thus 242 sq ft, compared with only 174 sq ft for the Bf 109E. This enabled the Spitfire to easily out-turn the German fighter at all altitudes. Although this was a tremendous advantage, large numbers of Spitfires, and even Hurricanes, were shot down simply because their pilots failed to fly to the limit. Tests with a captured Bf 109E-3 showed that, when the British fighter (either type) was on the tail of the German, it had no difficulty in keeping the 109 in its sights, with plenty of turn radius in hand. When the positions were reversed, the 109 could invariably get the British fighter in its sights because the RAF pilot was not pulling maximum g. Even the motivation of self-preservation was often not enough to make Spitfire and Hurricane pilots boldly haul back with all their might and use their tight-turn advantage.

The 109 did possess a distinct advantage in that it could climb at full power so steeply that the slats opened and the aircraft was close to the stall. A determined Spitfire or Hurricane pilot could hold this for a while, but it was very hard indeed to bring the reflector sight to bear on the 109, even for a moment. Curiously, this manoeuvre was seldom practised by 109 pilots. It was common, however, for a 109 to dive away, and as airspeed built up the German fighter became almost unmanageable. With the Hurricane there was plenty of control at all speeds, though the speeds were generally slightly lower than for its rivals. The Spitfire tightened up considerably more, and the ailerons in particular were very heavy at speeds around 400 mph (these comments do not apply to the much faster later marks of Spitfire). But at around 400 mph the 109 could hardly manoeuvre at all. The ailerons

**Messerschmitt Bf 109E-4**
By mid 1940 the lessons of the French campaign were being incorporated in the Bf 109E. The E-4 replaced the troublesome engine mounted MG FF/M cannon with two wing mounted MG FF cannon with an improved rate of fire
*Span:* 32 ft 4½ in *Length:* 28 ft 4 in *Engine:* DB 601A, 1100 hp *Max speed:* 354 mph at 12,300 ft *Max take-off weight:* 5523 lb *Ceiling:* 36,500 ft *Armament:* 3×7.9-mm MG 17 mg, 2×20-mm MG FF cannon

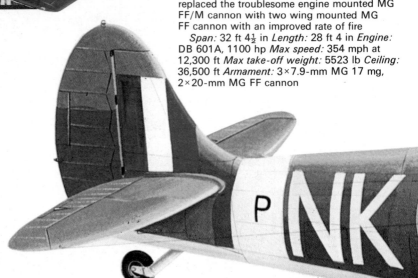

**Spitfire IIa**
During the battle a new version of the Spitfire, incorporating the first results of operational experience, began to reach front-line squadrons. The Spitfire II, with the Merlin XII engine driving a three-blade constant-speed airscrew, had increased armour protection and improved speed and ceiling. The type joined the battle in its closing stages and began the first offensive sweeps into occupied Europe
*Span:* 36 ft 10 in *Length:* 29 ft 9 in *Engine:* Rolls-Royce Merlin XII, 1236 hp *Max speed:* 354 mph at 17,550 ft *Ceiling:* 37,600 ft *Max take-off weight:* 6172 lb *Armament:* 8×.303-in Browning mg

*Hurricanes receive simultaneous refuelling from a bowser*

*Armourers work frantically on a Spitfire Ia at Duxford, October 1940. The eight .303-in Brownings were fed from 300-round belts*

were almost immovable, and in the narrow cockpit the pilot had no room to apply his full force to the stick except in the fore/aft direction. There was a further 109 deficiency that made combat an exhausting business. The 109 had no rudder trimmer and, as speed was increased, the pilot had to apply an increasing amount of left rudder, until at over 300 mph substantial force was needed. This became exhausting after a few minutes, and made gun sighting more difficult. Moreover, the slats snatched open and shut in tight turns, causing asymmetric drag and throwing the pilot off his aim.

It was remarkable that with so many deep-seated deficiencies the 109 remained in production until the end of the war. With a larger wing and improved high-speed lateral control it would have been a far more dangerous opponent, but the small 173–176-sq-ft wing remained unchanged. As a result the 109 was never far from the stall when pulling g, and though the stall was gentle in early versions, it could mean that a fight would be lost rather than won.

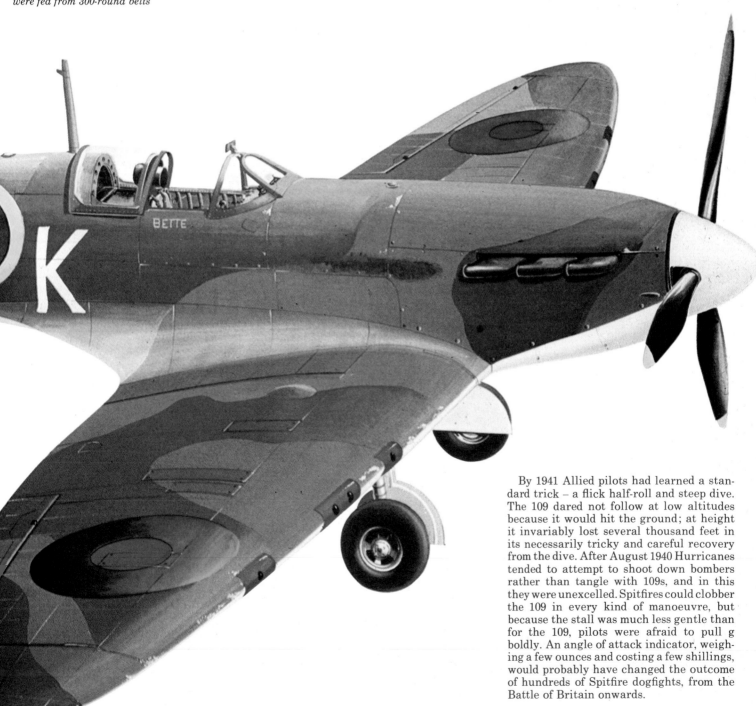

By 1941 Allied pilots had learned a standard trick – a flick half-roll and steep dive. The 109 dared not follow at low altitudes because it would hit the ground; at height it invariably lost several thousand feet in its necessarily tricky and careful recovery from the dive. After August 1940 Hurricanes tended to attempt to shoot down bombers rather than tangle with 109s, and in this they were unexcelled. Spitfires could clobber the 109 in every kind of manoeuvre, but because the stall was much less gentle than for the 109, pilots were afraid to pull g boldly. An angle of attack indicator, weighing a few ounces and costing a few shillings, would probably have changed the outcome of hundreds of Spitfire dogfights, from the Battle of Britain onwards.

Little need be said of the other Battle of Britain fighters. The Boulton Paul Defiant had a wing smaller than a Hurricane yet a weight considerably greater, mainly because it carried a gunner in a 680-lb power-driven turret with four rifle-calibre Brownings. Admittedly one could subtract from this the weight of the guns and ammunition, but the bare turret still weighed 361 lb and the gunner might add another 180, and such figures were important in 1940. The Defiant was a poor fighter in terms of speed, climb and turn radius, and though in May–June 1940 it had one or two moments of startling success, it was soon judged useless except by night.

The Blenheim fighter was also useful only at night. It was not a vicious machine, had a crew of two or three, and good endurance. But it was unable to catch even a Ju 88, and the hitting power of its four Brownings was totally inadequate.

By 1940 the fighter business had grown more serious. Bullet-proof windscreens, armour and self-sealing tanks were fast becoming universal, and the rifle-calibre gun was equally rapidly becoming inadequate. The 860-hp Hispano-Suiza, and similar engines of this power class, were no longer sufficient; even the 1000-hp Merlin and DB 601 had to be worked frantically to yield increased power at ever-greater heights. The beautifully tractable Fiat C.R.42 biplane fighters which accompanied the *Corpo Aereo Italiano's* Fiat B.R. 20 bombers over Britain in October and November 1940 were totally ineffective. Though they had the manoeuvrability to get out of harm's way, they could neither protect the bombers nor shoot down British fighters, and their value was no greater than the proposed last-ditch defence armed versions of British light aircraft, such as the Mew Gull and Tiger Moth, of July 1940.

Finally, mention must be made of a fighter that never reached the squadrons, although it was designed, built and flown in 65 days, and in some respects was better than the contemporary Hurricane and Spitfire. F G Miles had schemed the M.20/1 fighter in 1939, and in 1940 asked the British aircraft-production minister, Lord Beaverbrook, for permission to build a fighter to help defend the nation. The resulting M.20/2, designed by Walter Capley, was outstanding in many respects. A typical Miles wooden aircraft, with thick wing, it featured a Merlin XX 'power egg' of the form later standardised on the Lancaster and other aircraft, standard parts from Master trainers, eight Brownings with space for another four plus 5000 rounds, and 154 gallons of fuel. Even with fixed landing gear it reached 333 mph with full armament installed. Handling was excellent, the teardrop canopy years ahead of its time, and both ammunition and fuel capacity approximately double that of existing RAF fighters. It needed longer take-off and landing runs because it had a higher wing loading, although a production version might well have incorporated a hydraulic system, and thus had larger flaps and retractable landing gear. If the M.20 had been built in 1939, there would probably have been thousands; but in 1939 permission to build it would have been unlikely. Thus this excellent basis for development was rejected, as were the equally outstanding fighters by Martin-Baker Aircraft which appeared at intervals up to 1944.

### Spitfire Cockpit
The firing button for the eight Brownings is incorporated in the joystick handle. The seat accommodates the parachute and the gunsight base is fitted with a crash-pad

### Boulton Paul Defiant Mk I
The tactical concept of the turret fighter (originally as a bomber destroyer) was a disastrous failure in day fighting, although the Defiant had some success early in the battle when mistaken for a single-seater. The type soldiered on as night-fighter and target tug
*Span:* 39 ft 3 in *Length:* 34 ft 7 in *Engine:* Rolls-Royce Merlin III, 1440 hp *Max speed:* 312 mph at 10,000 ft *Ceiling:* 28, 100 ft *Max take-off weight:* 7510 lb *Armament:* 4×.303-in Browning mg

*The Miles M.20, the wooden 'utility' fighter which could outperform the Hurricane and carry twice as much fuel and ammunition as a Spitfire. It never entered production*

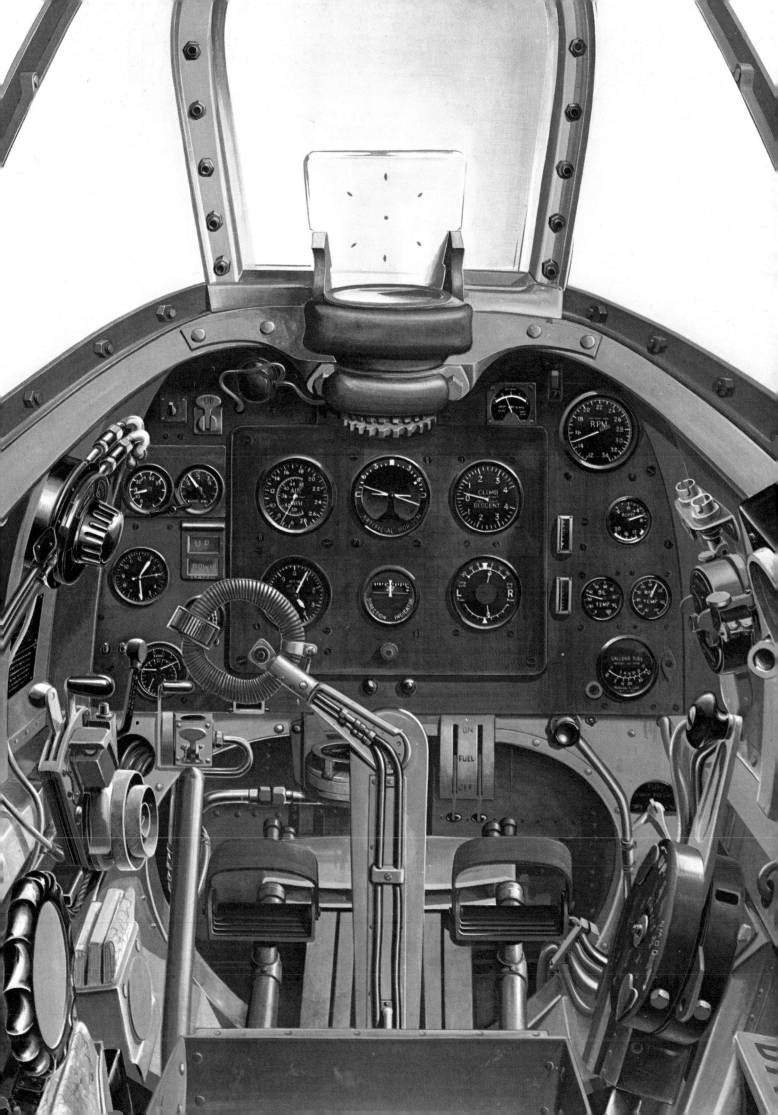

# THE SPREADING WAR

*Adlertag* (Eagle Day), the date of Göring's promised air offensive against Britain, happened as planned on 13 August, 1940. But the date planned for Operation *Seelöwe* (Sealion), the invasion of southern England, was held back because RAF Fighter Command had not been eliminated. But the invasion never came to pass. Though the Luftwaffe's near-total switch to the night-bombing 'Blitz' reduced its losses over England to vanishing point, it meant that, for the first time, it had failed completely in its objective. For the moment Britain was clearly not going to be defeated. Although Hitler promised to return in 1941, to finish off the obstinate islanders, he was forced to look for fresh fields to conquer.

The obvious direction was south through the Balkans to help his faltering ally, Italy, which could not even subdue the island of Malta, let alone conquer the Greeks or drive the British from North Africa. Rumania joined the German/Italian Axis on 23 November 1940, with Bulgaria following on 1 March, 1941. But the Greeks were pushing back the Italians, and the Yugoslavs overthrew their leaders and repudiated the Axis Pact. On 6 April, 1941, the Wehrmacht rolled into both countries.

The principal opposition to the Luftwaffe's Bf 109E was provided by the latest Yugoslav fighters, ironically Bf 109Es. They were valiantly flown, but were soon overcome. However, a few Yugoslav aircraft managed to escape to Greece and later Egypt, including two Do 17 bombers loaded with gold bullion. Yugoslavia also had a few Hurricanes, Hawker Fury biplanes and a handful of home-produced fighters. The parasol-winged Ikarus IK-2 was obsolescent, having been conceived in 1934, but the six or eight still in use saw intensive action and shot down several Luftwaffe aircraft, besides being heavily committed to ground-attack missions. The Rogozarski IK-3 was a later concept, first flying in 1938, and though it looked undistinguished, it proved to be remarkably manoeuvrable. On a 920-hp Czech-built Hispano engine it reached 327 mph, had the same armament of one cannon and two machine guns as the IK-2, and proved to be able to turn inside the Yugoslav Hurricanes. Pilots evaluated the IK-3 against both the Hurricane, which was being built under licence, and the Bf 109E-3, rating it distinctly preferable to either. By 1941 an IK-3 had demonstrated even better performance with a German DB 601 engine, but only six Hispano-powered machines were ready in April 1941, operated by the 52nd Squadron at Zemun. Estimates suggest that they destroyed more than 14 Luftwaffe aircraft before being overcome.

Operations against Greece had been started by Italy in 1940, but that unhappy nation soon found itself in deep trouble in Greece. In sharp contrast to the impressive and flamboyant spectacles of Italian air-power in the 1930s, the performance of the *Regia Aeronautica* was invariably undistinguished. Until mid-1941 Italian fighter philosophy adhered tenaciously to the tradition of seeking the greatest possible manoeuvrability at the expense of performance, firepower and protection. This in turn removed from the brilliant engineers in the famous engine companies – Fiat, Alfa Romeo, Isotta-Fraschini and Piaggio – the fierce pressure for more power that should have spurred them ahead. As in Japan, Italian fighters tended to opt for engines of 1000 hp or less, with armament of two 0.5-in machine guns and speed barely in excess of 300 mph. Though Italian pilots lacked little in the way of courage or skill, this outdated design philosophy hardly gave them a chance until, under German tuition, they received fighters with abundant horsepower and firepower in 1943. By then their will to fight, never strong, had almost vanished.

Best of the Italian fighter families were those that stemmed from the prototype Macchi C.200 flown in December 1937. C stood for designer Mario Castoldi, who had earlier planned the company's racing seaplanes; but the C.200 had to fit the requirements and was a lumpy machine, with only 840 hp, that could not even catch a Hurricane I, despite its light weight and small size. With just two heavy machine guns, its only favourable attributes were its short field lengths, rapid and steep climb, excellent manoeuvrability and extremely pleasant flying qualities. These qualities were not lost in the subsequent M.C.202, flown in August 1940 with the German DB 601A engine, and M.C.205, flown in April 1942 with the powerful DB 605, which by then was the standard engine of the Bf 109. Unlike the German fighter, the Macchis were to the end extremely likeable flying machines, and their armament crept up to an eventual standard of three 20-mm Mauser MG 151 cannon and two 0.5-in machine guns. But by this time, in late 1943, about half the M.C.205s had come over to the Allies.

Fiat's biplane fighters ought not to be dismissed as of no consequence in the air fighting of the Second World War; although they were obsolescent, they were so manoeuvrable that RAF fighters found it difficult to shoot them down. At the start of 1940 the most numerous fighter in the *Regia Aeronautica* was still the C.R.32, but it was fast being supplanted by the C.R.42. One of the last C.R.42 was fitted with a DB 601, and its speed of 323 mph may possibly have made it the fastest biplane ever (certainly much faster than the McGregor FDB-1 popularly reported as such). It is a telling reflection on Italian lack of determination and efficiency that the number of C.R.42s

**Avia B-534**
One of the last of the fighter biplanes, the B-534 constituted the Czech Air Force's front line when the Germans marched in during 1939. Three squadrons fought briefly with the Slovak Air Force on the Kiev Front in 1941
*Span:* 30 ft 10 in *Length:* 26 ft 7 in *Engine:* Hispano-Suiza HS 17Ycrs, 850 hp *Max speed:* 245 mph at 14,435 ft *Ceiling:* 38, 875 ft *Max take-off weight:* 4365 lb *Armament:* 4×7.7-mm mg

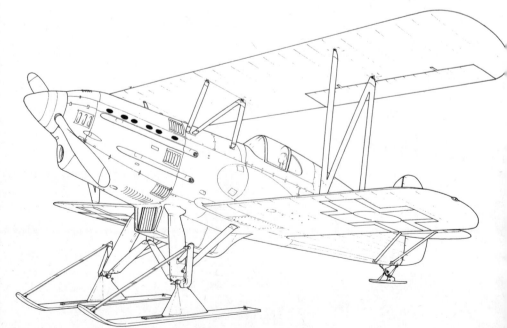

**Macchi C202 Folgore**
The availability of German liquid-cooled engines brought Italian fighter design up to world standard with the C202, an adaptation of the highly manoeuvrable radial-engined Macchi MC 200. Problems of engine supply kept deliveries of this most successful of Italian fighters to limited numbers, however
  *Span:* 38 ft 8½ in *Length:* 29 ft 0½ in *Engine:* Daimler-Benz DB 601A, 1150 hp *Max speed:* 360 mph at 20,000 ft *Ceiling:* 34, 490 ft *Max take-off weight:* 6400 lb *Armament:* 2×12.7-mm mg, 2×7.7-mm mg

**Reggiane Re 2000 Falco I**
Loser in a fighter competition with the MC 200, the Re 2000 was exported to Sweden, licence-built in Hungary and flown by the Italian Navy
  *Span:* 36 ft 1 in *Length:* 26 ft 2½ in *Engine:* Piaggio P XI(*bis*), 1025 hp *Max speed:* 329 mph at 16,400 ft *Ceiling:* 36, 745 ft *Max take-off weight:* 5722 lb *Armament:* 2×12.7-mm Breda-SAFAT mg

built, a mere 1781, was markedly greater than the total for any other Second World War Italian type. The fact that the C.R.42 was a biplane was immaterial; like the Royal Navy's Swordfish, this helped it continue in useful front-line service while later and faster machines faltered. By early 1942 all the *Regia Aeronautica*'s specialised ground-attack aircraft had failed, but the old C.R.42 filled the gap admirably, and remained in production until well into 1943, flying close-support and attack missions alongside the *Stormi* equipped with the German-built Ju 87.

The C.R.42's major defect was its limited engine power of 840 hp. Though well matched to the aircraft when it was designed, by 1942 this was almost useless. The lesson was learned by American designers in 1940 in time for the mass-produced US fighters of 1942–45 to have 2000 hp under their bonnets. But though the Italians had engines in this class under development, they progressed slowly and played virtually no part in the war. When Fiat planned a monoplane fighter in 1936, the blinkered vision of both the company and customer resulted in the engine being the same as that of the C.R.42, rated at 840 hp. The aircraft was the G.50, the first major design by Giuseppe Gabrielli, and it was uninspired and pedestrian. Fitted with the standard pair of synchronised heavy machine guns, its only advantages were manoeuvrability, durability and fair reliability, and it made little impact on the war. Another Fiat design, the last major effort by Rosatelli, was even less successful. The C.R.25, first flown in 1939, was a large twin-engined fighter for use in the 'Italian colonies', with long range and the ability to fly reconnaissance missions. Though its two engines were the same as those fitted to other Fiat fighters, it was slow and pathetically ill-armed, finishing its career as a transport, surely the ultimate indignity for any fighter!

Of Italy's many other fighters only the Reggiane series deserves mention. The Re.2000 of 1937 had excellent performance, and was widely used by Sweden and Hungary. The first model used by the *Regia Aeronautica* was the Re.2001 with a German DB 601 engine; the 252 built included versions capable of carrying a 1410-lb bomb or a torpedo. The most important variant was the 2001CN night fighter, with its synchronised heavy machine guns supplemented by two hard-hitting MG 151/20 cannon under the wings. Reggiane, a Caproni company, also developed the Re.2002 close-support fighter, with radial engine, heavy bomb load and armour. However, no more than 50 were built. The output of other Italian fighters did not reach double figures: the Caproni Ca 331B three-seat twin-engined night fighter, with up to six 20-mm cannon firing ahead; the Caproni Vizzola F.4 (DB 601), F.5 (Fiat A74 radial) and F.6 (licensed DB 605); IMAM Meridionali Ro 57 and 58 twin-engined fighters (the former with forward-firing armament of two machine guns and the latter with five cannon); the Piaggio P.119, with very advanced features and 1700-hp radial engine behind the pilot; the outstanding light fighters of SAI (Ambrosini), beginning with the tail-first S.S.4 and finishing with the SAI.403 *Dardo*, which achieved 403 mph on a mere 750 hp while carrying two cannon and two heavy machine-guns; and the extremely interesting and promising Savoia-Marchetti

Bristol Centaurus-powered Hawker Tornado prototype

The cockpit of the prototype Hawker Tempest V, powered by the Napier Sabre II, first flown in September 1942

S.M.91 and 92 with two DB 605 engines at the front of long tail booms, the 91 having a central nacelle and the 92 having the tandem-seat cockpit in the left boom. These little-known machines would have been formidable, had more than one of each been built.

Like France in 1935–40, the potentially important Italian industry frittered away its strength in badly managed programmes. By 1942 its only hope lay in advanced developments of existing conventional fighters redesigned with German engines. The standard engine was the DB 605, and the normal armament three MG 151 cannon and two heavy machine guns. It is remark-

able that Macchi, Fiat and Reggiane were able to fit almost double the horsepower and many times the armament into their fighters with only very slight increase in weight. Comparing such machines as the Macchi 205N, Fiat G.55 and Reggiane Re.2005 with the far more important Bf 109G series is very instructive. While the German fighter had about 174 sq ft of wing, all the Italian aircraft had considerably more than 200 sq ft, yet on the whole were comparable in weight. All three were evaluated against the 109G and proved to have dramatically better manoeuvrability, and at least equal performance. But by late 1942 Italy's production had fallen away sharply, and none

Imperial War Museum

John Batchelor

of these three excellent machines made any significant contribution to the war. In no case did deliveries to the *Regia Aeronautica* reach 100 aircraft.

At first the Allied fighters in the Mediterranean were an old and motley collection, but by late 1940 some Mk I Hurricanes were made available, backed up by Fulmar squadrons of the Royal Navy. The Fulmar was a tandem-seat carrier fighter based on a pre-war light bomber somewhat smaller than the Battle. Though a pedestrian machine, as was inevitable with 342 sq ft of wing, a gross weight exceeding 10,000 lb and a mere 1080 hp, the Fulmar met all the requirements and was quickly put into use. The prototype flew on 4 January, 1940, yet by August 806 Sqn was fully equipped and combat-ready aboard *Illustrious* in the Mediterranean. Altogether 600 of these useful machines were built, the last 350 having 1300 hp and consequently a speed raised from 244 to 272 mph, which almost kept it abreast of typical Italian aircraft. A much more formidable carrier fighter was the Grumman G.36, first ordered by the French *Aéronavale* in 1939 but transferred to Britain on France's capitulation. As the

Martlet it went into use with 804 Sqn of the Royal Navy in October 1940. On Christmas Day 1940 two Martlets destroyed a Ju 88 near Scapa Flow. Variously powered by the Wright R-1820 Cyclone or Pratt & Whitney R-1830 Twin Wasp, both of 1200 hp, the Martlet had a squarish wing of 260 sq ft. As it could comfortably exceed 300 mph and carry four 0.5-in guns (later six) it was extremely welcome in the Mediterranean and in many other battle areas.

In 1941 there were numerous important developments in Britain, which included both new marks and new types. The evergreen Hurricane grew versatile wings containing 12 Browning 0.303-in, four 20-mm British Hispano cannon or two 0.303-in (for sighting) and two underslung 40-mm Rolls-Royce or Vickers anti-tank guns, with racks for 250-lb (later 500-lb) bombs, rocket projectiles, small-bomb containers, drop tanks or smoke generators. The Merlins for Hurricanes rose in power to 1260, 1460 and finally 1635 hp, in each case the engine being low-blown for peak performance at low altitude. Following the 'Hooked Hurricanes' came a navalised Sea Hurricane which still lacked folding wings but gave vitally needed defence aboard escort carriers and merchant ships equipped with catapults in 1942. The CAM (Catapult – Armed Merchantman) enabled fighter cover to be provided for convoys without a

carrier, at the cost of the pilot having to bale out or ditch after every sortie, with loss of the fighter. A scheme was studied for a Hurricane carried on a pylon above a Liberator, for release upon hostile aircraft sighted possibly 1500 miles from the nearest RAF airfield, but the scheme remained only a study by Short Brothers. A related idea was the Hillson 'slip wing' Hurricane, with a jettisonable upper wing to help overload takeoffs.

Hawker's completely new generation of fighters began with the prototype Tornado, built to specification F.18/37 and flown on 6 October 1939. The forward-looking specification called for 2000 hp, and the chosen armament was 12 machine guns. The work was eventually directed towards two fighters, the R-type with the Vulture engine and the N-type with the Sabre. Two more unfortunate engines could hardly have been selected, but after long and painful development the N-type reached the RAF as the Typhoon IB in September 1941. Though powerful, it had a disappointing climb and altitude performance, and was also unable to turn as tightly as a Bf 109F or Fw 190A. Coupled with persistent engine failure and structural weakness of the rear fuselage, the Typhoon almost failed to make it. However, by 1944 it was the RAF's top low-level tactical attack aircraft, with rockets, 1000-lb bombs and four cannon.

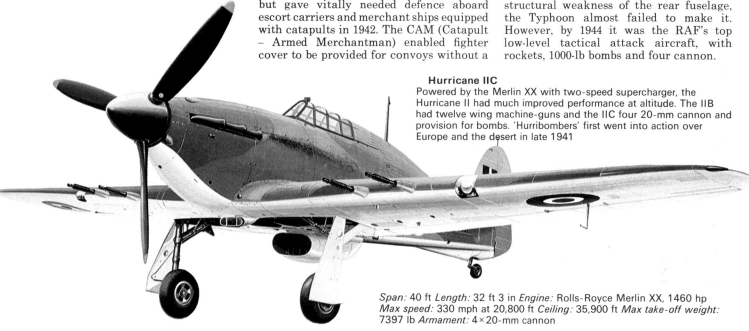

**Hurricane IIC**
Powered by the Merlin XX with two-speed supercharger, the Hurricane II had much improved performance at altitude. The IIB had twelve wing machine-guns and the IIC four 20-mm cannon and provision for bombs. 'Hurribombers' first went into action over Europe and the desert in late 1941

*Span:* 40 ft *Length:* 32 ft 3 in *Engine:* Rolls-Royce Merlin XX, 1460 hp
*Max speed:* 330 mph at 20,800 ft *Ceiling:* 35,900 ft *Max take-off weight:*
7397 lb *Armament:* 4×20-mm cannon

*Hurricane 11Cs in formation. Four cannon and a wing stressed for carrying up to 500 lb of bombs extended the Hurricane's life as a ground-attack aircraft*

When the might of the Luftwaffe was launched against the Soviet Union in Operation *Barbarossa* on 22 June, 1941, the objective was to kill or capture as much as possible of the Red Army in the shortest possible time. More than half the Soviet combat aircraft in the West were put out of action within the first week, mainly by attacks on airfields. Soviet losses of all kinds were grievous; yet the Russians were not to be entirely discounted in the air, as the German planning had tended to do. Though the invaders penetrated to the suburbs of Moscow before winter set in, the campaign was not over in six or eight weeks as had been hoped. By October 1941 the Soviet aircraft industry was being evacuated more than 1000 miles to the East, beyond the Urals, in an operation without parallel. At a cost of some five weeks' output, the whole manufacturing operation was re-started in completely new factories which proved, in the event, to be beyond the reach of the invaders. Whilst this was done, the types in production changed completely to the new designs which were first delivered in 1940–41.

Numerically by far the most important fighters in 1941 were still those from the Polikarpov bureau. These were of basic types dating from the 1930s when it seemed impossible to decide whether to build biplanes or monoplanes. The I-15 biplane and I-16 monoplane were in production side-by-side for years, and both proved to be fast, highly manoeuvrable, reliable and fitted with possibly the best machine gun of the day. Numerically the I-16 family was predominant, output exceeding 7000 in all. These were outstanding performers on primitive technology, and though extremely stubby and prone to poor longitudinal stability and tail-heaviness, they were fast, good at altitude and capable of very tight turns. They could absorb severe punishment, and were easy to maintain. By 1941 the most common fighter versions had the 1000-hp M-62 engine and two 20-mm cannon in place of the wing machine guns. Racks were standard for various external loads including six of the new RS-82 rockets for use chiefly against armour. The 50-kg (110-lb) PTAB hollow-charge anti-tank bomb was another common store.

These under-wing weapons were also carried by the biplane fighters which began with the I-15 of 1933. By 1941 most of these 700-hp four-gun machines had been replaced by the slightly faster and more capable I-15*bis*, with upper wing carried clear of the fuselage, and the 1000-hp I-153 with retractable landing gear. All were extremely manoeuvrable, and carried the same attack loads as later I-16 monoplanes, but faded fairly swiftly during 1942. In the course of this year the new monoplanes became available in gigantic numbers. Ilyushin's Shturmovik was not a fighter, though it often had to act like one. The final Poli-karpov fighter, the I-17, was outstandingly advanced for 1934 but was withdrawn in 1941 (the last of these 860-hp stressed-skin machines departed in 1942). First of the really important new fighters was the LaGG-1, flown as the I-22 in March 1939. Built almost entirely of wood, this new product of an untried team was startlingly good. It was light, immensely strong and very serviceable, though manoeuvrability would have been better with a larger wing. With 1100 hp, it carried adequate armament (such as a 20-mm cannon and two heavy

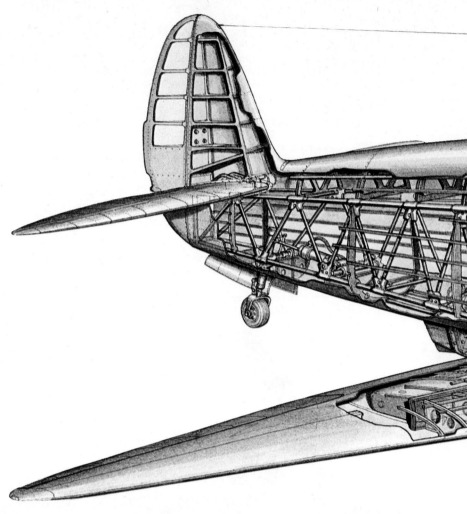

machine guns) and had excellent all-round performance, and the slightly modified LaGG-3 version was built in large numbers in 1941-42. In November 1941 an LaGG-3 flew with the much more powerful (1600-hp) M-82A radial engine, and this so greatly improved the fighter that it at once replaced the more vulnerable liquid-cooled machine, with designation La-5. Until the end of the Second World War the La-5 and its succes-sors were numerically the second most important of all Allied fighters, output exceeding 26,000. Nearly all could exceed 400 mph, and around the cowling were the blast tubes of two or three 20-mm cannon. Metal construction gradually replaced wood, pilot view was improved by cutting down the rear fuselage, and altogether the La-5FN of 1942 and La-7 of 1943 were equal in a close dogfight to the best Bf 109 or Fw 190.

Though today a household name, 'MiG' got off to a mediocre start with the MiG-1, first flown as the I-61 in March 1940. The 1200-hp engine conferred an excellent 390 mph, but armament was only two machine guns and handling was very poor. The MiG-3 of 1941 was even more powerful and reached 407 mph, but as a fighter it was not much of an improvement. Fortunately for the Soviet Union Alexander Yakovlev had flown his I-26 in the summer of 1940, and this entered service in the spring of 1941 as the Yak-1. To meet demands for reduced usage of light alloy it had a wooden wing and steel-tube and ply fuselage, but weight penalties were no worse than for the rival all-wood or mixed-construction fighters, and the

### Mikoyan-Gurevich MiG-3
The MiG-3, developed from the MiG-1 and carrying more fuel for the uprated engine, was introduced towards the end of 1941. Several thousand were built, but the type was less manoeuvrable than its opponents and was replaced by fighters from the rival Yakovlev bureau. The MiG-3 then flew reconnaissance missions but was soon withdrawn from front-line service

*Span:* 33 ft 9½ in *Length:* 26 ft 9 in *Engine:* Mikulin AM-35A, 1350 hp *Max speed:* 407 mph at 23,000 ft *Ceiling:* 39,400 ft *Max take-off weight:* 7700 lb *Armament:* 2×7.62-mm ShKAS mg, 12.7-mm Beresin BS mg, 6×RS-82 rocket projectiles or 2×110-lb bombs or 2×220-lb bombs or two chemical containers (VAP-6M or ZAP-6)

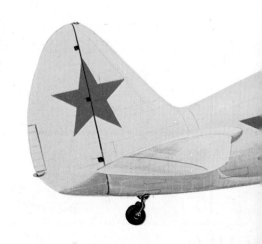

## Yakovlev Yak-3

The Yak-3 was developed alongside its Yak-9 stablemate specifically for ground-attack and low-altitude bomber escort. French and Polish units operating with Soviet forces also received the type, which was manoeuvrable and easy to fly although relatively lightly armed and lacking in protection.

*Span:* 30 ft 2 in *Length:* 27 ft 11 in *Engine:* Klimov M-105PF-2, 1222 hp *Max speed:* 403 mph at 16,400 ft *Ceiling:* 35,500 ft *Max take-off weight:* 5865 lb *Armament:* 20-mm ShVAK cannon, 2×12.7-mm Beresin BS mg

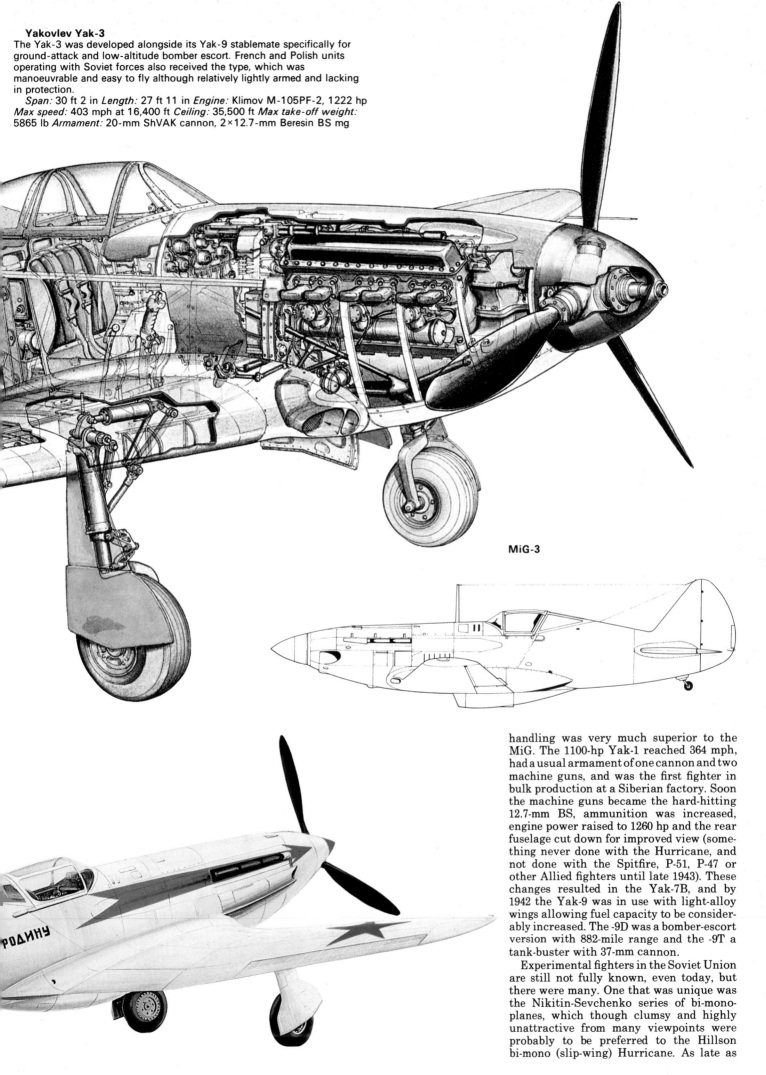

MiG-3

handling was very much superior to the MiG. The 1100-hp Yak-1 reached 364 mph, had a usual armament of one cannon and two machine guns, and was the first fighter in bulk production at a Siberian factory. Soon the machine guns became the hard-hitting 12.7-mm BS, ammunition was increased, engine power raised to 1260 hp and the rear fuselage cut down for improved view (something never done with the Hurricane, and not done with the Spitfire, P-51, P-47 or other Allied fighters until late 1943). These changes resulted in the Yak-7B, and by 1942 the Yak-9 was in use with light-alloy wings allowing fuel capacity to be considerably increased. The -9D was a bomber-escort version with 882-mile range and the -9T a tank-buster with 37-mm cannon.

Experimental fighters in the Soviet Union are still not fully known, even today, but there were many. One that was unique was the Nikitin-Sevchenko series of bi-monoplanes, which though clumsy and highly unattractive from many viewpoints were probably to be preferred to the Hillson bi-mono (slip-wing) Hurricane. As late as

1939 Soviet opinion hovered between mono-planes and biplanes. On the face of the argument, there is no point in having an aircraft that can behave as both, because this is bound to be complex, heavy and possibly potentially dangerous. There might be a case for it if biplane qualities were needed only at one time in each flight (such as take-off or landing) and monoplane speed at another; this is the reasoning behind the 'swing-wing' today. But in fact biplane qualities are needed all the time. Though the monoplane may go faster, the biplane will invariably climb higher and have a smaller radius of turn, so at what point does the pilot convert to a monoplane? Be that as it may, the two Soviet designers, V V Niki-tin and V Sevchenko, designed a fighter designated IS-1 (or NS-1) which flew in 1940. It was based on an I-153 but had landing gear folding inwards into the inner panels of the cantilever lower wing, which then folded into the sides of the fuselage and upper wing! A later prototype was designated IS-2 (NS-2) and flown in the same year. According to reports the trials were successful; probably this was one of those bright ideas which one knows all along will never go into general use no matter how successful the trials may be.

## Rocket Research
Another aircraft demonstrating the breadth of Soviet research, and its exceed-ingly advanced nature, was the BI-1 target-defence interceptor produced extremely quickly by A Ya Bereznyak and A M Isayev, under guidance of V F Bolkhovitinov. In 1933 the Soviet Union had set up the RNII, the world's first major rocket research establishment, and by 1941 this was produc-ing reliable engines that flew in many combat types, including the La-7 fighter. The D-1A-1100, apparently rated at 660-lb thrust on nitric acid and kerosene, was fitted into an extremely small and simple fighter of mainly wooden construction, with two 20-mm cannon. The first BI-1 was flown by Capt G Ya Bakhchivandzhi on 15 May 1942, having previously undertaken trials as a glider. No details of performance were disclosed, but the decision to stick to traditional fighters was dictated not by any faults in the BI-1 but by the inability of the short-endurance interceptor to fight a fluid war of movement over enormous distances.

Backing up these Soviet fighters were large numbers of Western types, of which the Bell P-39 Airacobra was one of the most important (4924 supplied). This unconven-tional machine was the first major product of the new Bell team, and at first it had a struggle. When 601 Sqn of the RAF equipped with it in July 1941 various factors combined to damn the aircraft, to which could be added some official resistance to the unusual layout. Bell put the 1150-hp Allison engine behind the pilot, to enable a battery of guns to be installed in the nose, to improve pilot view and improve manoeuvrability by put-ting the engine on the centre of gravity. In practice the P-39 proved a disappointment, and an uphill struggle for Bell in trying to make it a competitive fighter. It never did make a dogfighter, but 9584 were built and used with fair success in low-level attack missions, the 37-mm cannon especially appealing to the Russians.

An even more numerous and far more varied family were the P-40 series by Curtiss, which began with the airframe of the P-36 (Hawk 75A) fitted with the liquid-cooled Allison engine. The RAF used the French-ordered Hawk 81A under the name Toma-hawk I, IA and IB, though they were devoid of all protection and lacked most opera-tional gear. The Tomahawk II and USAAF P-40B did have some armour and self-sealing tanks, and reached about 350 mph carrying two synchronized 0.5-in and two rifle-calibre guns. The Tomahawk IIB had six 0.303-in, and served in various sub-types with 17 RAF squadrons in Britain (with the Army Co-operation Command, formed in late 1941) and many parts of Africa. From it were developed the P-40D to P-40M Warhawk family, called Kittyhawk by British Commonwealth air forces. Fitted with a revised type of Allison engine which shortened the nose and raised the thrust line, the Warhawks were much more for-midable with speeds exceeding 360 mph and armament of six 0.5-in in the wings. Just over 2000, with designations P-40F and L, had Packard-built Merlin engines. Strong and reliable, these later P-40 models were very widely used as low-level ground-attack

**North American A-36A**
The early Mustangs with Allison engines were used almost exclusively for low-altitude reconnaissance and ground attack. Five hundred A-36As (with the USAAF's Attack prefix) were delivered in 1943 equipped as dive-bombers and one (illustrated) was supplied to the RAF in March 1943 for experimental purposes
*Span:* 37 ft *Length:* 31 ft 11 in *Engine:* Allison V-1710-87, 1325 hp *Max Speed:* 356 mph *Ceiling:* 30,000 ft *Max take-off weight:* 10,700 lb *Armament:* 2×.5-in mg, 4×.3-in mg, 1000 lb bombs

machines – including the Soviet Union – with fair combat capability when needed.

They were useful, but totally outclassed as a basic aircraft by the Mustang, which the North American company had very quickly built in 1940 for the British in preference to building the P-40 on British account. North American had no experience of fighters, and before the British would sign a contract for the Mustang they insisted the firm should buy Curtiss P-40 wind-tunnel data. In retrospect this was like asking the Concorde builders to prove their capability by buying data on the 707, because the Mustang was vastly more advanced and in most ways dramatically superior. From the start it was a smooth and sweet performer, reaching about 390 mph on its 1200-hp Allison and having amazingly low drag from a laminar-profile wing, advanced-shape body and carefully ducted rear radiator that gave positive thrust instead of drag. Deliveries to the RAF began in 1941 with armament of four 0.5-in (two under the engine) and four 0.303-in, while subsequent sub-types had four 20-mm, four 0.5-in or six 0.5-in and two 500-lb bombs (USAAF A-36A attack model). Altogether 1579 Mustangs, P-51A and A-36A were built, before production switched to the high-rated Merlin.

### Bell P-39 Airacobra
The P-39 with its unconventional layout with the engine behind the pilot to accommodate heavy nose armament was a complete failure as an interceptor, but as a ground-attack fighter it excelled and many examples were supplied to the Soviet Union

*Span:* 34 ft *Length:* 30 ft 2 in *Engine:* Allison V-1710-35, 1150 hp *Speed:* 355 mph at 13,000 ft *Ceiling:* 29,000 ft *Armament:* 4×.3-in mg, 2×.5-in mg, 1×37-mm cannon

*RAF P-40 Tomahawk II. Underpowered and underarmed, the P-40B was soon supplanted by the E-series with considerably more horsepower*

Imperial War Museum

## Cockpit, Fw 190A-8: Key

### Focke-Wulf Fw 190A-8

The A-8 variant was introduced at the end of 1943 with increased internal fuel tankage. Sub types of the series included the A-8/R-1 with four 20-mm MG 151 cannon and the A-8/R3 ground support variant with wing-mounted MK 103 cannon.

*Span:* 34 ft 5½ in *Length:* 29 ft *Engine:* BMW 801D-2 1700 hp *Max speed:* 408 mph *Ceiling:* 37,400 ft *Armament:* 2 × 13-mm MG 151 mg, 4 × 20-mm MG 131 cannon

Opposition from the Luftwaffe intensified sharply during 1941, that year's crop of new fighters being among the best of the whole war as far as handling was concerned. By far the biggest, and least pleasant, Luftwaffe revelation was the Focke-Wulf Fw 190. The prototype of this totally new fighter had flown on 1 June, 1939, but the event escaped the notice of the Allies so that, when in the summer of 1941 RAF pilots reported radial-engined fighters, they were described officially as 'Mohawks captured from the French'. The RAF were soon disabused of this cosy notion, and in June 1942 an Fw 190A-3 landed by mistake in England. The lessons learnt in a technical inspection of it put a bomb under the RAF procurement machine. It taught the officials that a fighter can have an air-cooled radial and be dramatically faster than a pointed-nose liquid-cooled machine (this was obvious from the US Navy's Corsair, flown at over 400 mph in 1940, but this made no impact in Britain). It also taught that aircraft can be small and compact and still carry heavy armament and heavy external loads. Truly the 190 was an outstanding piece of engineering, and unlike the later Bf 109 models had no evident shortcomings.

Though the prototype had had the BMW 139 engine, initially with a ducted spinner that made the machine look like a jet, the production engine was the BMW 801, rated at 1600 hp. The big 18-cylinder radial was beautifully cowled, with a fast-revving cooling fan just behind the spinner and multi-stack ejector exhaust around the cooling exit slits. Above the engine were two 7.92-mm MG 17, replaced in many later models by 13-mm MG 131. Inboard in the wing were two of the hard-hitting MG 151 cannon, previously unknown in Britain, while early models had two 20-mm MG FF cannon outboard. Speed began at 389 mph, but soon the BMW 801D-2 engine with MW50 boost giving 2100-hp raised emergency speed to 416 mph, and armament schemes began to proliferate. The A-4 carried a 551-lb bomb and then a 1102-lb bomb and two drop tanks. The A-5/U3 carried a 2205-lb bomb load, and the A-5/U15 a torpedo. The A-8/R1 had six MG 151 cannon, while the A-10 was the first to carry the 3968-lb SC1800 heavy bomb. This bomb load, four times that of a Blenheim, was impressive for the smallest combat aircraft in the west with a span of 34 ft. In combat the 190 proved fast, small, elusive and unbreakable. Features included a beautiful clear-vision hood and very wide-track landing gear, and handling was well-nigh perfect. Even to sit in the cockpit without the chance to get airborne – as the author did in 1942 – was tremendously exciting.

*Fw 190A-5/U3 Trop, with special filters for desert operations and racks for up to 2200 lb of bombs for Jabo (fighter-bomber) operations*

John Batchelor

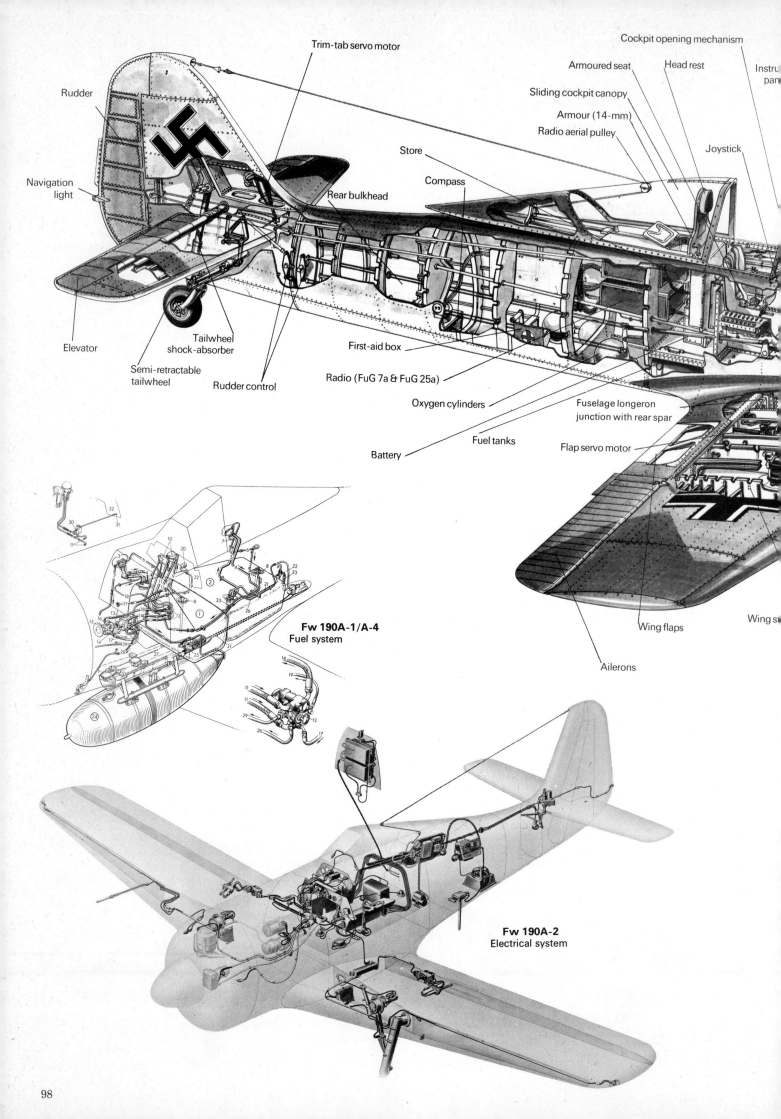

Rudder

Trim-tab servo motor

Cockpit opening mechanism

Armoured seat

Head rest

Instru
pan

Sliding cockpit canopy

Armour (14-mm)

Radio aerial pulley

Joystick

Navigation
light

Store

Compass

Rear bulkhead

Elevator

Tailwheel
shock-absorber

First-aid box

Semi-retractable
tailwheel

Rudder control

Radio (FuG 7a & FuG 25a)

Oxygen cylinders

Fuselage longeron
junction with rear spar

Battery

Fuel tanks

Flap servo motor

**Fw 190A-1/A-4**
Fuel system

Wing flaps

Wing s

Ailerons

**Fw 190A-2**
Electrical system

98

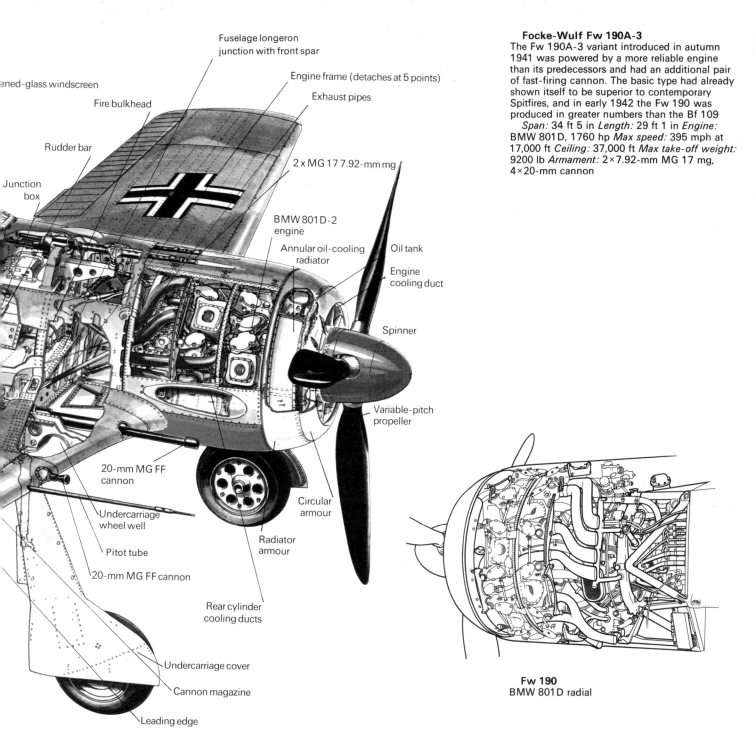

Fuselage longeron
junction with front spar

Engine frame (detaches at 5 points)

Exhaust pipes

ened-glass windscreen

Fire bulkhead

Rudder bar

Junction
box

2 × MG 17 7.92-mm mg

BMW 801D-2
engine

Annular oil-cooling
radiator

Oil tank

Engine
cooling duct

Spinner

Variable-pitch
propeller

20-mm MG FF
cannon

Undercarriage
wheel well

Pitot tube

20-mm MG FF cannon

Rear cylinder
cooling ducts

Circular
armour

Radiator
armour

Undercarriage cover

Cannon magazine

Leading edge

### Focke-Wulf Fw 190A-3

The Fw 190A-3 variant introduced in autumn 1941 was powered by a more reliable engine than its predecessors and had an additional pair of fast-firing cannon. The basic type had already shown itself to be superior to contemporary Spitfires, and in early 1942 the Fw 190 was produced in greater numbers than the Bf 109

*Span:* 34 ft 5 in *Length:* 29 ft 1 in *Engine:* BMW 801D, 1760 hp *Max speed:* 395 mph at 17,000 ft *Ceiling:* 37,000 ft *Max take-off weight:* 9200 lb *Armament:* 2×7.92-mm MG 17 mg, 4×20-mm cannon

**Fw 190**
BMW 801D radial

*Fw 190A-1, part of the initial production batch of 100 aircraft tested under operational conditions by Luftwaffe fighter units*

John Batchelor

In contrast Messerschmitt went one step forward with the Bf 109 and later took two steps back. In late 1940 the first evaluation models of Bf 109F were coming off the line, and these are generally judged the nicest 109 sub-types ever built. They had the 1300-hp DB 601E in a redesigned and more streamlined cowling, with a fat-spinner propeller of reduced diameter, and cooled by shallow redesigned wing radiators. Control surfaces, flaps and other aerodynamics were altered, the tailplane was restressed so that the bracing strut could be eliminated and the tailwheel was made retractable. Most obvious of all, the wing-tips were rounded, though span and area remained almost unchanged. Altogether the result was a substantial reduction in drag and improvement in control. Most F models were lightly armed, with a single engine-mounted cannon (MG FF, then MG 151/15 and finally MG 151/20) and two synchronized machine guns; yet it was to remain the preferred type of several of the top-scoring aces of the Luftwaffe. It was a superior dogfighter, and effective in the hands of an expert pilot. From January 1941 it gave all Allied fighters a hard time, and when the Fw 190 appeared alongside it there was a marked qualitative superiority on the side of the Luftwaffe.

In fact, as described later, the F was soon to give way to the G, and though this far outnumbered all other 109s and all other Luftwaffe fighters (unless all Fw 190s are grouped as one type) it was a thoroughly retrograde step that replaced the F's virtues by vices, all for the sake of piling on additional guns and equipment. All was not well on the twin-engined front, either. Though by 1940 the Bf 110C series were well-proven and reliable, they received such a mauling at the hands of Hurricanes and Spitfires

that they were themselves escorted by Bf 109s, and soon withdrawn entirely from the British theatre. There followed a major rethink of the once-proud *Zerstörer* (destroyer) wings, which Göring had regarded as the lead elements of the entire Luftwaffe, ranging across the breadth of Europe and clearing a path for the bombers. It was clear by October 1940 that the concept worked only in the absence of top-quality opposition. Many ZG *(Zerstörergeschwader)* were reassigned in the tactical attack role, and, though the Bf 110 gave good service in the Balkans, North Africa and on the Eastern Front, flying many kinds of mission, it was never again effective.

### Barbettes
In 1940–41 the poor showing of the 110 over Britain was not judged important because production was already running down and in 1941 was scheduled to terminate in favour of the completely new Me 210. The latter seemed an ideal successor, with 1395-hp DB 601F engines, an internal bomb bay, remotely controlled MG 131 barbettes covering the entire rear hemisphere and various arrangements of forward-firing armament. But no fighter ever suffered from more deep-seated faults than the Me 210, and though production deliveries did begin in early 1941 the result was a series of disasters. In April 1942, after countless faults and accidents, production was stopped, while the Bf 110 remained in increasing production. Eventually Messerschmitt developed the 210 into the DB 603-powered Me 410 for use as a day and night fighter, bomber, anti-tank machine, reconnaissance aircraft and anti-ship fighter, but it was never a success and only just over 1100 were built. Meanwhile, the old 110 strove to fill a widening gap.

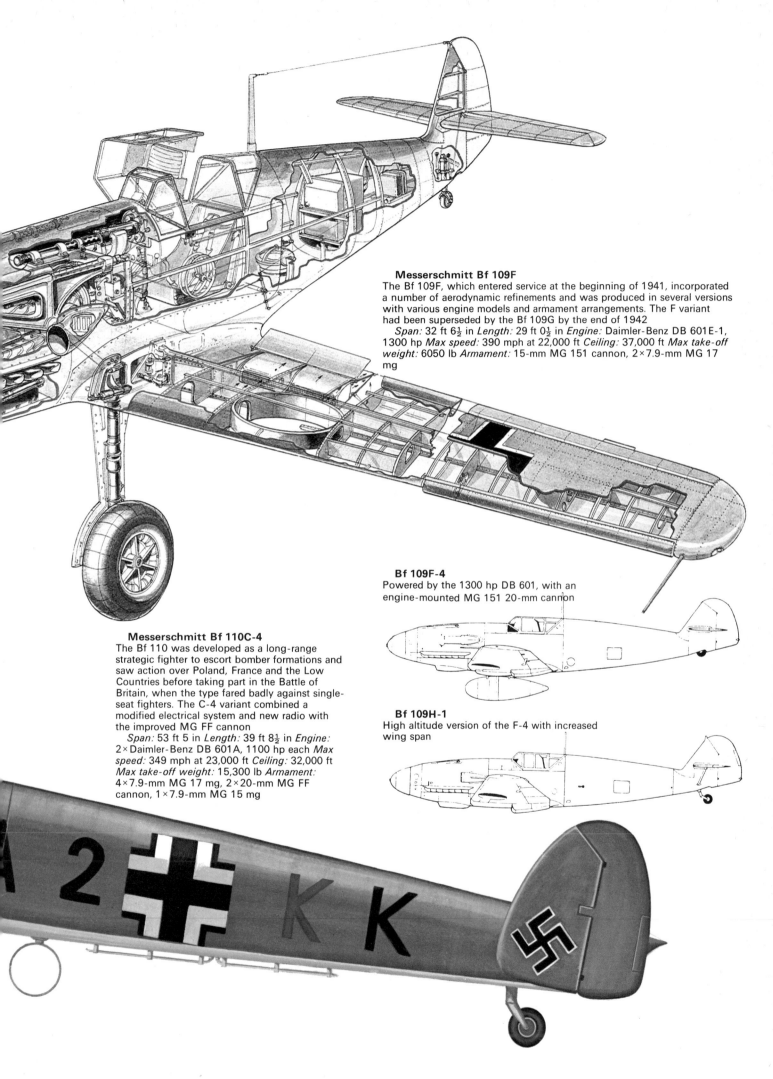

## Messerschmitt Bf 109F
The Bf 109F, which entered service at the beginning of 1941, incorporated a number of aerodynamic refinements and was produced in several versions with various engine models and armament arrangements. The F variant had been superseded by the Bf 109G by the end of 1942

*Span:* 32 ft 6½ in *Length:* 29 ft 0½ in *Engine:* Daimler-Benz DB 601E-1, 1300 hp *Max speed:* 390 mph at 22,000 ft *Ceiling:* 37,000 ft *Max take-off weight:* 6050 lb *Armament:* 15-mm MG 151 cannon, 2×7.9-mm MG 17 mg

### Bf 109F-4
Powered by the 1300 hp DB 601, with an engine-mounted MG 151 20-mm cannon

### Bf 109H-1
High altitude version of the F-4 with increased wing span

## Messerschmitt Bf 110C-4
The Bf 110 was developed as a long-range strategic fighter to escort bomber formations and saw action over Poland, France and the Low Countries before taking part in the Battle of Britain, when the type fared badly against single-seat fighters. The C-4 variant combined a modified electrical system and new radio with the improved MG FF cannon

*Span:* 53 ft 5 in *Length:* 39 ft 8½ in *Engine:* 2×Daimler-Benz DB 601A, 1100 hp each *Max speed:* 349 mph at 23,000 ft *Ceiling:* 32,000 ft *Max take-off weight:* 15,300 lb *Armament:* 4×7.9-mm MG 17 mg, 2×20-mm MG FF cannon, 1×7.9-mm MG 15 mg

# THE RISING SUN

In August 1940, at the height of the Battle of Britain, Chinese and Japanese fighters were locked in combat over the city of Chungking. Japan and China had been at war since 1937, and during that time there had been severe fighting in the air. However, in August 1940, in a dogfight 27,000 ft over Chungking, a swarm of Japanese fighters massacred the Chinese defenders. All the accounts of the engagement record that every one of the Chinese fighters was shot down. The few surviving pilots could suggest only that the Japanese aircraft appeared to be of a new type.

General Claire Chennault of the US Army Air Corps, leader of the American Volunteer Group (Flying Tigers), immediately sent urgent signals to Washington, warning of the new fighter. However, it appears that these signals were ignored, for on 7 December, 1941, the 'Zero' burst on the astonished defenders at Pearl Harbor. One early testimony to the Zero's quality is provided by the fact that during the 18 months between the dogfight over Chungking and the attack on Pearl Harbor not a single piece of one Zero fell into Allied hands, although the new fighter had been heavily engaged in battle over China almost daily.

The A6M *Zero-Sen*, designed by Jiro Horikoshi, chief designer of Mitsubishi, became a symbol to the Japanese of their air power. The aircraft's initial successes were so spectacular that it gained an aura of invincibility. However, just as with Hitler's Luftwaffe, belief in this myth proved disastrous. Though it was greatly developed over the years, the A6M remained in production to the end of the war, by which time it was outclassed. Its successor was started late and never saw front-line service. Meanwhile the ageing A6M remained in production, total output reaching at least 10,937, far outstripping all other Japanese aircraft before or since.

In the 1930s Japanese fighter philosophy had been based on attaining the greatest possible manoeuvrability. This had resulted in the two mass-produced fighters that dominated the Chinese war, the A5M of the Imperial Navy and the Army Ki-27. Both were stressed-skin monoplanes with open cockpits, powered by 710-hp Nakajima radial engines derived from the Bristol Jupiter; standard armament was two synchronised machine guns based on the Vickers. Their turn radius was outstanding, and with a time to 4000 m (16,400 ft) of $5\frac{1}{4}$ minutes, and speed at that height of around 280 mph, they were among the greatest performers of the 1936–40 era.

With this tradition of clean monoplane design and extremely lightweight construction, the next stage was predictable. Nakajima followed the Ki-27 with the Ki-43 *Hayabusa* (Peregrine Falcon) for the Army, while Horikoshi created the A6M. Both proved successful. The Ki-43 was slightly compromised by its small size and limited capability, but the A6M was just sufficiently endowed with horsepower to carry reasonable armament from the outset —two machine guns above the engine and two 20-mm Type 99 cannon in the wings immediately outboard of the tall landing gear. From the start provision was also made for two bombs of up to 60 kg (132 lb) each. After delivering sixty-four of the A6M2 Model 11 type, which immediately went to China in January 1940, Mitsubishi began volume production on the A6M2 Model 21 with folding wing-tips for carrier operation. These were the first aircraft to explore long-range flights on an operational basis, as distinct from special long-range flights formed for record-breaking purposes. With a small engine, clean design and drop-tank, the range was exceptional from the start, and pilots had no difficulty flying missions considerably longer than 2500 km (1553 miles). By the summer of 1940 careful cruise techniques had stretched the limit to almost 2000 miles, 1940 miles being the usual published figure.

This allowed the Model 21s in service at the time of Pearl Harbor – there were approximately 420 – to cover vast areas of the Pacific, China and SE Asia, often appearing at places where the presence of Japanese fighters had been thought impossible. During the crucial period after Pearl Harbor the entire Japanese ascendancy in the air rested on the achievements of the A6M. On countless occasions these lightweight fighters appeared several hundred miles from the nearest Japanese carrier or airstrip, inflicting crushing defeats on Allied pilots flying such machines as the Brewster Buffalo or P-40 near to their own bases. Though the stern dogfighting that ensued showed that the Zero could in fact be shot down, its marked ascendancy over the motley and obsolescent array of Allied fighters gave the Allies a respect for Japanese fighters that was perhaps greater than the latter deserved, until the winter 1942–43. By this time there were many new A6M versions, including the A6M2-N float seaplane (code-named 'Rufe') for use where land or carriers were not available, the A6M3 with clipped wings, and the A6M2-K tandem trainer.

Surprisingly, Mitsubishi's next Navy fighter was a complete about-face. The J2M *Raiden* (Thunderbolt) sacrificed manoeuvrability, and even good handling, in order to attain higher speeds and rates of climb. Fitted with an engine much larger than the A6M (the 1820-hp 14-cylinder *Kasei*) it had a much smaller wing, and at first even had a shallow cockpit with a curved, almost horizontal, windscreen like a racer. Pilot view was totally unsatisfactory, and when this was combined with poor manoeuvrability and unforgiving handling the result was not promising. The J2M entered service at the end of 1943 after an extremely troublesome period of development, receiving the Allied code-name 'Jack'. Subsequent variants had heavier armament, such as four 20-mm cannon of a modified type, and a much better canopy with proper forward view. Gradually the handling was improved, and late models captured in 1944–45 were judged to be excellent by Allied test pilots. Like so many Japanese combat aircraft, the J2M suffered such troubled development that only small numbers reached the squadrons, and these were mainly inadequate, undeveloped versions.

The design of the principal fighters of the Imperial Army also followed the philosophy of manoeuvrability. Parallel with the Navy A6M came the Nakajima Ki-43, which adhered more closely to earlier concepts; it was small, and even lighter than the Zero, and was armed with only two machine guns. The overriding need for less weight and better manoeuvrability resulted, in the early development in 1939, in much discussion about replacing the retractable landing gear by a fixed type.

Deliveries of this beautiful little fighter, the *Hayabusa*, code-named 'Oscar' by the Allies, began in 1941. It was always a sweet performer, with outstanding manoeuvrability. Most of the Army aces gained nearly all their victories on it, but they had an increasingly difficult task. Though the calibre of the two guns soon went up to 12-mm, this was insufficient to destroy the increasingly tough Allied machines, while the Ki-43 itself was poorly protected. It was by far the most important Army fighter, with 5878 delivered. However, it represented out of date technology for which no replacement was forthcoming in anything like comparable numbers.

Nevertheless, Nakajima did build later fighters. Like the Navy J2M, the Ji-44 discarded all the old goals in favour of more speed and climb, and the first Army pilots to receive it, in 1942, considered this a step backwards. Code-named 'Tojo' by the Allies, the Ki-44 had a more successful career than the J2M; eventually about 1223 were delivered, later models having heavier armament including two low-velocity 40-mm cannon.

After this quite effective machine came a new design that is generally rated the best Japanese fighter of the entire war,

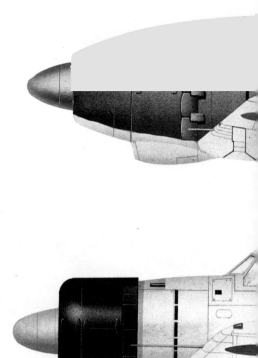

with the possible exception of the Ki-100. The Ki-84 *Hayate* (Gale), called 'Frank' by the Allies, was a thoroughly superior aircraft. No longer was the accent on any specific attribute; instead Nakajima's design team, under T Toyama, strove to produce an aircraft that was outstanding in all respects. Powered by the complex and troublesome *Homare* radial of 1900 hp, the Ki-84 was larger and heavier than earlier Japanese fighters, with an extremely strong structure (though the main legs were prone to snap from faulty heat-treatment of the steel). Eventually, after great efforts, no fewer than 3413 of these superb machines were built, armed with various mixes of 12.7-, 20- and 30-mm guns. Late variants included redesigned versions made of wood to conserve light alloys.

Another manufacturer, Kawanishi, flew a powerful float seaplane fighter in August 1942; from this they derived the N1K1-J *Shiden* (Violet Lightning) landplane. Had

*Mitsubishi A6M assembled by Americans from captured aircraft and flown in mock dog-fights against US types*

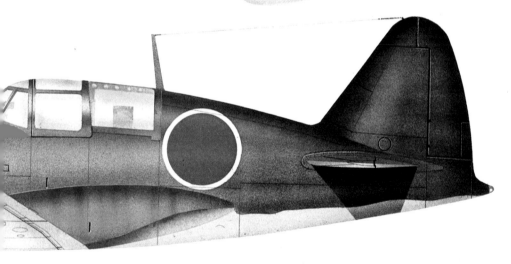

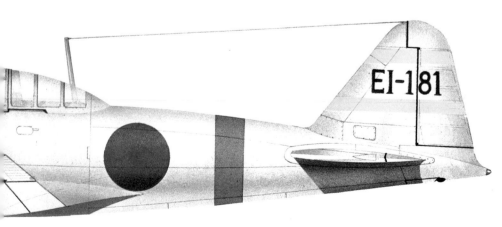

## Commonwealth CA-12 Boomerang

The Boomerang, using many components of the Wirraway trainer, was developed to counter Japanese forces in the Pacific and saw extensive action over New Guinea. The type was outclassed by its Japanese contemporaries but operated effectively as a bomber destroyer, on reconnaissance missions and in co-operation with ground forces

*Span:* 36 ft 3 in *Length:* 25 ft 6 in *Engine:* Pratt & Whitney R-1830-S3C4G Twin Wasp, 1200 hp *Max speed:* 296 mph at 7600 ft *Ceiling:* 29,000 ft *Max take-off weight:* 7600 lb *Armament:* 2×20-mm Hispano cannon, 4×.303-in Browning mg, 1×500-lb bomb

## Mitsubishi J2M3 Raiden

The Raiden (Thunderbolt) was the Japanese Army Air Force's first specialist intercepter, with emphasis placed on climb rate and speed rather than pure manoeuvrability. The J2M3 was the first variant with the armament doubled to four cannon and the fuselage-mounted machine-guns deleted

*Span:* 35 ft 5 in *Length:* 32 ft 7½ in *Engine:* Mitsubishi MK4R-A Kasei 23a, 1820 hp *Max speed:* 371 mph at 20,000 ft *Ceiling:* 38,400 ft *Max take-off weight:* 8700 lb *Armament:* 2×20-mm Type 99-I cannon, 2×20-mm Type 99-II cannon, 2×66-lb or 132-lb bombs

## Mitsubishi A6M2 Zero-Sen

The A6M2 was the first operational version of the Japanese Naval Air Force's Type 0 carrier fighter (hence Zero-Sen). The aircraft was blooded over Chungking in August 1940 and later variants were equally devastating in the Pacific war, although the Japanese policy of sacrificing protection for speed and firepower led to many Zeros being lost in combat

*Span:* 39 ft 4½ in *Length:* 29 ft 9 in *Engine:* Nakajima Sakae 12, 925 hp *Max speed:* 336 mph at 20,000 ft *Ceiling:* 33,800 ft *Max take-off weight:* 5310 lb *Armament:* 2×7.7-mm Type 97 mg, 2×20-mm Type 99 cannon, 2×66-lb or 132-lb bombs (all figures for A6M2 Type 21)

this fighter, called 'George' by the Allies, been designed as a landplane from the start, it might have been as good as the Army Ki-84. As it was, its ancestry left it with a complicated structure, mid wing, poor view and long landing gear which caused much trouble. The 1990-hp *Homare* gave excellent performance, even when carrying four cannon and two machine guns, but was hampered by its poor reliability and complexity. Just over 1000 were built, followed by 428 of the M1K2-J type in which the structure was redesigned to eliminate 23,000 parts and give shorter and better landing gear, improved view and even more outstanding combat performance. According to Imperial Navy records, an N1K2-J flown by Warrant Officer Muto once engaged 12 Hellcats and destroyed four, escaping almost unscathed.

Alone among Japanese fighters in having a liquid-cooled engine, the Army Kawasaki Ki-61 *Hien* ('Tony') was built in large numbers (2654) and served on all fronts. Superior to a P-40 or Bf 109E, it was at first powered by a refined Japanese version of the 1175-hp DB 601 and armed with various mixes of 12.7-mm heavy machine guns, 20-mm cannon (the German MG 151 or Japanese Ho-5) or 30-mm. Though possessed of fair performance and well protected, the Ki-61 suffered from engine and other troubles which were exacerbated by the introduction of the 1500-hp Ha-140 engine. However, the Ki-61 suddenly sprang into the limelight when a radial engine was fitted as a matter of necessity.

## Nakajima Ki-84 Hayate

The Hayate (Gale) represented a departure from normal Japanese fighter design methods, being much sturdier than its predecessors. Although marginally slower than its adversaries – Thunderbolts and Mustangs – the Ki-84 could outmanoeuvre and climb faster than its rivals. The strong construction also well suited the dive-bombing and close-support roles

*Span:* 36 ft 10 in *Length:* 32 ft 6½ in *Engine:* Nakajima Ha.45/11 Type 4, 1900 hp *Max speed:* 388 mph at 20,000 ft *Ceiling:* 34,500 ft *Max take-off weight:* 9200 lb *Armament:* 2×12.7-mm Type 103 mg, 2×20-mm Type 5 cannon, up to 1100 lb of bombs

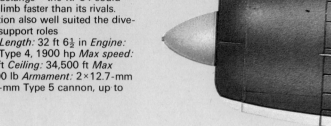

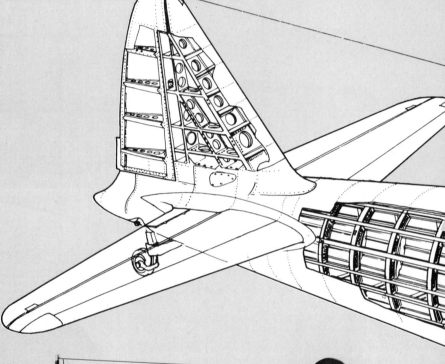

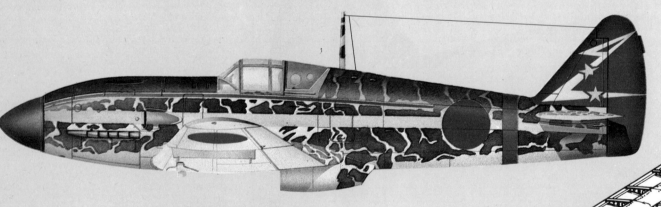

## Kawasaki Ki-61 Hien

The Hien (Swallow) entered service with the Japanese Army Air Force in the spring of 1943 and was at first mistakenly thought by the Allies to be a derivative of the Bf 109. The fighter did use a development of the Daimler-Benz DB 601A, however, and was the only Japanese fighter with a liquid-cooled engine to see service in the Second World War. The Ki-61 remained operational until the end of the war, serving in every Pacific theatre

*Span:* 39 ft 4 in *Length:* 29 ft 4 in *Engine:* Kawasaki Ha.40 Type 2, 1175 hp *Max speed:* 348 mph at 16,400 ft *Ceiling:* 32,800 ft *Max take-off weight:* 7650 lb *Armament:* 2×12.7-mm mg, 2×20-mm Ho-5 cannon, 2×550-lb bombs (all figures for Ki-61-Ic)

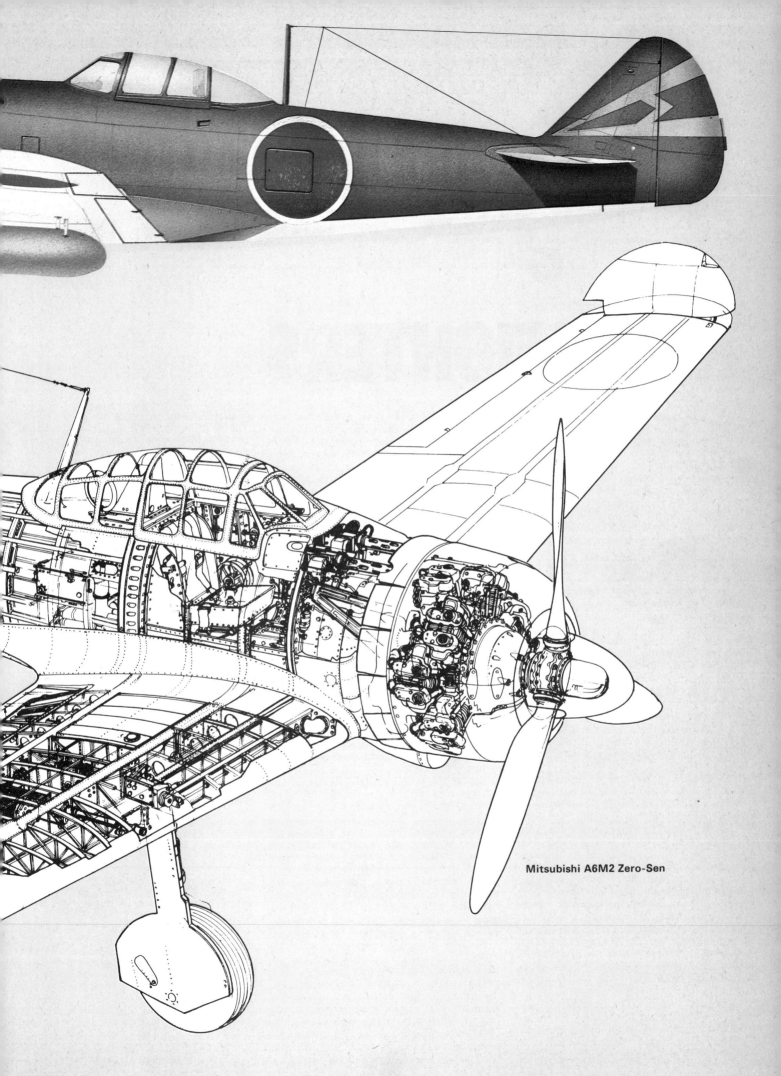

Mitsubishi A6M2 Zero-Sen

105

# NIGHT FIGHTERS

Until 1939 night fighters were usually merely day fighters equipped with cockpit lighting, flares to light up the airfield for landing, and similar minor additions. But by July 1939 an extremely difficult process of technical development in Britain had resulted in a primitive, though practical AI (airborne interception) radar set. The original 'Radio Direction Finding' equipment, built around Britain's coasts from 1936 onwards, was too large to fit into a fighter. It took completely new technology, greatly reduced operating wavelength, and the solution of problems of a kind never previously encountered, before the first Blenheims could be fitted with AI Mk III in the final weeks of peace. The early set was hopelessly erratic in operation, and even with a skilled operator (a crew-member assigned to watch the flickering cathode-ray tube and advise the pilot which way to steer) results were invariably most discouraging. In theory the AI radar gave pictorial indications on the CRT, like a dim television screen, from which the observer could work out the position and range of a hostile target. In practice, the targets tended not to be there, or merely to vanish into thin air.

GCI, Ground Control of Interception, played a vital part in the night battles over Britain, which began in earnest in the early summer of 1940. With the clearer VHF (very high frequency) voice communications, the controller could inform the patrolling night fighter of the location and approximate height of the nearest hostile bomber. The controller's objective was to guide the night fighter to a position astern of the enemy, where the marauder would betray its presence by casting a 'blip' as a small spike of light on the bearing and elevation CRTs in the fighter. On the nose or wing of the fighter were two sets of transmitter aerials forming a device resembling a double-headed harpoon. This sent out intense pulses of radio energy, a minute fraction of which would be reflected back by the enemy. A rapid rotary switch fed the the CRT displays with the reflected signals picked up by four sets of receiver aerials in turn: left and right azimuth (bearing)

aerials, looking like pairs of vertical wires on the sides of the fuselage or wing leading-edge; and upper and lower elevation aerials, like another harpoon projecting above and below the left wing. The signals picked up by these four combinations of receiver gave four sets of illuminated 'blips' on the CRTs, and skilled observers were taught to translate the lengths and positions of these faint traces into clear instructions to the pilot. The climax of the chase was meant to be closure of the range until the enemy suddenly appeared as a recognisable shape in front, even blacker than the night sky, where it could be shot down.

Blenheims had few successes, and their poor performance and armament did not help. Throughout 1940 they could not match ordinary Hurricanes and Spitfires flown by experienced pilots who had the same GCI to a position somewhere a mile or two behind the enemy but thereafter were on their own. Another successful night fighter was the Defiant, which had the advantage that, backing up the pilot, the observer in his turret could search constantly all over the upper hemisphere and sometimes could shoot down bombers from abeam or directly below.

By far the greatest night fighter of the early part of the war was the Bristol Beaufighter, first flown as a company private venture in July 1939 to make up for the glaring omission in the RAF's fighter inventory: a long-range twin. Powered by two Bristol Hercules 14-cylinder sleeve-valve radials, the early Beaufighters could reach about 320 mph despite their great weight and bulk. The somewhat fat body was oddly arranged, with the pilot in the nose and the radar observer well aft under a separate cupola which in some later versions was fitted with a flexible machine gun. Under the floor was the extremely welcome armament of four 20-mm Hispano cannon, and by 1941 there were also six machine guns in the wings, two on the left and four on the right. Through official stupidity the cannon had to be fed with 60-round drums by the observer despite the fact that a continuous belt feed had been offered by Bristol from the start. Eventually, in September 1941, a

### Bristol Beaufighter Mk IIF
The Mk IIF version of the Beaufighter, fitted with Mk IV AI (Airborne Interception) radar, succeeded Blenheims in the night-fighting role in 1941 and bore the brunt of bringing night interception techniques to the fine pitch achieved with later marks of AI. The observer was responsible for changing ammunition drums as well as operating the radar.
*Span:* 57 ft 10 in *Length:* 41 ft 8 in *Engine:* 2 × Bristol Hercules, 1460 hp each at 6250 ft *Max speed:* 301 mph at 20,200 ft *Ceiling:* 26,500 ft *Max take-off weight:* 19,190 lb *Armament:* 4 × 20-mm cannon, 6 × 0.303-in mg

### Junkers Ju 88G-6
The ever adaptable Ju 88 airframe was extensively redesigned to accept specialised night-fighter equipment, and the result was a potent and successful aircraft. The three seater G-series entered service in mid-1944 equipped with sophisticated interception radar
*Span:* 65 ft 10½ in *Length:* 54 ft 1½ in *Engines:* 2 × Jumo 213A, 1750 hp each *Max speed:* 344 mph at 19,700 ft *Ceiling:* 32, 800 ft

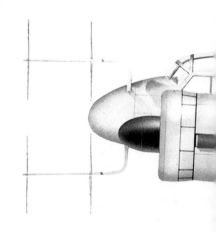

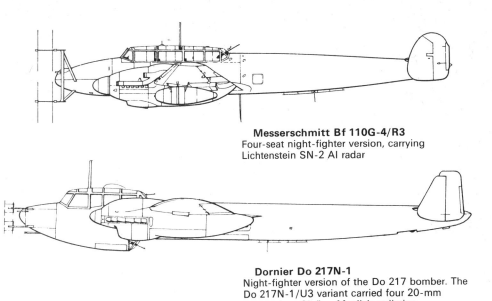

**Messerschmitt Bf 110G-4/R3**
Four-seat night-fighter version, carrying
Lichtenstein SN-2 AI radar

**Dornier Do 217N-1**
Night-fighter version of the Do 217 bomber. The
Do 217N-1/U3 variant carried four 20-mm
cannon in a *Shräge Musik* installation

belt feed identical to the Bristol pattern was adopted from the 401st aircraft. Thus, during the night Blitz over Britain in September 1940 to May 1941, the 'Beau' had to make do with the drum-fed guns. Moreover, it also lacked AI radar until November. By this time the AI. MkIV set was available, still rudimentary by later standards but much better than the Mk III.

The first Beaufighter kill was a Ju 88 on 11 November, 1940. As more aircraft, with more proficient crews, came into service kills mounted. There were 22 in March 1941, 48 in April, and 96 in the final 11 days ending on 11 May. Subsequently many marks of 'Beau' served in many roles on many fronts. Several hundred formed the backbone of the Allied night fighter force in the Mediterranean and Italy, being flown by all Allied air forces including the USAAF. Tough, highly manoeuvrable and with devastating firepower, the Beaufighter also performed an increasingly useful role against ground targets.

The only reason one cannot describe the

*A USAAF Liberator is literally chopped in half by anti-aircraft fire. The American massed daylight raids were severely mauled by fighters until the arrival of the Mustang long-range escorts, leaving the night raids to the unescorted RAF heavies and the waiting night fighters*

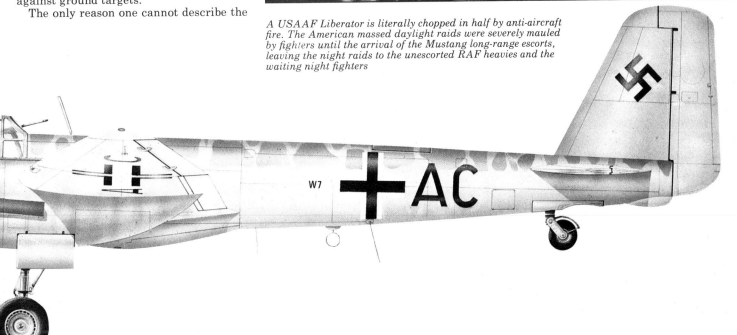

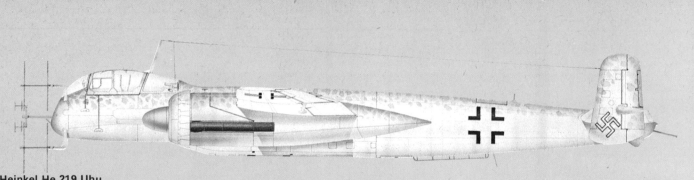

**Heinkel He 219 Uhu**
The *Uhu* (Owl) was designed as a night fighter, and in this role it was extremely successful. Converted prototypes were claimed to have destroyed 20 British bombers, including six Mosquitoes, in their first six missions during the summer of 1943. Many He 219s were fitted with upward-pointing 'Schräge Musik' cannon, but fewer than 300 entered service

*Span:* 60 ft 8 in *Length:* 60 ft *Engine:* 2 × Daimler-Benz DB 603G, 1900 hp each *Max speed:* 416 mph at 23,000 ft *Ceiling:* 41,600 ft *Max take-off weight:* 33,700 lb *Armament:* 4 × 30-mm MK 108 cannon, 2 × 30-mm MK 103 cannon, 2 × 20-mm MK 151 cannon

*Four .303-in Brownings blast a hail of fire from the rear turret of a Lancaster bomber. A night fighter equipped with* Schräge Musik *could attack from below – the bomber's blind spot*

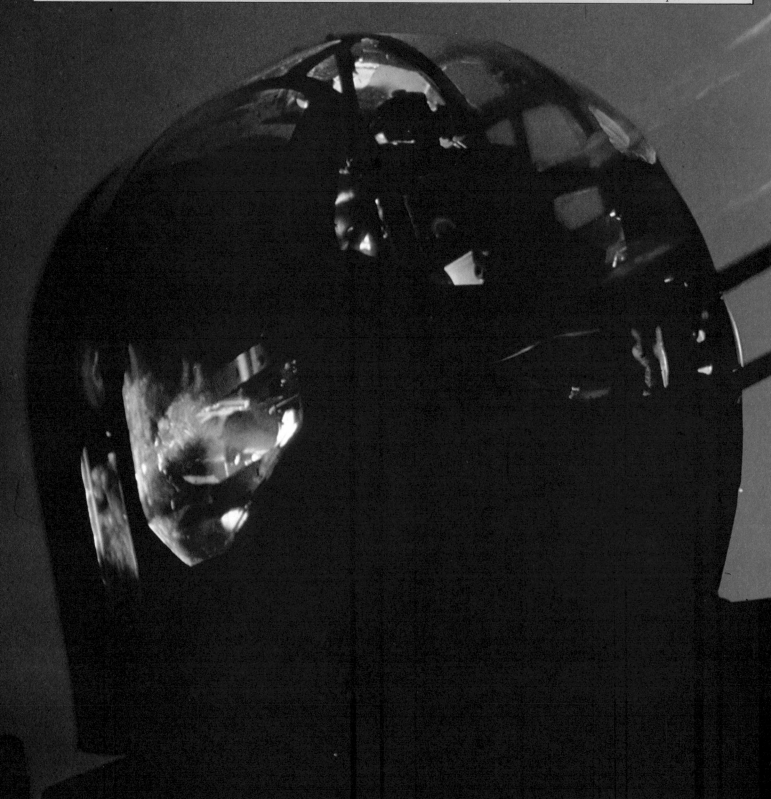

**Schräge Musik**
Radar-equipped Luftwaffe night-fighters were deadly opponents for RAF Bomber Command. Equipped with *Schräge Musik* ('Slanting Music' – Jazz) a night-fighter could attack from below and rake the unprotected belly of a bomber with 20-mm cannon fire

*The extraordinary Northrop P-61 was an enormous aircraft of some 28,500 lb gross weight, yet with excellent handling and landing characteristics. The shiny all-black finish, designed to conceal the aircraft even in a searchlight beam, earned the P-61 its sinister name*

'Beau' as the leading Allied night fighter is because de Havilland's 1938 proposal for an unarmed wooden bomber – contemptuously rejected by the officials – finally led not only to the Mosquito bomber but also to a fighter and many other versions. The first NF.II night fighter 'Mossie' flew on 15 May 1941, with AI.IV radar (fitted later), four 20-mm Hispano under the floor, four Brownings in the nose, a side crew door and flat bullet-proof windscreen. Though unnecessary effort was wasted on such fitments as a four-gun dorsal turret and a Turbinlite searchlight in the nose, the basic NF.II was a superb aircraft; 466 were built in 1941–42. Capable of over 370 mph with full equipment at medium altitudes, the Mosquito was the first aircraft to combine AI radar, heavy armament and fighter-like performance. The robust Beaufighter had come close, but the 'Mossie' was almost 50 mph faster, and could be flung round the sky with its fighter-type stick in a way that made it a dangerous opponent even in a day-time dogfight.

The NF.II was followed by many more powerful versions. Some of these were day multi-role fighter/attack machines (notably the mass-produced FB.VI) while others exchanged the nose machine guns for AI radar of the new centimetric kind. The first sets had worked on wavelengths of about one metre, but by 1942 another British breakthrough in technology, the Magnetron valve, had opened the way for new families of radars with wavelengths that began at about 10 cm. The new sets had transmitting and receiving aerials combined into a single unit, backed by a parabolic 'dish' reflector which could be aimed to search the sky ahead in a spiral fashion. The first of the new sets, AI.VII, was flown in November 1941 in Beaufighter X7579, the first aircraft to carry the excrescence today familiar as a radome. In this case the dome was on the nose, and described as of thimble shape. In

1942 the AI.VIII, with minor improvements, was fitted to the Mosquito II, the result being called NF.XII (197 were later converted). The Mk VIII radar could be used at low level, because its screen was less prone to obliteration by the reflection from the ground, which had hampered the old radars. Range was increased, and there were other improvements, but the arc of sky scanned was considerably reduced.

By 1943 hundreds of NF Mosquitoes had been committed to production, including the II, XII, XIII (new build with AI.VIII and underwing tanks), XVII (II converted with American SCR-720 radar) and XIX (with universal 'bull nose' able to enclose any centimetric radar). They ranged far and wide, though centimetric-equipped marks were forbidden to cross enemy coasts until D-Day on 6 June, 1944.

With little to do over Britain, the Mosquitoes were used increasingly as long-range intruders, ranging all over Europe and often making bomb and rocket attacks on enemy airfields. Other Mosquito night fighters served with 100 Group as counter-measures and special-electronic aircraft. The final wartime mark was the NF.30 powered by a high-altitude Merlin 72 or 76 with new flame-damped exhausts and paddle-blade propellers.

In 1940 the RAF had produced a remark-ably good night fighter and intruder by converting the Douglas DB-7 (Boston) light bomber. Operations began in early 1941 with a version called Havoc I, with glazed bombardier nose carrying four Brownings and internal bomb load of 2400 lb. They reached about 295 mph on 1200-hp Twin Wasp engines, and once crews had got used to the tricycle landing gear and the un-familiarity of an American aircraft, they became very popular. The Havoc II was powered by 1600-hp 14-cylinder Cyclones, reaching about 330 mph. Though a consider-able number were wasted in experiments

with the Turbinlite and LAM, many had a 12-gun nose. The LAM stood for Long Aerial Mine, a scheme for trailing or releasing explosive charges with long cables intended to catch the wings of hostile bombers. The Turbinlite was even more hare-brained; a searchlight aircraft was required to com-plete an interception and then, having lined up on the target, switch on a searchlight so that other aircraft (usually Hurricanes) could somehow get past the searchlight carrier and despatch the enemy. The chief result was a high wastage rate in Havocs and Hurricanes.

In 1940 British scientists and engineers visited the United States and handed over details of centimetric radar. At once an American development was started, with British engineers (the leader of the team was Australian) giving advice. While this produced the excellent SCR-720 radar in fairly quick time, the aircraft planned to carry it, the Northrop P-61 Black Widow, took over three years to develop and almost missed the war. The first purpose-designed radar-equipped night fighter, the Black Widow was powered by two 2000-hp Double Wasp engines; it was a large and extremely complex machine with a broad wing with advanced high-lift features, twin tail booms and a nacelle accommodating a pilot, radar observer (above and behind) and rear observer/gunner. In the nose was the SCR-720, under the floor were four 20-mm M-2 cannon, and on top (in some versions) was a remarkable remote-sighted electric turret mounting four 0.5-in guns, aimed by either the front or rear observer, or slewed to fire ahead with the cannon. The P-6IA finally became operational with the USAAF in May 1944 in the Pacific, and in June in Europe. Until then the USAAF had used the Beaufighter and, to a lesser extent, the Douglas P-70 Havoc in various versions with radar and belly cannon, or without radar and with nose guns.

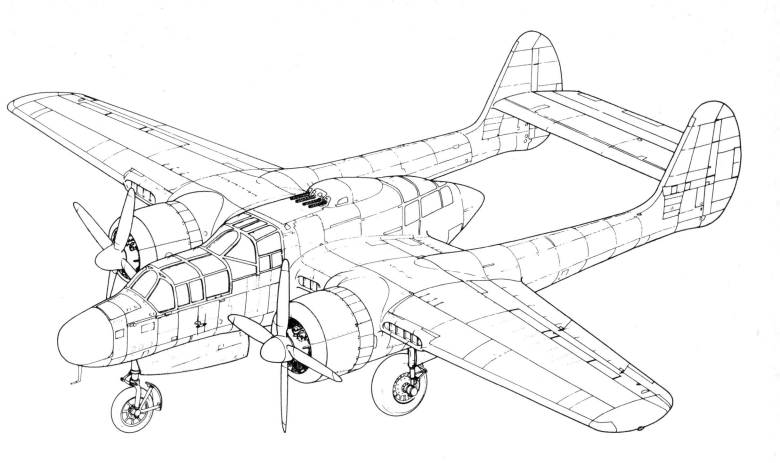

**Northrop P-61**
The twin-boom layout of the P-61 Black Widow
grouped all the specialised functions of the
night-fighter in the central nacelle, housing AI
radar, pilot, radar operator and rear gunner with
his own four-gun barbette
*Span:* 66 ft *Length:* 40 ft 11 in *Engine:* 2×
Pratt & Whitney R-2800-10 Double Wasp, 2000
hp *Max speed:* 362 mph at 20,000 ft *Max take-
off weight:* 28,500 ft *Armament:* 4×.5-in mg,
4×20-mm cannon

In the US Navy, development was boldly
aimed at an AI radar operating on the short
wavelength of 3 cm, and in 1943 this materialised as the APS-4 and APS-6. The set was
small enough to fit a single-seater, and
first went into action with the F6F-3 Hellcat.
Subsequently it equipped the F4U-4 Corsair.

In the Luftwaffe two aircraft designed for
other purposes came to dominate the night
sky: the Ju 88 and Bf 110. The first Ju 88
fighter prototype flew in September 1938,
and in 1940 some of the first NJG (night-
fighter wings) were equipped with converted
A-1 bombers designated C-2. There followed
many night fighters in the C series, as well
as day fighters in the R series, with Jumo 211
or BMW 801 engines and heavy nose armament. From 1942 some C models, starting
with the C-6b, had been fitted with AI radar
of the FuG 212 Lichtenstein type, operating
on a wavelength of around a half-metre and
notable for a large array of dipole aerials on
nose and wings. Though a conversion, as
was the Blenheim, the Ju 88 was destined
to be one of the greatest warplanes in
history.

During 1943 work began on a version
tailored from the start for night fighting.
Burdened by extra equipment and weapons,
the earlier models had placed heavy demands on pilots who – operating under
extreme stress caused by fear of the omni-
present Mosquito – suffered high attrition
through accidents. In the Ju 88G there was

a complete revision of systems and equipment, and handling was improved by fitting
the large tail of the Ju 188. From 1944 almost
all Ju 88 output was of the G-6 or G-7 series,
powered by the Jumo 213 or BMW 801, and
with various radars and devastating armament in ventral blister or *Schräge Musik*
installations. The latter, meaning Slanting
Music (Jazz), comprised cannon mounted
at about 70°–80°, allowing the pilot to
formate under the bomber, where the RAF
crew could not even see or fire on their
enemy. Skilled pilots could aim precisely
at the wing spars between the engine, well
clear of the bomb bay, and then avoid the
stricken monster as it tumbled downwards.
Five or six kills a night became common,
interceptions being made childishly simple
by the Naxos radar, which homed on the
bomber's $H_2S$ mapping radar, and the
Flensburg, which homed on the bomber's
Monica tail-warning radar. Monica had
been added to protect the RAF heavies, but
in practice it gave so many warnings of
other bombers that its indications were
often ignored, and it served merely as an
aerial lighthouse to guide the Luftwaffe
night fighters.

The other important NJG aircraft was
the old Bf 110, much more restricted than
the Ju 88 but docile and pleasant and, in
Bf 110G and H forms, specially planned for
night fighting with DB 605 engines. The
Luftwaffe's lack of forward planning is well
demonstrated by the fact that the 110 was
intended to be withdrawn from production
in 1941, but in the absence of anything else
its output was tripled in 1942–43 and held
at the same level in 1944, virtually all being
night fighters.

Other German night fighters included
special versions of Bf 109G and Fw 190, used
chiefly in the free-lance *Wilde Sau* (Wild
Boar) role, searching visually for RAF
bombers illuminated by searchlights,
ground fires or flares. They enjoyed much

greater success than RAF night fighters in
1940, partly because there was more illumi-
nation and far greater intensity of bombers.

One of the finest combat aircraft of the
entire war was the Heinkel He 219 *Uhu*
(Owl), which combined powerful armament
and good sensors with outstanding performance and handling. However, only small
numbers of this aircraft were built. Night
fighter versions of the Do 335 and Me 262
were flying at the end of the war, but these
few leaders of the Luftwaffe's last fighter
generation appeared too late.

Though Japanese airborne interception
radar seldom got into action before the final
weeks of the war, the chief twin-engined
fighters all served in the vital night role.
The Kawasaki Ki-45 *Toryu* (Dragon Killer),
code-named 'Nick' by the Allies, served the
Imperial Army well, but, like the Bf 110,
was obsolescent. Many combinations of
12.7-, 20-, 30-, 37-, and even 75-mm guns were
fitted; night fighters had an oblique pair of
12.7- or 20-mm weapons and on several
occasions managed to shoot down the B-29,
which many single-seaters could not even
reach. The main Navy twin-engined fighter
was Nakajima's JINI, flown in May 1941
as a long-range escort with remotely controlled rear barbettes. There were several
subsequent versions, called 'Irving' by the
Allies, of which the most important was the
JINI-S *Gekko* (Moonlight) night fighter,
with oblique pairs of 20-mm cannon firing
above and below, and often with nose
searchlight or radar. Though successful
against the B-17 and B-24, the JINI lacked
the performance to reach the B-29. Best of
all the Japanese night fighters was probably
the Kawanishi PIY2-S *Kyokko* (Aurora),
derived from a Yokosuka attack bomber in
the class of the Ju 88. Powered by two
1850-hp *Kasei* engines, the *Kyokko* carried
AI radar and three cannon (four if a twin-
20-mm dorsal turret was fitted), but arrived
too late to make an impact on the war.

# FIGHTERS' WAR

In the dark winter of 1941–42, dozens of Allied fighter types battled against enemy machines that were generally superior. In the Far East and SW Pacific the A6M and Ki-43 found little difficulty in mastering the early marks of Curtiss P-40, the Dutch Curtiss-Wright CW-21B (1000-hp Cyclone and four machine guns), the Brewster Buffalo (1100-hp Cyclone and four guns) and a collection of other mediocre types.

The CW-21B was nimble but unprotected. The Buffalo was nimble at the start of its career, in 1938, but by 1941 it had been loaded with armour, self-sealing tanks, extra equipment and four 0.5-in guns. The result was a sluggish performance—it took almost half an hour to reach 21,000 ft, and manoeuvrability was pathetic. Often guns were removed, or replaced by small 0.303-in Brownings, in an attempt to restore some performance. In these circumstances the Boomerang, designed and flown in fourteen weeks by Commonwealth Aircraft in Melbourne, was of great value. The only engine

available was the 1200-hp Twin Wasp, yet the Australians created a tough and 'operable' tactical machine with two cannon and four 0.303-in, plus 500-lb bomb, which remained outstanding in the close-support role until the end of the Pacific war.

An identical armament had been adopted in 1940 for the so-called 'B' wing of the Spitfire. In early 1941 this went into production as the Spitfire V, with a strengthened fuselage and more powerful Merlin 45 or 50-series engine. The VB was the standard front-line fighter in 1941–42, flown by all Allied air forces in all theatres. Many had clipped wings, bomb racks, tropical filter under the engine or other changes, and Air Service Training fitted arrester hooks to produce the deck-landing Seafire IB. Large numbers were fitted with the 'C' wing, which could carry four 20-mm Hispanos (but usually did not), and this Spitfire VC in turn led to the Seafire IIC with catapult spools and stronger landing gear. In 1942 a way was found to make the wing fold,

*The development of the Spitfire from the first prototype in 1936 right until the end of the war was continuous, making it one of the most 'stretched' aircraft designs in history. The Griffon-engined Mk IV of 1941 was renumbered the Mk XX to avoid confusion with the reconnaissance PR IV*

Imperial War Museum

resulting in the excellent Seafire III of 1943; but by this time more performance was needed. While leading-edge tankage and many other changes led to long-range unarmed PR (photo-reconnaissance) Spitfires, the great pressure on Rolls-Royce to deliver more power at greater height resulted in the Merlin 60-series. Supermarine, helped by the Royal Aircraft Establishment, devised a pressure cabin, and in 1941 flew the Mk VI as an interim high-altitude machine with Merlin 47 and extended wing-tips. Design went ahead on the definitive high-altitude fighter, the VII, and the all-round version, the VIII.

Meanwhile, Merlin 60-series engines went into production in 1942 and a quick lash-up conversion of the VC went into production as the IX. This had the two-stage engine, twin radiators and four-blade Rotol propeller, but was otherwise unchanged. Possibly a limit should have been placed on the numbers built. However, it stayed in production, and to the end of the war was by far the most numerous Spitfire, 5664 being built. There were LF, F and HF versions, depending on the low, medium or high altitude rating of the engine, often accompanied by clipped, normal or extended tips. A closely related model was the XVI powered by the Packard-built engine; by

1944 the 'E' wing moved the cannon outboard and added 0.5-in guns inboard, the LF.XVI often having clipped wings and later a cut-down rear fuselage and teardrop hood. Other changes included broader, pointed rudder and rear-fuselage fuel tanks. Only modest numbers were made of the pressurized HF.VII and so-called 'definitive' Mk VIII, which had many refinements, including retractable tailwheel, and had been meant to replace the IX. The standard photo-reconaissance machine became the XI, with unpressurized cockpit but greatly augmented fuel capacity, and numerous other changes, including shiny Cerulean Blue finish and plain red/blue roundels.

*Heavily-armed P-47D Thunderbolts prepare for a strike mission over France just prior to D-Day*

### Republic P-47D Thunderbolt

Heavy, ungainly and with a massive deep fuselage to accommodate supercharger and ducting, the P-47 nevertheless made up in brute power and at altitude its performance was exceptional. Later models had a teardrop hood and the late-model R-2800 Double Wasp, giving no less than 2550 hp emergency power

*Span:* 40 ft 9½ in *Length:* 36 ft 1¾ in *Engine:* Pratt & Whitney R-2800-59 Double Wasp, 2300 hp *Max speed:* 429 mph at 30,000 ft *Ceiling:* 42,000 ft *Max take-off weight:* 17,500 lb *Armament:* 8×.5-in mg, 1500 lb bombs/rockets

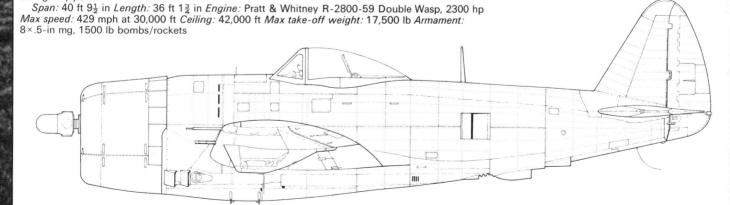

In 1940 Rolls-Royce discussed installation of the Griffon engine, which had a capacity of more than 36 litres compared with the Merlin's 27. A few were fitted in early airframes, and in 1942 production was urgently started on the Spitfire XII, with 1735-hp low-blown Griffon giving a speed at low altitude of around 370 mph, an increase of some 50 mph and fast enough to catch the Luftwaffe's Fw 190 and Bf 109 fighter-bombers making sneak raids on English coastal towns. The Griffon resulted in a bigger nose, with fairings over the projecting cylinder blocks, together with a four-blade propeller and large spinner. In 1943 the more refined Mk XIV entered production with the two-stage Griffon 65, deep twin radiators and five-blade propeller married to an airframe based on the Mk VIII retractable tailwheel and broad rudder. This was a superb combat machine that saw much action in 1944 and restored ascendancy of the Spitfire over the best Luftwaffe machines. It caught flying bombs with ease, reaching over 400 mph near the ground and almost 450 mph at 26,000 ft, and was an impressive dogfighter. The final batches were fitted with the teardrop hood, and small numbers of the long-range Mk XVIII and ultra-long-range PR.XIX were also built. Many of the Spitfires built from 1942 were of the FR (fighter-reconnaissance) type with one or more oblique cameras as well as full armament. Unfortunately, little was done to improve armament, and the RAF ended the war with the same guns, of 1916 design, which it had used in 1939.

Hawker followed the Hurricane through its design office with a totally new fighter, boldly begun in 1937–38, to carry 12 machine guns or four cannon on 2000 hp. After many difficulties and false starts, Gloster Aircraft got into limited production in 1941 with a version called the Typhoon IB, with 2200-hp Napier Sabre engine and four cannon. Though basically extremely strong, this machine at first had a tendency to shed its tail. Coupled with the extreme unreliability of the complex engine and disappointing performance at altitude, this led to the otherwise impressive Typhoon being written off as virtually useless. However, the Typhoon gradually demonstrated its tremendous value as a low-level tactical machine. It was able to catch any Luftwaffe fighter-bomber at low level, and after D-Day wrought havoc among Panzer divisions with cannon, bombs and rockets. Eventually 3330 were built, the majority of them having the odd car-type doors and clumsy canopy replaced by a neat teardrop sliding

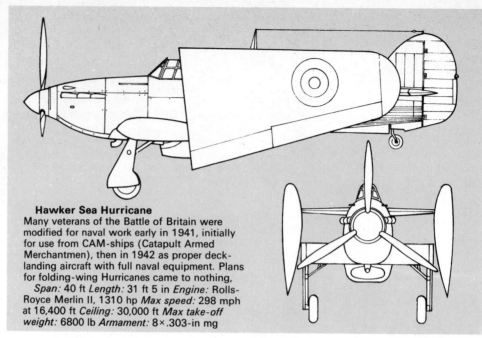

**Hawker Sea Hurricane**
Many veterans of the Battle of Britain were modified for naval work early in 1941, initially for use from CAM-ships (Catapult Armed Merchantmen), then in 1942 as proper deck-landing aircraft with full naval equipment. Plans for folding-wing Hurricanes came to nothing.
*Span:* 40 ft *Length:* 31 ft 5 in *Engine:* Rolls-Royce Merlin II, 1310 hp *Max speed:* 298 mph at 16,400 ft *Ceiling:* 30,000 ft *Max take-off weight:* 6800 lb *Armament:* 8×.303-in mg

hood. Early in Typhoon development buffeting was traced to transonic airflow around the thick wing. A proposed thin-wing Typhoon led to the Tempest, with thin elliptical wing and longer fuselage housing fuel previously carried in the thick wing. Armed with four short-barrel Mk V cannon, the Tempest V went into production in the summer of 1943, proving approximately 40 mph faster than the Typhoon at most altitudes. In the summer of 1944 it destroyed more flying bombs than any other fighter. Later versions, with more powerful Sabre and Bristol Centaurus engines, entered service after the war.

Though they were never adopted, for reasons unconnected with the aircraft, the Martin-Baker M.B.3 and M.B.5 were outstanding fighters, in many ways superior to those used by the RAF. The M.B.3, powered by the Sabre, had six cannon with belt feeds, and features that promised exceptionally easy maintenance. Sadly, the prototype crashed as a result of engine-failure. The Griffon-engined M.B.5 is widely judged the best piston-engined fighter ever built. The official report described the general design as 'infinitely better, from the engineering and maintenance aspects, than any other similar type of aircraft'. Had a positive decision been taken in time, the M.B.5 could have seen war service.

Other British fighters built only as prototypes included the high-altitude Mosquito XV and Vickers 432, the former

having four machine guns and the latter six cannon. The official high-altitude choice fell on the long-span Westland Welkin, powered like its two rivals by two high-blown Merlins and with a wing of no less than 70-ft span. Though 67 were built, including a two-seat night fighter, none reached a squadron.

**Hawker Typhoon 1B**
The Thyphoon was not suited to its original role as an intercepter but proved to be a devastating ground-attack fighter. The Mk 1B variant carried cannon in place of the earlier machine-guns, but the most effective armament was the rocket projectile. Rocket-firing Typhoons destroyed large numbers of German tanks in the weeks following D-Day

*The Martin-Baker MB5, last of a series of remarkable fighter prototypes from this small company*

*Span:* 41 ft 7 in *Length:* 31 ft 10 in *Engine:* Napier Sabre IIA, 2180 hp *Max speed:* 405 mph at 18,000 ft *Ceiling:* 34,000 ft *Max take-off weight:* 11,400 lb *Armament:* 4×20-mm Hispano cannon, 2×1000-lb bombs or 8×60-lb rockets

Testing the Browning machine-guns in the nose of a Mosquito FB.VI, the fighter-bomber sub-type produced in larger quantities than any other variant of this remarkable wooden warplane

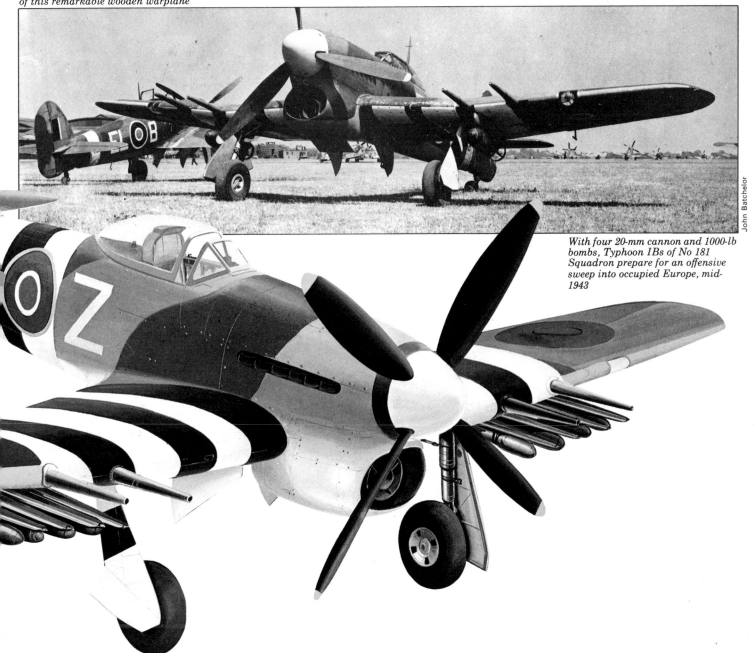

John Batchelor

With four 20-mm cannon and 1000-lb bombs, Typhoon IBs of No 181 Squadron prepare for an offensive sweep into occupied Europe, mid-1943

In the United States, Curtiss relentlessly pursued the P-40 design through 15 basic variants; most variants had the advanced Allison engines, but a few were powered by the Packard V-1650 Merlin. Nearly all carried six 0.5-in machine-guns in the wing, giving reliable service as tactical support and fighter-bomber aircraft in many theatres. Handicapped by the fact that even at the most effective altitude the speed seldom reached 350 mph, the P-40 Warhawk/Kittyhawk family took the total of P-40 production to 13,738. None of the shoal of other Curtiss prototypes reached the production stage, though the XP-55 Ascender (Arse-ender), XP-60 family, giant XP-62 with up to eight cannon, and shipboard XF14C were technically of intense interest.

Although the extremely unconventional P-39 Airacobra had been rejected by the RAF, Bell fought to get the aircraft into wide service, eventually succeeding to the tune of 9558 aircraft. Of these just over half went to the Soviet Union, where the effective armament in the ground-attack role and ability to absorb heavy punishment brought the Airacobra a good reputation. Bell also built 3303 P-63 Kingcobras, which looked similar (apart from the tail) but were actually a complete redesign. Armed with the same 37-mm cannon and four 0.5-in as most P-39s, the P-63 saw action almost entirely on the Russian front; 332 P-63s were converted by the USAAF into armoured target aircraft to be shot at by

frangible (easily shattered) practise bullets.

Lockheed's first fighter, the Model 22, was a masterpiece of originality. It was designer Hall L. Hibberd's answer to a challenging Army Air Corps requirement for a speed of 360 mph and endurance at full throttle of one hour. The prototype flew in January 1939, startling observers by its size, twin liquid-cooled Allison engines, twin tail booms and remarkably slim design. By 1941 deliveries began to the RAF, with the turbochargers and handed propellers (rotating in opposite directions) removed. As predicted by Lockheed, these aircraft were poor performers, and the RAF never used the Lightning, as it was named. But the US Army P-38 went from strength to strength, and eventually 9923 were delivered, most having nose armament of one 20-mm and four 0.5-in, with an exceptional amount of ammunition. Most had a combat setting of 8° flap to improve manoeuvrability, and later models also had an extra electrically driven dive-flap under each wing to counter strong nose-down pitch at very high speeds. Comfortable on long missions, and flown by a wheel instead of a stick, the popular P-38 served as an escort and ground-attack fighter. At the end of the war the radar-equipped P-38M night fighter saw action, and many were converted as two-seat 'droop snoot' lead ships for accurate level bombing. The monster XP-58, with 3000-hp double engines, was built only as a prototype.

In the Pacific war two US Navy fighters of

**North American P-51B Mustang**
The P-51's airframe had proved itself an outstanding and advanced design, and with the decision to mass-produce the Merlin engine in the United States the two were married to produce a completely revitalised fighter of outstanding quality. The introduction of the long range P-51B at the end of 1943 gave tremendous impetus to the US daylight bombing offensive over Europe

*Span:* 37 ft *Length:* 32 ft 3 in *Engine:* Packard Merlin V-1650-3, 1450 hp *Max speed:* 445 mph at 30,000 ft *Ceiling:* 41,800 ft *Max take-off weight:* 9800 lb

**North American P-51D Mustang**
Teardrop canopy and a fin fillet to compensate for the reduced side area were the main changes in the D-model. With engines rated for low altitude and equipped with rockets, Mustangs made many successful marauding raids into Europe

*Span:* 37 ft *Length:* 32 ft 3$\frac{5}{16}$ in *Engine:* Packard-Merlin V-1650-7, 1790 hp *Max speed:* 443 mph at 25,000 ft *Ceiling:* 41,900 ft *Max take-off weight:* 10,100 lb *Armament:* 6×.5-in mg, 1000 lb bombs/rockets

*Seafire XV comes in to land aboard the escort carrier* Pretoria Castle

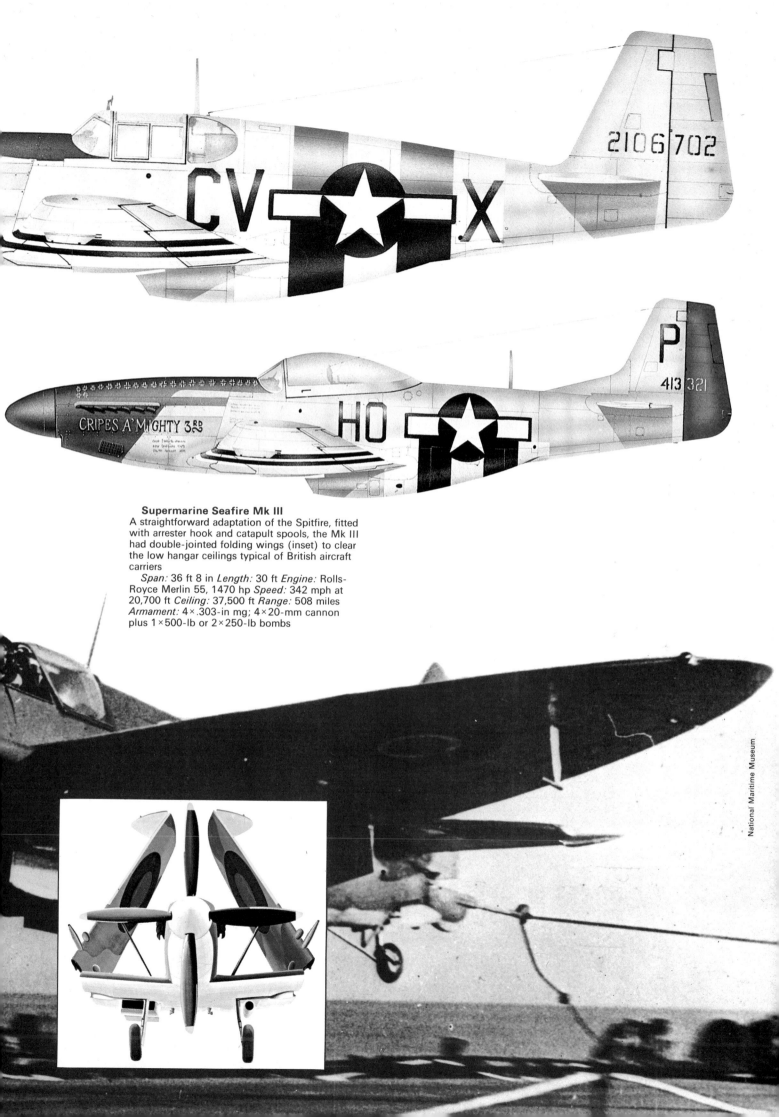

**Supermarine Seafire Mk III**
A straightforward adaptation of the Spitfire, fitted
with arrester hook and catapult spools, the Mk III
had double-jointed folding wings (inset) to clear
the low hangar ceilings typical of British aircraft
carriers
   *Span:* 36 ft 8 in *Length:* 30 ft *Engine:* Rolls-
Royce Merlin 55, 1470 hp *Speed:* 342 mph at
20,700 ft *Ceiling:* 37,500 ft *Range:* 508 miles
*Armament:* 4×.303-in mg; 4×20-mm cannon
plus 1×500-lb or 2×250-lb bombs

National Maritime Museum

similar design and power came into service in 1942–43 and at once turned the tables on the previously victorious Japanese. First to be designed was Chance Vought's F4U Corsair, planned in 1938. At that time there was not the mass of information which was later gleaned from warfare in Europe, and American fighters tended to have 1000 hp. Rex Beisel and his team boldly selected the monster R-2800 Double Wasp aircooled radial, then experimentally giving 1850 hp, and as a result laid the basis for an aircraft many consider the best fighter of the Second World War. In October 1940 the prototype became the first American aircraft to exceed 400 mph, but development was protracted. Fuel was added to the fuselage, moving the pilot aft where his view was poor. The Navy insisted that the F4U could not be used from carriers, but Corsairs of the British Fleet Air Arm, with wings clipped to fit below-decks, successfully operated from small escort carriers from mid-1943 onwards. After a considerable number of engineering changes the F4U-1 was built in large numbers, 4102 by Vought, 3808 by Goodyear (FG-1) and 735 by Brewster (F3A). The F4U-2 interim night fighter was in service in early 1943, followed by the F4U-4E and 4N, all of which had the APS-4 or -6 radar with the aerial in a pod on the right wing. Most Corsairs had six 0.5-in guns and up to 2000 lb of bombs or rockets; some versions had four 20-mm cannon. By 1945 there were further versions, some having the 3000-hp R-4360 Wasp Major; the last F4U was completed when No. 12,571 came off the line in December 1952.

The Grumman F6F Hellcat, partner and rival to the Corsair in defeating the Japanese, was designed in 1941–42, flown in 1942 with two types of engine, and then built with great rapidity after extremely quick development. Like the F4U it was a very large fighter, with an even bigger wing of 334 sq ft housing the same armament of six 0.5-in (in some sub-types, two 20-mm and four 0.5-in). By 1945 no fewer than 12,274 Hellcats had been produced, and their capability is underlined by the fact that, of 6477 enemy aircraft credited to US Navy carrier-based fighters, 4947 were gained by the Hellcat. Again several versions were night fighters, with the aerial system in a pod on the right wing, and most versions could carry two 1000-lb bombs or rockets. Just over 1200 were supplied to the Fleet Air Arm (which originally considered the name Gannet) and there were also high-altitude reconnaissance versions. Though few Hellcats could reach 400 mph, they combined so many attributes that from 1943 onwards the Allies dominated the Pacific skies.

In the Soviet Union the Yakovlev fighters, powered by liquid-cooled M-105 engines, continued to be built in huge numbers, reaching approximately 37,000 by 1945. The mass-produced Yak-7B was replaced on the Urals assembly line in late 1942 by the Yak-9, in which metal wing spars allowed a considerable increase in fuel capacity. Total output of all versions of Yak-9 reached approximately 20,000; despite their light armament of one engine-mounted cannon and one or two machine guns, their all-round performance and manoeuvrability were so good that they were at least equal to Bf 109G or Fw 190 fighters flown by pilots of similar experience to the Russians.

There were many sub-types, with different marks of engine, the gun being 20-, 37- or even 75-mm; the -9D was a long-range model. The trim Yak-3 of 1944 was very similar but had a smaller wing for low-level duties, which increased speed from about 365 to 405 mph. The Yak-9U, a total redesign with all-metal structure, came too late to see action.

The rival Lavochkin bureau achieved great success by switching to the M-82A radial engine, the resulting La-5 going into production in 1942. By the end of that year the -5FN was in production with a more powerful 1640-hp engine and cut-down rear fuselage giving all-round vision. Further development led to increasing use of metal construction, and during 1943 to the La-7 with a third ShVAK cannon on the left side supplementing the original two above the fuselage. Most -7s also had an engine boosted to 1775 hp, and with an aerodynamic clean-up their speed rose considerably above 400 mph. Despite their small wings, which were fully slatted, the Lavochkin fighters were excellent in close combat and were flown by most of the leading Soviet aces, including Ivan Kozhedub (62 victories) and Alexander Pokryshin (59). Total production amounted to about 26,000, the completely new La-9 with all-metal airframe of different shape and four fuselage cannon not entering service uutil near the end of the European war.

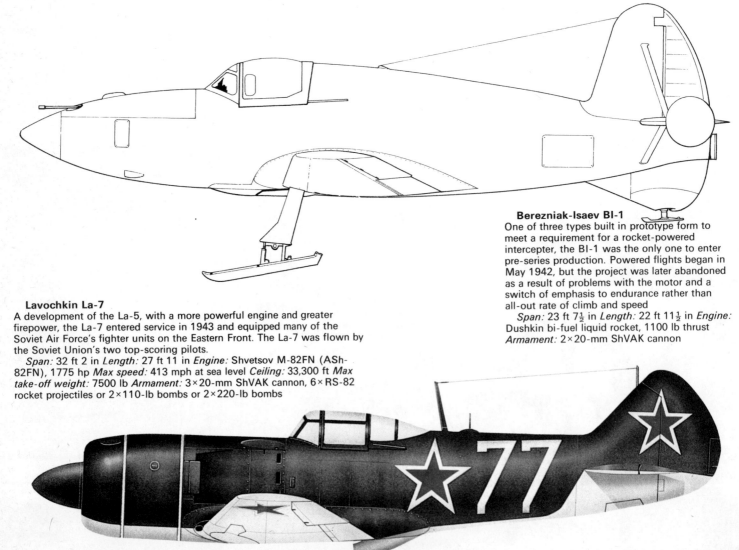

**Lavochkin La-7**
A development of the La-5, with a more powerful engine and greater firepower, the La-7 entered service in 1943 and equipped many of the Soviet Air Force's fighter units on the Eastern Front. The La-7 was flown by the Soviet Union's two top-scoring pilots.
  *Span:* 32 ft 2 in *Length:* 27 ft 11 in *Engine:* Shvetsov M-82FN (ASh-82FN), 1775 hp *Max speed:* 413 mph at sea level *Ceiling:* 33,300 ft *Max take-off weight:* 7500 lb *Armament:* 3×20-mm ShVAK cannon, 6×RS-82 rocket projectiles or 2×110-lb bombs or 2×220-lb bombs

**Berezniak-Isaev BI-1**
One of three types built in prototype form to meet a requirement for a rocket-powered intercepter, the BI-1 was the only one to enter pre-series production. Powered flights began in May 1942, but the project was later abandoned as a result of problems with the motor and a switch of emphasis to endurance rather than all-out rate of climb and speed
  *Span:* 23 ft 7½ in *Length:* 22 ft 11½ in *Engine:* Dushkin bi-fuel liquid rocket, 1100 lb thrust *Armament:* 2×20-mm ShVAK cannon

# THE FALLEN EAGLE

*Like the Spitfire, the Fw 190 underwent continuous development. The long-nosed 190D series were powered by the liquid-cooled Jumo 213 and not a radial engine as the annular radiator suggests*

Despite having an excellent engineering capability, and at the lower and medium levels good management, the essentially flawed nature of the RLM *(Reichsluftministerium)* procurement machine ensured that, right up to the final defeat in May 1945, the Luftwaffe was equipped in the main with modified versions of the aircraft in service before the war.

Fighters were no exception – the newest type in widespread service was the Fw 190, first flown in June 1939. Though a superb combat aircraft, and technically more advanced than almost all contemporary machines, the 190 was increasingly hard-pressed by the later Spitfires, Mustangs and Tempests; nor did it have things all its own way on the Eastern front. The Fw 190F and G were offshoots from the mainstream of development, intended for various close-support and attack roles and built in large numbers. Heavily armed and armoured, they could carry bombs of up to 3968 lb (1800 kg), the heaviest weapon load of any comparable aircraft until almost 20 years later. Some indication of what had been done to the 190 – dimensionally a small fighter – can be deduced from the fact that the loaded weight of the first prototype was 3968 lb, compared with 12,900 lb in some F and G models.

In 1942 several high-altitude prototype 190s were built with turbochargers under the rear fuselage, some of these being powered by DB 603 liquid-cooled engines. Though not built in quantity, these led to a completely new 190, the D, with a 1776-hp Jumo 213 boosted by MW50 injection to 2240-hp.

By early 1944 the D-9, called Dora 9 by

the Luftwaffe and the long-nosed 190 by the RAF, was proving itself faster than radial-engined versions, though carrying fewer guns in most instances (often only two MG 151 and two MG 131). The D-12, however, had a 30-mm MK 108 on the centreline and a useful bomb load, yet in clean condition could reach 453 mph at height with MW50 injection. The RLM rewarded designer Kurt Tank by allowing future aircraft from his team to have the prefix 'Ta', and the Fw 190D series thus became the Ta 152. By late 1944 the effort was becoming dissipated into numerous prototypes, but the formidable Ta 152C and H did reach production, the former having a 2100-hp DB 603L and heavy armament (usually one 30-mm MK 108 and four 20-mm MG 151) and the latter being a high-altitude version with Jumo 213E, long-span wing and lighter armament (e.g. MK 108 and two MG 151). All were great performers. Tank himself simply opened the throttle when bounced by a flight of Mustangs on a test-flight, leaving the straining American aircraft, the fastest Allied fighters in the high-level sky, at a relative speed of about 30 mph, the 152H being capable of about 472 mph. The shortage of both fuel and pilots meant that only very few saw service.

The final mark of Bf 109 also saw only limited combat, though production until 1945 was on an unprecedented scale, 14,212 being delivered in 1944, despite Allied bombing. All the 109G series were powered by various types of DB 605, rated at 1475–1550-hp. This was insufficient to compensate for the increase in weight, and the all-round performance was unimpressive. Many of the inherent virtues of the 109 became

submerged, while the shortcomings of terrible discomfort, limited endurance, and flight controls that seized almost solid at high speeds, persisted or worsened. Despite the wealth of outstanding guns and rocket missiles fitted, the 109G was steadily outclassed even over Germany itself; this process was exacerbated by the failure of the training schools to replace experienced fighter pilots. By mid-1944 the shortage of pilots and fuel was critical, and the task of Allied pilots – notably those flying the P-47 and P-51 – became much easier. The final Bf 109 versions introduced an improved canopy, often called the 'Galland' hood after the Luftwaffe General of Fighters, and a wooden tail of improved shape. These features were seen on many late sub-types of both the G and the K, the latter tending to replace the G in December 1944 with small structural changes. The high-altitude 109H had a long-span centre section but was abandoned, as was the completely new and radical Me 155, which was passed to Blohm und Voss, becoming the Bv 155. The 155 gradually ceased to resemble the 109, and with a wing of various spans from 61 to 69 feet carrying enormous radiators, was intended to fight at up to 56,000 ft.

Messerschmitt's greatest achievements were the Me 163 and Me 262. The former had a history extending back to a 1938 DFS project for a rocket test aircraft. Professor Alex Lippisch made the DFS 194 tailless, and flight tests in 1940 encouraged the RLM to sanction a high-speed aircraft designated Me 163. Glider champion Heini Dittmar reached 570 mph in this in August 1941, and on 2 October, 1941, was towed to 13,000 ft. He cast off, ignited the Walter

## Messerschmitt Bf 109G-6

The G-6 version of this famous fighter, which remained in production in one form or another throughout the Second World War, was the first to have a pair of 20-mm cannon mounted beneath the wings. The G series was being built at the rate of 725 a month by mid-1943 but, despite having the more powerful DB 605 engine, was in some ways considered inferior to its Bf 109F predecessor

*Span:* 32 ft 6½ in *Length:* 29 ft 8 in *Engine:* Daimler-Benz DB 605A-1, 1475 hp *Max speed:* 387 mph at 23,000 ft *Ceiling:* 38,500 ft *Max take-off weight:* 7500 lb *Armament:* 2×13-mm MG 131 mg, 3×20-mm MG 151 cannon

## Messerschmitt Me 309

Intended as a successor to the Bf 109, the Me 309 was cancelled after its development was delayed so long by technical problems – especially with the undercarriage – and the fact that the new generation of Focke-Wulf fighters proved a better proposition

*Span:* 36 ft 1 in *Length:* 31 ft 4 in *Engine:* Daimler-Benz DB 603A-1, 1750 hp *Max speed:* 496 mph at 26,200 ft *Ceiling:* 39,300 ft *Max take-off weight:* 9050 lb *Armament:* 30-mm MK 103 or MK 108 cannon, 2×13-mm MG 131 mg (light version); 30-mm MK 103 or MK 108 cannon, 2×20-mm MG 151 cannon, 4×13-mm MG 131 mg (heavy version)

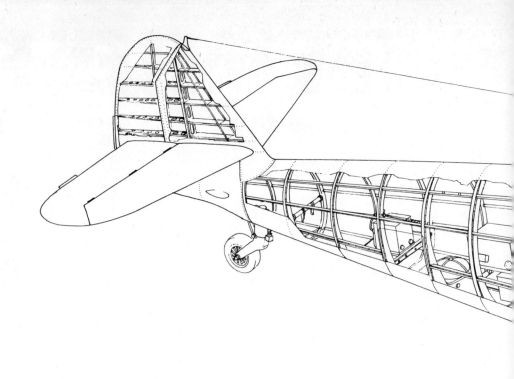

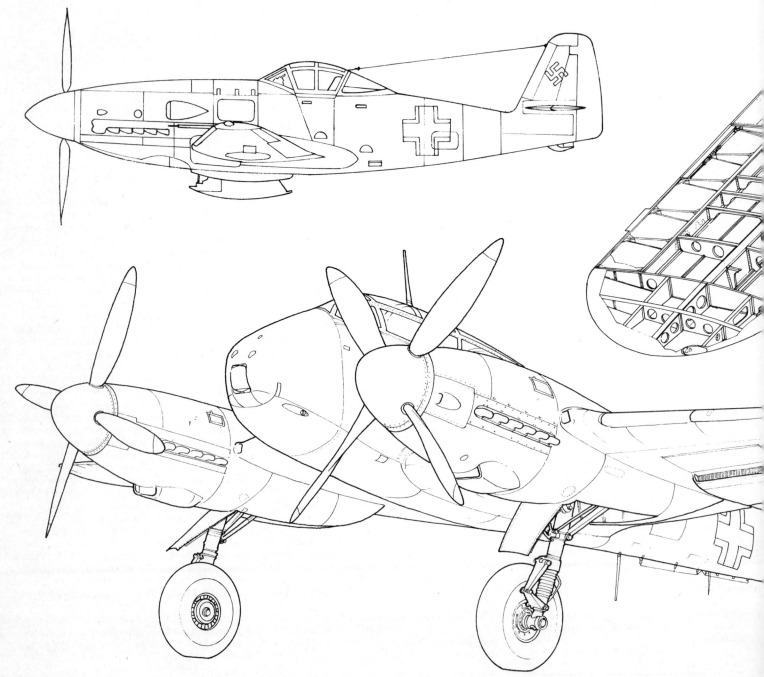

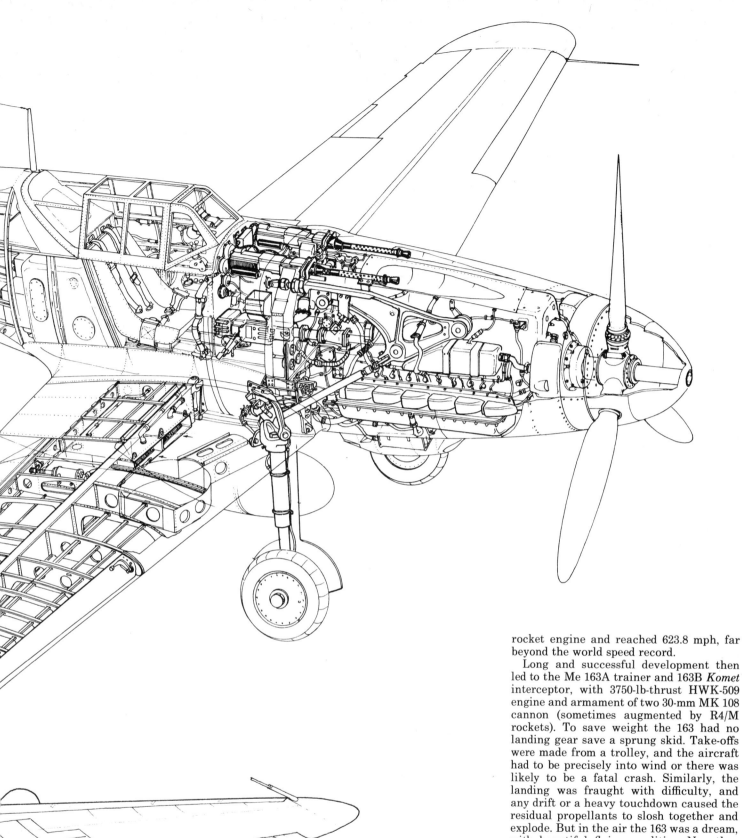

**Messerschmitt Me 210**

The Me 210, designed as a successor to the Bf 110, had an unhappy career largely as a result of being rushed into production before the basic design deficiencies had been corrected. An order for 1000 Me 210s was placed even before the first prototype had flown, but in the event only 350 were completed and the type was succeeded by the Me 410 Hornisse

*Span:* 53 ft 7 in *Length:* 40 ft 3 in *Engine:* 2×Daimler-Benz DB 601F, 1395 hp each *Max speed:* 385 mph *Ceiling:* 23,000 ft *Max take-off weight:* 17,850 lb *Armament:* 2×20-mm MG 151 cannon, 2×7.9-mm MG 17 mg, 2×13-mm MG 131 mg

rocket engine and reached 623.8 mph, far beyond the world speed record.

Long and successful development then led to the Me 163A trainer and 163B *Komet* interceptor, with 3750-lb-thrust HWK-509 engine and armament of two 30-mm MK 108 cannon (sometimes augmented by R4/M rockets). To save weight the 163 had no landing gear save a sprung skid. Take-offs were made from a trolley, and the aircraft had to be precisely into wind or there was likely to be a fatal crash. Similarly, the landing was fraught with difficulty, and any drift or a heavy touchdown caused the residual propellants to slosh together and explode. But in the air the 163 was a dream, with beautiful flying qualities. No other aircraft could get near it, and the 370-odd delivered achieved great success in steep interceptions of US 8th Air Force heavy bombers, followed by a swift glide back to base.

In contrast the Me 262 was a conventional fighter, with the same outstanding flying qualities (quite different from the 163 but exceptionally pleasant and safe) combined with normal range and endurance. What made the 262 different was that it was a jet. Development had begun in late 1938, and the first prototype flew with a nose piston engine in April 1941. In July 1942 it flew with two Jumo 004 turbojets under the wings, eventually receiving a tricycle landing gear, full combat equipment and the

## Messerschmitt Me 262

The first jet fighter to enter service, the Me 262 could have altered the course of the Second World War if it had been employed in the role for which it was intended. On Hitler's instructions, however, the type was used primarily as a ground-attack aircraft, and fewer than one-sixth of those built ever saw service.
*Span:* 40 ft 11½ in *Length:* 34 ft 9½ in *Engine:* 2×Junkers Jumo 004B-1, -2 or -3, 1980 lb thrust each *Max speed:* 583 mph at 30,000 ft *Ceiling:* 37,500 ft *Max take-off weight:* 14,100 lb *Armament:* 4×30-mm MK 108 cannon, 24×R4M rocket projectiles

Cockpit is a totally enclosed unit for pressurisation purposes, although no service machines were ever pressurised

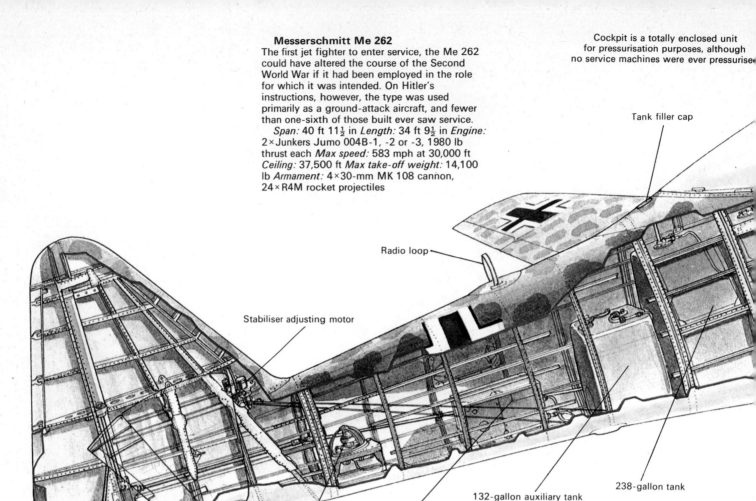

Tank filler cap

Radio loop

Stabiliser adjusting motor

238-gallon tank

132-gallon auxiliary tank

Flaps

Radio

Master compass

Variable orifice 'bullet' moves in and out to vary exit area

devastating armament of four 30-mm MK 108 cannon. Production was scheduled to get under way in 1943, but Allied bombing, and Hitler's insistence that all the Me 262s should be used as bombers instead of fighters, seriously delayed deliveries. EK 262, the operational-reasearch and development group, received their first aircraft in June 1944, followed in September by III/EJG2 training unit and 8/ZG26 fighter squadron. The first RAF Meteor squadron had been equipped two months previously. There were many variants of the 262, but nearly all the 1433 completed were of the A-1a *Schwalbe* (Swallow) fighter type or the A-2a *Sturmvogel* (Stormbird) bomber with two 551-lb bombs. The A-2a/U2 had a prone bombardier in place of the guns, and among the other versions was the B-family of tandem-seat radar night fighters. All were extremely formidable, though many pilots were killed in training accidents through failure to appreciate that a dual conversion trainer was needed. Most of the Allied victories over these aircraft were gained by patrols orbiting over the Luftwaffe airfields.

Fastest of the non-jets was the Dornier Do 335, which in the only form built in numbers (the A-0 and A-1) attained 477 mph with MW50 boost. Fitted with a 1900-hp DB 603 ahead of and behind the cockpit, driving a propeller at each end, the 335 *Pfeil* (Arrow) was a substantial machine first flown in the autumn of 1943. Many sub-types were planned, including night fighters, big-winged high-altitude versions and the Do 635 twin-fuselage model with 'an engine at each corner'.

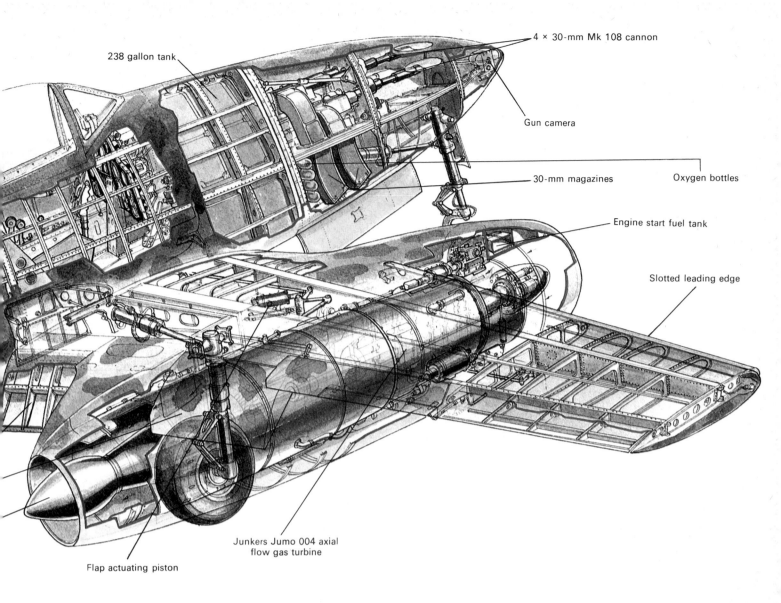

238 gallon tank

4 × 30-mm Mk 108 cannon

Gun camera

30-mm magazines

Oxygen bottles

Engine start fuel tank

Slotted leading edge

Junkers Jumo 004 axial
flow gas turbine

Flap actuating piston

**Dornier Do 335**

The unique push-pull engine configuration of the
Do 335 had been patented by Dr Claude Dornier
in 1937, but a prototype warplane did not fly
until 1943. Performance was exceptional but
delays and diversification into night-fighter and
reconnaissance variants kept production down

*Span:* 45 ft 3⅓ in *Length:* 45 ft 5¼ in *Engine:*
2 × Daimler-Benz DB 603G, 1900 hp *Max speed:*
473 mph at 26,200 ft *Ceiling:* 37,400 ft *Max
take-off weight:* 25,800 lb *Armament:* 2 × 15-mm
MG 151 cannon, 1 × 30-mm MK 103 cannon,
1100 lb bombs

**Heinkel He 162 Salamander**
The *Volksjäger* (People's Fighter) was a desperate attempt to develop a high-performance fighter which used commonly available materials, could be built by relatively unskilled labour and flown by novice pilots. The planned output of a thousand a month was — hardly surprisingly — not achieved, and total production barely exceeded 100

*Span:* 23 ft 8 in *Length:* 29 ft 8½ in *Engine:* BMW 003A-1, -2, or E-1, 1760 lb thrust *Max speed:* 522 mph at 19,700 ft *Ceiling:* 39,500 ft *Max take-off weight:* 5940 lb *Armament:* 2×30-mm MK 108 cannon

*The RAF's first jet fighter, the Gloster Meteor, entered service on a test basis in mid-1944 and by August one had claimed a victory over a V-1. Tactical sorties began over Europe in April 1945 but no Meteor ever met a Luftwaffe jet fighter*

Imperial War Museum

**Bell P-59 Airacomet**

[Th]e first turbojet-powered aircraft to be built in [the] United States made its maiden flight in [co]nditions of great secrecy in late 1942. Stability [pro]blems made it unsuitable for operational use [and] the P-59 was used solely for research and [trai]ning

*Span:* 45 ft 6 in *Length:* 38 ft *Engine:* 2 × [Ge]neral Electric J 31-GE-5, 2000 lb thrust *Max [spe]ed:* 413 mph at 30,000 ft *Ceiling:* 46,200 ft [Ma]x take-off weight: 12,700 lb *Armament:* [1×]37-mm cannon, 3×.5-in mg

Focke-Wulf's glued-wood Ta 154 *Moskito* was a failure; the extraordinary Bachem Ba 349 *Natter* (Viper) and Blohm und Voss Bv 40 target-defence interceptors never reached service; and the impressive Heinkel He 162A Salamander, popularly called the *Volksjager* (People's Fighter) was designed, flown and put into mass production in the final three months of 1944, but the Luftwaffe was by then denuded of pilots and even jet fuel. Built of wood and metal, the 162 was an excellent answer to the challenge of finding a truly modern fighter 'to be built as a piece of consumer goods' at the rate of 4000 per month. It is also an indication of the depths to which planning had fallen that this extremely advanced machine was to be flown by Hitler Youth previously

**Messerschmitt Me 163 Komet**

The Me 163 was originally intended as a high-speed research aircraft, but the encouraging trials results led to the type being adopted as a home-defence intercepter. The Me 163B Komet, the

only pure-rocket aircraft to see service in the Second World War, first went into action against B-17 Fortresses in August 1944. The type was of limited operational use, despite its exotic design, and only about 350 were built

*Span:* 30 ft 7 in *Length:* 18 ft 8 in *Engine:* Walter HWK 109-509A-2 bi-fuel rocket motor, 3750 lb thrust *Max speed:* 596 mph at up to 30,000 ft *Ceiling:* 40,000 ft *Max take-off weight:* 9500 lb *Armament:* 2×30-mm MK 108 cannon, up to 24× R4M rocket projectiles

given a few quick towline hops in primary gliders.

The last German fighter to fly was the Gotha Go 229, designed as the Ho IX V2 prototype by the Horten brothers as the first jet-propelled member of a long family of flying-wing machines. On two Jumo 004B engines it reached 497 mph in early trials in March 1945. A production version would have had four 30-mm MK 108 or 103 cannon and two 2205-lb bombs.

In the final year of the war German designers explored a number of bizarre configurations for fighters, most of them of the short-range target-defence type. Several attempted to save time and resources by using the Argus 014 impulse-duct engine of the Fi 103 flying bomb, while others used a completely new kind of jet engine, the ramjet. One of the most unusual of all these experiments was intended to have a ramjet fuelled by powdered coal, or a mixture of finely divided carbonaceous fuel and a liquid hydrocarbon. This was one of the Lippisch shapes, with wings and fin of delta (triangular) form. Many projects were tailless, while the Ta 183 had a sharply swept vertical tail carrying a swept tailplane on top (today this looks normal, but in 1943 it was utterly futuristic). Messerschmitt explored many wing shapes, and in the P.1101, almost ready for flight in 1945, arranged for the wing sweep to be adjusted to different settings on the ground. In 1951 this

aircraft flew in the United States as the Bell X-5, with swing-wings adjustable in flight.

Possibly the strangest fighter of all was the Focke-Wulf *Triebflügel,* a target-defence interceptor which stood on its tail and achieved VTO (vertical take-off) by three rotary wings mounted on bearings around the fuselage and spun by tip-ramjets. In normal flight it was to reach 620 mph with the fuselage in the conventional attitude, and two Mk 103 and two MG 151 were to be fitted in the nose. With skill and luck the pilot was to regain the vertical attitude, gently descending onto a landing wheel at the rear tip of the fuselage with outrigger wheels extended from the tips of the four fins. Though scarcely credible, the same concept was used in 1948–56 by two turbo-prop VTOL interceptors of the US Navy intended for use from small ship platforms. Thus, while the Luftwaffe soldiered on with the 190 and 109, German research mapped out tracks leading far into the future but irrelevant to the defence of the crumbling Reich during the last two years of the Second World War.

In contrast, British designers kept their feet firmly on the ground and seldom built anything that did not relate directly to the operating needs of the front-line squadrons. The outstanding de Havilland 100 'Spider Crab', later named Vampire, flew in September 1943, but just missed the war. The

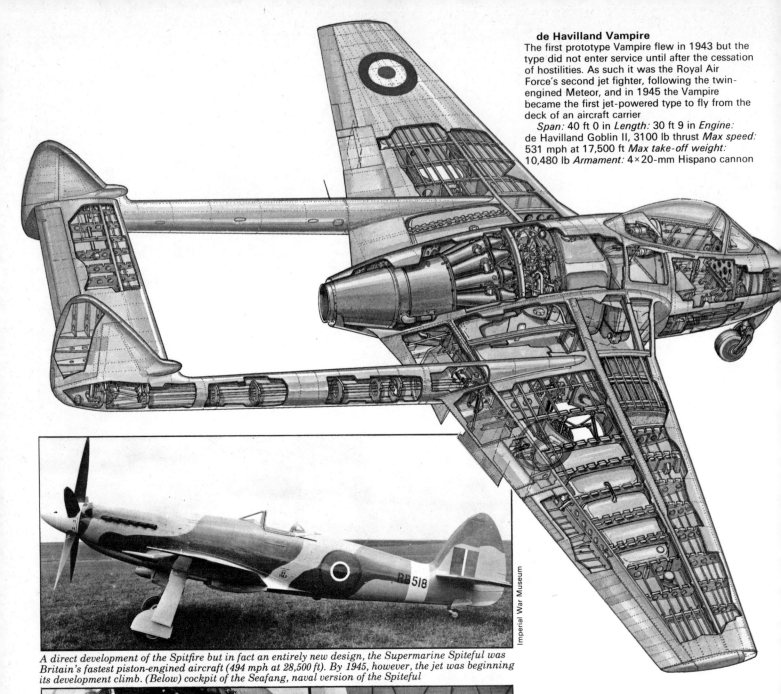

Imperial War Museum

## de Havilland Vampire

The first prototype Vampire flew in 1943 but the type did not enter service until after the cessation of hostilities. As such it was the Royal Air Force's second jet fighter, following the twin-engined Meteor, and in 1945 the Vampire became the first jet-powered type to fly from the deck of an aircraft carrier

*Span:* 40 ft 0 in *Length:* 30 ft 9 in *Engine:* de Havilland Goblin II, 3100 lb thrust *Max speed:* 531 mph at 17,500 ft *Max take-off weight:* 10,480 lb *Armament:* 4×20-mm Hispano cannon

*A direct development of the Spitfire but in fact an entirely new design, the Supermarine Spiteful was Britain's fastest piston-engined aircraft (494 mph at 28,500 ft). By 1945, however, the jet was beginning its development climb. (Below) cockpit of the Seafang, naval version of the Spiteful*

John Batchelor

Gloster F.9/40, however, later named Meteor, flew in March 1943, and after development with seven types of British turbojet entered RAF service with 616 Sqn in July 1944, becoming the first fully operational jet aircraft in the world. The production Meteor I was powered by the 1700-lb Rolls-Royce Welland and reached 410 mph in fighting trim, sufficient to catch and destroy flying bombs, becoming operational on 4 August. By April 1945 the more powerful (2000-lb Derwent I) Meteor III was in service from 2nd Tactical Air Force Continental bases with 616 and 504 Sqns, reaching 493 mph and fitted with an improved canopy that slid instead of being hinged. In the United States the Bell P-59A Airacomet remained a jet conversion trainer, but the outstanding Lockheed YP-80A did see evaluation service in Italy in 1945, but no combat however.

Fighter development in Japan in the Second World War ran broadly parallel to that in Germany: while the front-line pilots had to make do with improved models of old aircraft, the design teams created numerous exciting fighters which never got into action. In both countries the situation was to a large degree a vicious circle; the need to concentrate on modifications to the old types interfered seriously with development of the new ones, and continuing delay to the replacement types kept the pressure on improvement of the existing aircraft. Nowhere was this more apparent than in the great Mitsubishi company, where not even the brilliance of Jiro Horikoshi could produce a replacement for the *Zero-Sen* in time to save defeat.

Most of the effort went into extremely urgent work on the A6M *Zero-Sen,* which by 1943 was demonstrably hard-pressed, if not outclassed, by such aircraft as the F4U and F6F. By late 1943 the A6M5 family was in production, and like the 109G this late model soon outnumbered all its predecessors. It was little changed at first, having the clipped wings rounded and the engine installation improved with individual ejector exhaust stacks. Along with aerodynamic detail refinement, the 5a then introduced belt-fed cannon, with 25 extra rounds per gun; in the 5b one of the fuselage machine guns was replaced by a 12.7-mm weapon and bullet-proof canopy; and the grossly underpowered 5c received new armament of two 20-mm Type 99 Mk 4 and two 13.2-mm Type 3 in the wings and a 13.2-mm in the fuselage. In late 1944 the

# THE SETTING SUN

water/methanol-boosted *Sakae* 31 resulted in the 6c version, which also introduced self-sealing wing tanks. The 7 was a fighter-bomber made in trivial numbers in the final desperate weeks of the war, and production did not even begin on the A6M8 with the bigger 1560-hp *Kinsei* engine.

Though the total of 10,049, 10,449 or 10,937 of all models of *Zero-Sen* (the records are confusing) is impressive, this was the only Japanese warplane to be built in anything like such quantity. From the start of 1944 loss of carriers and skilled pilots greatly diminished the effectiveness of the Imperial Navy Air Force, and rapidly changing tactical situations resulted in many fighters being modified as home-defence night fighters (such as the A6M5d-S with oblique 20-mm cannon in the rear fuselage) or *Kamikaze* suicide bombers. Nakajima built 327 of the A6M2-N float-seaplane version, which was intended to

clear the skies for amphibious landings but was forced into defensive duties where it was easy meat for the Allied fighters.

The fighter which failed to replace the Zero was the A7M *Reppu* (Hurricane), which received the code-name 'Sam' though it never got into action. Horikoshi was certain that the required performance demanded the powerful MK9 engine, and tried to scheme his new fighter around it, but the Navy insisted on the lower-powered *Homare* NK9. There is abundant evidence that the Navy did not appreciate the vital need to introduce a much more powerful fighter at the earliest possible date, and at one time decided not to go ahead with development at all until April 1943. As it was, the prototype A7M did not fly until 6 May 1944, and though by the final defeat there were four major versions, the total number of aircraft built was a mere ten. The same pattern can be seen in the great array of new or proposed Japanese fighters. Though of immense technical interest, they never fired a shot against the Allies, except, perhaps, for the Kawasaki Ki-102b (Randy), which during a test flight is alleged to have shot an engine clean out of a B-29 with a single shell from its 57-mm gun! In fact, quite a few of these potent high-altitude, ground-attack and night fighter Ki-102 versions were built, the 102b sub-type reaching operational units.

A development of the Ki-102 was the Ki-108 high-altitude fighter, which demonstrated that the Japanese found the design of pressure cabins as difficult as did engineers in Europe. The Ki-109 was a

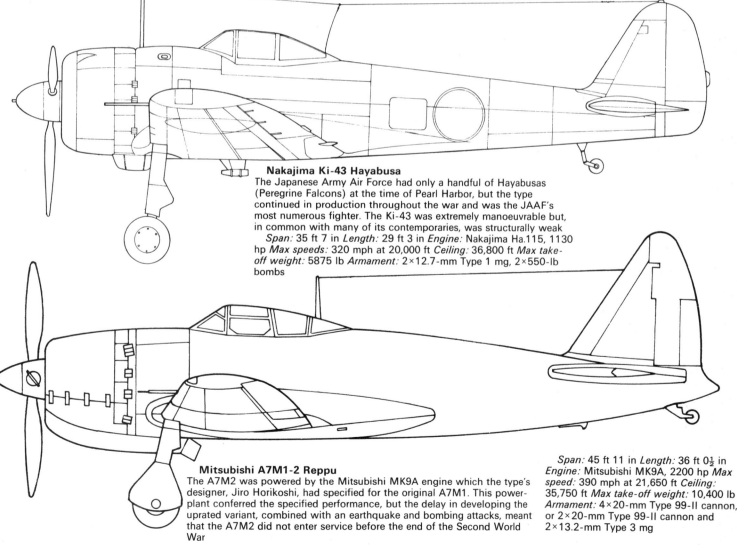

**Nakajima Ki-43 Hayabusa**
The Japanese Army Air Force had only a handful of Hayabusas (Peregrine Falcons) at the time of Pearl Harbor, but the type continued in production throughout the war and was the JAAF's most numerous fighter. The Ki-43 was extremely manoeuvrable but, in common with many of its contemporaries, was structurally weak
*Span:* 35 ft 7 in *Length:* 29 ft 3 in *Engine:* Nakajima Ha.115, 1130 hp *Max speeds:* 320 mph at 20,000 ft *Ceiling:* 36,800 ft *Max take-off weight:* 5875 lb *Armament:* 2×12.7-mm Type 1 mg, 2×550-lb bombs

**Mitsubishi A7M1-2 Reppu**
The A7M2 was powered by the Mitsubishi MK9A engine which the type's designer, Jiro Horikoshi, had specified for the original A7M1. This power-plant conferred the specified performance, but the delay in developing the uprated variant, combined with an earthquake and bombing attacks, meant that the A7M2 did not enter service before the end of the Second World War

*Span:* 45 ft 11 in *Length:* 36 ft 0½ in *Engine:* Mitsubishi MK9A, 2200 hp *Max speed:* 390 mph at 21,650 ft *Ceiling:* 35,750 ft *Max take-off weight:* 10,400 lb *Armament:* 4×20-mm Type 99-II cannon, or 2×20-mm Type 99-II cannon and 2×13.2-mm Type 3 mg

remarkable conversion of the Ki-67 heavy bomber, resulting in one of the largest fighters of the Second World War. More than 20 of these unusual machines were built, most having the unique but carefully considered armament of a single 75-mm Type 88 anti-aircraft gun firing ahead, with 15 rounds loaded individually by the co-pilot. This enabled the monster interceptor to cruise in formation with a B-29 but out of range of smaller guns. Great efforts were made to increase altitude performance, and one aircraft was flown with a battery of rocket motors to reduce the time taken to reach B-29 altitude. A far more deadly Mitsubishi Army fighter was the Ki-83, one of the best warplanes built in Japan but started so late that only four flew. Powered by two 2200-hp turbocharged Ha-211Ru engines, the -83 ('Steve') looked like a Grumman F7F Tigercat, another design that virtually missed the war, and like the US Navy fighter had both power and punch. With a speed of 438 mph, range exceeding 2000 miles and armament of two 3-mm and two 20-mm, plus light bombs internally, the Ki-83 demonstrated that Japan could create competitive aircraft for the 1945–46 warfare that never came to pass.

Other high-altitude Army fighters include the Nakajima Ki-87 and Tachikawa Ki-94, both powered by the 2400-hp Nakajima Ha-44 with turbocharger and pressurised cockpit. Both were armed with two 30-mm and two 20-mm cannon, plus a heavy bomb, and reached about 440 mph. The only Army fighter that paid off was a conversion of the indifferent Ki-61 'Tony', large numbers of which were lying about without engines in late 1944 because of lack of liquid-cooled power plants. The only available alternative was the 1500-hp Mitsubishi Ha-112-II radial, and though this hardly fitted the slim fuselage, a quick lash-up conversion was flown. It proved to be an amazing success, with comparable speed and significantly improved handling. Over 270 conversions were rushed into service, proving safe in the hands of inexperienced pilots

and able to tackle anything from a Hellcat to a B-29. Production urgently began on a version specially designed to the radial engine, but this did not reach service.

One of the Army fighters was the Kawasaki Ki-64, with two liquid-cooled Ha-40 engines combined in a single 24-cylinder unit driving a contraprop, with the cockpit placed between the front and rear banks of cylinders. Similarly, the Navy Yokosuka R2Y had coupled V-12 engines, but this was an even bigger aircraft and was converted into an attack bomber. A bizarre Navy fighter was the Kyushu J7W, a canard (tail first) with efficient variable-camber fore-plane resembling those flown on some fighters of the 1970s. Powered by a 2130-hp MK9D driving a six-blade pusher propeller, the J7W1 *Shinden* (Magnificent Lightning) might have been outstanding, with a speed of 466 mph and armament of four 30-mm cannon, but the prototype did not fly until two weeks before the surrender. Aichi built the S1A *Denko* (Bolt of Light) as a specialised night fighter, with two fan-cooled *Homares* and a radar-operator in a broad but shallow dorsal turret with two 20-mm cannon, presumably to hit whatever the two 30-mm in the nose missed.

The Navy operated Japan's only Second World War jets. The Mitsubishi J8M1 *Shusui* (Swinging Sword) was intended to be a copy of the Me 163B, but the submarine bringing the drawings was sunk and the Mitsubishi team were left with only a Walter rocket engine and a rough idea of the German interceptor's features. Quite quickly a design was prepared, trials begun with gliders and a powered version built (it crashed). Nakajima's *Kikka* (Orange Blossom) was a downgraded Me 262, with only 1047-lb thrust from each Ne-20 turbojet. Influenced by German design, the *Kikka* was conceived as an attack fighter carrying bombs; because payload was limited, the proposed 30-mm guns were then left out. The war was over before the Japanese had made engines with enough thrust (Ne-130 of 1984 lb) for them to be put back.

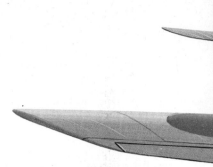

**Kawanishi N1K2-J Shiden-Kai**
The Shiden-Kai resulted from an extensive redesign of the N1K1-J Shiden (Violet Lightning) and has been described as the Japanese Naval Air Force's finest Second World War fighter. Fewer than 500 were built, however, because teething troubles delayed the build-up of production until late in the war, when extensive US bombing was disrupting the supply of components
*Span:* 39 ft 4 in *Length:* 30 ft 8 in *Engine:* Nakajima NK9H Homare 21, 1990 hp *Max speed:* 369 mph at 18,400 ft *Ceiling:* 35,300 ft *Max take-off weight:* 9040 lb *Armament:* 4×20-mm Type 99-II cannon, 2×550-lb bombs

**Grumman Tigercat**
This twin-engined single-seater could be flown off carriers and served with the US Marines flying close-support missions as the Pacific war was ending. Of the 364 machines built, most were post-war night-fighters
*Span:* 51 ft 6 in *Length:* 46 ft 11 in *Engine:* 2×Pratt & Whitney R-2800-34W, 2800 hp *Max speed:* 429 mph at 22,100 ft *Ceiling:* 37,300 ft *Max take-off weight:* 22,091 lb *Armament:* 4×20-mm cannon, 8×5-in rockets/2000-lb bomb/torpedo

**Goodyear FG-1D**
Goodyear-built version of the Vought F4U Corsair, the ungainly but powerful fighter-bomber that gained a legendary reputation in the Pacific war
*Span:* 41 ft *Length:* 32 ft 10 in *Engine:* Pratt & Whitney R-2800-18W Double Wasp, 2000 hp at take-off *Max speed:* 415 mph *Ceiling:* 36,900 ft *Armament:* 6×.5-in mg/2000-lb bombs/rockets

John Batchelor

*The Grumman F6F Hellcat incorporated the operational recommendations gathered from US Navy pilots who had met the Zero in combat. With excellent speed, climb and manoeuvrability, the Hellcat could outmatch the Zero in a dogfight*

# FIGHTER TECHNOLOGY

The superb Merlin 61 engine in a Mustang. Installing the Packard-built Rolls-Royce engine in the P-51, originally designed around an Allison engine, produced a first-class intercepter

John Batchelor

Always dictating fighter design, and more often than not success or failure in combat, were the guns and engines around which the fighters of the Second World War were designed. Fighters were often planned in one form and later emerged in another, while many fighters were actually designed as bombers (for example, the Ju 88). Nevertheless, running through most major programmes there was a coherent thread, which may have been occasionally disrupted by technical snags or changes in official requirements but usually resulted in the development taking place in sensible steps. Nowhere in aviation history is this more evident than in the case of the Spitfire, which began with a simple airframe with 242 sq ft of wing and a gross weight of 5800 lb, designed to a load factor of 10, and finished the war with a markedly different airframe with 244 sq ft and a gross weight of 12,750 lb, designed to a load factor of 11. Such a change transformed the character of the aircraft. Unlike most combat aircraft, which over the years progressively became slower and more sluggish as weight increased, the 120% increase in weight of the Spitfire/Seafire was accompanied by dramatic increase in performance, solely because

of Rolls-Royce's impressive progression from the 990-hp Merlin I to the 2375-hp Griffon 85.

Few fighters could enjoy a comparable increase in available power. The power of some aircraft actually decreased, while most increased by about ten per cent. Britain farsightedly planned in 1937 for fighters with engines of 2000 hp, but unfortunately picked the wrong ones. While the Bristol Centaurus radial was ignored, concentration on the Rolls-Royce Vulture (a failure) and Napier Sabre (a success after five costly years of effort) killed the Hawker Tornado and delayed the Hawker Typhoon and Tempest, delaying their operational serviceability until near the very end of the war. In the United States in 1935–40, the emphasis was on the powerful aircooled radial, but the fact that virtually every fighter in the Battle of Britain had a liquid-cooled engine prompted the US Army to launch big liquid-cooled engine programmes at Pratt & Whitney, Curtiss, Lycoming, Chrysler and Continental, backing up the existing work by Allison. Many fighters flew with these new engines, but all eventually fell by the wayside. The final winners for US fighter propulsion, apart from the

Packard-built Merlin, were the same big radials that were used at the start, by far the most important of which was the R-2800 Double Wasp. This drove a Corsair at over 400 mph in 1940, no mean achievement for so large a warplane, and by 1945 was driving much heavier versions of the Corsair, Thunderbolt and other American fighters at speeds exceeding 450 mph.

There was no substitute for sheer horsepower. Frequently, in comparisons between 'streamlined' liquid-cooled engines and 'unstreamlined' radials, the powers are dissimilar. For example, the Italian fighters of 1940–41 had radials of around 840-hp, and it is hardly surprising that their performance was improved by fitting German liquid-cooled engines of 1475 hp. A closer comparison was seen in the Curtiss P-36A, of 1150 hp and 327 mph speed, which when re-engined with a 1160-hp Allison, reached 337 mph. Much later a P-47H Thunderbolt reached 490 mph with the Chrysler XIV-2220, one of the experimental liquid-cooled units, but another reached 504 mph with the bluff-fronted Double Wasp. Perhaps the supreme expression of the piston-engined fighter can be seen in two late-wartime models that just failed to see combat duty. Britain's

Supermarine Spiteful XIV was a beautiful little fighter with four 20-mm guns, a new laminar-section wing of only 210 sq ft, and a high-blown 2375-hp Griffon engine, giving a speed of 483 mph. America's Republic XP-72 was equally fine in its lines, armed with various guns including six 0.5-in or four big 37-mm cannon, plus two 1000-lb bombs, and fitted with the same 300-sq-ft wing as the P-47; this had a 3450-hp R-4360-13 Wasp Major and reached 490 mph. They represented different philosophies, but both were supremely effective.

It is popularly supposed that the capture of a Fw 190 in 1942 led to the sudden realisation in Britain that good fighters might have radial engines. There is some truth in this, but the fixation with the liquid-cooled engine was of quite brief duration. Most of the RAF's fighters between the wars – Grebe, Gamecock, Siskin, Bulldog, Gauntlet and Gladiator – had radial engines, and only the Fury had the in-line. The supposed superiority of the in-line may have been reinforced by the lack of top-quality radial-engined fighters in Western Europe in 1938–41, until the emergence into combat of the Focke-Wulf 190. No such attitude afflicted the Soviet Union, where there were good engines of both types. Although the Yak fighters continued to use the VK-105 and -107-type in-line engines, the other mass-produced fighter family, by Lavochkin, did not hit its stride until it switched to a radial.

*RATO (Rocket Assisted Take-Off) installation on a Seafire giving a burst of power for take-offs from escort carriers*

John Batchelor

### German Aero Engines

By the outbreak of war the rearming German aircraft industry had a range of powerful in-line and radial engines available from a variety of manufacturers. Liquid-cooled in-lines were favoured for fighters (with the very notable exception of the Fw 190) and radials for transport and bomber aircraft but the precision products of the German aero-engine industry could not compete in volume output with the industrial might of the United States

**Daimler-Benz DB 601**

**BMW 801**

*The evolution of the fighter-bomber was one of the most important developmental themes in Second World War aircraft design. Many types were adapted – some, like the Fw 190G and Hurricane II, with outstanding success. Equally successful was the massive P-47 Thunderbolt, seen here being refuelled and armed with .5-in machine-gun belts and 250-lb bombs on a forward airstrip in France, 1944*

# Fighter Armament

Except for the Lewis, virtually all the guns used in air combat in the First World War had been originally designed for use by armies or navies. But rifle-calibre weapons are so numerous, and often so closely inter-related, that it is difficult to be dogmatic about which were new designs, or planned from the start for air use. Certainly in the 1920s and 1930s the Vickers, used in fighters all over the world, would have been immediately recognized by Hiram Maxim in the 19th century, and it retained its popularity by refinement and reliability. Few of the other rifle-calibre guns of the period were dramatically different, though some were gas-operated. The American fighters backed up the Lewis with the belt-fed Colt-Browning, an outstanding weapon fitted to a few fighters in 1917, and a variant of it called the Marlin. The Browning was developed in the 1920s with a much harder-hitting calibre of 0.5-in, and in Italy, and several other countries, the same 12.7-mm calibre was adopted as being well worth the extra size and weight; paradoxically, Germany, which had been the pioneer of the 12.7-mm machine gun for tanks and aircraft, stayed with rifle calibre until 1938.

Germany had also done more than any other country to develop fast-firing cannon, fed with explosive or other special types of ammunition. The best cannon actually used in numbers in Second World War air combat was the German 20-mm Becker, which was obviously more lethal than the cumbersome and slow-firing Hotchkiss used by the French. But in 1919 Becker sold his patents to Oerlikon Machine Tool Works in Switzer-land, and when the reborn Luftwaffe wanted a cannon in the mid-1930s Rheinmetall had to buy the old German patent back. As the MG FF the Oerlikon was mass-produced in Germany from 1938, at first for the Bf 109 and subsequently for many fighters, bombers and other aircraft. It was a fairly short and light weapon, its main drawback being that it was fed by a 60-round drum. But so was the only other fully developed, and related, aircraft cannon of the pre-1939 era, the 20-mm Hispano. The growing Hispano-Suiza company had developed some of the best and most powerful aircraft engines of the First World War, and their chief designer M Marc Birkigt had spent much time trying to perfect a *moteur canon* with the gun lying between the cylinder blocks and firing through the hub of the propeller. Of course, this could be done only with a geared engine, and the geared Hispano gave a lot of trouble (as related earlier in describing the S.E.5 fighting scout, and the Caudron R.11). The first *moteur canon* in combat was the old 37-mm Hotchkiss, and though Spad XII fighters fitted with this were often deadly (Guynemer achieved four victories in one, and Fonck 11 in another) the heavy slow-firing cannon caused severe recoil problems and choked the pilot with cordite fumes. Birkigt master-minded the design of an extremely efficient new cannon of only 20-mm calibre but firing automatically, and at 550–600 rounds per minute. Moreover, the muzzle velocity was high, so that the trajectory was flat, accuracy high

and hitting power (in terms of kinetic energy of a solid shot) actually greater than for the old Hotchkiss. The new 20-mm cannon was mounted in the traditional propeller-hub position in the Dewoitine D.501 fighters of the early 1930s, and Dewoitine boldly fitted two in the wings of many D.37-series (373) fighters of the same period. The 20-mm *moteur canon* was fitted to the main French fighters of 1940, the M.S.406 and Dewoitine D.520, and two wing-mounted examples were fitted to the Bloch 152. By 1940 there were also manually aimed Hispano instal-lations and the first power turrets in French bombers. After years of delay, the same gun was also adopted by the ailing RAF, to make up for lack of a British cannon, and first used in the twin-engined Whirlwind, designed to carry four in the nose. Later it was the standard British fighter gun in the Second World War.

Curiously, no fighter in operational service had a gun firing through the propeller shaft of a radial engine, though such an arrangement was tried experiment-ally. With a geared engine having the propeller shaft offset from the crankshaft, a *moteur canon* was not difficult to install, and with a liquid-cooled in-line engine the gun was sometimes mounted on the engine itself. Designers sometimes adopted un-conventional aircraft layouts to facilitate the provision of powerful nose armament, examples being the American P-38 and P-39 of 1939. In fact there was no need for unusual layouts, and there were other reasons, in addition to armament, for the unconventionality of the P-38 and -39.

Nevertheless, the variety of armament in the Second World War was exceptional. Some nations adopted more or less standard armament and stuck to it. Britain began with eight rifle-calibre machine guns and quite soon switched to four licence-built 20-mm Hispanos, while the Soviet Union seldom departed far from one 20-mm and one or two 12.7-mm (in Yaks) or two 20-mm (later three) in Lavochkin. In Italy the standard armament of two 12.7-mm machine guns was gradually recognised as hope-lessly inadequate and supplemented by three of the excellent German MG 151 cannon. But in Germany and Japan such rigid standardisation did not apply. Though a few aircraft, including the most numerous Japanese Army and Navy fighters, tended to have uniform armament, a great variety of guns and other armaments were available and were fitted in every conceivable com-bination.

Originally Japanese guns were based on Western patterns such as the Vickers 0.303-in and Browning 0.5-in, but by 1940 guns of Japanese design were coming into use. One of the first of these was the Army Ho-5 cannon, virtually a Browning scaled up to 20-mm calibre and very soon giving good service. The standard Navy gun of this calibre was the Type 99; this was a large family with many lengths, weights, muzzle velocities and feed systems. Develop-ment in Japan had begun in 1930, making use of the original Becker and Oerlikon drawings, but it soon became distinctive and was used in both fixed installations and in many hand-aimed or turreted forms. By 1944 production was beginning on a 30-mm version of the same family, the Type 5, with a muzzle velocity of 2460 ft/sec, high for a weapon of so large a calibre. The Army 30-mm gun was lighter, the Ho-105 having lower velocity. Quite large numbers of

*The Vickers 'K' gun superseded the Lewis as a flexible observers gun on British aircraft*

37-mm Ho-203 were used, these being short-barrel, slow-firing, low-velocity guns of ancient origin. The 40-mm Ho-301 was new, but rather cumbersome and limited in range by the low velocity of its rocket-propelled shells. By far the hardest-hitting Japanese gun was the 57-mm Ho-401, with rate of fire of only 90 shots/min but high velocity and with extremely destructive projectiles. The 401 was used on several heavy fighters in 1944–45.

Britain was the only major Second World War power to have virtually no home-developed aircraft guns, apparently because in the vital 1930s nobody thought about producing any. The standard guns were the American Browning of 1916, altered to fire rimmed 0.303-in ammunition, and the hard-hitting Hispano of 1917, which by 1940 was in licence-production by British MARC Ltd and was later developed to have belt-feed, a shorter barrel and other changes. Almost all US fighters were armed with the 0.5-in Browning or M-2 cannon (Hispano-derived), though several almost identical patterns of 37-mm cannon were also used in the P-38, P-39 and some other aircraft.

Soviet guns were almost entirely home-developed from 1931, and their quality and performance were outstanding. In fitness for hard service, ease of mass-production, reliability and firepower they had few equals. The rifle-calibre gun was the ShKAS of 7.62-mm, with rate of fire up to 1800 rds/min and weight appreciably less than the rifle-calibre Browning. The heavy machine gun was the Beresin BS, firing 12.7-mm ammunition similar to the Browning but at 900 rds/min instead of 650–850, and with a gun weight of 55 lb (fighter installation) instead of 64 lb. The 20-mm gun was the ShVAK, much shorter and lighter than a Hispano but with almost as much hitting power and a rate of fire of 800/min instead of 600. The remarkable 23-mm V-Ya had the extremely high muzzle velocity of 3020 ft/sec, coupled with 600/min rate of fire, but this tank-killing gun was seldom fitted to fighters.

German guns were by far the most adventurous. The Rheinmetall MG 17 was the standard machine gun for fighters in 1939, and despite the development from the MG 34 infantry weapon of the simple and light MG 81, the old MG 17 remained the chief fixed machine gun to the end, though it was almost entirely replaced by heavier weapons. The standard heavy machine gun was the excellent Rheinmetall MG 131 of 13-mm calibre, first used in 1938 and soon the most common synchronised fuselage gun on the Bf 109 and Fw 190. Weighing only 38 lb, it was extremely reliable, firing high-velocity ammunition at 900–960/min with detonation accomplished electrically instead of by a mechanical pin fired into the percussion cap; this greatly facilitated synchronisation.

Electrical firing was selected in 1935 by the Mauser company for a completely new gun, the MG 151, firing 15-mm ammunition at the extremely high velocity of 3150 ft/sec. Though a superb gun, it was used in only a few aircraft, being replaced by the most important gun of the Luftwaffe, the MG 151/20. Identical except for calibre, this fired 190 lb/minute at 750 rds/min, compared with only 90 lb/min at a slightly lower rate of fire, and its muzzle velocity of 2600 ft/min gave it excellent ballistics and hitting power. The MG 151 was used in about 80% of all German fighters built after 1941, and some Fw 190s, Ju 88s and Me 410s had as many as six or eight, all firing ahead. It completely replaced the old Ikaria MG FF, of original Oerlikon design, which fired different low-velocity ammunition at a lower rate from a 60-round drum.

By 1939 development was well advanced on a large 30-mm cannon derived from the Solothurn 20-mm S-18 anti-tank gun. It emerged in 1941 as the Rheinmetall MK 101, but with a length of almost 10ft and bare weight of 287–396 lb it was too cumbersome for any except anti-tank aircraft. Rheinmetall's MK 108 was a complete contrast. This was the stubbiest little 30-mm gun imaginable, with length of 45 in, weight of 133 lb and rate of fire of 450/min of low-velocity (1640 ft/sec) ammunition by simple

*The quest for a heavy concentrated nose armament brought many design solutions, but the twin-engined single-seat Westland Whirlwind with four 20-mm Hispano cannon in the nose had only a modestly successful career*

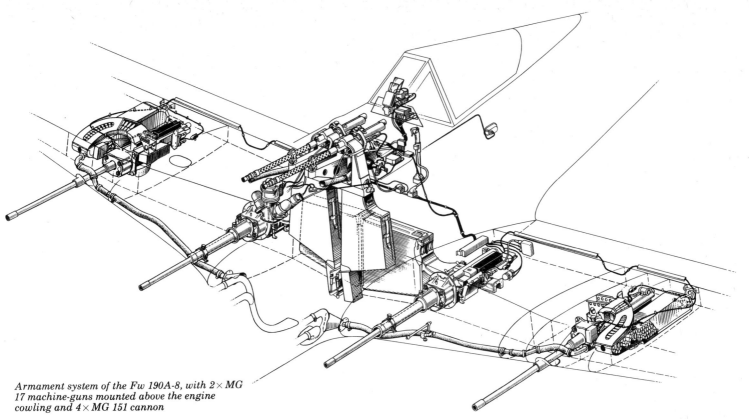

*Armament system of the Fw 190A-8, with 2 × MG 17 machine-guns mounted above the engine cowling and 4 × MG 151 cannon*

blow-back action. Easily made from sheet stampings, it was mass-produced for the Bf 109G, Fw 190, Ta 152, Me 163, Me 262 and many other fighters, making up for its low velocity (which limited effective range except with a very experienced pilot) by the devastating explosive power of its HE ammunition. In 1944 a third, and again totally different, 30-mm cannon completed development and came into service on selected fighters. The Rheinmetall MK 103 was a large (91-in, 320-lb) high-velocity gun with gas operation and rate of fire of some 420 rds/min. Among the aircraft fitted with this tremendous weapon were the Ta 152C-3, the Bf 109K (selected models), most types of single-seat Do 335 and several sub-types of He 219.

Mauser's brilliant MG 213 revolver cannon, with five-chambered feed drum allowing rates of fire up to 1200 rds/min, did not see action in the Second World War but served as the basis for almost all the aircraft cannon of the immediate post-war era such as the British Aden, French DEFA, American M-39 and certain Soviet cannon. Still further off were a great array of recoilless cannon, rocket launchers and other unconventional weapons. The SG 116, for example, was a gun which fired inert masses to the rear to counteract the kinetic energy of the projectile, as in the British Davis gun of 1915. The more promising MK 115 was a 55-mm cannon firing large rounds with thick paper cases which burned through, allowing gas to escape from recoil-neutralising ejector nozzles at the rear; rate of fire was 300 rds/min.

One weapon which was received into large-scale service was the R4/M spin-stabilised air-to-air rocket, with calibre of 55-mm (about 2.2-in), fired from underwing racks on several kinds of fighter including the Me 163B and 262A (the Ba 349 *Natter* would have had a battery of 24 in the nose). This folding-fin projectile spurred development of the post-war American Mighty Mouse of 2.75-in calibre.

A weapon which just failed to complete its development was the Ruhrstahl X-4 air-to-air guided missile, with spools of wire to transmit command signals. With this, as in so many other aspects of air combat, the Germans were years ahead of their enemies. Had their basic planning been better, the outcome of the war might have been very different.

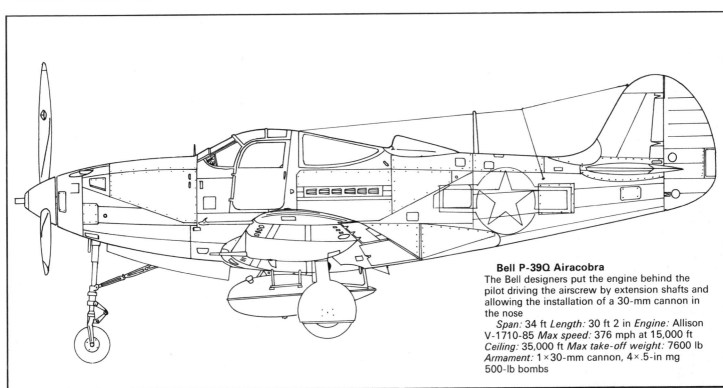

**Bell P-39Q Airacobra**
The Bell designers put the engine behind the pilot driving the airscrew by extension shafts and allowing the installation of a 30-mm cannon in the nose
*Span:* 34 ft *Length:* 30 ft 2 in *Engine:* Allison V-1710-85 *Max speed:* 376 mph at 15,000 ft *Ceiling:* 35,000 ft *Max take-off weight:* 7600 lb *Armament:* 1 × 30-mm cannon, 4 × .5-in mg 500-lb bombs

# JET FIGHTERS
## *David Anderton*

This chapter covers the development of the jet fighter from the early days of Heinkel and Whittle, through the Second World War and all the wars since then, up to the present day and beyond. The author, a technical consultant who specialises in forecasting trends in technical developments, includes his ideas on the future of jet fighters and rounds off the story with fighter armament.

The dazzling full colour illustrations by John Batchelor combine with a lavish use of photographs to show virtually every important jet fighter of the past 35 years, right up to the YF-16 and its contemporaries that will be the fighters of the future.

# CONTENTS

*RAF McDonnell Phantom intercepts a Russian Bear*

This narrative follows the common threads of a single subject: the jet fighter. Here a fighter is defined as an aircraft designed to destroy other aircraft as a primary mission. This rules out those planes which were designed primarily for strike or ground attack, even though they may have remaining capability as fighters to claw their way back from the target area.

The chapters that deal with the development of the jet fighter are divided to follow that development in one country at a time. Further, the arrangement is chronological, with the key time being the date of the first flight of either a fighter prototype or of the research aircraft that immediately preceded development of a jet fighter:

Two types of basic jet propulsion powerplants are considered here. One is the rocket engine, a self-contained motor burning a liquid fuel and oxidiser, and generating its thrust by combustion of the two in a suitably shaped chamber and exhaust nozzle. The other is the aircraft gas turbine for jet propulsion, an engine which takes in outside air, mixes it with fuel, burns the mixture, and exhausts the hot gases at high speed through a suitable nozzle.

Neither engine type was new when it was first applied to aircraft. The rocket engine, albeit in the form of a solid-propellant system, had been known for years, and was familiar to millions in the form of fireworks. They had been used in warfare and, in the daring years of flight during the 1920s, they had been adapted to launch a glider into the air in the first recorded flight of any aircraft anywhere using jet propulsion of any type.

But that experiment, demonstrated by the Opel-Sander Rakete 1 on 30 September 1929 at Frankfurt, led to a dead end. Neither the state of the solid-propellant rocket art nor that of aeroplane design was ready then for the use of such a novel form of propulsion.

The gas turbine is also an old form of powerplant. The first patent for such an engine

# "THIS IS HEINKEL. WE'VE JUST FLOWN THE WORLD'S FIRST JET FIGHTER!"

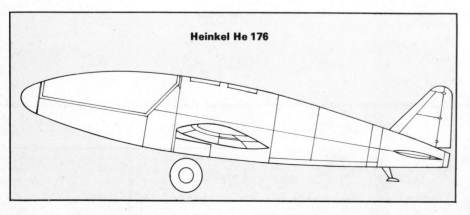

Heinkel He 176

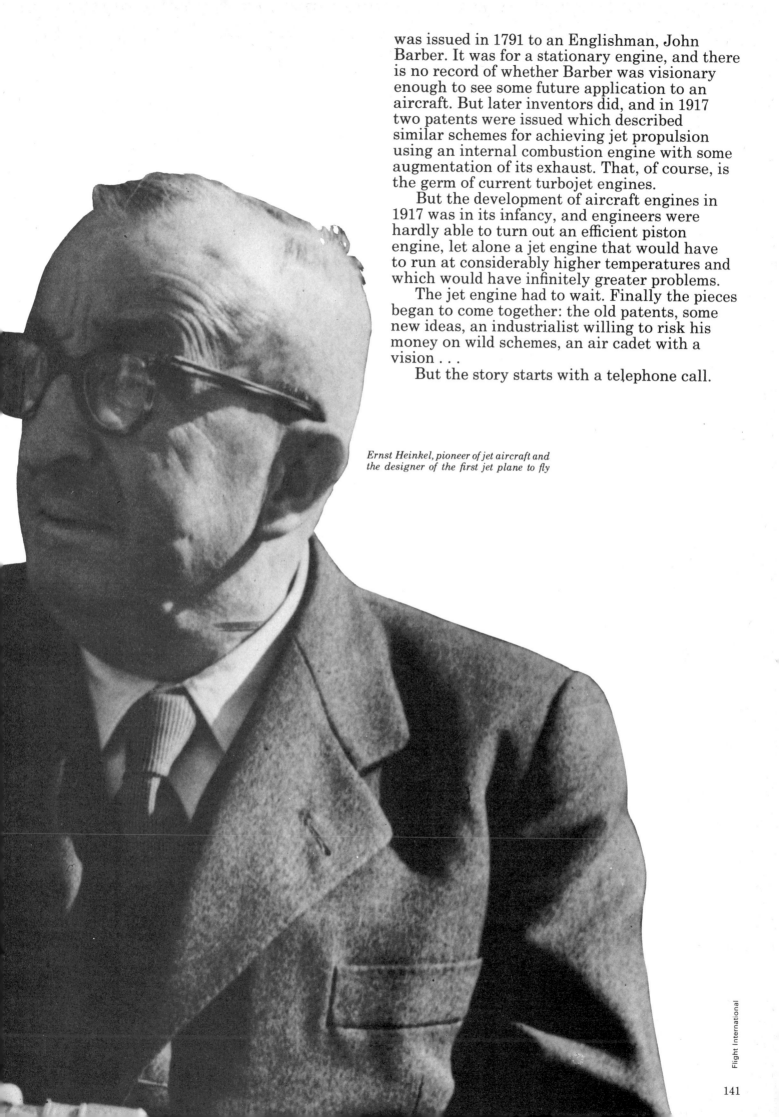

was issued in 1791 to an Englishman, John Barber. It was for a stationary engine, and there is no record of whether Barber was visionary enough to see some future application to an aircraft. But later inventors did, and in 1917 two patents were issued which described similar schemes for achieving jet propulsion using an internal combustion engine with some augmentation of its exhaust. That, of course, is the germ of current turbojet engines.

But the development of aircraft engines in 1917 was in its infancy, and engineers were hardly able to turn out an efficient piston engine, let alone a jet engine that would have to run at considerably higher temperatures and which would have infinitely greater problems.

The jet engine had to wait. Finally the pieces began to come together: the old patents, some new ideas, an industrialist willing to risk his money on wild schemes, an air cadet with a vision . . .

But the story starts with a telephone call.

*Ernst Heinkel, pioneer of jet aircraft and the designer of the first jet plane to fly*

Flight International

At half-past four on the morning of 27 August 1939, the telephone's insistent ringing woke Ernst Udet, the top-ranking living ace of the former German Imperial Air Service, and head of the new Luftwaffe's Technical Department.

Udet growled into the phone, still sleepy. The voice from the distance was jubilant. 'This is Heinkel. We've just flown the world's first jet airplane!'

'Fine,' said Udet, grumbling. 'Now let me get back to sleep.'

Germany was only a few days from its invasion of Poland; and Udet could reasonably be expected to be on edge, wondering if each phone call would bring news of the march to the East. But the news that he

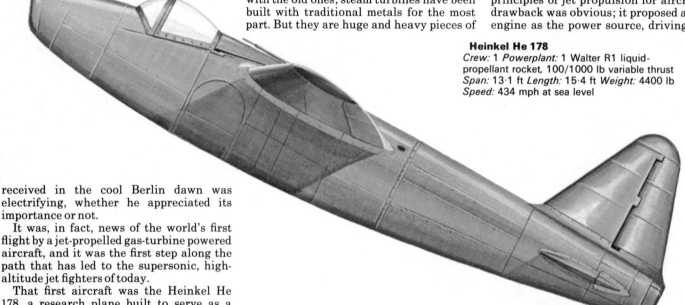

**Heinkel He 178**
*Crew:* 1 *Powerplant:* 1 Walter R1 liquid-propellant rocket, 100/1000 lb variable thrust *Span:* 13·1 ft *Length:* 15·4 ft *Weight:* 4400 lb *Speed:* 434 mph at sea level

received in the cool Berlin dawn was electrifying, whether he appreciated its importance or not.

It was, in fact, news of the world's first flight by a jet-propelled gas-turbine powered aircraft, and it was the first step along the path that has led to the supersonic, high-altitude jet fighters of today.

That first aircraft was the Heinkel He 178, a research plane built to serve as a flying test-bed for the new form of power-plant. It was tiny, spanning less than 24 ft and weighing less than two tons, fully loaded with fuel. Its turbojet engine produced about 1000 lb of thrust. Its performance was modest; it probably never flew faster than 350 mph, and never got very far out of sight of the aerodrome at Marienehe. It was overtaken by events. The German Air Ministry ignored it, and eventually it was dropped from active development only a few months after its historic first flight.

In a way, the German Air Ministry couldn't be blamed then for its lack of interest. There were other tasks of a higher priority. Hitler was about to invade Poland, and the air support needed for the job was marginal. If the Polish resistance lasted longer than planned, if the offensive strikes did not knock out enough of the Polish air force, if another country should side with the Poles, there was serious doubt about the ability of the Luftwaffe to maintain its part of the push eastward.

Further, the concept of war held by Hitler and his staff was built around the Blitzkrieg – lightning war – which would move so fast and so powerfully that it would steam-roller the opposition before any countering forces had time to realise what was happening. There was no time for development that might take years to come to fruition. New weapons were needed, yes; but they were needed tomorrow. The jet fighter was an interesting new idea, but of no immediate practical value.

It is not exactly correct to speak of jet propulsion as a new idea. The idea of

propulsion by reaction is as old as propulsion itself. The classical example is the aeolipile, which is supposed to have been demonstrated by the Alexandrian philosopher Hero before the time of Christ. His little sphere, whirled on its bearings by two steam jets driven by the boiling of water inside the sphere, was what we would now call an interesting laboratory experiment. It served to illustrate the principle, but did nothing else.

Jet propulsion was an idea that had to wait for its time to come, primarily because it depended on the generation of high temperatures within an engine. High temperatures and the attendant high pressures required containment in shells that were strong when red-hot. And that required new lightweight metals. True, it could be done with the old ones; steam turbines have been built with traditional metals for the most part. But they are huge and heavy pieces of stationary machinery, hardly suitable for powering aircraft in flight.

And that was the toughest requirement of all. Whatever form of engine was to be built, if it were to fly it had to be light as well as strong. There was no way the jet engine could have been built and flown by the Wright brothers. The state of the art in 1903 simply would not have permitted it.

Jet propulsion is a very general term, because everything that flies under power is propelled by a jet. A propeller, a helicopter rotor, a turbojet or a rocket engine all move an aircraft by jet propulsion. Each of those engines generates a mass of air moving at a higher speed than its surroundings. The difference in momentum is applied as thrust to the airframe, and the aircraft moves through the air, propelled by a jet of faster air.

### Toughest problems

The gas turbine for jet propulsion presents some of the toughest design problems ever faced by engineers. The basic problem is to increase the momentum – the mass times the velocity – of the air going through the engine. That is done by burning fuel and adding its energy to that of the incoming air, which is not too difficult to accomplish, in engineering terms. But it is difficult to accomplish with economy of fuel, with safety of operation, and with long life of the moving and stationary parts of the engine.

Building jet engines and getting them to

work was, in retrospect, the simple part. The hard part was getting them to stand up to high performance for hour after hour, while a white-hot exhaust roared out of the tailpipe and the engine glowed from orange to dull red.

The story of today's jet fighters starts, truly, with a young British air cadet, Frank Whittle, studying at the RAF College at Cranwell preparatory to getting his commission. In his fourth term, Whittle wrote a paper, 'Future Developments in Aircraft Design', pointing out some of the possibilities of rockets and of gas turbines driving propellers, but not of the gas turbine producing only a jet of hot air for propulsion.

That came later, about 18 months into his RAF career. Whittle's research had found a basic patent, dated 1917, that covered the principles of jet propulsion for aircraft. Its drawback was obvious; it proposed a piston engine as the power source, driving a fan, or a shrouded propeller within the fuselage, and with the addition of afterburning for additional thrust from the engine.

The principle was sound, but its failings were obvious to Whittle. He hit on the idea of using a gas turbine instead of the piston engine, proposed it to the Air Ministry, and was politely turned down. Materials just didn't exist that could do the job, said the Ministry rejection. Whittle persisted, and filed a patent application for an aircraft gas turbine on 16 January 1930. Because the Air Ministry was not officially interested, the patent was openly published after it was granted about 18 months later. The secret was out.

Another jet pioneer was German aircraft designer Ernst Heinkel, a brilliant innovator. Heinkel was always looking for ways to improve performance, to go faster and higher.

He had become interested in jet propulsion while doing research for an article he was writing in 1935, called 'An Inquiry into Engine Development'. It was a look at future trends desirable for aircraft power-plants. Heinkel saw 500 mph as a practical limit to the speed of propeller-driven aircraft. To get over that hurdle, he reasoned, it would be necessary to have some new kind of propulsion, and he thought in terms of jet engines, primarily those driven by gas turbines.

Late in 1935, Heinkel met Wernher von Braun, then a young engineer testing rocket engines of his own design and longing for an airplane with which to do flight tests. Heinkel loaned von Braun a fuselage from an He 112 fighter, and eventually supported experiments in flight with a modified He 112 driven by an auxiliary rocket engine.

In early 1936, Heinkel's interest was further sparked by a letter from a colleague, Professor Pohl, head of the Science Institute at the University of Göttingen. Pohl's

assistant was a 24-year old scientist named Pabst von Ohain, who had been working on a new kind of aeroplane engine, said Pohl, which did not need a propeller. Pohl believed in the young man and his ideas, and urged Heinkel to investigate.

Heinkel met von Ohain, hired him and his assistant Hahn, and set the team to work in a special building across the aerodrome at Marienehe, a former Mecklenburg state park which Heinkel had bought as the site of a new factory for Luftwaffe production.

Marienehe is rolling country, lying along the Warnow river in the north of Mecklenburg province, now part of the German Democratic Republic. There were farms and estates in the vicinity, and it was a quiet backwater of rural Germany in the years before the war.

But in September 1937 the stillness of the night was disturbed by a low humming that built quickly to a scream and then to a sudden roar. A tongue of flame shot out of the building at the Heinkel field, hot and red, jutting toward the river, scattering the first leaves of fall, lighting the area around the hangar. Von Ohain's strange engine had just been fired for the first time. Hahn telephoned Heinkel with the news, and within a few minutes Heinkel himself was at the building to see and hear the birth of the jet age.

By the next spring the engine was advanced enough to produce repeatable performance. Its fuel was gasoline, instead of the hydrogen used during development work. It produced about 1100 lb of thrust, and it seemed to Heinkel to be ready to power an aircraft.

The design of the aircraft began, with engineers, draftsmen and technicians sworn to secrecy. With that project in capable hands, Heinkel turned back to rocket aircraft, and to experiments aimed at getting one into the air. The first flight of the modified He 112 lent to von Braun's team had been made in April 1937, with the rocket engine operated during flight only. On later tests, pilot Erich Warsitz made takeoffs on the combined power of the He 112's Junkers Jumo piston engine and the von Braun rocket engine in the tail.

Finally, in the summer of 1937, Warsitz made the takeoff on rocket power alone, climbed to altitude, circled the field and landed, using only the rocket engine for propulsion during the entire flight. The point had been proven; Heinkel now was keen to build a special aeroplane for rocket propulsion instead of attempting to modify the He 112 or any other aircraft.

He and Warsitz agreed that the design should aim for the round number of 1000 kilometres per hour speed (621 mph). The airplane was designed around Warsitz. It had a tiny wing span, only 13·1 ft, and the top of the fuselage came up to Warsitz' waist. The nose was detachable in the event

Caproni-Campini N 1
*Crew:* 2 *Powerplant:* 900 hp Isotta-Fraschini piston engine driving 3-stage ducted fan with afterburner *Span:* 50·3 ft *Length:* 42·9 ft *Weight:* 9229 lb *Speed:* 233 mph at 9800 ft

of an accident, and used a drogue parachute to cut its speed to a point where Warsitz would have been able to bail out and use his own parachute.

Design and construction of the He 176, as the rocket research aircraft was designated, took about one year. It was trucked to the experimental airfield at Peenemünde, later to become world-famous for its development of the A-4 rocket weapon under von Braun. But it was to take another year of slow development before the He 176 was ready for flight. Meanwhile, Warsitz made taxi runs, extending the speed range and distance covered. The Peenemünde runway was lengthened by nearly one mile to accommodate the tests, and Warsitz occasionally took advantage of the long runway to lift the little plane into the air for a few seconds.

## Official interference

Meanwhile the Air Ministry did develop some interest in the He 176. They saw it as a potential rocket-powered interceptor, heavily armed and able to slash through bomber formations with great effectiveness. Their insistence on that role for the He 176 caused Heinkel to install small blisters on the fuselage, alleging they were the provisions for armament, but actually filling them with test instrumentation.

Finally Warsitz and the He 176 were ready. On 30 June 1939 the tiny plane blasted off the runway on its first full flight. It lasted less than one minute, but Warsitz and the rest of the Heinkel team were

jubilant. Rocket-powered aircraft had flown before, for short distances and times; but they had been modified aircraft, flying test-beds for the rocket and only a means for getting in-flight data on the powerplant. But none had been considered as a candidate for development into a fighter, although that thought was not exactly uppermost in Heinkel's mind.

On 1 July, Heinkel himself saw the He 176 fly. So did Udet, Milch and others in the Air Ministry. After Warsitz landed, Udet shook his hand and then forbade any further flying with the He 176. 'That's no aeroplane' was his verdict. Warsitz and Heinkel argued for continuance of the tests, and eventually they were successful. Then they were refused permission again. That was rescinded, and permission reinstated, because there was to be a demonstration of new aircraft for Hitler, and the Air Ministry wanted to include the He 176.

Hitler saw it fly, passed a few compliments around, and left the field. Warsitz later had a personal talk with Hitler, during which the subject of the He 176 never came up. Obviously there was to be no official support for the project.

The He 176 and He 178 had been hangar mates, both having been built in a special hangar erected for the purpose away from the rest of the Marienehe plant. Now work concentrated on the He 178, aiming for an early flight.

On the morning of 27 August 1939, Warsitz climbed into the He 178 and started the engine. It whined up to speed; he taxied out, roared down the runway and lifted off. The undercarriage could not be retracted and Warsitz, after trying every trick in the trade, finally resigned himself to circling the field about 1500 ft altitude with the gear hanging. After about six minutes in the air, he swung into the approach pattern, sideslipped on final, and touched lightly down on the grass, rumbling across the field to a stop.

For Heinkel and the rest of his design team, it was the justification for the hours of work on two pioneering aircraft, both jet-propelled, one by a rocket and one by a gas turbine. Under the circumstances, the Heinkel organisation would have had a tremendous technological lead on the rest of industry. But the events of September 1939 caught up with them. After only a few flights by both airplanes, and despite long hours of arguing, cajoling and pleading by Heinkel, both aircraft projects were stopped

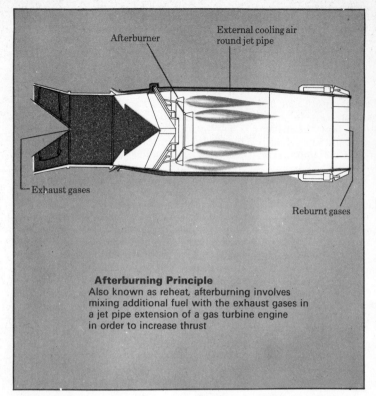

**Afterburning Principle**
Also known as reheat, afterburning involves mixing additional fuel with the exhaust gases in a jet pipe extension of a gas turbine engine in order to increase thrust

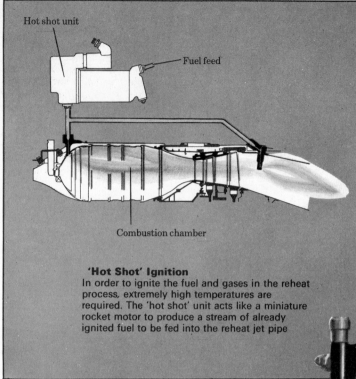

**'Hot Shot' Ignition**
In order to ignite the fuel and gases in the reheat process, extremely high temperatures are required. The 'hot shot' unit acts like a miniature rocket motor to produce a stream of already ignited fuel to be fed into the reheat jet pipe

dead. The word was production, and there was no time for research and development. Besides, they would not be needed for the short war that would be over before winter.

Sadly Heinkel saw both the He 176 and He 178 crated and sent off for display in the Berlin Air Museum. Their fate was predictable from the day they left Marienehe. In 1943, one of the many bombing raids that were to destroy Berlin pounded the Air Museum and its priceless collection of aeronautical history into rubble.

Soon after the pioneering flights of the two Heinkel jet-propelled aircraft in 1939, an Italian research aircraft flew from Milan to Rome under jet power. It was the first cross-country flight of any length by a jet-propelled aircraft, and even though the run included a stop for fuel, the flight was epochal.

The airplane was the Caproni-Campini N 1, a low-winged monoplane with a cylindrical fuselage. A circular inlet at the nose and an exhaust nozzle at the tail gave the only clues to the internal arrangement, which was unusual. In some ways, it predated the most advanced jet engines operational today; but it was then only a makeshift approach to jet propulsion.

Inside the N 1, an Isotta-Fraschini piston engine, which developed about 900 hp, drove a three-stage variable-pitch fan in an application of the principle of the ducted fan, today the most economical and advanced type of turbojet development. Downstream of the ducted fan discharge was a ring burner fed with fuel to augment the thrust of the engine-driven fan; it was, in essence, an afterburner.

Designer Secondo Campini had been working on the idea of a jet-propelled aeroplane for eight years, and had hit upon this particular scheme, perhaps as a result of finding earlier research and patents along this line.

The N 1 made its first flight on 27 August 1940 from Forlanini aerodrome near Milan, and was airborne for about ten minutes. The cross-country flight to Rome was basically a ferry flight to get the aircraft to the Italian aeronautical test establishment at Guidonia. With maximum publicity planned, the N 1 took off from Forlanini and headed south. A fuelling stop was made at Pisa, and the final landing at Rome.

It was found to be a not very efficient method of achieving jet propulsion. The programme was abandoned about two years after the first flight, although Campini continued to press for adoption of his basic ideas as a possible auxiliary powerplant for fighter aircraft.

*Frank Whittle, British pioneer of jet aircraft who patented his first jet engine in 1930*

Flight International

# THE WHITTLE ENGINE

*Whittle W 1 Turbojet:* 2-sided centrifugal compressor, 10 reverse-flow interconnected combustion chambers *Fuel:* Paraffin with atomised burners *Specific fuel consumption:* 1·4 lb/lb thrust/hour *Performance:* 850 lb static thrust at 16,500 rpm *Thrust/Weight ratio:* 1:0·66

# TOO FAST TO FIGHT

The early experiments with jet engines and aircraft led to both Britain and Germany having jet fighters operational during the Second World War. Me 262s terrorised British bombers, while Gloster Meteors brought down several V-1 flying bombs. But against conventional aircraft the jet planes' superior speed could also be a handicap, hindering accurate fire

The real impetus to the development of jet aircraft was the start of the Second World War in September 1939. As country after country realised the ugly truth, industry, science and engineering teams girded for their parts in the struggle to come.

Germany had a head start, thanks to the pioneering work by Ernst Heinkel. So it should not be surprising that the world's first turbojet aircraft to be designed as a fighter from the beginning was yet another Heinkel effort, the He 280. It was also the world's first twin-jet aircraft.

Heinkel visualised a twin-engined jet fighter, with engines slung in individual nacelles under the wing, minimising the length of the intake and exhaust ducting. For ground clearance, and to avoid blasting loose huge chunks of the runway surface, he decided on tricycle landing gear, the first on a German aircraft.

Design of the He 280 started in late 1939. In March 1940, the Air Ministry awarded a contract to the Messerschmitt organisation for prototypes of a twin-engined aircraft. A few days later, Heinkel got essentially the same sort of contract.

The first prototype He 280 was completed by September, lacking only airworthy engines. Heinkel had the prototype flown first as a glider, testing basic aerodynamic characteristics. When the Heinkel HeS 8A engines were ready for flight the following spring, the airplane was already well understood.

On 2 April 1941, at Marienehe airfield, Fritz Schaefer climbed into the cockpit of the He 280. He taxied out and took off, climbing to 900 ft or so for a circle of the field. He did not attempt to retract the landing gear, or to do anything exceptional with the aircraft.

Three days later it was flown again, and demonstrated to Udet and others from the Air Ministry. Their indifference was annoying to Heinkel, who could not understand why his advanced ideas were continually rejected.

He thought the He 280 had proved its point and that it should be considered for production. So he arranged a series of tests against the Luftwaffe's top fighter of the time, the Focke-Wulf 190. It was no contest; the jet-propelled fighter outperformed the Fw 190 in every way. The Ministry bent a little, and awarded Heinkel a contract for 13 pre-production aircraft.

His designers put together a further development, with the unusually heavy armament of six 20-mm cannon, and proposed it to the Air Ministry. To everybody's surprise, the Ministry awarded a contract for the production of 300, but Heinkel's facilities, strained as they were by existing production programmes, were bypassed by this order and the He 280 was scheduled to be built by another firm.

But by then the Me 262 had flown under jet power; it appeared so promising that the Ministry cancelled the He 280.

The rocket-powered jet fighter arrived, in prototype form, in 1941. On 13 August Messerschmitt test pilot Heini Dittmar strapped himself into a prototype Me 163A, started the rocket engine, and blasted across the turf at Peenemünde-West, the experimental Luftwaffe airfield. The Me 163A was held in a climb until the fuel was burned; then Dittmar turned and began a circling letdown to a landing. It was the first flight by a rocket-powered interceptor prototype, and it began a long, frustrating and ultimately unsuccessful development programme.

It had begun some years earlier, as a project to power a tailless glider designed by Dr Alexander Lippisch. Working at the German Research Institute for Soaring (DFS), Lippisch's team had brought along the design of their DFS 194 to the point where it obviously required industrial support.

Messerschmitt was designated, and Lippisch's team went to Augsburg. The aircraft turned out well, but its rocket powerplant did not, and the DFS 194 was never flown under power. It was used instead for ground tests of the rocket.

The baulky rocket was replaced by a new design with controllable thrust, other changes were made, and the result was the Me 163A series, prototypes used for development of the interceptor version.

It was one of this first batch of 13 that Dittmar first flew in August 1941. But there was a long time between that first flight and the first operational sortie. The Me 163 did not see action until 13 May 1944, and even that attempt to seek combat was made in a development aircraft, one of the Me 163B prototypes. By the time the Luftwaffe had had production versions of the Me 163B in service, the war was running down and the visions of hundreds of the tiny rocket fighters slashing through disrupted bomber formations had been reduced to the actual-

**Messerschmitt Me 163**
*Crew:* 1 *Powerplant:* 1 Walter RII liquid-propellant rocket *Span:* 30·5 ft *Length:* 17·8 ft *Weight:* 5291 lb *Speed:* 558 mph

ity of a few sporadic intercepts and some hideous operational accidents.

A dispassionate examination of the concept led to one conclusion: it was possible to be too fast for effective combat. The Me 163s were designed to be used as interceptors of daylight bombing raids. They were to take off and climb rapidly (they could get to bomber height in less than three minutes), attack the bombers with their paired 30-mm cannon, and break away for the return to base.

In practice, the speed of the rocket fighter was so much greater than that of

its bomber target that a pilot only had two or three seconds to aim and fire. It proved to be nearly impossible. The Me 163 was not suitable for combat against slow-flying bombers.

Those that did get into combat managed to shoot down a few bombers, but it was too late. The factory producing one of the essential fuel components was bombed in

*May 1973: 'Vintage Pair' of the RAF's Historic Flight, a De Havilland Vampire T 11, the last Vampire still flying with the RAF, and a Gloster Meteor T 7.*

August 1944. Ground transportation was under constant attack, and several complete shipments of rocket fuel were lost to Allied gunnery. As winter neared, the weather worsened – and the Me 163 was not suitable for bad weather or night operations. The whole programme ground to a halt, with only a few intercepts flown against special targets such as high-altitude photo flights.

There was only one truly successful jet fighter developed and brought to operational status during the Second World War: the Messerschmitt Me 262. In spite of setbacks to the smooth development of the programme caused by such diverse factors as Hitler's dreams and bombing realities, the project maintained and even gained momentum.

It began in late 1938 with an Air Ministry contract with Messerschmitt for a twin-engine jet fighter. By March 1940, both Messerschmitt and Heinkel were told to go ahead with the development of their respective twin-jet fighters.

The first Messerschmitt prototype was completed well before its jet engines were ready for flight. The first alternative, to fit

rocket engines in the nacelles for flight tests, was ruled out because the engines weren't considered safe enough.

So Messerschmitt installed a standard Junkers Jumo piston engine in the nose, and the first flight of an Me 262 was made on 18 April 1941, with a piston engine and propeller providing propulsion, and empty jet nacelles under the wings. By March of the following year airworthy jet engines were available, and on 25 March the prototype was flown on the combined power of its piston engine and the two new jets. It nearly ended in disaster for the pilot, Fritz Wendel, because both turbojets failed shortly after takeoff, and he had a tough time keeping the Me 262 in the air.

Wendel made the first flight on jet power only with the third prototype, which had been fitted with a pair of Junkers Jumo 004A-0 turbojets producing about 1850 lb of thrust each. On 18 July 1942 he took off from the hard-surfaced runway at Leipheim for a flight of about twelve minutes. He completed a second flight that day, and was delighted with the way the plane handled.

But he had had to use brakes momentarily during the takeoff roll, in order to get the tail up into the slipstream so that the elevators would be effective. The braking served to rotate the aircraft nose down and had to be done carefully, gently and at exactly the right time.

This must have been one of the reasons that Messerschmitt decided to redesign the Me 262 with a new type of landing gear – the tricycle type with nosewheel – that became the standard for all subsequent Me 262s.

Happily for the Allied cause, the decision-making machinery broke down on the Me 262 programme. Production schedules were changed almost monthly. Variations on the theme were developed on request and the Me 262 was built as a fighter, an all-weather fighter, a reconnaissance aircraft, a ground-attack aircraft, a fighter with reconnaissance capabilities, a fighter-bomber, and an interceptor with rocket booster engines in the nacelles, all in a single-seat version. Two-seat models were developed as trainers and night fighters. They were built in small batches of only a few of most of the versions, and only one model was produced in any quantity.

It was July 1944 before the Me 262 engaged in combat, the first recorded instance being an encounter with a re-connaissance Mosquito flown by Flt Lt Wall, RAF. Wall reported that an Me 262 made five passes at his Mosquito, but in each case he was able to break away and finally dove into clouds to escape his persistent adversary.

Time and the losing position of Germany caught up with the Me 262. By tremendous industrial effort, mass production of the aircraft had been achieved under mountains of difficulties. The first production aircraft had been delivered in March 1944, and by February 1945 production had peaked at 300 completed aircraft per month. Factory delivery data show that 1320 were rolled out of the doors for delivery to the Luftwaffe during the 13-month production programme.

The most famous unit to operate the Me 262 was JV 44, formed and commanded by General Adolf Galland. The unit arrived

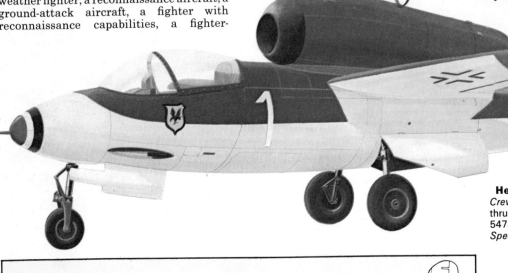

**Heinkel He 162A-2**
*Crew:* 1 *Powerplant:* 1 BMW 003E-1, 1760 lb thrust *Span:* 23·6 ft *Length:* 29·7 ft *Weight:* 5478 lb *Armament:* 2×20-mm cannon *Speed:* 518 mph at 19,680 ft

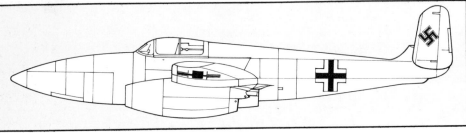

Cockpit is a totally enclosed unit for pressurisation purposes, although no service machines were ever pressurised

**Heinkel He 280**
*Crew:* 1 *Powerplant:* 2 HeS 8, 1100 lb thrust each *Span:* 40 ft *Length:* 34·1 ft *Weight:* 9500 lb *Armament:* 3×20-mm cannon *Speed:* 558 mph at 19,680 ft

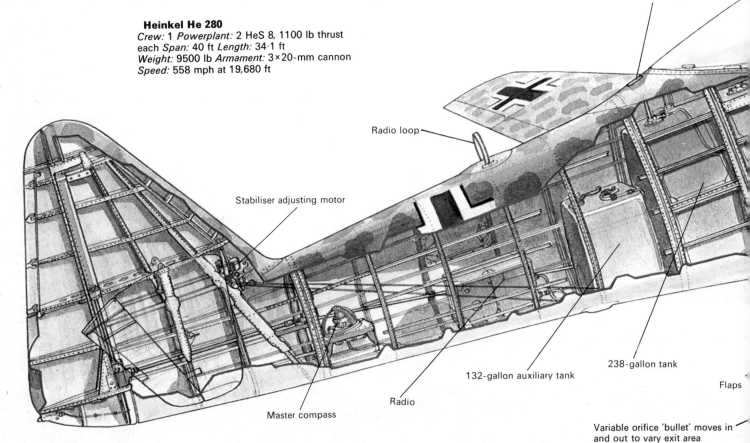

Tank filler cap

Radio loop

Stabiliser adjusting motor

238-gallon tank

Flaps

132-gallon auxiliary tank

Radio

Master compass

Variable orifice 'bullet' moves in and out to vary exit area

at its base near Munich on the last day of March 1945, and operated for only about one month, finally making its sorties from the autobahn between Munich and Augsburg. But in that time, they terrorised bomber crews, made about 50 kills, and established once and for all the value of the jet fighter.

As Germany's position grew more desperate, so did attempts to develop new weapons to stave off the inevitable. One of these was the Heinkel 162, a tricky single-engined jet fighter. Its specification, issued in September 1944, called for a lightweight fighter, using an absolute minimum of strategic materials, and capable of being put into rapid mass production. It was to be flown into combat by the loyal Hitler Youth,

after they had been given a brief training period on gliders.

Heinkel was awarded the contract on 30 September. By 29 October the He 162 had been designed, and construction had begun. The first prototype was flown on 6 December, with Flugkapitän Peter at the controls. One month later, the first He 162s were delivered to a test unit, and in February 1945 I/JG-1 began conversion to the type.

Few German records remain of those frantic last days, but there is at least one reported incident of combat between an He 162 and a USAAF P-51 Mustang. The jet was able to turn and climb with the Mustang, but it was much faster and had greater acceleration. The combat was inconclusive; neither claimed victory.

By the end of the war, about 275 had been built and another 800 were in various stages of assembly. It was a formidable accomplishment by the Heinkel organisation. They designed a contemporary jet fighter in one month, flew it nine weeks after starting design, and delivered 275 in less than seven months.

There was one more last-gasp defence effort to fly: the Bachem 349 Natter. Work had begun in the spring of 1944, to a specification for a target defence interceptor. Bachem's first proposal was rejected in

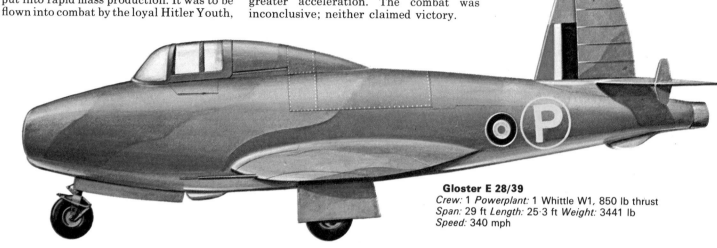

**Gloster E 28/39**
*Crew:* 1 *Powerplant:* 1 Whittle W1, 850 lb thrust *Span:* 29 ft *Length:* 25·3 ft *Weight:* 3441 lb *Speed:* 340 mph

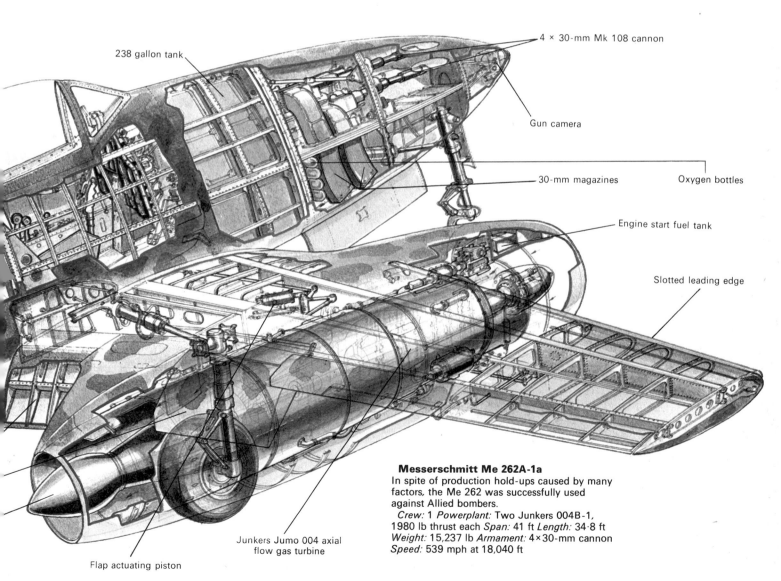

238 gallon tank

4 × 30-mm Mk 108 cannon

Gun camera

30-mm magazines

Oxygen bottles

Engine start fuel tank

Slotted leading edge

Junkers Jumo 004 axial flow gas turbine

Flap actuating piston

**Messerschmitt Me 262A-1a**
In spite of production hold-ups caused by many factors, the Me 262 was successfully used against Allied bombers.
*Crew:* 1 *Powerplant:* Two Junkers 004B-1, 1980 lb thrust each *Span:* 41 ft *Length:* 34·8 ft *Weight:* 15,237 lb *Armament:* 4 × 30-mm cannon *Speed:* 539 mph at 18,040 ft

favour of a Heinkel design; but Erich Bachem knew the sources of power and had an interview with Heinrich Himmler. The decision of the Ministry was immediately changed to support the Bachem proposal as well.

It was a tiny wooden airframe powered by a single rocket engine, boosted by four solid-propellant rockets, launched from a vertical tower, and armed by a nose full of air-to-air rockets. The attack over, the pilot was expected to bail out. He and the valuable engine were to be saved by parachutes.

The Natter was tested as a glider in November 1944, launched unmanned in December under boost power only, and was successful in both tests. But the first piloted flight ended in disaster. On 28 February 1945 Oberleutnant Lothar Siebert, a volunteer for the test flight, was killed when the canopy came off during launch, ap-

parently knocking him unconscious as it left. The Natter crashed out of control. But the next three manned launches were successful, and the programme moved ahead. Seven manned flights were made in all, and the production programme continued to grind out the wooden airframes which took only a few hundred man-hours each to build.

In April 1945, a squadron of 10 Natters was set up ready to launch near Stuttgart, waiting for the next bomber raid for its initiation into combat. But before the aerial assault, Allied armoured units rolled into the area, and the Natter crews destroyed their aircraft to keep them from falling into enemy hands. That was the effective end of the Natter programme.

The turbulence of war was a major factor in the establishment and cancellation of aircraft programmes. It was the beginning of war that must have been one of the events prompting the issuing of a British Air Ministry specification, E 28/39, for a single-seat fighter prototype aircraft powered by a gas turbine for jet propulsion.

Earlier, the Ministry had contracted with Power Jets, a firm headed by Frank Whittle, for development of an airworthy jet engine. Power Jets received its first Ministry support in March 1938; the engine contract was received on 7 July 1939.

The aircraft contract, issued to Gloster Aircraft on 3 February 1940, described a design based on the need for an interceptor. Top speed was to be about 380 mph, and armament was to be four machine-guns. The primary purpose of the aircraft was to obtain flight data on the engine, but it was also to be a prototype fighter.

The first run of an engine in the E 28/39 airframe was made on 6 April 1941, using an unairworthy engine. The next day, Flt Lt P E G Sayer began taxi tests at Brockworth. The plane rolled across the green field, picking up speed and slowing again as Sayer felt out the handling. Three times during the taxi runs, Sayer lifted the plane off the ground briefly. It seemed ready to fly.

The first prototype was trucked to the airfield at Cranwell, home of the RAF College where Whittle had spent his cadet days. There were practical as well as sentimental reasons for selecting that field. It had a long runway, with clear approaches, and was one of the best available fields for test work.

On 14 May Sayer repeated some of the taxi tests and planned to fly the following day. Low clouds hid the sky on the morning of May 15, but towards evening they began to lift. The camouflaged E 28/39 with Sayer in the cockpit trundled out to the starting area. There was a rising howl from the

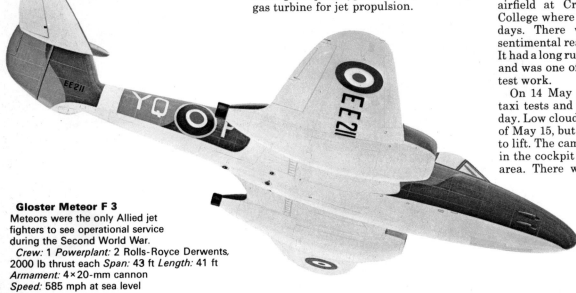

**Gloster Meteor F 3**
Meteors were the only Allied jet fighters to see operational service during the Second World War.
*Crew:* **1** *Powerplant:* **2 Rolls-Royce Derwents,** 2000 lb thrust each *Span:* **43 ft** *Length:* **41 ft**
*Armament:* **4 × 20-mm cannon**
*Speed:* **585 mph at sea level**

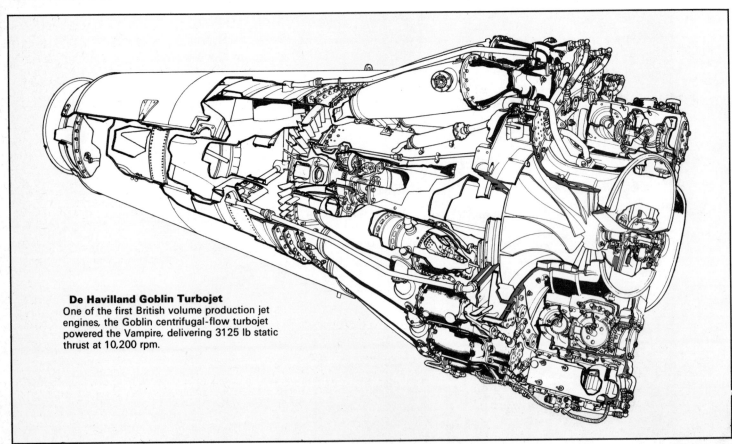

**De Havilland Goblin Turbojet**
One of the first British volume production jet engines, the Goblin centrifugal-flow turbojet powered the Vampire, delivering 3125 lb static thrust at 10,200 rpm.

engine, the plane began to move, and with darkness already gathering, Sayer lifted the plane off on its first flight.

He stayed aloft 17 minutes. It was the first flight ever made by a British jet-propelled aircraft.

Official support came soon. The flight programme was pushed to learn more about this new form of propulsion. Gradually the aircraft was taken to 25,000 ft and 300 mph in less than 10 hours flying.

Later, Rolls-Royce took over development of the engine, and raised its basic thrust to 1400 lb. Then the plane was flown to a maximum speed of 466 mph, and to an altitude above 42,000 ft. Gloster completed its portion of the programme in late June 1943, and turned the E 28/39 over to the care of the Royal Aircraft Establishment at Farnborough.

Britain's first true jet fighter was the Gloster Meteor, begun as an answer to specification F 9/40. It was planned as a twin-engine craft, because one engine of the type then available was hardly sufficient to obtain performance better than that of contemporary piston-engined fighters. Further, there was a supposed advantage of twin-engine reliability and safety.

The Gloster design team laid out their twin with the jet engines buried in the wings, and with the rear spars built around large holes for the jet pipes to pass through. Tricycle landing gear and a high tail were other basic decisions. Armament was to be four 20-mm cannon in the nose, and the cockpit was to be pressurised. Design began some time around August 1940.

About a year later, problems arose with the specified engines; the Power Jets W 2B engines had not been declared airworthy. One prototype was converted to take the Halford H 1 engines then in advanced development, and another to take the Metropolitan-Vickers F 2 engines. The H 1 engines were first cleared for flight, and the fifth prototype Meteor was trucked to the aerodrome at Cranwell, where Gloster pilot Michael Daunt made the first flight on 5 March 1943.

With Rolls-Royce in the engine programme, the final choice for the Meteor powerplant was the Welland W 2B, basically the Whittle/Power Jets engine. Wellands powered the first 20 production F Mk 1 Meteors, a fighter rushed into production

and action near the end of the war. Issued to 616 Squadron RAF, based at Culmhead and later at Manston, they first saw action on 27 July 1944, on 'Diver' patrol against the German V-1 buzz-bombs. Sqdn Ldr Watts was the first Meteor pilot to contact one; but his guns jammed and the flying bomb continued on course. First kill of a V-1 was made on 4 August by F/O Dean, whose guns also jammed. So Dean closed the distance, eased the Meteor's wingtip under that of the V-1, and banked sharply away. The Meteor's wingtip slammed against the V-1's wing and sent it into a spiral dive and a crash in open country.

F/O Rogers, almost at the same time, was having more conventional success. His guns fired, and he became the first RAF pilot to shoot down an enemy aircraft from a jet fighter.

Britain's only other jet fighter of the war years, the de Havilland DH 100 Vampire, was very different from the Meteor. Designed to specification E 6/41, which defined an experimental aircraft rather than

the fighter required by the Gloster Meteor specification of F 9/40, the Vampire started to take shape on the drawing boards at Hatfield in May 1942.

The single jet engine was enclosed in an egg-shaped fuselage, with inlets for the air at the root of each wing, and the exhaust discharging directly aft on the centre line of the egg. De Havilland designers used twin tail booms, perhaps borrowing the idea from the piston-engined Lockheed P-38 Lightning.

The Vampire was all metal, but there was one holdover from earlier DH designs; the cockpit section was constructed of a plywood and balsa sandwich material.

It was an all-DH project. The engine was the Halford H 1, designed by Maj Frank Halford and built by de Havilland. Geoffrey de Havilland, Jr, made the first flight on 30 September 1943, at Hatfield, six months after the Meteor had flown. The time differential was critical; the Meteor just barely saw action near the end of the war, but the Vampire was too late to be tested under combat conditions.

About a year earlier, the first flight of the first US jet fighter, the Bell XP-59A, had taken place. The site was a remote desert area, part of the USAAF Muroc Bombing and Gunnery Range located on a dry lake bed about 100 miles north of Los Angeles, California. (That site later became Edwards Air Force Base.)

Robert M Stanley, then chief pilot for Bell, fired up the twin General Electric I-A turbojets, which had been closely but not completely copied from the British W 2B engines. A few minutes later, on the after-

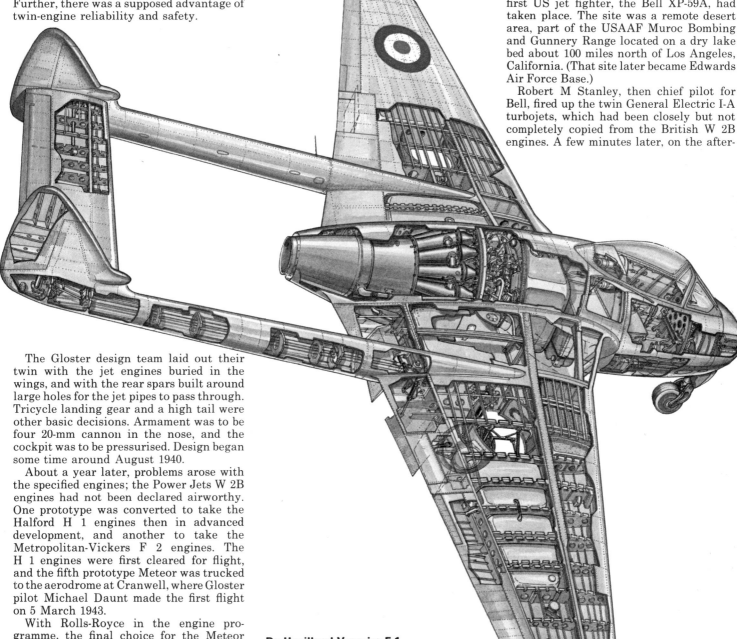

**De Havilland Vampire F 1**
*Crew:* 1 *Powerplant:* DH Goblin, 3125 lb thrust
*Span:* 40 ft *Length:* 30·8 ft
*Armament:* 4×20-mm cannon
*Speed:* 525 mph at 25,000 ft

noon of 1 October 1942, the XP-59A lifted off the dry lake bed into the California sky.

It would not have made such progress without British help. Major General Henry H Arnold, then Chief of the USAAC, visited Britain in the spring of 1941, saw the Whittle engine and the E 28/39, and was impressed. After follow-up meetings, it was agreed that the US should copy the Whittle engine and develop a twin-engine fighter around it. Bell were chosen as the airframe company to be responsible, and General Electric were chosen to build the engines. Bell were given eight months from the date of the contract approval to have their first aircraft ready for flight.

Construction stayed on schedule, but the timetable for GE engine deliveries slipped. They were not ready until August 1942, and they were never trouble-free. Their performance did not meet expectations, because

**Bell XP-59A Airacomet**
More of a research aircraft than a service fighter, the XP-59A first flew in October 1942

Smithsonian Institution Photo No A516A

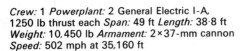

*Crew:* 1 *Powerplant:* 2 General Electric I-A, 1250 lb thrust each *Span:* 49 ft *Length:* 38·8 ft *Weight:* 10,450 lb *Armament:* 2×37-mm cannon *Speed:* 502 mph at 35,160 ft

the British data used as the basis for their design proved optimistic. Neither the original nor later production engines ever developed the predicted thrust. Consequently, the P-59 Airacomet never served with front-line units; it served instead with a squadron training pilots and mechanics on the new aircraft.

It was obvious early in the Bell programme that its performance was not going to be earth-shaking. Everybody had ideas about what to do, but Lockheed's Clarence L (Kelly) Johnson was to see his ideas take tangible form.

Lockheed had done earlier work on a jet fighter proposal, had been rejected, but had persisted. On one of Johnson's periodic visits to Wright Field, then the technical headquarters of the USAAF, he was asked to consider designing a jet fighter around a British engine. Within a few days the first sketches were ready, and Johnson got the go-ahead in June 1943. At the far side of the Lockheed airport at Burbank, California, a temporary building was erected, the 'Skunk Works', named after a mythical factory in the popular comic strip, 'Li'l Abner'.

The contract gave Lockheed 180 days to design, build and fly the XP-80. They beat the construction deadline and had the plane ready to go in 143 days. But the first flight was delayed by engine availability, and it was not until 9 January 1943, that Lockheed chief test pilot Milo Burcham made the first flight with the XP-80 from the dry lake bed at Muroc.

Then Lockheed had to repeat the whole performance. It was decided that the production P-80 would be powered by the new General Electric I-40 engine, based on British designs. Back to the Skunk Works went Johnson's team, to emerge 139 days later with another prototype, the XP-80A. It first flew on 11 June 1944, and by the time the war ended, 45 had been delivered to USAAF squadrons. A few had even been tested at operational bases in England and Italy, but had been kept from any area where combat might have been possible.

Early in the development of the jet fighter, the navies of Great Britain and the US had studied the new type and wondered how best to adapt it to carrier operations. In the US, the Navy Bureau of Aeronautics were sponsoring the development of a series of axial-flow turbojets by Westinghouse Electric Corp. These small-diameter engines promised much better overall installed performance than did the bulkier centrifugal-flow engines pioneered by Whittle, Rolls-Royce, and General Electric.

The Navy, Westinghouse and the McDonnell Aircraft Corp got together in early 1943 to discuss the design of a Naval fighter built around two or more of the Westinghouse engines. McDonnell designers investigated a wide range of possibilities, guided by basically conservative design policies. They checked eight-, six-, four- and twin-engined schemes and settled on the twin as the basis for their design of the XFD-1. It was to be a fighter

with a defensive mission of combat air patrol at 15,000 ft above a carrier task force. Two years and a few days after the contract was signed, the first prototype XFD-1 took to the air on 26 January 1945. Two months later McDonnell received a production order.

But the war was to be over by almost two years when the first McDonnell Phantoms, redesignated FH-1, were delivered to the fleet. By then, it was apparent that the Phantoms were only an interim type serving to accumulate some fleet experience with jet fighters.

To the East, the Russians had been working for several years to develop their own rocket-powered interceptor. Two designers – Bereznyak and Isaev – planned a tiny aircraft around a single rocket engine rated at 2420 lb of thrust. They designed a conventional fighter, armed with a pair of 20-mm cannon in the nose, and intended for the same kind of mission as the Me 163.

The expected bomber raids against Russia never happened; the rocket-powered interceptor would not have been needed. But it was a fatal accident during a test flight that put an end to the development programme. The first flight had been successful. Test pilot Grigori Bakhchivandzhe flew the BI for a little longer than three minutes on its maiden trip on 15 May 1942.

It was slow development. A second plane was added, but the rocket engine proved troublesome. Only six flights were logged in ten months. Bakhchivandzhe was killed

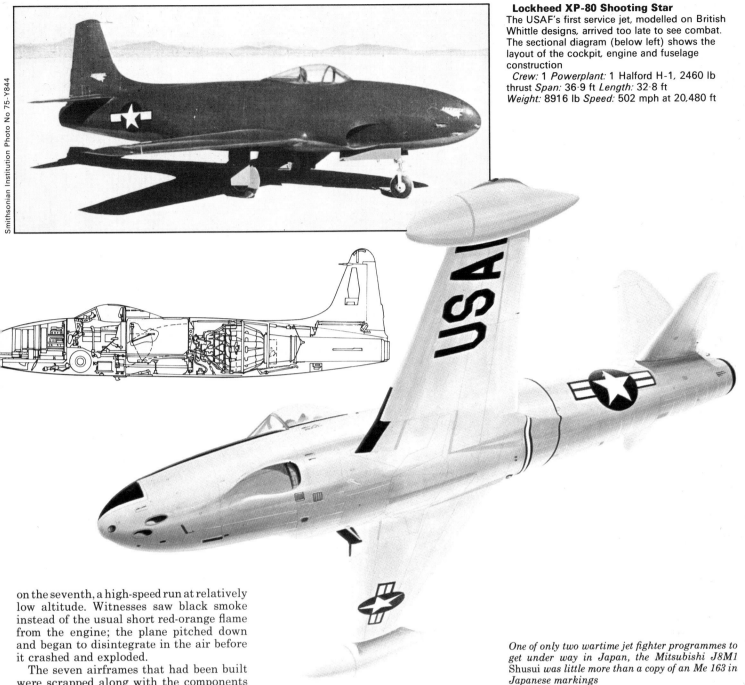

**Lockheed XP-80 Shooting Star**
The USAF's first service jet, modelled on British Whittle designs, arrived too late to see combat. The sectional diagram (below left) shows the layout of the cockpit, engine and fuselage construction
*Crew:* 1 *Powerplant:* 1 Halford H-1, 2460 lb thrust *Span:* 36·9 ft *Length:* 32·8 ft *Weight:* 8916 lb *Speed:* 502 mph at 20,480 ft

*One of only two wartime jet fighter programmes to get under way in Japan, the Mitsubishi J8M1 Shusui was little more than a copy of an Me 163 in Japanese markings*

on the seventh, a high-speed run at relatively low altitude. Witnesses saw black smoke instead of the usual short red-orange flame from the engine; the plane pitched down and began to disintegrate in the air before it crashed and exploded.

The seven airframes that had been built were scrapped along with the components for another 20 or so, on orders from the Kremlin banning all further work on rocket fighters.

Only two jet fighter projects ever got under way in Japan, and both were inspired by German developments. Japan acquired licence rights to the Me 163 and its rocket engine. But delivery of a sample Me 163 and a complete set of blueprints was not completed; the submarine carrying them to Japan was sunk. Japan received only a single rocket engine and an Me 163 manual a Japanese naval officer had brought back from a visit to Germany.

In July 1944 the Japanese Navy issued a specification for a rocket-powered interceptor. The Army joined the programme, and the first prototype of a training glider was completed by December 1944. It flew successfully, after being towed to altitude and released. But the aircraft itself, the Mitsubishi J8M1, was not as successful. The first prototype was finished in June 1945, and its first flight was scheduled for 7 July. The engine failed soon after takeoff, and the J8M1 smashed into the ground, killing Lt Cdr Toyohiko Inuzuka, the test pilot. Although production had started,

and other J8M1 aircraft were available, no more flights were made before the Japanese surrender.

The success of the Me 262 programme sparked Japanese interest in a twin-jet fighter, and the Navy issued an order to Nakajima for development of such a fighter, based on the German twin-jet craft but smaller. Data were limited. The turbojet engines were designed using, among other

sources, photographs of the German BMW 003 turbojet. The first prototype was completed in August 1945, just days before the final bell rang for Japan. On 7 August it made its first flight from the Naval air base at Kisarazu, with Lt Cdr Susumu Tanaoha at the controls. On his second flight, Tanaoha had to abort during the takeoff run because of engine failure. It was the last attempt to fly the Nakajima J8N1 *Kikka*.

Britain and Germany were the first countries to develop jet fighters. Unfortunately, Britain was slow to adapt to the development of jet aircraft technology. The USA and USSR adopted the axial flow engine and sweptback wing configuration (both based on German research) much earlier: seizing the lead in the postwar years, they never let it go

The victorious Allied armies that steam-rollered through Germany towards the end of the Second World War liberated filing cabinets and desk drawers crammed with documents on aerodynamic, structural and powerplant ideas, designs and tests. It was a bonanza, a major foundation for the postwar development of jet fighters.

In spite of the intense pressures of war, German scientists working in university and government research institutes were able to develop ideas at their own pace. They had time to calculate, sketch, build models and test them in flight or a wind tunnel. Combined with practical experience from operational use of a wide variety of unusual weapons and aircraft, this wealth of data fell almost intact into Allied hands at the end of the war.

Many of these ideas had been discovered or developed earlier by scientists in other countries. But in Great Britain and the United States, there was more pressure to produce aircraft in quantity than there was to improve the breed with new and revolutionary ideas, even though both countries did develop and produce jet fighters during the war. But they did not do so on the scale of Germany, and their operational experience was very limited compared to that of the Luftwaffe.

One major contribution made by the Germans was their standardisation of axial-flow jet engines of reduced diameters, compared to the centrifugal-flow types pioneered by Heinkel and Whittle. The axial-flow jets were better suited to installation in a slim fighter fuselage, or under the wings. It took several postwar years before other nations realised that the axial-flow engine was really the best way to do the job.

Another German contribution was sweep-back. Known as early as 1935, sweepback reduces the drag of the wing by aerodynamically thinning the wing section. German wind tunnel tests proved and evaluated this, and almost every late wartime German design featured a swept-back wing.

The combination of these two basic concepts – the axial-flow engine and sweep-back – produced a long series of combat aircraft after the war, spilled over into civilian designs in the late 1950s, and remains as the basic configuration of many military and civil aircraft today.

The Germans spent a tremendous amount of their scientific resources on guided missiles. Even though they were not all applicable to fighter design, or to exploitation for fighter use, the basic technology developed for them furnished valuable background experience.

Unguided missiles and air-to-air rockets were developed and used operationally by

Smithsonian Institution Photo No 75-Y843

*The Republic XF-91 mixed-powerplant (rocket and turbojet) sweptwing interceptor*

## POSTWAR DEVELOPMENTS

# THE SWEEP TOWARDS MACH 1

the Germans, and that type of weapon was destined to become an important part of the striking power of fighters to come.

Airborne radar systems, primitive though they were, had been used by both sides during the war. Postwar, they blossomed as technology advanced. Combined with ground-based long-range radars and improved communications, they formed the beginnings of the highly effective command and control systems now in operation.

Gradually these developments began to come together. A new design might evaluate one or two new ideas; later designs might add a third or fourth. And, piece by piece, the unsophisticated jets of the war's end built to advanced designs that broke through the speed of sound, and could fly and fight in an all-weather environment.

Those few postwar years were exciting. There was money to spend on unusual concepts, and designers had a wealth of data to draw on. The jet fighters we talk about here are only those that were

significant during that period. For every one of these, there were others that led only to a dead end in development.

During the five years between the end of the Second World War and the start of the Korean conflict, the design lead was seized and exploited by the United States and Russia, and they have never let go of it. The British, whose truly pioneering efforts contributed so much to the early development of the jet fighter, never exploited their position with advanced technology. They stayed too long with the straight wing and the centrifugal engine and – with one exception – never again became a technological competitor in fighter design.

This period of time also saw the emergence of strong jet fighter design teams in France and Sweden. Both countries continued to improve their position, the French more rapidly and on a broader scale because of their greater size and wealth. Today they stand on a technical par with the United

States and Russia in advanced aeronautics.

**28 February 1946:**
### Republic XP-84 Thunderjet
This American aircraft was the first significant jet fighter to fly in the postwar years. Sleek, powered by a new General Electric axial-flow engine, the XP-84 was designed as an interceptor, but was destined to spend most of its long career in the USAF as a fighter-bomber. It was later blooded in the Korean war, and was the mainstay of tactical airpower for many Allied countries and the United States during the years of cold war.

**24 April 1946:**
### Yakovlev Yak-15 and Mikoyan MiG-9
Both these Russian jet fighters, first of that country's postwar types to fly, were powered by originals or copies of German jet engines that had been captured in quantity by the Russians. The Yak-15 was a single-engined modification of a piston-engined interceptor that saw much service during the war. It was the first of the pair to fly, followed into the air within minutes by the MiG-9, a bulkier, twin-engined fighter. Both types went into production, although the MiG-9 faded from the scene early and the Yak-15 stayed on in service in Russia and some of its allied countries, and was later developed further.

**27 July 1946:**
### Supermarine Attacker
This was the first jet fighter to serve with the Royal Navy on carriers, but it had started life as a land-based interceptor design for the Royal Air Force. Developed late in the war years, it was built around the ubiquitous centrifugal-flow engine and a straight wing adapted from the last of the piston-engined Spitfire line. With the Attacker, the Royal Navy learned the operational problems of jet fighters.

**11 November 1946:**
### SNCASO 6000-01 Triton
French daring developed their first jet aircraft, designed under the noses of occupying German troops. Work began in 1943, in spite of the lack of contact with other countries developing jet aircraft, the systematic despoiling of the industry and German labour drafts that decimated its personnel. The Triton was a single-engined test-bed, built to be able to handle a variety of jet engines. The first prototype flew – on the anniversary of Armistice Day – under the power of a German jet Junkers 004B. It was the harbinger of dynamic French fighter designs to come.

**27 November 1944:**
### North American XFJ-1 Fury
This stubby, straight-winged aircraft was North America's first jet fighter. It was designed as one of two successors to the Navy's first jet, the McDonnell Phantom, and was bought only in small quantity because other and newer developments were coming along rapidly. It served with only one squadron on one carrier, and might have been forgotten but for one thing: its rugged airframe was the basis that led to NAA's sweptwing XP-86 the following year.

**11 January 1947:**
### McDonnell XF2H-1 Banshee
A bigger and more powerful brother-in-arms to the Phantom then in fleet service with the US Navy, the Banshee was a linear, almost scaled-up development of the earlier McDonnell jet fighter. It had more of

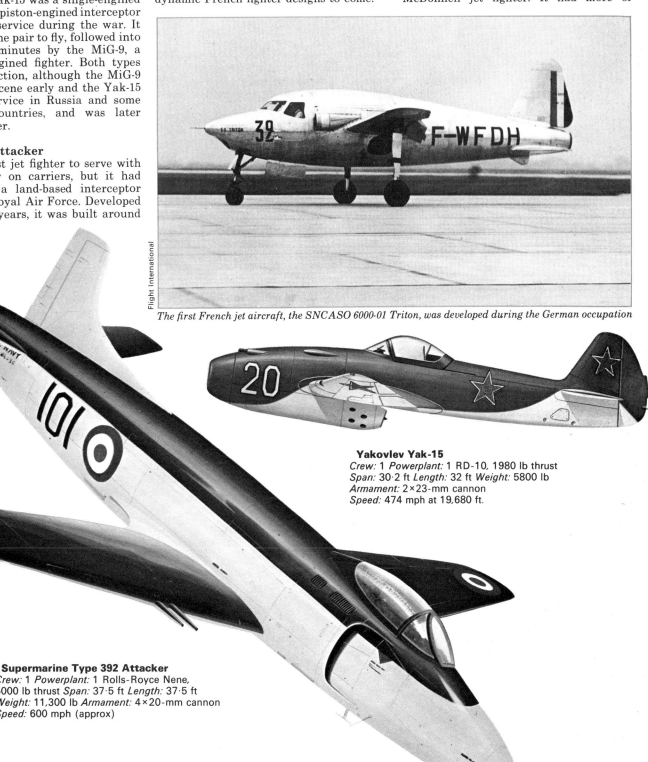

Flight International

*The first French jet aircraft, the SNCASO 6000-01 Triton, was developed during the German occupation*

**Yakovlev Yak-15**
*Crew:* 1 *Powerplant:* 1 RD-10, 1980 lb thrust
*Span:* 30·2 ft *Length:* 32 ft *Weight:* 5800 lb
*Armament:* 2×23-mm cannon
*Speed:* 474 mph at 19,680 ft.

**Supermarine Type 392 Attacker**
*Crew:* 1 *Powerplant:* 1 Rolls-Royce Nene,
5000 lb thrust *Span:* 37·5 ft *Length:* 37·5 ft
*Weight:* 11,300 lb *Armament:* 4×20-mm cannon
*Speed:* 600 mph (approx)

everything, including range – one of the more elusive performance characteristics of early, fuel-guzzling jets. That goal achieved, the Banshee stayed in the fleet to serve as a potent fighter-bomber during the Korean war some years later.

*10 March 1947:*

## SAAB 21R

Sweden, later to become known for superlative combat aircraft, built its first jet fighter by converting a piston-engined type. This is the only known case where the same basic configuration served in both a piston-engined and jet-engined form. The SAAB J 21R could not have been a very efficient aircraft, and it was not produced in large quantity. But it served a very useful purpose, furnishing both industry and the Royal Swedish Air Force with valuable experience they could not get otherwise.

A McDonnell F2H Banshee of the US Navy, photographed in 1957 with an armament of Zuni missiles. The Banshee was virtually just a scaled up version of the earlier McDonnell Phantom

Smithsonian Institution Photo No 75-480

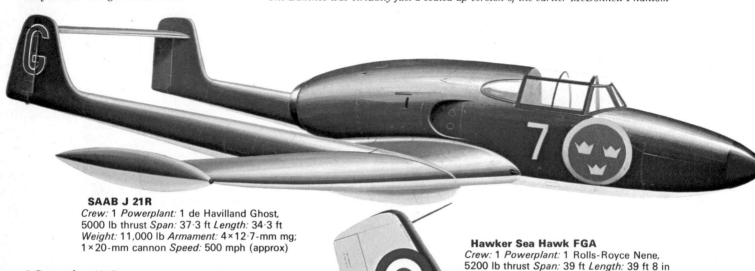

**SAAB J 21R**
*Crew:* 1 *Powerplant:* 1 de Havilland Ghost, 5000 lb thrust *Span:* 37·3 ft *Length:* 34·3 ft *Weight:* 11,000 lb *Armament:* 4×12·7-mm mg; 1×20-mm cannon *Speed:* 500 mph (approx)

**Hawker Sea Hawk FGA**
*Crew:* 1 *Powerplant:* 1 Rolls-Royce Nene, 5200 lb thrust *Span:* 39 ft *Length:* 39 ft 8 in *Weight:* 13,785 lb *Armament:* 4×20-mm cannon *Speed:* 560 mph at 36,000 ft

*2 September 1947:*

## Hawker P 1040 Sea Hawk

The elegant P 1040 was Hawker's first jet aircraft, and it was developed into the Sea Hawk interceptor that served with the navies of Great Britain and other countries. It was laid out around the straight-wing, centrifugal-flow engine formula, but it featured an unusual exhaust pipe that must have given fits to metal fabricators in England. It was described as bifurcated, which means that the tailpipe divided into two, instead of extending straight aft in a single cylinder from the exhaust nozzle of the engine. The Sea Hawk was a sprightly aircraft, with fine flying qualities, and it was shown off beautifully by No 738 Naval Training Squadron at the 1957 Farnborough show in England.

*1 October 1947:*

## North American XP-86 Sabre

The immortal Sabre started as a parallel design to the Fury, with a similar layout, including the straight wing. But the exciting wartime German data on swept wings led to a complete rethinking of the design concept and the alteration of the Sabre proposal to sweptwing geometry. The wings were angled back at 35°, measured at the quarter-chord line, the tail was matched to that sweep angle, and the Sabre was born.

It was an advanced fighter for its day, was built in Australia and Canada in licensed – and improved – versions, and later racked up an astonishing combat record in the high skies over MiG Alley in the Korean war.

*24 November 1947:*

## Grumman XF9F-2 Panther

Grumman's first jet fighter followed the established and conservative formula of British jet fighters. Not that it was a bad formula to follow at that time; given the state of both engine and aerodynamic technology, the Grumman choice was as justifiable as that of North American.

Events were to prove that it was the conservative choice, but the Panther was built in quantity, maintained the Grumman reputation for fine and rugged combat aircraft, and acquitted itself well in Korea.

*30 December 1947:*

## Mikoyan MiG-15

The thrust of an exported Nene engine from England urged this little Russian sweptwing jet fighter into the air on its first flight. Copied versions of that engine formed the basis for a later Russian industry and the powerplant for many thousands of the Red fighters. The MiG-15 became one

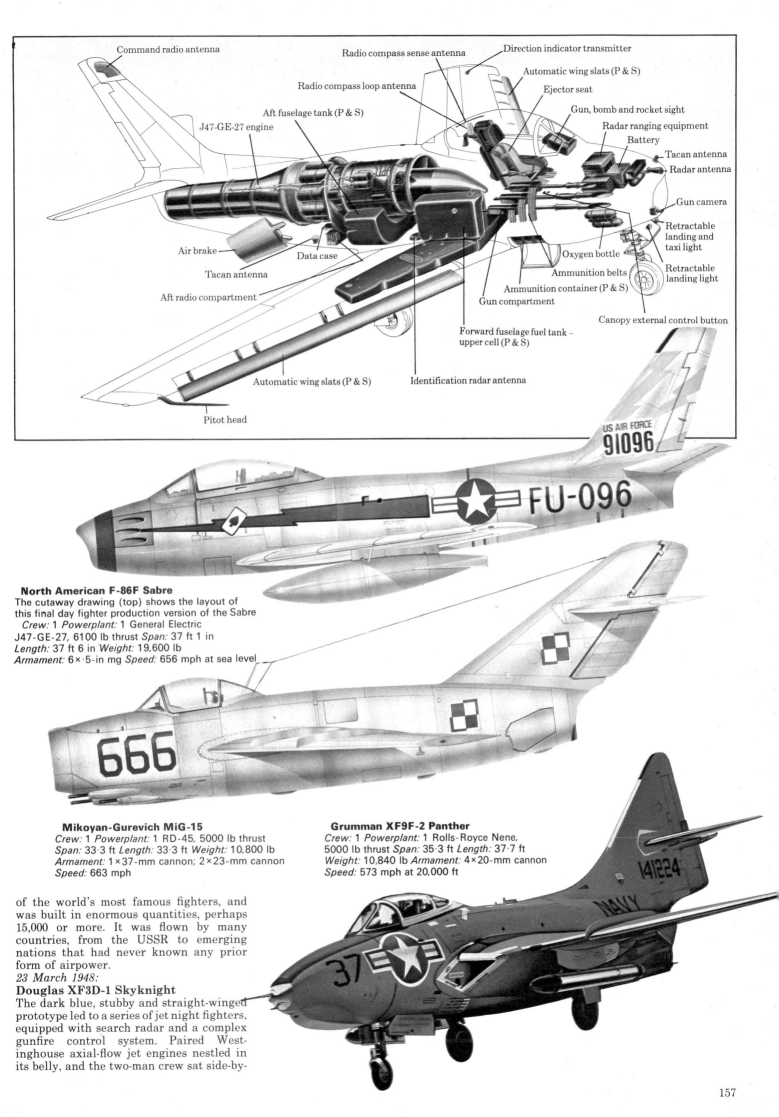

Command radio antenna

Radio compass sense antenna

Direction indicator transmitter

Radio compass loop antenna

Automatic wing slats (P & S)

Aft fuselage tank (P & S)

Ejector seat

J47-GE-27 engine

Gun, bomb and rocket sight

Radar ranging equipment

Battery

Tacan antenna

Radar antenna

Gun camera

Air brake

Retractable landing and taxi light

Data case

Tacan antenna

Oxygen bottle

Retractable landing light

Aft radio compartment

Ammunition belts

Ammunition container (P & S)

Canopy external control button

Gun compartment

Automatic wing slats (P & S)

Identification radar antenna

Forward fuselage fuel tank – upper cell (P & S)

Pitot head

US AIR FORCE
91096

FU-096

**North American F-86F Sabre**
The cutaway drawing (top) shows the layout of
this final day fighter production version of the Sabre
*Crew:* 1 *Powerplant:* 1 General Electric
J47-GE-27, 6100 lb thrust *Span:* 37 ft 1 in
*Length:* 37 ft 6 in *Weight:* 19,600 lb
*Armament:* 6×·5-in mg *Speed:* 656 mph at sea level

**Mikoyan-Gurevich MiG-15**
*Crew:* 1 *Powerplant:* 1 RD-45, 5000 lb thrust
*Span:* 33·3 ft *Length:* 33·3 ft *Weight:* 10,800 lb
*Armament:* 1×37-mm cannon; 2×23-mm cannon
*Speed:* 663 mph

**Grumman XF9F-2 Panther**
*Crew:* 1 *Powerplant:* 1 Rolls-Royce Nene,
5000 lb thrust *Span:* 35·3 ft *Length:* 37·7 ft
*Weight:* 10,840 lb *Armament:* 4×20-mm cannon
*Speed:* 573 mph at 20,000 ft

of the world's most famous fighters, and
was built in enormous quantities, perhaps
15,000 or more. It was flown by many
countries, from the USSR to emerging
nations that had never known any prior
form of airpower.
*23 March 1948:*
**Douglas XF3D-1 Skyknight**
The dark blue, stubby and straight-winged
prototype led to a series of jet night fighters,
equipped with search radar and a complex
gunfire control system. Paired Westinghouse axial-flow jet engines nestled in
its belly, and the two-man crew sat side-by-

side under the cockpit canopy. Like the Sabre, Panther and MiG, the Skyknight was to go on to gain combat laurels in the Korean war.

*16 August 1948:*
## Northrop XF-89 Scorpion
The night fighter has always been a specialised weapon, working with a combination of electronic and human sensors to seek out and destroy its prey under cover of darkness or bad weather. It needs a two-man crew to fly and to operate the complex radar, plus endurance and heavy armament. Thus it tends to be big and heavy. The Northrop XF-89 featured the same basic layout as the Skyknight, except that its crew sat in tandem positions, and it dispensed with guns as the primary

*The Douglas XF3D Skyknight was a radar-equipped night fighter. Two Westinghouse engines gave it a maximum speed of 500 mph, and its armament consisted of four 20-mm cannon*

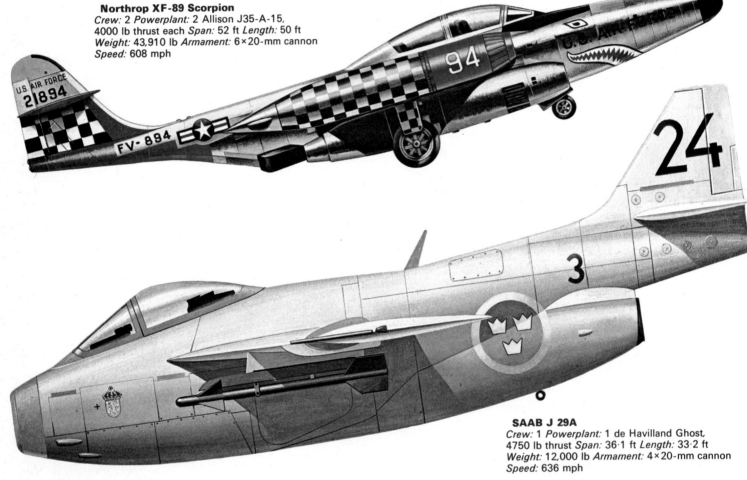

**Northrop XF-89 Scorpion**
*Crew:* 2 *Powerplant:* 2 Allison J35-A-15, 4000 lb thrust each *Span:* 52 ft *Length:* 50 ft *Weight:* 43,910 lb *Armament:* 6×20-mm cannon *Speed:* 608 mph

**SAAB J 29A**
*Crew:* 1 *Powerplant:* 1 de Havilland Ghost, 4750 lb thrust *Span:* 36·1 ft *Length:* 33·2 ft *Weight:* 12,000 lb *Armament:* 4×20-mm cannon *Speed:* 636 mph

weapons. It was armed with 104 small folding-fin air-to-air unguided rockets, housed in wingtip pods. They were fired in a devastating ripple pattern, rocket after rocket bursting from the pod at split-second intervals.

*1 September 1948:*
## SAAB J 29
Europe's first sweptback wing fighter design, the barrel-shaped prototype was the first in a large quantity of the speedy Swedish fighters. Some were delivered as low-level reconnaissance aircraft, with batteries of cameras in the forward fuselage. Many years later, the J 29s served in the Congo with a United Nations force.

*18 September 1948:*
## Convair XF-92A
The delta wing, a major technical innovation based on German experiments, first flew on the Convair XF-92A. Since there was no previous flight research experience with the new wing form, the Convair design was developed as both a flight-test aircraft and the possible prototype for an interceptor. It featured trailing-edge elevons, control surfaces that combined the functions of elevators and ailerons. Experience with the XF-92A led to Convair's later successes with the more advanced F-102A and F-106A delta-winged supersonic interceptors.

*29 September 1948:*
## Vought XF7U-1 Cutlass
Another tailless design, but based on German sweptback wing technology rather than that of the delta shape, the Cutlass was developed for the US Navy as a carrier-based, twin-engined interceptor. Its performance was based on the use of afterburners for its jet engines, which increased the thrust substantially by adding and burning additional fuel downstream of the engine in the tailpipe. Trouble dogged the series, and the Cutlass never achieved the expected performance, what with its problematic Westinghouse jet engines and its tricky aerodynamics.

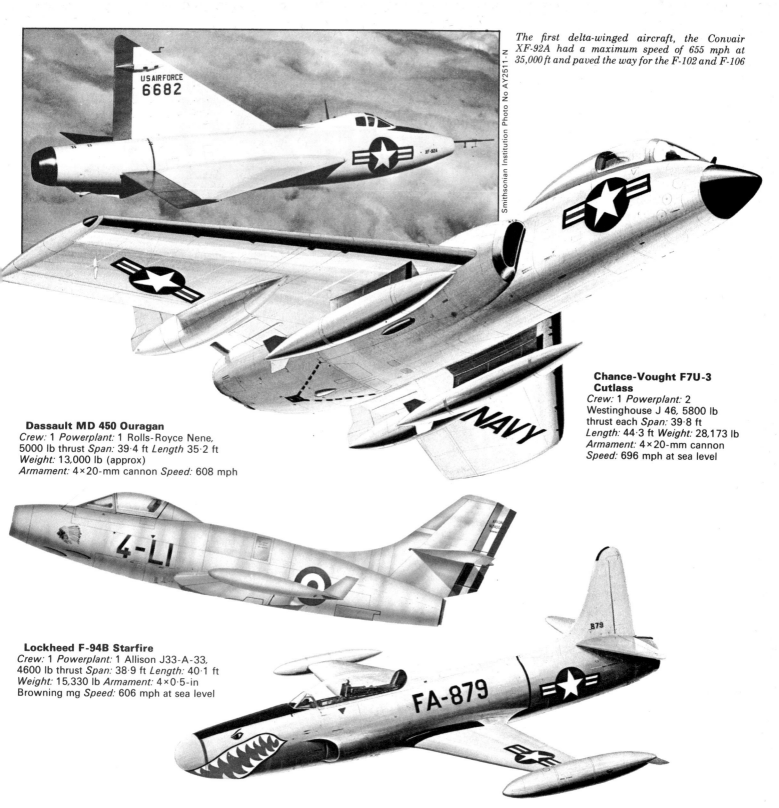

Smithsonian Institution Photo No AY2511-N

*The first delta-winged aircraft, the Convair XF-92A had a maximum speed of 655 mph at 35,000 ft and paved the way for the F-102 and F-106*

**Chance-Vought F7U-3 Cutlass**
*Crew:* 1 *Powerplant:* 2 Westinghouse J 46, 5800 lb thrust each *Span:* 39·8 ft *Length:* 44·3 ft *Weight:* 28,173 lb *Armament:* 4×20-mm cannon *Speed:* 696 mph at sea level

**Dassault MD 450 Ouragan**
*Crew:* 1 *Powerplant:* 1 Rolls-Royce Nene, 5000 lb thrust *Span:* 39·4 ft *Length* 35·2 ft *Weight:* 13,000 lb (approx) *Armament:* 4×20-mm cannon *Speed:* 608 mph

**Lockheed F-94B Starfire**
*Crew:* 1 *Powerplant:* 1 Allison J33-A-33, 4600 lb thrust *Span:* 38·9 ft *Length:* 40·1 ft *Weight:* 15,330 lb *Armament:* 4×0·5-in Browning mg *Speed:* 606 mph at sea level

*28 December 1948:*
## Supermarine Type 510
This experimental prototype led directly and eventually to Britain's first sweptwing service fighter, but there were to be other prototypes with many changes before the final production configuration had been adopted. Then more than five years were to elapse between the first flight of the prototype and the early deliveries to Royal Air Force squadrons. In the end, the Swift – as the service fighter development was named – never achieved a full measure of success. It was outpaced by the advances of technology and the complexities of high-speed aircraft design.

*28 February 1949:*
## Dassault MD 450 Ouragan
After more than two years of experimentation and trial of a variety of designs, the French produced the first in a long series of jet fighters that were to establish that country as a major exporter of aircraft. Like so many early jet designs, the Ouragan had a straight wing and a centrifugal-flow engine. But it was a major breakthrough for the French, and particularly for Marcel Dassault himself, then and still the leading exponent of private enterprise in the nationalised French aircraft industry. The Ouragan was the first French jet fighter to go into large-scale production, and was also sold abroad.

*16 April 1949:*
## Lockheed XF-94 Starfire
The USAF sponsored continuing development of night- and all-weather fighters, hoping to counter the trend toward larger, heavier and more costly aircraft of this type. One of the successful attempts was the Starfire. Single-engined, using its afterburner for bursts of power during takeoff and at altitude, the F-94 had excellent performance for its time. Its design was an adaptation of the basic Lockheed F-80/T-33 series, so its cost was lower than it would have been if developed as a new type. Its unusual feature was the rocket armament. Like the F-89, it was also armed with folding-fin rockets, named 'Mighty Mouse' after a cartoon character of the day, and they were housed in mid-wing pods and in an annulus around the blunt nose radome.

*9 May 1949:*
## Republic XF-91
Designed as a mixed-powerplant interceptor, the XF-91 was built around the combined thrust of a turbojet engine and a powerful four-barrelled liquid-propellant rocket engine. The theory was that the

rocket engine would provide super-performance at high altitudes, long after the thrust of the turbojet had fallen off to a fraction of its sea-level value. The Republic design was the first aircraft to fly with such a combined powerplant, although it was not until much later in the programme that it did so, and then not with the proposed production engine. Among the unusual technical features of the design were the inverse-taper wings, broader at the tips than at the root. This improved the low-speed performance. Additionally, the wings had variable incidence. Tandem landing gear was another innovative feature. The XF-91 was both the first and last attempt in the United States

to follow the mixed-powerplant formula, and it never progressed beyond the experimental prototype.

*2 September 1949:*
**De Havilland DH 112 Venom**
Britain's first all-weather interceptor came out of this successful fighter-bomber design. It was rather like an enlarged Vampire, with its twin tail boom layout and the egg-shaped fuselage carrying the jet engine. Its bulbous radome housed airborne intercept radar, and the two-man crew sat side by side under the broad canopy. It pioneered the use of wingtip fuel tanks in the RAF. Its wings had a modest degree of sweepback to give a straight-across trailing edge.

*22 December 1949:*
**North American YF-95A Sabre**
Originating as a modification of the Sabre line, the YF-95A (later redesignated F-86D) was designed as a night- and all-weather fighter, armed only with rockets. It required an entirely new fuselage to house a more powerful engine with an afterburner, and the nose radome changed the contours of the straight-in nose inlet of the standard Sabres. Wings, tail and landing gear came unchanged from the Sabre production line.

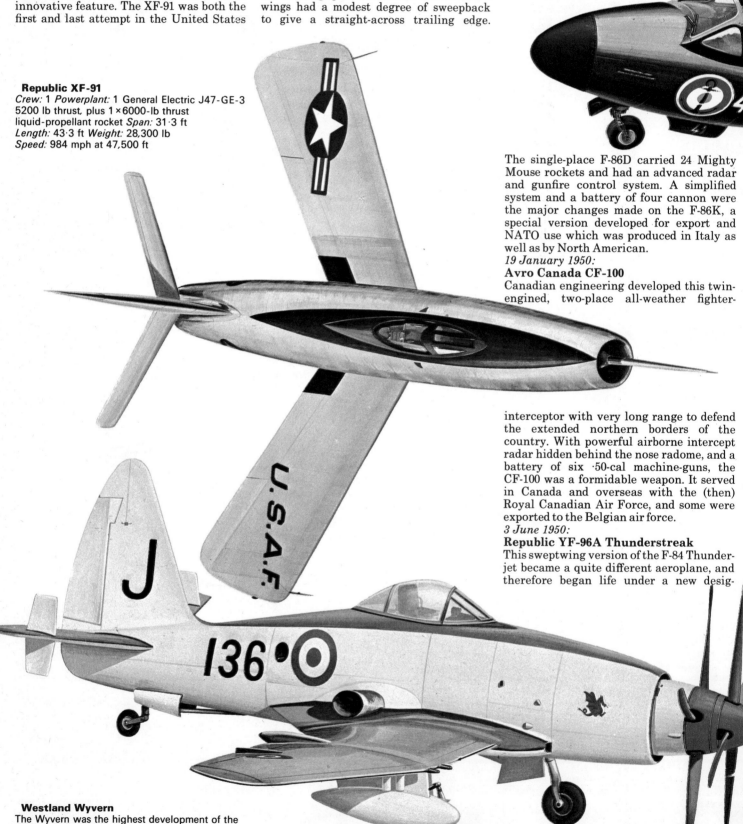

**Republic XF-91**
*Crew:* 1 *Powerplant:* 1 General Electric J47-GE-3 5200 lb thrust, plus 1 × 6000-lb thrust liquid-propellant rocket *Span:* 31·3 ft *Length:* 43·3 ft *Weight:* 28,300 lb *Speed:* 984 mph at 47,500 ft

The single-place F-86D carried 24 Mighty Mouse rockets and had an advanced radar and gunfire control system. A simplified system and a battery of four cannon were the major changes made on the F-86K, a special version developed for export and NATO use which was produced in Italy as well as by North American.

*19 January 1950:*
**Avro Canada CF-100**
Canadian engineering developed this twin-engined, two-place all-weather fighter-interceptor with very long range to defend the extended northern borders of the country. With powerful airborne intercept radar hidden behind the nose radome, and a battery of six ·50-cal machine-guns, the CF-100 was a formidable weapon. It served in Canada and overseas with the (then) Royal Canadian Air Force, and some were exported to the Belgian air force.

*3 June 1950:*
**Republic YF-96A Thunderstreak**
This sweptwing version of the F-84 Thunderjet became a quite different aeroplane, and therefore began life under a new desig-

**Westland Wyvern**
The Wyvern was the highest development of the torpedo strike fighter concept, but with a turbo-prop driving contra-rotating propellors it could not compete with the pure-jet naval fighter

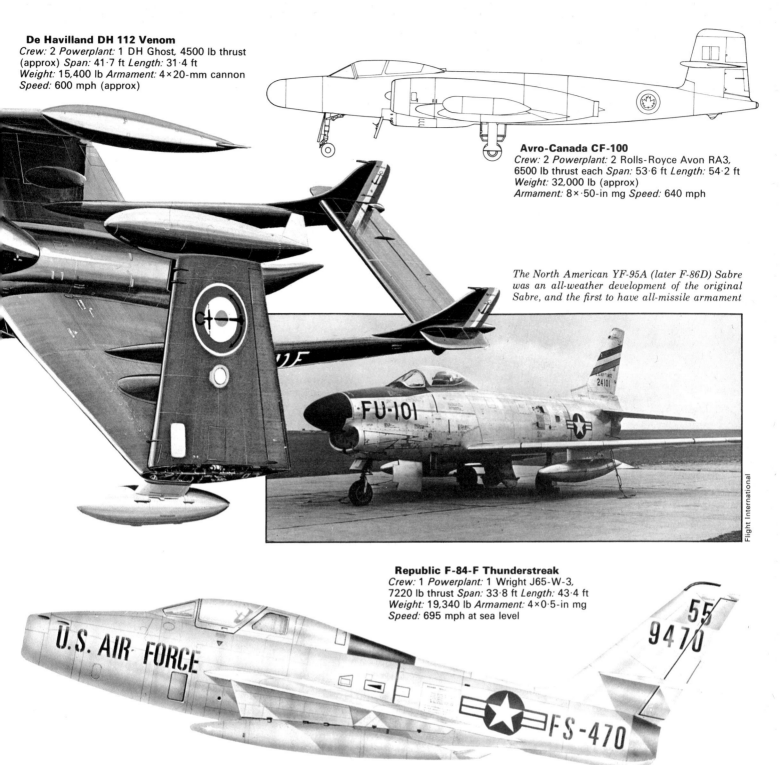

**De Havilland DH 112 Venom**
*Crew:* 2 *Powerplant:* 1 DH Ghost, 4500 lb thrust (approx) *Span:* 41·7 ft *Length:* 31·4 ft *Weight:* 15,400 lb *Armament:* 4 × 20-mm cannon *Speed:* 600 mph (approx)

**Avro-Canada CF-100**
*Crew:* 2 *Powerplant:* 2 Rolls-Royce Avon RA3, 6500 lb thrust each *Span:* 53·6 ft *Length:* 54·2 ft *Weight:* 32,000 lb (approx) *Armament:* 8 × ·50-in mg *Speed:* 640 mph

*The North American YF-95A (later F-86D) Sabre was an all-weather development of the original Sabre, and the first to have all-missile armament*

Flight International

**Republic F-84-F Thunderstreak**
*Crew:* 1 *Powerplant:* 1 Wright J65-W-3, 7220 lb thrust *Span:* 33·8 ft *Length:* 43·4 ft *Weight:* 19,340 lb *Armament:* 4 × 0·5-in mg *Speed:* 695 mph at sea level

nation, later changed to F-84F. It was powered by an American-built model of a British jet engine, the Armstrong Siddeley Sapphire. But the Sapphire must have suffered in the translation, because it took an abnormally long time before the F-84F was accepted for service. This author remembers a visit to Edwards AFB during the accelerated service testing of the early production Thunderstreaks, and hearing the almost uniformly bad comments from pilots and technicians alike about ,the short-lived engines (as much as 25 hours between overhauls, when they were lucky) and its flying qualities (it was called the Hog, the Lead Sled, and other uncomplimentary names). But eventually the troubles were licked, and the F-84F went on to serve well in the tactical air arms of many countries as well as that of the United States.

The F-84F, like many of the aircraft developed since the end of the Second

World War, was a basis for a major export programme. It happened, in almost every case, because somebody's air force wanted to convert from old-fashioned piston-engined fighters to the new jet breed, and didn't have a local industry that could develop the relatively complex airplanes. In fact, their local industry often could hardly cope with simpler types, and many a jet was to be delivered later, flown a few times until something went wrong, and left to stand on the ramp, deteriorating, unfixed and unfixable by the local talent.

The export pattern solidified early. It was clear that exporting was one way of recovering some of the high development costs of these new jet fighters and, at the same time, of exerting a powerful political and military influence on the customer country. The early postwar years saw British dominance in the marketplace with its Meteor and Vampire fighter and trainer lines. Between the two companies, they sold aircraft to 26

countries, alphabetically beginning with Argentina and ending with Venezuela. The planes were manufactured under licence in eight other countries besides Great Britain.

The United States did not, at first, sell its new jet aircraft abroad, preferring to unload some of its vast stock of piston-engined Mustangs and Thunderbolts. It was to be a while before the US industry began to take the export market very seriously.

The postwar years were characterised, then, by the maturing of wartime designs into a number of basic types of operational jet fighters. The discoveries and wild ideas of wartime were exploited in new designs and used as the foundation for further advances in technology, and the advantages of an export market became very apparent.

But the events of June 1950 were to play a very important part in the future development of jet fighters, and in the concepts of their design and operational use by opposing air forces.

# THE FIRST JET FIGHTER ACES

The experiments with jet aircraft in the years after the Second World War, and the planes that were developed, did not have to wait long to be put to the test. In 1950 the Korean War began, and soon Russian and American Jet fighters were mixing it over war-torn Asia – not for the last time

When the North Koreans struck across their borders against South Korea early on the morning of 25 June 1950, they set in motion events that became major factors in the maturing of military jet aircraft design.

In the conflict that followed, the aerial warfare quickly became a war of interdiction, with primary roles assigned to bombers and fighter-bombers. Whatever fighter-to-fighter combat resulted was subordinate to those primary missions. This is not to downgrade the extremely valuable role of fighters in that war, but to emphasise that their missions rose out of the USAF's need to protect its bomber forces, and the North Korean and Chinese need to destroy those forces.

Some significant milestones of the jet fighter war in Korea should be recorded here. Within a week of the invasion of South Korea, USAF jets scored their first victories. Four Lockheed F-80Cs from the 35th Fighter-Bomber Squadron tangled with eight Ilyushin piston-engined attack aircraft, and shot down four of the Russian-built planes.

The first-ever combat between jet fighters was on 8 November 1950. Lt Russell J Brown, pilot of an F-80C from the USAF's 51st Fighter-Interceptor Wing, blasted a Russian-built MiG-15 that unwisely tried to out-dive the Lockheed plane.

The first victory by a North American F-86A Sabre over a MiG took place on 17 December, gained by Lt Col Bruce H Hinton. On 20 May 1951, USAF Capt James Jabara became history's first jet ace, downing his fifth and sixth MiG-15s during a single combat.

The first jet night victory was achieved by a USMC Douglas F3D-2 Skyknight, vectored by ground radar to locate a Yak-15 in the night skies over Sinui-ju.

Near the end of the war, United Nations air superiority had been established without challenge. During June 1953, USAF Sabres sighted 1268 MiG-15s, and engaged 501 of them in battle. They destroyed 77, probably destroyed another 11, and damaged 41, without losing a single Sabre all month. Those kinds of scores helped increase the highly publicised kill ratio which, near the end of the war, averaged out to better than ten to one. USAF pilots gunned down 792 MiG-15s for the loss of 78 Sabres, according to final official US figures.

What was learned? Early in the fighting, the Sabre pilots wanted more thrust – they were tired of being bounced from above by MiGs with superior altitude performance. They wanted heavier-calibre guns with a high firing rate; the ·50-cal machine-guns, so effective against German and Japanese designs during the Second World War, often failed to destroy a MiG because of the lack of striking power. And they wanted a radar-ranging gunsight, because the gyro types in the Sabres were not suitable for holding the large leads required during deflection shooting in turning combat.

Conceptually, the idea of air superiority was again tested in the skies, and the final ability of the UN air forces to fly almost anywhere without serious challenge was proof of the value, and the attainment, of that concept. Jet fighters also made good fighter-bombers, it was found, able to deliver their ordnance loads with speed and accuracy.

*The end of a MiG-15, photographed with a camera mounted in the nose of an F-86 Sabre*

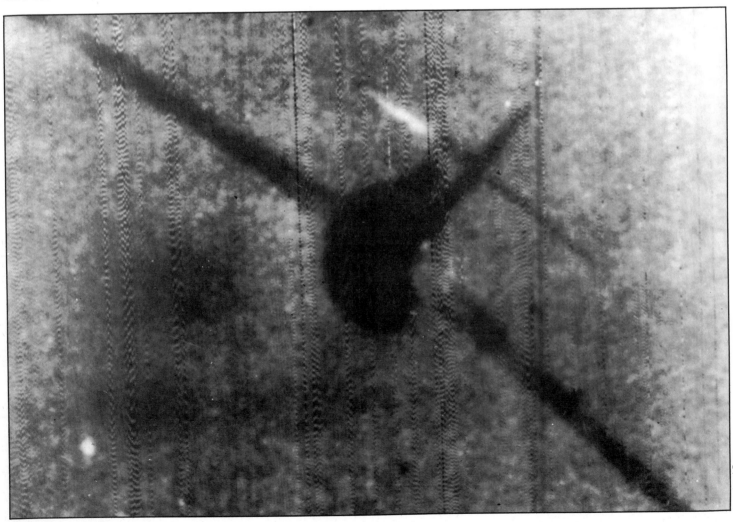

Associated Press

The idea of the long-range escort fighter grew out of early experience of the accompanied bomber raids. The B-29s operated by Strategic Air Command were slow; they were escorted by a top cover of F-86 Sabres and lower elements of Thunderjets. Typically, the attacking MiGs would streak through the top cover and go after the bombers; they were able to avoid combat with the Thunderjets by virtue of superior speed and manoeuvrability. One answer seemed to be a supersonic long-range fighter that could both escort and fight.

By the time the Korean war was seriously under way, one basic form of future air action had been further emphasised. Long-range bomber forces would continue to be one component of any future threat. The defence would be by a mix of interceptor aircraft and missiles, able to reach out with electronic senses to see, attack and destroy the bombers at either long or short range and in all kinds of weather.

This scenario, born during the Second World War, gave rise to a continuing series of fighter-interceptor designs during the decade which began with the Korean war. In the sequential descriptions that follow, note how many of the new aircraft fit into this single category.

*23 January 1951:*
### Douglas XF4D-1 Skyray
This tailless interceptor was designed for Fleet defence under the conditions of the last months of the Second World War. It emphasised climb performance at the expense of range. Although it demonstrated its potential by setting a series of world

*US jets scored heavily over MiGs in Korea, in spite of atrocious climatic conditions*

*Major James Jabara (with cigar), first ace of the Korean War, with two fellow pilots at their base*

records, it was delayed from operational use by engine problems. Later, with those problems solved, it served well as a fighter on US carriers.

*23 February 1951:*
### Dassault MD 452-01 Mystère
Basically a sweptwing version of the Ouragan, the Mystère prototype was the first of a series that went through extensive development, including major aerodynamic, structural and powerplant changes. It was built in quantity under off-shore procure-

ment contracts from the US, and became one of Europe's top fighter aircraft. An advanced version, the Mystère IVA, saw service in the combat in the Suez war of 1956.

*20 July 1951:*
### Hawker P 1067
The pale green prototype was an aircraft of classic beauty. It was developed into the Hawker Hunter, first-line Royal Air Force interceptor and later ground-attack fighter. It was widely used in a variety of roles by many countries, and is regarded by many as the peak development of the subsonic jet fighter. Originating as an interceptor study built around the then-new Rolls-Royce Avon engine, it featured heavy armament and outstanding flying qualities.

*7 August 1951:*
### McDonnell XF3H-1 Demon
The US Navy wanted this high-performance interceptor to give its carrier forces the same kind of defensive protection that land-based interceptors afforded. But the dismal failure of its Westinghouse engine to live up to requirements hamstrung the Demon from the start. It was only after long and trouble-filled delays that the type was cleared for Fleet service. The developed models served on carriers as night-fighters and as missile-armed interceptors.

*20 September 1951:*
### Grumman F9F-6 Cougar
First sweptwing fighter in the US Navy, the Cougar was essentially a sweptwing Panther with a new horizontal tail. Development time was shortened by this fairly simple modification of the straight-winged F9F series, and the Cougars went on to serve with the Navy and the Marines as a fighter, a reconnaissance aircraft, and a trainer.

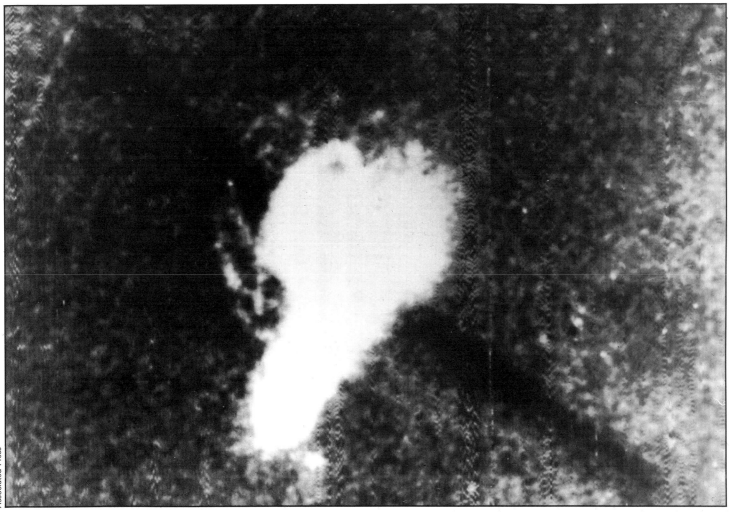

**26 September 1951:**
**De Havilland DH 110**
This twin-engined, two-place, twin-boomed, trans-sonic all-weather fighter was developed in response to a Royal Air Force requirement. After the structural failure and crash of the first prototype during the 1952 Farnborough show, the DH 110 was rejected in favour of the Gloster Javelin. The Royal Navy then funded the development programme and, after major redesign, the modified DH 110 was named the Sea Vixen. The first production version of that Naval all-weather interceptor flew in March 1957. It was the first British gunless interceptor, armed instead with four DH Firestreak missiles and 28 unguided rockets.

**3 November 1951:**
**SAAB 32 Lansen**
Swedish defence policy requirements for a strike fighter were the origin of this graceful and efficient two-place aircraft. Developed for a multi-role mission, the Lansen was produced as a night- and all-weather fighter in addition to its attack and reconnaissance versions. It was the first Swedish aircraft to attain supersonic speeds, flying in shallow dives to reach them for brief periods.

**26 November 1951:**
**Gloster GA 5 Javelin**
Conservatively designed as a tailed delta to improve its landing characteristics for night operations, the Javelin was the first twin-jet delta-winged aircraft to fly. It had a huge wing, in area and volume, and could carry a large load of fuel as well as the bulky and heavy radar needed for its mission. It went through a long development period and several production versions, used both British and American radars, and lasted in service with the Royal Air Force until 1967.

**27 December 1951:**
**North American XFJ-2 Fury**
With Sabres matching and beating MiG-15 performance in Korea, the US Navy concluded that its fastest way to get similar performance was to get a similar airplane. The XFJ-2, first of the Fury line, was a modified and navalised standard F-86E Sabre. Changes included arresting gear and heavier armament. But its production was slowed by increasing demands of higher priority for the F-86s, which were being built on the same production lines in the same factories. Eventually, the Fury was modified to a strike fighter with nuclear capability.

**16 October 1952:**
**SNCASO SO 4050 Vautour**
This versatile, twin-engined, two-place aircraft was produced in three versions, but

**Dassault Super Mystère B-2**
Last production version of the Mystère
*Crew:* 1 *Powerplant:* SNECMA Atar 101G, 7480 lb thrust *Span:* 34·5 ft *Length:* 46·1 ft *Weight:* 19,840 lb *Armament:* 2×30-mm cannon; 35×68-mm air-to-air rockets *Speed:* Mach 1·25 at altitude

*November 1952: prototype of the Hawker Hunter, probably the best subsonic jet fighter ever built*

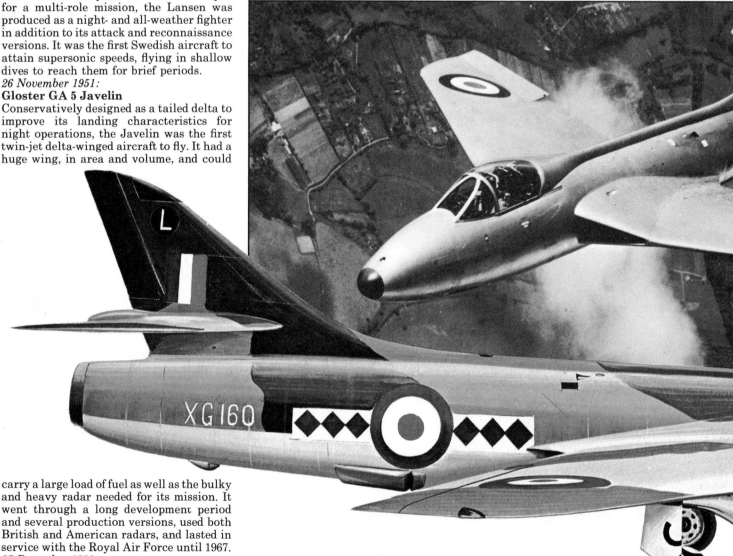

*Above: Grumman F9F-6 Cougar, a sweptwing development of the F9F-2 Panther, and the US Navy's first sweptwing fighter. Armed with four 20-mm cannon, it had a speed of about 550 mph*

**Hawker Hunter 6**
*Crew:* 1 *Powerplant:* Rolls-Royce Avon 200, 10,500 lb thrust *Span:* 33·7 ft *Length:* 45·8 ft *Weight:* 17,600 lb *Armament:* 4 Aden cannon *Speed:* 715 mph at 36,000 ft

the greatest number were night and all-weather fighters. They were armed with a powerful battery of four 30-mm cannon and 232 air-to-air rockets in paired belly trays. This was France's only all-weather fighter for several years, and it also served that country's defences as a two-place light bomber and a single-seat strike aircraft. Some of the latter were exported to Israel, and fought with the Israeli air force in the Suez war.

*2 March 1953:*
### SNCASO SO Trident I
Another in the periodic appearances of the super-performance manned interceptor, the Trident was a mixed-powerplant aircraft with turbojets at the wingtips and a powerful rocket motor in the fuselage. Its rate of climb was comparable to that of contemporary guided missiles, and it had high supersonic speed in level flight. Pre-production aircraft were built, but the planned production programme never materialised because of the lack of French government support.

*19 May 1953:*
### Grumman XF10F-1 Jaguar
The date was eagerly anticipated and now is happily forgotten by senior personnel at Grumman. The Jaguar was the world's first variable-sweep aircraft and – in that respect – it performed very well, without any trouble during the entire flight-test programme. But the Jaguar was another casualty of the Westinghouse jet engine fiasco. There is one former Grumman engineering test pilot who wears a pair of Jaguar cufflinks, given by his wife to remind him that whenever things seem bad, they once were worse.

*25 May 1953:*
### North American YF-100A Super Sabre
First of the Century Series of fighters for the USAF, the Super Sabre was designed as a tactical day fighter based on the lessons of Korea. The first approach had been to plan a new wing for the old Sabre; but a powerful new engine – the Pratt & Whitney J57 – became available, and it was decided to design around it. The combination was a winner: it was the world's first supersonic combat aircraft, and the progenitor of a long-lived line of fighters and fighter-bombers. It pioneered the low-set horizontal tail to eliminate a disastrous form of instability appearing on high-speed aircraft. Other innovations included the one-piece 'slab' horizontal tail, and the use of titanium metal for some components.

*September 1953:*
### Mikoyan MiG-19
Opposite number to the Super Sabre, this Russian fighter became the second supersonic combat aircraft in the world. It used a twin-engined layout and contemporary aerodynamic features to develop a high-performance fighter and night-fighter that served well and long with the Soviet Union and their allied countries and friends.

*24 October 1953:*
### Convair YF-102A Delta Dagger
Designed as an all-weather interceptor, the Delta Dagger did not, at first, meet its expected supersonic speed performance because of high drag. Happily, a development at the laboratories of the National Advisory Committee for Aeronautics (now NASA) produced a way of reducing trans-sonic drag. The YF-102 was speedily redesigned to take advantage of the 'area rule' developed by NASA's Richard T Whitcomb, and it easily slipped through the speed of sound in

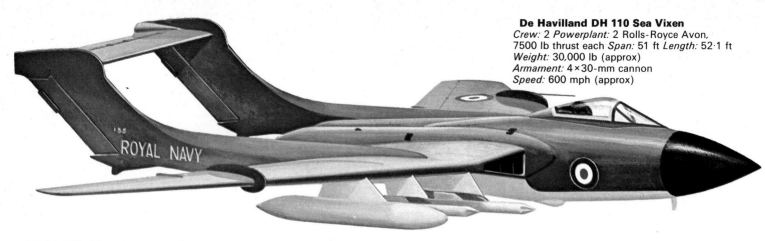

**De Havilland DH 110 Sea Vixen**
*Crew:* 2 *Powerplant:* 2 Rolls-Royce Avon,
7500 lb thrust each *Span:* 51 ft *Length:* 52·1 ft
*Weight:* 30,000 lb (approx)
*Armament:* 4×30-mm cannon
*Speed:* 600 mph (approx)

**SAAB A32-A Lansen**
*Crew:* 1 *Powerplant:* 1 SFA RM 5, 8050 lb thrust
*Span:* 42·7 ft *Length:* 48 ft *Weight:* 16,535 lb
(empty) *Armament:* 4×20-mm cannon
*Speed:* 700 mph at sea level

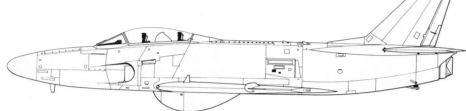

**Gloster Javelin FAW 8 (foot of page)**
The FAW 8 was the final production version of
the Javelin, and was equipped with American
radar equipment. Below left: Gloster Javelin
taking off
　(FAW 8) *Crew:* 2 *Powerplant:* 2 Bristol
Siddeley Sapphire 203/204, 11,000 lb thrust each
*Span:* 52 ft *Length:* 56·3 ft *Weight:* 38,000 lb
*Armament:* 2×30-mm cannon; 4 Firestreak
missiles *Speed:* 695 mph at 10,000 ft

Flight International

*Above: The SNCASO SO 4050 Vautour was France's only all-weather fighter for several years, and was powerfully armed with four 30-mm cannon plus 232 68-mm unguided rockets. Above right: Engine trouble with the Grumman XF10F-1 Jaguar made it a major disappointment to its sponsors. Right: The SNCASO SO Trident was powered by two Armstrong Siddeley Vipers and a 6600-lb thrust liquid-fuel rocket. It was France's first plane capable of supersonic speeds in level flight, but in spite of its exceptionally good performance it never achieved production status*

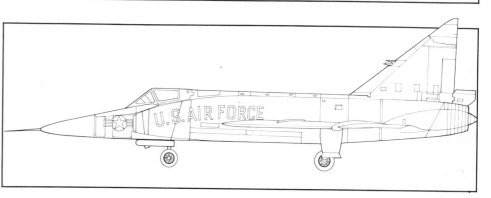

**Convair YF-102 Delta Dagger**
Above, and, left, in flight
 *Crew:* 1 *Powerplant:* 1 Pratt & Whitney
J57-P-11, 9700 lb thrust *Span:* 37 ft
*Length:* 52·5 ft *Weight:* 25,000 lb (approx)
*Armament:* 6 Falcon guided missiles; 12×2·75-in
rockets *Speed:* 780 mph

167

**North American F-100D Super Sabre**
*Crew:* 1 *Powerplant:* 1 Pratt & Whitney
J57-P-21A, 11,700 lb thrust *Span:* 38·8 ft
*Length:* 54·3 ft *Weight:* 29,762 lb
*Armament:* 4×20-mm cannon
*Speed:* 864 mph at 35,000 ft

Flight International

*F-100D Super Sabres, final production version of the F-100 Sabre, at the point of takeoff*

**Mikoyan MiG-19**
*Crew:* 1 *Powerplant:* 2 AM-5, 4850 lb thrust
each *Span:* 29·5 ft *Length:* 42·9 ft
*Weight:* 15,000 lb (approx)
*Armament:* 1×37-mm cannon; 2×23-mm cannon
*Speed:* 900 mph (approx)

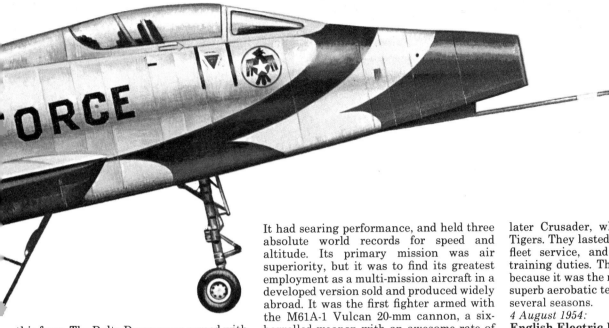

this form. The Delta Dagger was armed with Falcon guided missiles and a battery of air-to-air unguided rockets.

*16 December 1953:*
### Dassault Mystère IVB
Just about two months after its first flight, the Mystère IVB joined the level-flight supersonic club and became the first European fighter to do so. It was one of several progressive developments of the original Mystère prototype, and used after-burning on its jet engine to improve its takeoff, climb and speed performance.

*7 February 1954:*
### Lockheed XF-104 Starfighter
One answer to the unofficial requirements of the Korean war was the Starfighter, designed as an uncomplicated day fighter.

It had searing performance, and held three absolute world records for speed and altitude. Its primary mission was air superiority, but it was to find its greatest employment as a multi-mission aircraft in a developed version sold and produced widely abroad. It was the first fighter armed with the M61A-1 Vulcan 20-mm cannon, a six-barrelled weapon with an awesome rate of fire. It also carried Sidewinder missiles for air-to-air combat.

*30 July 1954:*
### Grumman YF9F-9 Tiger
The area rule that benefited the Convair F-102 had been applied earlier to the Grumman YF9F-9, a major modification of the Panther/Cougar series of Naval fighters. It was later redesignated F11F-1. It preceded the area-ruled YF-102A into the air by several months, and was the first aircraft to fly with this new applied principle of aerodynamics. As one result, it became the Navy's, and the world's, first supersonic carrier-based fighter. But its delivery to the fleet was delayed by engine problems, and it began to arrive at the same time as the

later Crusader, which soon replaced the Tigers. They lasted only about two years in fleet service, and then went ashore to training duties. The Tiger's fame remains, because it was the mount for the US Navy's superb aerobatic team, the Blue Angels, for several seasons.

*4 August 1954:*
### English Electric P 1
The angular shape of the P 1 prototype and the over-and-under arrangement of its twin engines looked like power personified. It was: the P 1 was the basis for development of the outstanding Lightning interceptor, still in active service with the Royal Air Force. It was Britain's first fighter capable of level-flight supersonic speed. Designed for the specific conditions of defence of the British Isles, the Lightning and its later

**Lockheed F-104G Super Starfighter**
*Crew:* 1 *Powerplant:* 1 General Electric J79-GE-11A, 10,350 lb thrust *Span:* 21·9 ft *Length:* 54·8 ft *Weight:* 20,900 lb *Armament:* 1 × 20-mm Vulcan cannon; 2 Sidewinder missiles *Speed:* Mach 2·2

*A Lockheed F-104G Starfighter during instant takeoff tests for the German air force at Edwards Air Force Base. The solid fuel rocket motor propelled it into the air, and was then jettisoned*

Associated Press

developments were characteristically short on range but long on performance. The astounding rate of climb, coupled with automatically controlled weapons, make the Lightning a formidable fighter, even by today's high standards.

*29 September 1954:*

## McDonnell F-101A Voodoo

USAF's Strategic Air Command, drawing on its Korean experience, wanted a long-range fighter capable of escorting bomber fleets to distant targets. The F-101 Voodoo was the result. This twin-jet, two-place

**Grumman F11F-1 Tiger**
*Crew:* 1 *Powerplant:* 1 Wright J65-W-18, 7800 lb thrust *Span:* 31·6 ft *Length:* 44·9 ft *Weight:* 21,035 lb *Armament:* 4 × 20-mm cannon; 4 Sidewinders *Speed:* 740 mph at 35,000 ft

fighter was developed from an earlier prototype, the XF-88, designed as a fighter able to strike deeply into enemy territory. SAC cancelled its requirements before the Voodoo flew, but the design was adopted by Tactical Air Command, was developed as a fighter-bomber, and later was further developed into a long-range all-weather interceptor of high performance, and a low-level photo-reconnaissance aircraft. In the latter role, Voodoos furnished many photographs of the missile sites emplaced by the Russians in Cuba in 1961.

*1954:*

## Yakovlev Yak-25

This twin-engined, two-place night- and all-weather fighter was first seen publicly in 1955, and therefore probably flew late in 1954. As the first Russian aircraft of its type, the layout and systems were a bit behind the state of the art, a deficiency that was remedied with surprising speed in later designs from the Yakovlev design bureau. The large nose radome hints at a radar dish dimensioned for long-range detection.

*2 March 1955:*

## Dassault Super Mystère B-2

For a quick and effective survey of the state of the French jet fighter art in the 1950s, look at the Dassault Mystère series. This model, the end of the line, was a major redesign of the basic format, featuring a thinner wing with a higher sweep angle, a redesigned windshield for lower drag, and other refinements. It easily went supersonic in level flight on its first flight.

*25 March 1955:*

## Vought XF8U-1 Crusader

The Navy, drawing on its Korean experience, asked for a supersonic day fighter for fleet defence. The Crusader was the answer. It flew supersonically on its first flight, was the first carrier-based aircraft to exceed 1000 mph in level flight, and crossed the United States at supersonic speed. Its technical innovations included a variable-incidence wing for superb visibility during approaches to carrier landings, and full application of the area rule. Armed with

*English Electric P1A, prototype Lightning, whose performance has kept it in front line service*

**Chance Vought F8U-2 Crusader**
*Crew:* 1 *Powerplant:* 1 Pratt & Whitney J57-P-16, 10,700 lb *Span:* 35·2 ft *Length:* 54·5 ft *Weight:* 28,000 lb *Armament:* 4 × 20-mm cannon; 4 Sidewinders *Speed:* Mach 1·7

**English Electric Lightning F1**
*Crew:* 1 *Powerplant:* 2 Rolls-Royce Avon 200,
11,250 lb thrust each *Span:* 34·9 ft
*Length:* 50 ft *Weight:* 40,000 lb (approx)
*Armament:* 2 Aden cannon; 2 Firestreak missiles
*Speed:* Mach 2·1 at 40,000 ft

cannon and Sidewinder, it packed a powerful punch. At its maximum deployment, it
equipped about half of the Navy and Marine
fighter squadrons. It was further developed
with a boundary-layer control system for
the French Navy. Most recently, it fought
in the Vietnam war where it acquired a
reputation as the 'best gun fighter' in the
theatre. It was redesignated as the F-8
Crusader in 1962.
*25 June 1955:*

**Dassault MD 550-01 Mirage I**
Like the Trident before it, the Mirage I
featured a mixed powerplant. But its paired
turbojets were in the fuselage, and the
rocket motor was slung in a droppable
package under the belly. It was the ancestor
of the current Mirage III line, and was
developed through a series of engine and
wing changes to become France's most
successful fighter, one of its best export
programmes, and one of the world's best
fighters, proven in combat against top-
notch Russian-built aircraft in the Middle
East wars.
*18 July 1955:*

**Folland Gnat**
Even though USAF pilots in Korea argued
loud and long for a simple, light fighter,
nobody took them seriously. All the fighters
inspired by that conflict were heavier and
more complex than the Sabres and MiG-15s

except for one: the Gnat. It was a small and
light fighter designed to carry the optimum
minimum in armament and fuel while still
being an effective interceptor. The Gnat
was not accepted in Britain until much
later, and then only as a trainer. But
Finland bought them and India built them,
and they fought in the Indo-Pakistan wars,
earning the nickname of 'giant-killer'.
*25 October 1955:*

**SAAB 35 Draken**
The Draken's unusual double-delta layout
was SAAB's answer to a Swedish requirement for a supersonic interceptor with
short takeoff and landing performance. The
unusual planform was first tested on the
SAAB 210, a little aeroplane with similar
aerodynamics, and then translated into the
full-scale Draken. Armed with cannon and
missiles, the Draken has a phenomenal rate
of climb and is highly manoeuvrable at low
and high altitudes. It has been operated
from ordinary stretches of highway, one
indicator of its handling qualities and its
runway requirements.

**McDonnell F101A Voodoo**
*Crew:* 1 *Powerplant:* 2 Pratt & Whitney J57-P-13
10,200 lb thrust each *Span:* 39·7 ft
*Length:* 67·4 ft *Weight:* 48,000 lb
*Armament:* 4×20-mm cannon; 3 Falcon
guided missiles; 12×2·75-in unguided rockets
*Speed:* 1000 mph at 35,000 ft

Flight International

*McDonnell F-101C Voodoos of the USAF. The F-101C is a modified and structurally strength-
ened version of the Voodoo for low altitude close support missions*

**Folland Gnat T1**
Trainer version of the Gnat, shown here in the
colours of the RAF Red Arrows aerobatic team
*Crew:* 2 *Powerplant:* 1 Bristol Siddeley
Orpheus 107, 4400 lb thrust *Span:* 24 ft
*Length:* 37·8 ft *Weight:* 8077 lb
*Armament (fighter version):* 2×30-mm cannon
*Speed:* Mach 1·15

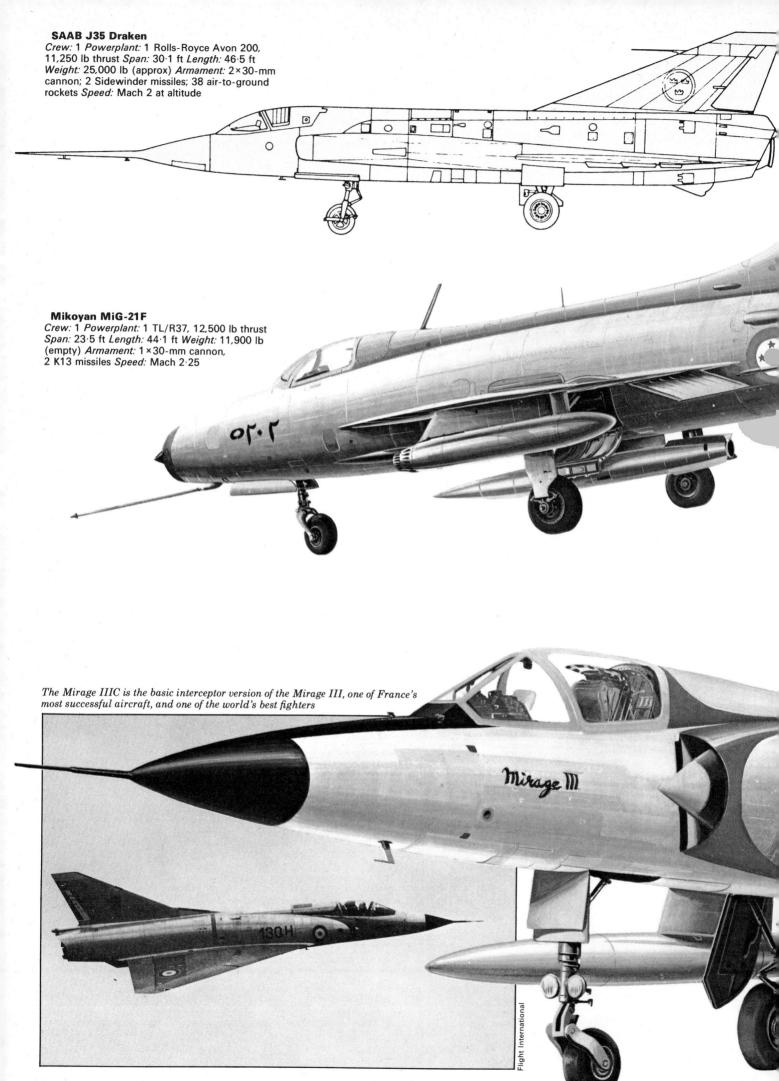

**SAAB J35 Draken**
*Crew:* 1 *Powerplant:* 1 Rolls-Royce Avon 200,
11,250 lb thrust *Span:* 30·1 ft *Length:* 46·5 ft
*Weight:* 25,000 lb (approx) *Armament:* 2×30-mm
cannon; 2 Sidewinder missiles; 38 air-to-ground
rockets *Speed:* Mach 2 at altitude

**Mikoyan MiG-21F**
*Crew:* 1 *Powerplant:* 1 TL/R37, 12,500 lb thrust
*Span:* 23·5 ft *Length:* 44·1 ft *Weight:* 11,900 lb
(empty) *Armament:* 1×30-mm cannon,
2 K13 missiles *Speed:* Mach 2·25

*The Mirage IIIC is the basic interceptor version of the Mirage III, one of France's
most successful aircraft, and one of the world's best fighters*

Flight International

172

### 1955:
### Mikoyan MiG-21

First publicly seen in 1956, the MiG-21 must have made its first flight during the previous year. Primarily an all-weather interceptor with secondary ground-attack capability in some models, the MiG-21 has been widely distributed among the allies and friends of the Soviet Union. Its design is based on a thin delta with a swept horizontal tail to improve altitude performance and landing characteristics. The MiG-21 is armed with both cannon and air-to-air missiles. Its defence of North Vietnam in later years was regarded with almost universal admiration and even some envy by its adversaries.

### 1955:
### Sukhoi Su-9

This single-engined all-weather fighter is also a tailed delta, like its contemporary, the MiG-21. But the Sukhoi design is larger, and its afterburning turbojet has a considerably higher thrust. Armament is based on missiles only, rather than on the combined cannon and air-to-air missile weaponry of the MiG-21.

### 20 January 1956:
### Supermarine Type 544 Scimitar

This single-place sweptwing fighter for the Royal Navy was area-ruled, had power controls and blown flaps, all innovations for a Fleet Air Arm fighter. It was also the FAA's first sweptwing fighter, their first able to top supersonic speed in a shallow dive, and their first equipped to carry nuclear weapons. Its high performance was a great advance over the straight-winged Sea Hawk which it replaced in service.

### 23 July 1956:
### Dassault Etendard

The French went to a smaller and lighter concept for their first carrier-based jet fighter. The Etendard was a loser in a NATO competition for a light fighter, but it became the basis for further development into the only true supersonic carrier-based fighter in European naval service at that time. Its design is aerodynamically similar to that of the long Mystère line, but it features layout modifications that make it more suitable for carrier use.

*The Convair F-106 Delta Dart high altitude fighter has been continuously updated, and is expected to continue in front-line service until the late 1970s*

### Dassault Mirage IIIC
*Crew:* 1 *Powerplant:* 1 SNECMA Atar 9B, 9370 lb thrust *Span:* 27 ft *Length:* 47 ft *Weight:* 32,630 lb *Armament:* 2×30-mm cannon; 1 Nord AS 30 plus 2 Sidewinder missiles *Speed:* Mach 2 at 36,000 ft

173

**McDonnell F4-E Phantom**
*Crew:* 2 *Powerplant:* 2 General Electric
J79-GE-10, 11,870 lb thrust each *Span* 38·4 ft
*Length:* 63 ft *Weight:* 59,000 lb (max)
*Armament:* 1 × 20-mm Vulcan cannon
*Speed:* Mach 2·4 at 40,000 ft

MOD

USAF

*17 November 1956:*
**Dassault Mirage III**
This was a redesigned Mirage I, and more nearly the true prototype of the contemporary line of Mirage III fighters. It used a single turbojet engine, setting the powerplant style that is maintained today in the latest of the Mirage fighters.

*26 December 1956:*
**Convair F-106A Delta Dart**
This delta-winged interceptor started life as the F-102B, but incorporated so many changes that it was redesignated with the later number. It is an automatically directed and fired weapon system; the pilot is along mostly to monitor the complex and advanced avionics systems that cram every cubic inch of this all-weather aircraft. By continuing modification programmes, this elderly design has been kept current, electronically speaking, and can handle the contemporary threat of high-altitude jet bombers. Like the F-102, it relies on both unguided and guided missiles for weapons.

*16 May 1957:*
**Saunders-Roe SR 53**
It's tempting to dismiss this as another mixed-powerplant interceptor, but its concept was a good one that ignored the ridiculous official requirements in favour of a logical design that would do the job envisioned. Unfortunately, it was caught in Britain's myopic White Paper of the late 1950s, which said that there was no foreseeable need for a manned fighter programme beyond the English Electric P 1. That bureaucratic decision knocked out the Saunders-Roe programme as well as some other innovative British designs of the time, and set the stage for the final decline of British fighter technology.

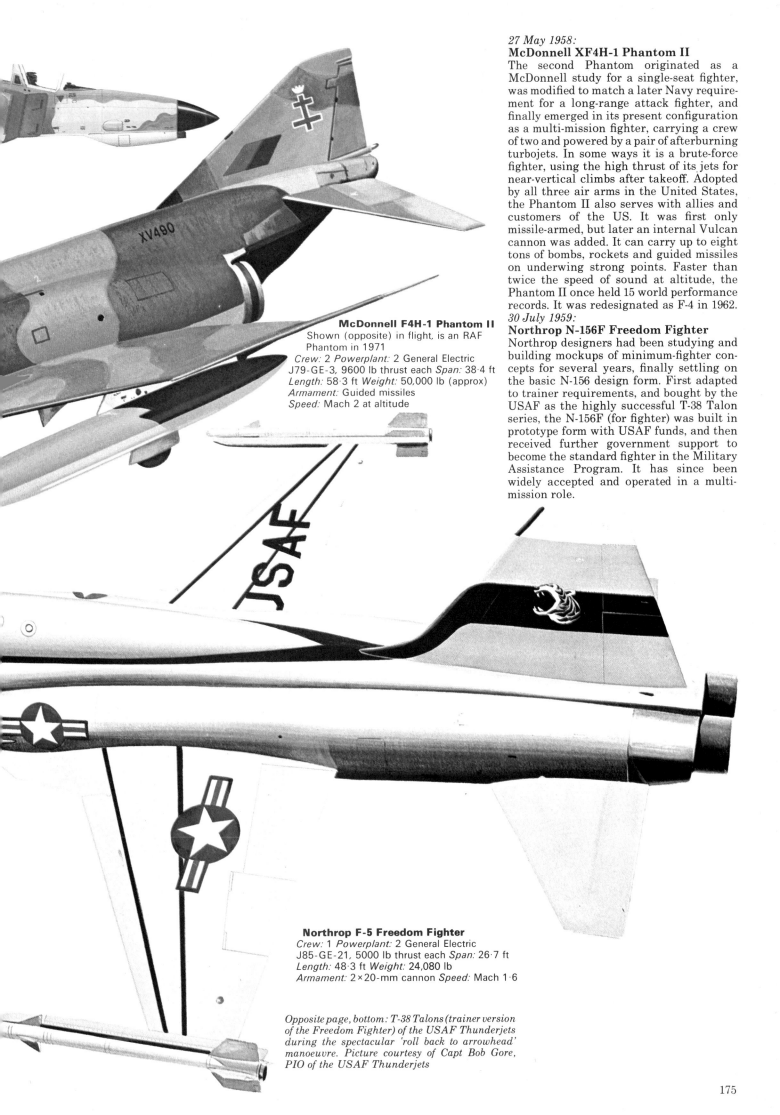

**27 May 1958:**

### McDonnell XF4H-1 Phantom II

The second Phantom originated as a McDonnell study for a single-seat fighter, was modified to match a later Navy requirement for a long-range attack fighter, and finally emerged in its present configuration as a multi-mission fighter, carrying a crew of two and powered by a pair of afterburning turbojets. In some ways it is a brute-force fighter, using the high thrust of its jets for near-vertical climbs after takeoff. Adopted by all three air arms in the United States, the Phantom II also serves with allies and customers of the US. It was first only missile-armed, but later an internal Vulcan cannon was added. It can carry up to eight tons of bombs, rockets and guided missiles on underwing strong points. Faster than twice the speed of sound at altitude, the Phantom II once held 15 world performance records. It was redesignated as F-4 in 1962.

**30 July 1959:**

### Northrop N-156F Freedom Fighter

Northrop designers had been studying and building mockups of minimum-fighter concepts for several years, finally settling on the basic N-156 design form. First adapted to trainer requirements, and bought by the USAF as the highly successful T-38 Talon series, the N-156F (for fighter) was built in prototype form with USAF funds, and then received further government support to become the standard fighter in the Military Assistance Program. It has since been widely accepted and operated in a multi-mission role.

**McDonnell F4H-1 Phantom II**
Shown (opposite) in flight, is an RAF Phantom in 1971
*Crew:* 2 *Powerplant:* 2 General Electric J79-GE-3, 9600 lb thrust each *Span:* 38·4 ft
*Length:* 58·3 ft *Weight:* 50,000 lb (approx)
*Armament:* Guided missiles
*Speed:* Mach 2 at altitude

**Northrop F-5 Freedom Fighter**
*Crew:* 1 *Powerplant:* 2 General Electric J85-GE-21, 5000 lb thrust each *Span:* 26·7 ft
*Length:* 48·3 ft *Weight:* 24,080 lb
*Armament:* 2×20-mm cannon *Speed:* Mach 1·6

*Opposite page, bottom: T-38 Talons (trainer version of the Freedom Fighter) of the USAF Thunderjets during the spectacular 'roll back to arrowhead' manoeuvre. Picture courtesy of Capt Bob Gore, PIO of the USAF Thunderjets*

*A mid fifties revival of the parasite fighter concept, launched from the belly of an airship, teamed the giant Vultee B-36 strategic bomber with the F-84 fighter. The concept was already obsolete, and surface-to-air missiles and advanced interceptors gave the escort fighter a doubtful utility*

# DEAD END DEVELOPMENTS

During the 1940s and 1950s, boom years for jet fighter development, many strange ideas were put into practice: on these pages we illustrate some of those that were built

**McDonnell XF-85**
Parasite fighter of 1948, designed to be carried in the bomb-bay of a B-36 bomber. Powerplant was a Westinghouse J34 of 3400 lb thrust; span was 21 ft and length 15 ft. The unusual tail configuration was adopted for maximum flight stability

**Saunders-Roe SRA-1**
Saunders-Roe's postwar flying boat
development ranged from the giant Princess to
the diminuitive SRA-1 fighter, whose fuselage
incorporated a planing hull
*Crew:* 1 *Powerplant:* 2 Metro-Vick F2/4A
Beryl MVB1, 3850 lb thrust *Span:* 25·1 ft
*Length:* 45 ft *Weight:* 16,255 lb
*Armament:* 4×20-mm cannon *Speed:* 516 mph

*The Convair XF2Y-1 Sea Dart of 1953 was the first and last hydroski jet fighter to be built. Its Westinghouse J46 engine gave it a maximum speed of 724 mph; armament was a single cannon; dimensions were 33·7 ft span and 52·6 ft length*

# LESSONS FROM SMALL WARS

**The wars in Vietnam and the Middle East, coupled with rapid advances in missile technology, had a profound effect on jet fighter design. Outside Russia, the emphasis switched away from pure interceptors; guns came back into fashion as armament; and the nature of local wars emphasised the usefulness of vertical takeoff aircraft**

By the beginning of the 1960s, the design of jet fighters had begun to turn away from continued emphasis on interception. During the late 1950s, strategic and tactical missiles were being developed and deployed, and the nature of any offensive threat was changing. Instead of relying on a high-altitude bomber force exclusively, major powers were switching to a strike force mix of bombers and missiles. Further, bombers were being modified and crews retrained for low-level missions, to get to their targets under radar coverage and below effective anti-aircraft missile height.

This had a profound effect on fighter development. The interceptor was no longer the be-all and end-all of fighter design. There were defence planners who seriously questioned the need for any further development of manned interceptors at all. In the United States, for example, no interceptors have been designed and built since the Convair F-106A, first flown in 1956. Almost 20 years later, that aircraft is still the USAF's only all-weather interceptor.

The Russians, on the other hand, apparently still clung to the belief that the major threat against them would include strikes by manned bombers. They have continued to develop interceptors, and during the 1960s turned out five new types.

But missile strikes by major powers were considered as the ultimate recourse in some future apocalyptic event. Meantime, there were some smaller wars that had happened, that were happening, and that were about to happen. From these, too, came useful lessons for fighter designers.

The Suez crisis in 1956 clearly showed that the primary use for fighters in wars of that kind was in ground attack. They still needed residual ability to fight their way home if they were jumped by an enemy counter-air strike, but the primary job was that of airborne artillery.

Continuing aerial engagements between Israeli aircraft and adversaries from one or another of the several Arab air forces in the area taught other lessons. They re-emphasised the importance of heavy calibre cannon with high rates of fire. They redirected some

of the earlier thinking that had concentrated on missiles as the only air-to-air weapon. It became apparent that mixed armament was better than a single type of weapon, and that the ideal mix for a fighter expecting trouble was cannon and missiles, the latter being able to home on infra-red signatures or radar returns.

The vulnerability of airfields to missile strikes and to bombing – the latter point made brilliantly by the Israelis in the Six-Day War of 1967 – gave emphasis to the concept of the vertical-takeoff and vertical-landing fighter. This novel type would be able to operate from any small cleared

Keystone

*Mirage kills MiG: during the 1973 Middle East War an Israeli Mirage III downs an Arab MiG-21. Both planes first flew in the late 1950s, proved their worth during the wars of the 1960s, and continue in front line service in many parts of the world during the 1970s*

area – a pasture, a crossroads, a clearing in a forest – and would not be restricted to the long and very vulnerable concrete runways. It was a prototype of this new class of fighters that first flew during the first year of the time period we are considering here.

*21 October 1960:*

### Hawker Siddeley P 1127

This prototype, built to explore the concept of vertical takeoff and landing for a tactical fighter, was developed into the first, and still the only, such fighter known to be in active service with military forces. The first flight was a captive one, made while the

P 1127 was tethered by cables to the ground. But it was the first time the aircraft left the ground under the power of its vectored-thrust turbofan engine. The P 1127 was developed further into the Kestrel and then the Harrier, a tactical fighter now in service with both the Royal Air Force and the United States Marine Corps. It is one of the truly pioneering designs in the development of fighter aircraft.

*1960:*

### Tupolev Tu-28P

This huge, all-weather interceptor was first shown publicly in 1961, and then was lost to Western view until the 1967 display at Domodedovo, the airport near Moscow chosen for the public display that year of many new and different Russian types. The Tu-28P is a two-place, twin-engined aircraft, weighing an estimated 100,000 lb. It is very obviously capable of supersonic performance, and is armed with four air-to-air missiles externally mounted, and possibly others in an internal bay. The likelihood is that it was designed as a specific counter to high-altitude strikes by the USAF's B-58A supersonic delta-winged bomber, then in very conspicuous service with the Strategic Air Command.

*1960:*

### Yakovlev Yak-28P

This twin-engined, two-place Russian interceptor seems at first glance to be an enlarged Yak-25. But all the indications are that it is basically a new design aerodynamically and internally, with new powerplants. It carries a pair of air-to-air missiles externally, and has trans-sonic performance. It would be capable of intercepting any subsonic bomber force targeted against Russia. The Yak-28P has been fitted with progressively updated avionics and more powerful turbojets since its service introduction.

*The Hawker Siddeley Harrier provides a complete contrast to the YF-12 concept. A subsonic V/STOL aircraft, it is designed specifically to meet the conditions encountered in local wars*

*The Lockheed YF-12A long-range interceptor, capable of sustained flight at Mach 3+ at altitudes of over 80,000 ft, was the peak development of supersonic cruise aircraft in the early 1960s. It was largely redundant as a fighter, however, and became the SR-71 strategic reconnaissance aircraft*

*26 April 1962:*
## Lockheed YF-12A
There have been USAF requirements for advanced manned interceptors from time to time, and the development of the YF-12A may have resulted from one of them. Or it might have been simply a cover operation to conceal the development of the SR-71 strategic reconnaissance aircraft. In either case, the dark blue-black YF-12A prototypes were built as potential interceptors, optimised for cruise speeds three times that of sound at altitudes above 80,000 ft. They carried missiles in belly bays. The unusual

layout, with its lifting-body aerodynamics and canted twin tails, represented the peak of the development of supersonic cruise design features at that time. The configuration was later modified slightly to produce the unarmed SR-71, a global reconnaissance aircraft that collects data supplementing that obtained from satellite photography.

*21 December 1964:*
## General Dynamics F-111
Variable-sweep wings had been tried and studied, with varying degrees of success, for nearly twenty years when a technical

innovation developed by the US National Aeronautics and Space Administration offered promise for a practical application. The idea was to pivot the wing sections outboard, rather than on the aircraft centreline. Technically, it worked. Then USAF and Navy fighter requirements and a new Secretary of Defence combined with the NASA innovation to give birth to the TFX, a multi-role combat aircraft that was intended to become the standard tactical fighter with all three US air arms. In the event, only the USAF got the developed TFX, or the F-111, as it was later designated after four design competitions and evaluations, and a controversial production contract. But the performance of the variable-sweep plane has marked it as one of the outstanding fighter designs of any era, and assured it of a place in aeronautical history. Missile-armed and loaded with advanced avionics, the F-111 series serves with Tactical Air Command and the Australian air force. In its FB-111A version, it equips two Strategic Air Command medium-bomber wings.

*1964:*
## Sukhoi Su-11
On display at Domodedovo in 1967, the Su-11 attracted notice as an obvious further development of the same designer's Su-9, then in widespread service as an all-weather fighter with the Red air forces. The Su-11 follows the same formula of the single-engined, single-seat all-weather fighter, but it carries improved missiles under its delta wings, has a longer fuselage nose, and an engine with perhaps a 20% thrust increase. It is supersonic, reaching speeds close to Mach 2 at altitude.

*1964:*
## Sukhoi Su-15
Second of the new Sukhoi designs to be shown in 1967 at the Domodedovo display, the Su-15 all-weather fighter forsakes the single-engine scheme for the power of twin turbojets with afterburning for super performance. Its aerodynamic layout draws heavily on the design bureau's experience with the Su-9 and Su-11, but the basic wing has been modified in one prototype to produce a compound-sweep planform. Another version of the Su-15, modified for short takeoff and landing operations, was also displayed and flown at Domodedovo.

*1964:*
## Mikoyan MiG-23
The Russian application of variable-sweep geometry to a fighter produced the MiG-23, a smaller and lighter aircraft than the F-111 series, although designed for approximately the same missions. It is a single-seat, single-engined tactical fighter and fighter-bomber, armed with a twin-barrelled 23-mm cannon. Performance estimates place it in the Mach 2 class at altitude. Apparently the MiG-23 ran into some development troubles, because it was not until several years after the first public display at Domodedovo in 1967 that it was reported in active service with front-line squadrons. It has since been exported to Russian allies and friends, including those in the Middle East arena.

*1964:*
## Mikoyan MiG-25
An all-weather interceptor with phenomenal performance, the MiG-25 has been the standard of comparison and of the fighter threat in almost any consideration of strategic or tactical aircraft design in the West during the past decade. The speedy

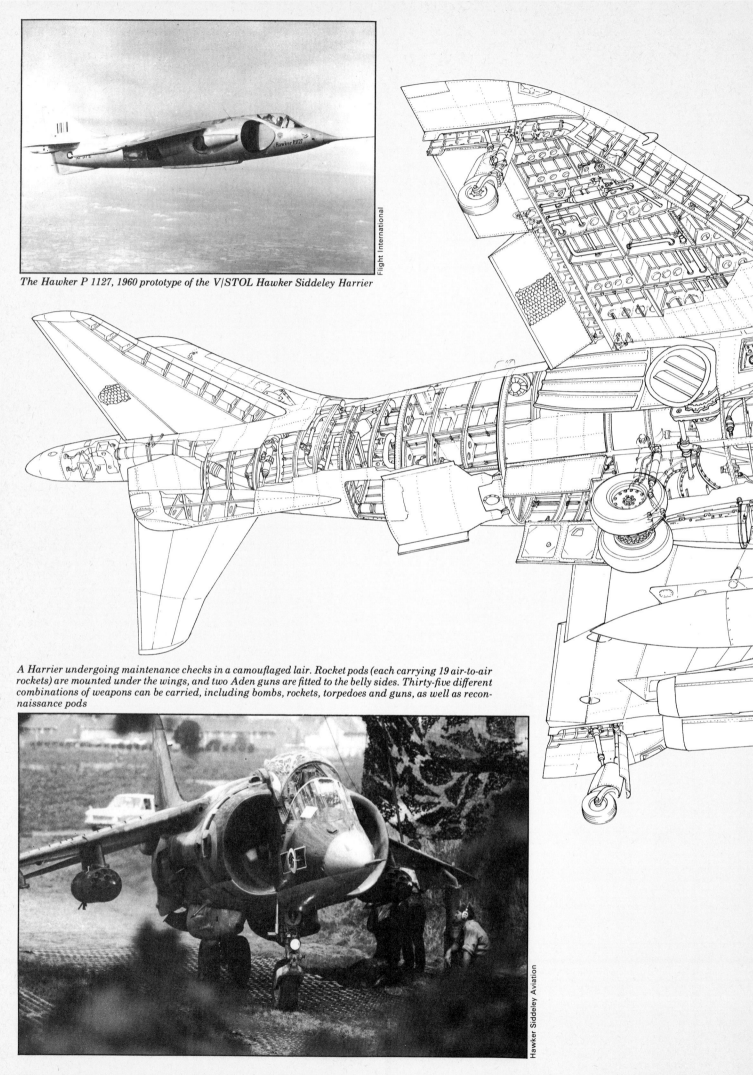

*The Hawker P 1127, 1960 prototype of the V/STOL Hawker Siddeley Harrier*

*A Harrier undergoing maintenance checks in a camouflaged lair. Rocket pods (each carrying 19 air-to-air rockets) are mounted under the wings, and two Aden guns are fitted to the belly sides. Thirty-five different combinations of weapons can be carried, including bombs, rockets, torpedoes and guns, as well as reconnaissance pods*

# THE HARRIER

The Hawker Siddeley Harrier V/STOL is unique among serving combat aircraft. The prototype P 1127 first flew in 1960; the Harrier itself in August 1966. At present serving with the RAF and US Marine Corps, the Sea Harrier version will shortly enter service with the Royal Navy.
*Crew:* 1 *Powerplant:* 1 Rolls-Royce Pegasus 11, 21,500 lb vectored thrust *Span:* 25·3 ft
*Length:* 45·5 ft *Speed:* 680 mph at sea level

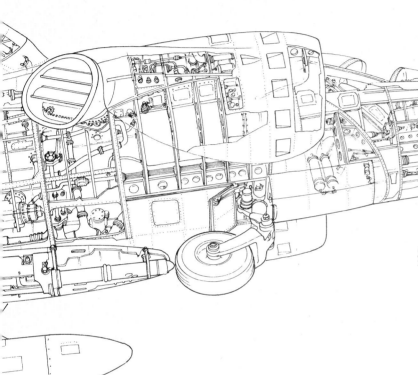

*The Harrier can lift one-third of its maximum load in vertical takeoff (below). In a short takeoff run (approximately a quarter of that required by conventional combat aircraft) it can lift a fuel and weapons load of 13,050 lb – greater than its own empty weight of 12,200 lb*

Hawker Siddeley Aviation

Pilot Press

**Yakovlev Yak-28P**
*Crew:* 2 *Powerplant:* 2 RD11, 13,100 lb thrust
each *Span:* 42·5 ft *Length:* 71 ft
*Weight:* 30,000 lb (approx)
*Armament:* 2 air-to-air missiles
*Speed:* 730 mph (approx) at altitude

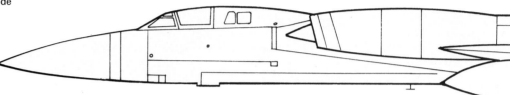

**Tupolev Tu-28P**
*Crew:* 1 *Powerplant:* 2 afterburning jets,
27,000 lb thrust each (estimate) *Span:* 65 ft
*Length:* 85 ft *Weight:* 100,000 lb
*Armament:* 4 air-to-air guided missiles
*Speed:* Mach 1·75 at altitude

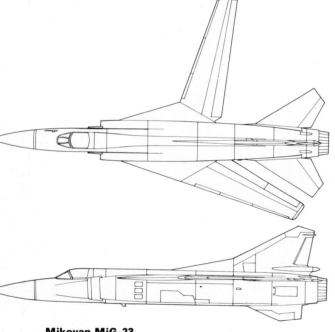

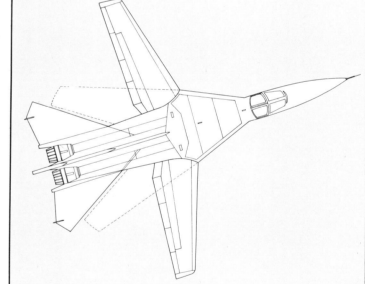

**Mikoyan MiG-23**
*Crew:* 1 *Powerplant:* 1 afterburning turbojet,
9300 lb thrust (estimate) *Span:* 46·7 ft
*Length:* 55·1 ft *Weight:* 12,700 lb
*Armament:* 2×23-mm cannon
*Speed:* Mach 2·3 at altitude

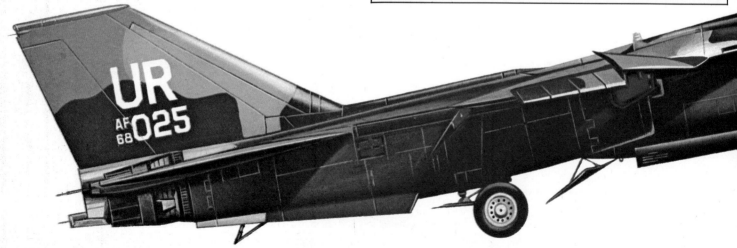

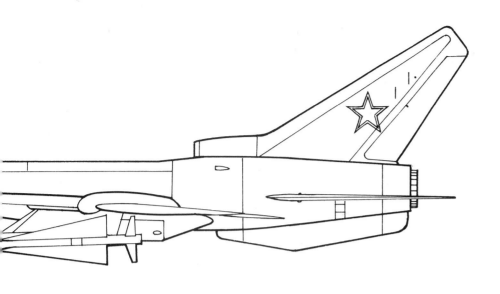

*The variable-sweep F-111 was dogged by trouble during its development*

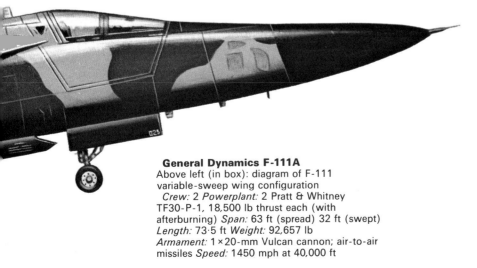

### General Dynamics F-111A
Above left (in box): diagram of F-111 variable-sweep wing configuration
*Crew:* 2 *Powerplant:* 2 Pratt & Whitney TF30-P-1, 18,500 lb thrust each (with afterburning) *Span:* 63 ft (spread) 32 ft (swept)
*Length:* 73·5 ft *Weight:* 92,657 lb
*Armament:* 1 × 20-mm Vulcan cannon; air-to-air missiles *Speed:* 1450 mph at 40,000 ft

twin-jet design held several absolute world speed and time-to-climb records, and held some of them unbeaten for nearly ten years. (In early 1975, the time-to-climb records, the last of the batch held by the MiG-25, were topped, and substantially, by a USAF McDonnell F-15A Eagle.) It carries four air-to-air missiles of a new type beneath its stubby swept wings, and can do better than Mach 3 at altitude. In a reconnaissance version, the MiG-25 has been observed, but not intercepted, in high-level flights in the Middle East. It is a single-place aircraft.
*23 December 1966:*

### Dassault Mirage F 1
Like so many of the series of advanced Dassault designs, this one grew out of earlier attempts to meet an entirely different requirement. The F 1 was developed into a multi-mission fighter, with its greatest strength in the air-superiority role. Yet its genesis was as a flying test bed aircraft built for development work with a new engine planned for a Dassault VTOL fighter. It is a single-seat fighter, powered by a single SNECMA Atar engine, probably the final development of that long line that traces its ancestry all the way back to the Junkers 004B. The F 1 wing design uses advanced aerodynamic features, tried and tested on other Dassault aircraft and refined for optimum performance in the specific F 1 configuration. Rugged landing gear gives the F 1 the ability to operate out of grass strips, or from unimproved airfields. It carries very heavy armament; as an interceptor, for example, it is armed with a pair of 30-mm cannon, a pair of Matra R 530 radar-homing missiles, and another pair of air-to-air missiles at the wingtips. It has been ordered by the French air force and will undoubtedly be sold abroad as well.
*8 February 1967:*

### SAAB System 37 Wiggen
The Wiggen is the major component of a complete air defence system, and the product of a most ambitious effort by Sweden that can stand comparison to any such effort by any country. The Wiggen aircraft was designed to be a flying platform capable of carrying a variety of sub-systems into the air for a variety of missions. Four major missions were chosen for the development: strike, reconnaissance, interception, and training. These requirements were all built into the airplane as far as possible, so that the final result is a multi-mission aircraft with cross-capabilities. Its unique aerodynamic layout with its main wing and the forward, separate, auxiliary wing produces low approach speeds. With thrust reversers, the Wiggen can land in less than 1700 ft of runway. Its low-speed characteristics give it STOL (short takeoff and landing) performance and it can operate from highways. At altitude it can streak along at twice the speed of sound. It carries missiles and advanced avionics for navigation and attack.

The SAAB Wiggen, Dassault F 1 and General Dynamics F-111, all multi-mission fighters, are one product of their times. Capable of a wide range of performance, from STOL to supersonic, and able to arm with missiles for dog-fighting or bombs for ground attack, their versatility assures them of continued use in the air arms of their respective countries.

But they begin to show a trend which will be accentuated in the next time period by the Grumman F-14A and the McDonnell F-15A. That trend will force another look at the philosophy of fighter design.

*The SAAB J37 Wiggen, another competitor in the race to replace Europe's combat aircraft*

*Marcel Dassault, designer and builder of the Mystère/Mirage series, started his career in aviation during the First World War*

*Dassault Mirage F1, the French candidate to replace the obsolescent NATO Starfighters*

### Dassault Mirage F1
*Crew:* 1 *Powerplant:* 1 SNECMA Atar 09K-50,
11,023 lb thrust *Span:* 27·5 ft *Length:* 49·3 ft
*Weight:* 24,000 lb *Armament:* 2×30-mm cannon;
2 Sidewinders plus other missiles
*Speed:* Mach 2·2 at altitude

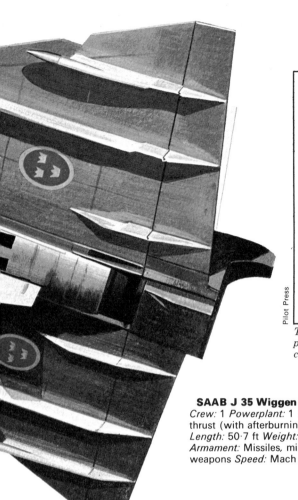

*The Mikoyan MiG-25 set the standard of jet fighter performance for a decade, and its phenomenal performance was only recently bettered by the F-15 Eagle. It carries four air-to-air missiles, and is capable of Mach 3·2 at altitude*

**SAAB J 35 Wiggen**
*Crew:* 1 *Powerplant:* 1 Svenska RM8, 26,500 lb thrust (with afterburning) *Span:* 34·8 ft
*Length:* 50·7 ft *Weight:* 35,000 lb (approx)
*Armament:* Missiles, mixed external stores and weapons *Speed:* Mach 2 at high altitude

*The Dassault Mirage G 8 swing-wing fighter, from which Dassault are developing the Super Mirage fighter*

# SOARING COST AND COMPLEXITY

The escalating costs of fighters like the Grumman Tomcat are causing much heart-searching among aircraft firms and Defence Ministries, and it begins to seem that the General Dynamics YF-16 is the fighter concept of the future

The aerial war in Vietnam continued to teach the same truths about fighter design. Versatility was one such lesson. A fighter-bomber might have to carry bombs on one sortie, rockets on the next, napalm on its third and guns on its fourth. Or it might have to take off with a mixture of all these and, after dumping them, fight its way back home through a curtain of intense flak and enemy missiles.

The concept of a multi-mission fighter was reinforced by Vietnam experience. It also added new emphasis to a lesson from earlier wars: the importance of electronic countermeasures. Electronic warfare had come to the battlefields of Southeast Asia with a vengeance; their environment was criss-crossed with invisible beams of electronic devices for finding, fooling and helping to destroy aircraft.

A major impetus to the growth of electronic warfare was the advance of the guided missile as an anti-aircraft weapon. It could not be shot down; it had to be evaded or avoided or decoyed some way. Evasion techniques had been developed, but – as missile technology improved – the dependence on spotting and manoeuvring grew riskier. The next stop: decoy the missile by giving it a false target to detect, track and intercept. This could be done electronically or physically.

Electronic warfare systems added to fighter capabilities. They could warn a pilot that his plane had been detected; that it was under fire from the rear; and that the missile was beginning its final tracking toward a hit on his plane. These passive indications could be augmented by electronic countermeasures activity. They could, in effect, create a cloud of electronic noise in which the fighter might hide. They could create a completely spurious fighter that would have a stronger attraction for the oncoming missile. They could confuse the missile, decoy it, send it to another quadrant of the sky.

Brains, even electronic ones, are not built cheaply. Further, like human brains, they function best in friendly environments; they tend to be troubled by excessive heat, cold, or physical shocks. Making electronic brains that can withstand those deteriorating factors and still perform involves more complexity, and that equates to more cost.

*The US Navy's best and most expensive fighter, the swing-wing Grumman F-14A Tomcat undergoing flight tests over San Clemente Island*

And so, as each side added an offensive technique or countered one with a defensive technology, the inevitable happened. Fighters got more complex, and consequently more costly to build.

When the Vietnam aerial action really began in the early 1960s, a North American F-100, then a standard day fighter, could be bought for about $600,000 flyaway factory. If one wanted the greater capability of a McDonnell F-4C, the price was tripled to $1·8 million.

These kinds of cost figures were impressive, especially to those who remembered 1958, when it was possible to get a North American F-86F, then the top-of-the-line day fighter, for $230,000.

But the price of an F-4C was only an indication of the trend. The data were coming in on the then-new General Dynamics F-111 series, with costs – according to testimony in the bitter but fascinating Congressional hearings – of more than $16 million per airplane. Admittedly this was a special case, but it served to send up warning flags. In the depths of the Pentagon in Washington, DC, fighter analysts and planners began to look for alternatives to the high-cost development and production of today's fighters.

## Costs and contracts

The final straw, perhaps, was the sobering experience shared by the Navy and Grumman in the development of the variable-sweep F-14A Tomcat series. Cost escalations, caused by inflationary factors and other problems, threatened the continued existence of Grumman as a company. The Navy wanted to hold Grumman to the fixed prices guaranteed in its contract; Grumman said that doing that would bankrupt the company, and claimed that the rules under which the contract had been written had made assumptions that were no longer valid. It was unfair to hold Grumman to those rules, they said.

After some recriminations and refinancings, the programme continued and the Tomcat entered the fleet, where it is serving with great distinction. It is costing the US Navy, and the American taxpayer, about $11 million per airplane, based on a system of accounting which produces agreement between Navy and Air Force cost figures.

The McDonnell-Douglas F-15A Eagle, a fighter, costs about 10% less than the variable-sweep Tomcat.

These amounts of dollars, it might be noted, could have bought a complete Boeing B-52 towards the end of that bomber's production life in 1962, at about the time that the top fighter, the F-4, was purchased for less than $2 million.

It should be realised that costs for aircraft may or may not include all the programme costs amortised out over each airplane. At one end of the cost scale is simply the amount of money the military

pays to the manufacturer for building and assembling the parts that make up one complete airplane. At the other end is the total cost of research, development, tooling, flight test, problem-solving, and all the rest of the programme, including production, for each plane.

The figure of $16·6 million, quoted in Congressional testimony for the F-111, is the total programme cost divided by the number of airplanes produced. In the Grumman F-14A and McDonnell-Douglas F-15A figures cited above, the programme costs are not all included in the airplane's price. To do so might double the figures.

There is black humour used by proponents of low-cost aircraft to make a point. They plot the cost of a fighter (or bomber or whatever) against time, and show how that curve has climbed upward more and more rapidly until, at some date not very far away, the cost of a single fighter exceeds the total US budget. They then speak of the Air Force as flying its fighter, or its bomber, against an enemy strike.

The use of such a simplistic approach dramatises the central fact: fighter costs are escalating out of sight. And a major portion of that cost is chargeable to inflation. By mid-1974, inflation in the price of metals that go into aircraft construction had accounted for more than a 60% rise in materials costs since 1967, or more than 8% annually.

Even stronger was the impact of world-wide inflation during the last few months of 1973. A Department of Defense estimate attributed a 36% cost rise in more than 50 major weapons systems to the financial near-panic of that period.

In some ways, the cost growth of fighter aircraft has been a result of evolution, rather than revolution, in design. It is common to base a new fighter on the latest successful one operational, because the fastest and lowest-cost way has been thought to be by that route. What this meant in practice is that the latest technology, which might have been able to reduce costs if effectively applied, was not used. An existing design was adapted and, although it often brought an apparent low cost with it because of its status as a high-production, amortised programme, it could wind up costing more because of fix after fix necessary to bring the ageing design up to date.

In the early 1970s, the USAF funded a development programme that aimed at reversing the trend to higher costs. The purpose was to evaluate the latest in fighter technology in minimum-sized aircraft, with the primary goal of seeing what could be done with new technology when a preconceived military mission was not a factor in design.

The General Dynamics YF-16 and Northrop YF-17 were the interesting results of that programme. Two different airplanes, they reflect their company's approaches to the design of a lightweight fighter type of aircraft, with carte blanche to use any new idea as long as it appeared promising.

Both these refined designs have won high marks from observers for their adherence to requirements and their low-cost approach.

In early 1975, prices for both were estimated at about $4 million each, should a large-scale production programme follow. This is still an expensive fighter, but it is a remarkable achievement to reach cost figures about one-third those of the immediate predecessors.

187

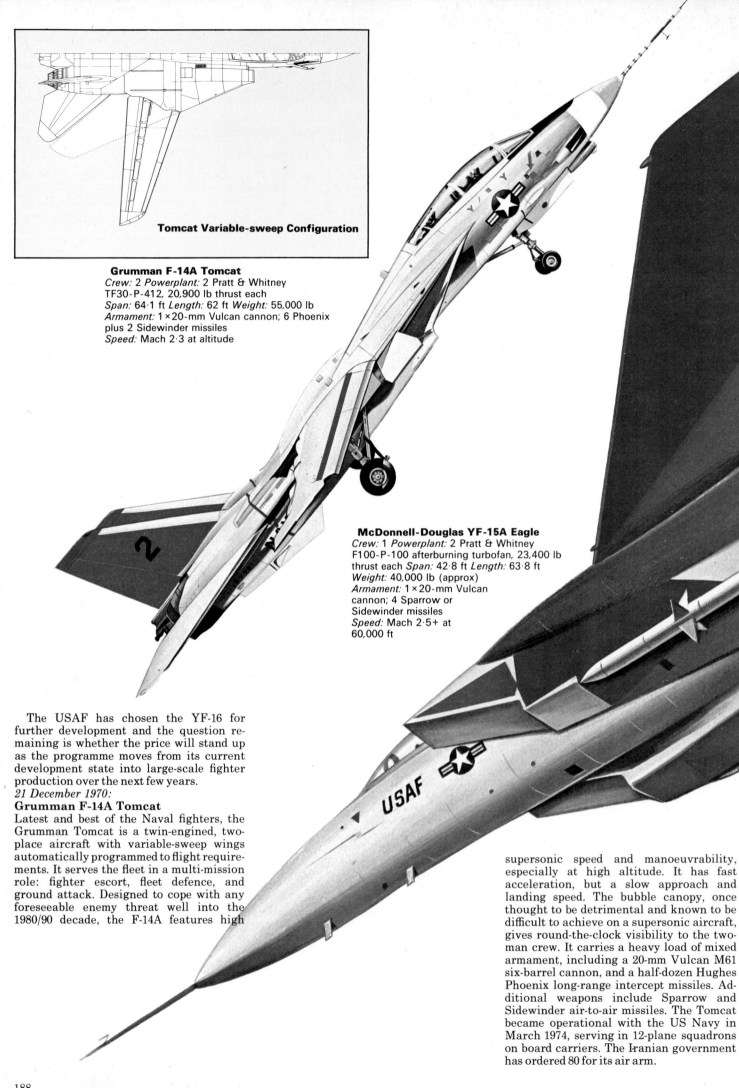

**Tomcat Variable-sweep Configuration**

### Grumman F-14A Tomcat
*Crew:* 2 *Powerplant:* 2 Pratt & Whitney
TF30-P-412, 20,900 lb thrust each
*Span:* 64·1 ft *Length:* 62 ft *Weight:* 55,000 lb
*Armament:* 1×20-mm Vulcan cannon; 6 Phoenix
plus 2 Sidewinder missiles
*Speed:* Mach 2·3 at altitude

### McDonnell-Douglas YF-15A Eagle
*Crew:* 1 *Powerplant:* 2 Pratt & Whitney
F100-P-100 afterburning turbofan, 23,400 lb
thrust each *Span:* 42·8 ft *Length:* 63·8 ft
*Weight:* 40,000 lb (approx)
*Armament:* 1×20-mm Vulcan
cannon; 4 Sparrow or
Sidewinder missiles
*Speed:* Mach 2·5+ at
60,000 ft

The USAF has chosen the YF-16 for further development and the question remaining is whether the price will stand up as the programme moves from its current development state into large-scale fighter production over the next few years.
*21 December 1970:*

**Grumman F-14A Tomcat**

Latest and best of the Naval fighters, the Grumman Tomcat is a twin-engined, two-place aircraft with variable-sweep wings automatically programmed to flight requirements. It serves the fleet in a multi-mission role: fighter escort, fleet defence, and ground attack. Designed to cope with any foreseeable enemy threat well into the 1980/90 decade, the F-14A features high supersonic speed and manoeuvrability, especially at high altitude. It has fast acceleration, but a slow approach and landing speed. The bubble canopy, once thought to be detrimental and known to be difficult to achieve on a supersonic aircraft, gives round-the-clock visibility to the two-man crew. It carries a heavy load of mixed armament, including a 20-mm Vulcan M61 six-barrel cannon, and a half-dozen Hughes Phoenix long-range intercept missiles. Additional weapons include Sparrow and Sidewinder air-to-air missiles. The Tomcat became operational with the US Navy in March 1974, serving in 12-plane squadrons on board carriers. The Iranian government has ordered 80 for its air arm.

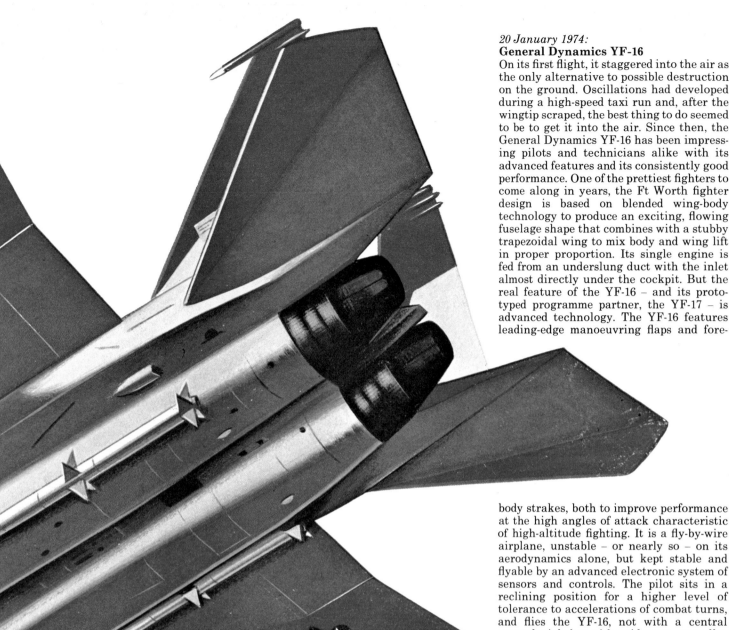

**General Dynamics YF-16**

On its first flight, it staggered into the air as the only alternative to possible destruction on the ground. Oscillations had developed during a high-speed taxi run and, after the wingtip scraped, the best thing to do seemed to be to get it into the air. Since then, the General Dynamics YF-16 has been impressing pilots and technicians alike with its advanced features and its consistently good performance. One of the prettiest fighters to come along in years, the Ft Worth fighter design is based on blended wing-body technology to produce an exciting, flowing fuselage shape that combines with a stubby trapezoidal wing to mix body and wing lift in proper proportion. Its single engine is fed from an underslung duct with the inlet almost directly under the cockpit. But the real feature of the YF-16 – and its proto-typed programme partner, the YF-17 – is advanced technology. The YF-16 features leading-edge manoeuvring flaps and fore-body strakes, both to improve performance at the high angles of attack characteristic of high-altitude fighting. It is a fly-by-wire airplane, unstable – or nearly so – on its aerodynamics alone, but kept stable and flyable by an advanced electronic system of sensors and controls. The pilot sits in a reclining position for a higher level of tolerance to accelerations of combat turns, and flies the YF-16, not with a central control stick, but with a side-arm controller mounted at the right of the cockpit. Armed with a single Vulcan cannon and a pair of Sidewinder dog-fighting missiles, the YF-16 also can carry external stores and weapon pods on seven underwing strong points.

**Northrop YF-17**

The other half of the lightweight fighter programme flew about six months later, and revealed a different approach from the one chosen by General Dynamics. The YF-17 designers used a pair of smaller engines, citing the dependability and reliability of a twin-engined installation, verified in practice with their F-5/T-38 line. Aero-dynamically, the YF-17 builds on advanced technology, using a refinement of the basic Tiger II wing and a forward modification of that surface. A shoulder-wing design, the Northrop fighter uses twin canted vertical tails. Engine intakes are what have been inelegantly called armpit types, located between wing root and the fuselage side. Armament is identical to that of the YF-16. Northrop designers optimised their aircraft around the turning rate performance, which they had concluded was the single most important factor in aerial combat. The radius of turn, the ability to pull high loads during the turn, and all the other arguments of pilots and technicians were boiled down to a shape and a powerplant that would get the nose of the airplane turned into a fight as rapidly as possible.

**McDonnell-Douglas F-15A Eagle**

For the first time in a quarter of a century, the USAF has a fighter optimised specifically for the air-superiority mission. The single-seat Eagle is built around a stubby sweptwing of low aspect ratio and high area, for extra manoeuvrability at high altitude. It carries mixed armament: a 20-mm Vulcan cannon, and Sidewinder and Sparrow missiles, updated specifically for dog-fighting at closer ranges. Early in 1975, the Eagle set eight world time-to-altitude records in flights under the project name of Streak Eagle. In timed climbs from three to 30 kilometres, the Eagle broke mark after mark in a staggering performance with numbers that almost defy the imagination. On the climb to three kilometres, the F-15A lifted off the ground after a roll of 400 ft, about seven lengths of the fuselage. On a single flight which broke three of the existing records, the F-15A accelerated to sonic speed within 19 seconds after takeoff. In the climb to the highest altitude, the pilot accelerated in climb to Mach 2 in less than two minutes from takeoff, and the subsequent energy climb got him to 102,000 ft before the Eagle slacked off. In less than one minute, the Eagle reached 12 kilometres (39,360 ft); its average rate of climb to 30 kilometres (98,400 ft) was better than 144 metres per second, or 28,438 ft per minute.

The rapid increases in cost and complexity are not the only considerations in assessing the future of the jet fighter. Most fighters of the last 20 years have had some capacity as strike/attack/reconnaissance aircraft, while some high performance interceptors such as the YF-12 and F-111 have failed to get into service. History is not encouraging, and the future is far from clear

Are the YF-16 and YF-17 the trend-setters for future fighter design? Or are they slated to be remembered as brilliant design efforts that gradually turned into heavy, complex and costly multi-purpose fighters?

History does not provide much encouragement for answering the first question positively. The one recent example – the Folland Gnat – was a successful approach to the problem, and it offered some unique and innovative solutions, but alone it could not stem the tide.

The avowed purpose of the YF-16/YF-17 programme was to develop a fighter-type aircraft without the need to meet a specific operational requirement for one. For that reason, the design teams at General Dynamics and Northrop were free to apply any late technology, almost without regard for its suitability in a combat aircraft.

But neither company got where it is by

## THE FUTURE
# WHERE DO WE GO FROM HERE?

being unaware of military requirements. Consequently, both aircraft reflect the extensive experience of their manufacturers in the development of recent high-speed combat aircraft. Further, along the line the Air Force began to have second thoughts about the programme, and let it be known that they might consider developing the better of the two into a combat-ready fighter, and that it might then be purchased in substantial quantity.

That is what happened. There was a competitive fly-off between the two, which gave the nod to the General Dynamics entry. Soon after, the USAF announced plans to purchase up to 650 of the little fighters. They would become part of the hi-lo mix, the compound word referring not to mission profiles, but to cost. Future USAF fighter fleets would be mixed, with a small number of versatile, expensive fighters and a larger number of less versatile, less-expensive fighters.

Current cost estimates for the YF-16 as developed into a production fighter vary between $4 and $5 million, with the majority view clustering around the lower end of

that range. Undoubtedly, escalation, inflation and the other enemies of constant cost will lay heavy influence on the development programme. Further, every fighter that has started out in pristine form has soon had all kinds of external and internal modifications added. Still further, the production cost of a large number of McDonnell-Douglas F-15 Eagles has been estimated as not too far above the $5 million mark, throwing another factor into the YF-16 equation.

But it is natural to expect that the YF-16 and YF-17 will be developed, and that they inexorably will grow heavier and more expensive. But at the same time, the same process will be happening to the F-15, and at some future point it may be possible to see an F-15 costing $20 million and an F-16 costing $8 million. And then, the Air Force will announce a competition for a revolutionary new concept in aircraft development: a fighter-type aircraft to use the very latest technology to achieve optimum performance at minimum cost.

There is no way of knowing certainly what the future will bring in technology,

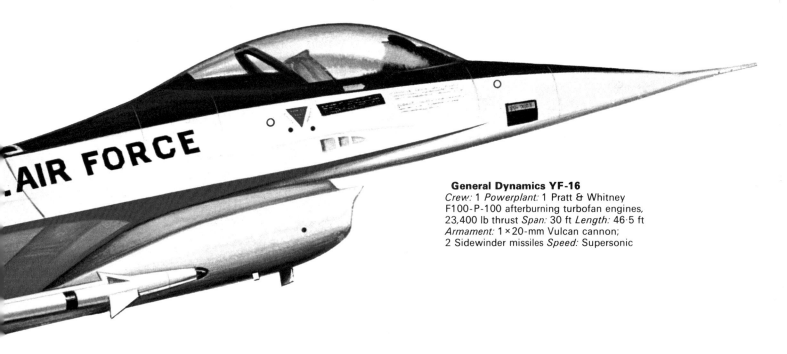

**General Dynamics YF-16**
*Crew:* 1 *Powerplant:* 1 Pratt & Whitney
F100-P-100 afterburning turbofan engines,
23,400 lb thrust *Span:* 30 ft *Length:* 46·5 ft
*Armament:* 1 × 20-mm Vulcan cannon;
2 Sidewinder missiles *Speed:* Supersonic

but it may be instructive to look at the YF-16 and YF-17 to see what the present offers.

General Dynamics wanted to build the smallest and lightest aeroplane possible, using low-risk developments. They emphasised the air-superiority mission almost to the exclusion of any other possibility, and wound up with a single-place, single-engine airplane built primarily of conventional materials.

Northrop, with the experience of a long line of small, light combat aircraft to draw upon, carried that series one step further in applying what they had learned in extrapolation of the technology. But they stressed the concept of an operational prototype, one that could make the transition from development to production and service with a minimum of change and difficulty.

The YF-16 is a blended wing-body configuration, in which the contours of wing and fuselage are aerodynamically melded to a smooth and continuous fairing of surfaces. This was done to draw on body lift; the resulting wing size is smaller, and therefore lighter, and therefore cheaper. The wing has leading-edge manoeuvring flaps, which are programmed to function automatically as Mach number and angle of attack change in flight. They increase the camber of the wing and the lift at altitude.

Strakes – long, thin, horizontal fences ahead of the wing leading edge – are placed there to generate vortices. These rotating streams of energised air then move aft, and keep the boundary-layer flow from breaking away in the intersection area between

*Flight International*

*The YF-17, Northrop's latest idea for a low-cost lightweight fighter, has similar performance and armament to the YF-16, and in spite of its twin-jet configuration the two designs are strikingly similar*

wing and fuselage. All these features – the blended wing-body shape, the leading-edge manoeuvring flaps, the strakes – work to improve the lift characteristics of the aircraft, especially at high altitudes and high angles of attack that are typical of dog-fighting. And they save weight also – General Dynamics say that the wing-body blending saved them about 320 lb in the fuselage, and another 250 on the wing. The strakes made a smaller wing possible, saving another 490 lb.

Weight saved means money saved. A pound of aircraft weight represents many dollars in direct labour and material costs, plus indirect research, development, design, engineering, and tooling costs. In aircraft, smaller means lighter means cheaper.

In order to get a very agile fighter, General Dynamics engineers deliberately reduced static stability, because a stable aircraft – while desirable for cruise flight and other portions of a mission – is not the best for manoeuvring. Next, they designed

191

*An interceptor version of the MRCA (Multi-Role Combat Aircraft) is scheduled to replace RAF Phantoms and Lightnings in the air-superiority role for the 1980s. Different versions of the MRCA will be equipped with extremely advanced avionics for a wide range of different combat roles*

in a fly-by-wire system, making the YF-16 the first aircraft planned from the start around such a control system. Fly-by-wire uses electronics to transfer the input signal from the pilot's control stick to the moveable surface; it reduces weight and vulnerability of the airplane. But it does more; it can be used to give unusual groupings of control motions, something no conventionally rigged system could do. In a combat situation, a pilot could easily pop up his airplane above the enemy on his tail by using fly-by-wire signals to move all of the horizontal surfaces to generate lift simultaneously. On an enemy's tail, he could aim the fuselage independently of what flight path his plane was following.

Further recognising that the pilot is under maximum physical stress during combat, the YF-16 design team slanted the pilot's seat back to a 30° angle, to increase his tolerance to high-g turns, and to make it easier for him to see and to look around in those situations. They gave him a side-arm controller instead of a central control stick.

By using a simple engine inlet under the belly, General Dynamics engineers saved another major weight increment. The inlet duct is short, and its leading edge is fixed, rather than built with the sharp-edged

variable geometry of most supersonic fighters. In the speed range where the YF-16 will fight, the complexity is unnecessary.

Finally, they built the airplane out of familiar aluminium alloys, with minimum use of steel and titanium. There is some use of graphite epoxy skins on the tails for stiffness with reduced weight, but that is the only area where new and different materials are used.

One novel feature of the overall design is that many of these features can be removed and replaced if they should not prove to be what they were predicted to be. The entire wing is quickly removeable, and can either be replaced, or moved back along the fuselage to restore some of the static stability designed out of the layout. The forward strakes can come off, the inlet can be simply changed, the canopy can be replaced, and even the side-arm controller can be removed and replaced by a central control stick in the conventional manner.

Northrop were faced with the same set of conditions, and it is interesting to note that their solutions differ in detail but not so much in concept. The twin-tailed YF-17

uses a pair of engines, rather than the single jet of the YF-16. Paired engines present unique problems, among which is 'base drag', that portion of the total airplane drag due to the blunt end of the jet exhaust. Northrop has been working on the problem a long time and the YF-17 is claimed to have the lowest base drag of any twin-jet fighter.

The wing is basically the wing of the Tiger II, with extra leading-edge extensions which reach well forward along the fuselage and which serve basically the same purpose as the YF-16's strakes: generation of strong vortex flow to re-energise the wing root boundary layer flow. But rather than relax the stability standards, Northrop chose to handle the vortex flow by using twin tails, canted outboard, and placed far forward on the fuselage. Here they also serve to close an aerodynamic gap between the wing trailing edge and the horizontal tail, and

they do not require a carry-through structure that pierces the rear fuselage. That means a lighter fuselage back there, and also an engine bay free of internal obstructions. The engines can be easily dropped out for maintenance or replacement.

The horizontal tail is larger than usual, and was sized for manoeuvring to give the best turning rate at high speeds. It is set as far back as possible, which means a slightly heavier and longer fuselage, but Northrop think that the manoeuvring advantages outweight the objections in this case.

The YF-17 also uses fly-by-wire, but on the ailerons only. The control stick is the conventional central one, and the pilot's seat is tilted aft to an 18° angle. Like its competitor, the YF-17 is largely aluminium alloys with some graphite composite skins. One nice detail is the use of fireproofing paint in the engine bay. The paint expands under the heat and creates an air-filled insulating layer which effectively can contain and delay the fire for longer periods of time than usual.

These two aircraft characterise the best aerodynamic features and the best structural techniques available to designers today, given the restraints of the programme.

### Heart of the fighter

Powerplants are the heart of the fighter. The technology of today's afterburning turbofans is a far cry from the inventions of von Ohain and Whittle. New metals, much higher temperatures, improved fuels and better internal flow characteristics have all added their incremental improvements to those early engines. The general feeling in the engine industry is that future changes are going to continue to be incremental, as better materials come along to permit higher operating temperatures. There should be no major changes in powerplant design for perhaps another decade.

Weaponry, another major factor in fighter design, is considered in the succeeding chapter.

*The Israel Aircraft Industries Kfir, essentially a revamped Mirage III, was revealed in April 1975. Maximum speed is around Mach 2·2; armament is a 30-mm cannon plus missiles and stores of various types*

Today, the over-riding considerations of costs are in the minds of every planner and designer of fighter aircraft. It can be expected that there will be at least one round of lightweight and low-cost fighters of the YF-16/YF-17 type. The success of translating those prototypes into operational fighters will determine to a large extent whether there will be a second round.

The future is not clear. It is difficult to see a follow-up as being more complex, more versatile and with higher performance than today's breeds. Besides, the fighter cannot be considered alone, but must be treated as part of an overall strategic battlefield system, working with ground troops and weaponry, airborne command and control systems, and perhaps in the context of long-range missile strikes as well as bomber and fighter-bomber assaults.

Consequently, the future trend of fighter design will be affected primarily by the overall defence programme of any country. It has been decided in the United States, for example, that there will be no more manned interceptors. The British once decided that there would be no more manned fighters at all after the Lightning.

It is possible that the current round of fighters could be the last. The average life of today's types is predicted to be about ten years at the beginning of their programme lives. Yet as the ten-year mark approaches, ways are found to stretch that life. Look at two fighters in large-scale service today. The Russian MiG-21 first flew in 1955; the McDonnell-Douglas F-4 prototype in 1958. Both are still very much the front-line fighters of their respective countries and allies, far from the end of their useful lives.

Applying that kind of a life span to the F-14 through F-17 series gets close to the end of this century. What will be the state of international affairs then? Will multilateral agreements have been worked out that will systematically reduce armaments? Or will an apocalypse of natural forces – lack of pure air and water, a starved earth and stagnant oceans – be controlling our international relations?

It's safer to look at the short-term approach. For the foreseeable future, the YF-16/YF-17 will be the route of future fighter development.

**SEPECAT Jaguar S2**
The first prototype of this tactical strike fighter flew in 1968. It will equip several RAF and *Armee de l'Air* squadrons
*Crew:* 2 *Powerplant:* 2 Rolls-Royce Turboméca RT172 Adour 102 turbofans, 4620 lb thrust each
*Span:* 28·5 ft *Length:* 53·8 ft *Weight:* 23,000 lb
*Apartment:* 1 × 30-mm cannon; 10,000 lb ordnance load *Speed:* Mach 1·6 at 33,000 ft

# EJECTOR SEATS

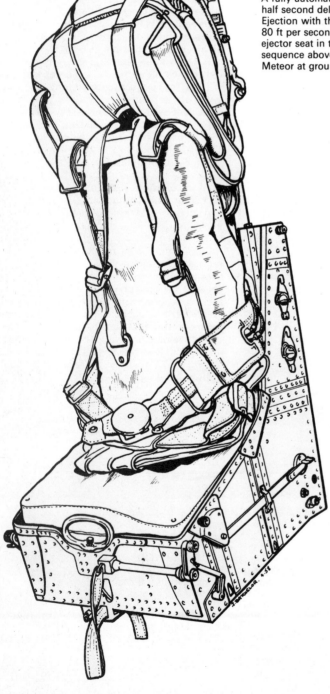

**Martin-Baker Mk IV Ejector Seat**
A fully automatic ejector seat incorporating a half second delay and a duplex drogue system. Ejection with this seat took place at the rate of 80 ft per second; it was the standard British ejector seat in the mid-1950s. The photo sequence above shows a pilot ejecting from a Meteor at ground zero

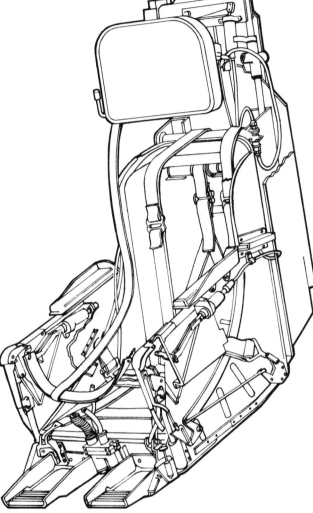

## US Ejector Seat

This is the model of ejector seat fitted to the Sabre fighter; ejection was effected by means of a cartridge exploding under the seat to propel it into the air. The photo sequence below shows a Crusader making a bad landing on the deck of an aircraft carrier; as the plane goes overboard into the sea the pilot is shot to safety by his seat

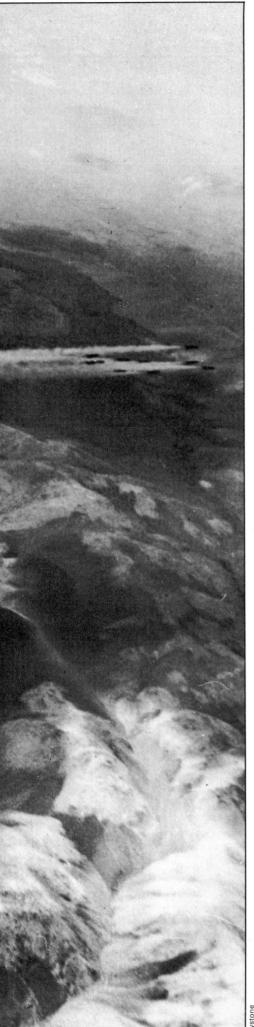

Keystone

# CHOOSE YOUR WEAPON

**The first jet fighters carried cannon, with some experimental air-to-air rockets on German planes. Since then, cannon, for weight of fire, and machine-guns, for maximum rpm, have vied for popularity, while missiles almost took over**

Guns to missiles to guns and missiles is the short history of fighter armament.

At the end of the Second World War, most of the standard fighters on the Allied side were armed with machine-guns, characterised by a high rate of fire. Cannon had been introduced on the German fighters, and on some of the Allied types, and there was some use of unguided air-to-air rockets by the Germans.

This mixed weaponry set a trend that has continued to this day. Since then, fighters have been armed with machine-guns, cannon and unguided rockets, plus a newer development, the guided missile.

Most early jet fighters were armed with cannon. The value of the heavier weapon had been proved during the Second World War. Only in the United States was there a holdout position for the machine-gun, and not until the Korean war did the military finally make an all-out switch away from those weapons.

It would seem obvious that a cannon is better than a machine-gun; it fires a heavier shell, and therefore has greater striking power. It also has a generally higher muzzle velocity and greater range, both desirable attributes. But there is a strong argument for the 'buzz-saw' theory of aircraft armament. A cluster of six or eight heavy calibre machine-guns can bring an effective weight of fire upon an enemy aircraft that is sufficient to cut it to pieces – literally. There is a problem, though; the firing has to be done within effective range, and it must be accurately aimed.

Korean combat of jet against jet introduced a new element. That kind of combat stretched the combat range of the machine-gun to the point where it was no longer effective. The ranges were increased substantially from those the USAF pilots had grown used to in their previous war. And eight machine-guns at those distances were often totally ineffective. Strikes could be seen on the MiGs, but the Russian-built planes absorbed them and continued to fly and fight.

*A Lockheed F-94C Starfire fires a salvo of rockets over the California Desert in 1952. Inset: 'Phantom' effect of heat condensation as a Lightning fires its three 30-mm cannon*

Actually, two things were happening: the range had opened up, and the airplanes had grown more rugged. The speed of jet aircraft dictated thicker skins, approaching the thickness of armour plating in some cases. And, hit at a shallow angle in the classic tail-chase, those surfaces were almost as good as armour in deflecting machine-gun bullets.

Korea, then, spelled the end of the machine-gun as an armament for jet fighters. As cannon batteries took their place, certain national characteristics began to emerge. The United States standardised on the 20-mm cannon; Europe built around the heavier 30-mm weapon. The Russians tried several – they have used 20-mm, 23-mm and 37-mm cannon in their fighters.

Conventional designs of aircraft cannon lacked the really high rate of fire that fighter pilots wanted. A fierce combat gives very little time to a pilot to aim and fire. He wants to be able to fire a maximum weight of slugs in minimum time.

There are two apparent ways to solve this problem, and the United States and Europe took different paths. The European developments generally were based on the revolving chamber concepts that were evolved by Mauser in Germany during the Second World War. The United States reached back into history to the Gatling gun, a 19th Century field weapon with a multiple barrel. The six barrels rotated and fired as they came opposite the breech mechanism. The Gatling patents were a basis for the modern 20-mm Vulcan cannon with an awesome rate of fire which reaches 6000 rpm.

But while these developments were going on, significant changes were being made in overall fighter armament concepts. For one thing, the guided and unguided rockets were beginning to come into their own. By the mid-1950s, just a few short years after the start of the Korean war, the first air-to-air guided missiles began to appear. Early guidance methods were improved, and homing missiles evolved, able to fly unerringly towards an enemy by reading his radar reflection, or his infrared signature, and homing on it.

It was only a matter of time until the nuclear weapons that had once filled bomb bays were reduced in size to fit into the warheads of air-to-air missiles. Unguided, because their lethal radius was so great that guidance would have been gilding the lily, these weapons were carried as standard armament on USAF interceptors for many years. They since have been replaced by a further updating, a smaller nuclear device in a guided weapon.

In fighter-against-bomber combat, the fighter held the speed advantage for many years. The standard way to shoot down a bomber was to approach from the rear in a tail chase, or a curve of pursuit as it was more elegantly called. The fighter pilot kept the nose of his aircraft bearing on the enemy target, and when he was within range of the bomber, the bomber was within range of him, and he was often met with a blast of fire from the enemy.

Obviously the curve of pursuit had drawbacks as a fighter tactic. But to replace it with anything else would seem to call for superhuman skill in piloting, because from any other approach angle, the speed differentials in the approach and the rapidly changing bearing of the target made hitting it more a matter of luck than skill.

The Germans developed a different tactical manoeuvre: they flew their jet interceptors in a line abeam at right angles to the bomber stream, and fired salvos of unguided rockets into the stream. They had to be successful; the densities of the bomber stream and the rocket salvos were so high that missing was impossible.

This technique was refined for single interceptors after the war. Armed with unguided rockets, the interceptor would fly against its target on an intersecting path, called a lead-collision course, heading for the point in the sky where the bomber would be when the interceptor's missiles got there. This called for skill in piloting and firing, and it was inevitable that a simple computing sight would be developed for this particularly difficult situation.

But the unguided rocket was short-lived, and was soon replaced by a guided missile with a brain of its own. It could change course to match any last-second evasion by the bomber, and it was infinitely more accurate than the unguided rockets.

In a sense, then, armament for the fighter-against-bomber combat evolved from an airplane carrying guns to a two-stage missile. The first stage, of course, is the

A US Navy F-14 Tomcat with its load of Phoenix missiles. In 1973 a world record was set when a Phoenix scored a hit on a supersonic jet drone at 126 miles range. They were introduced into service in 1973/74

Associated Press

### Republic F-105D Thunderjet
This long-range fighter-bomber typifies the weight of armament carried by modern combat planes
  *Crew:* 1 *Powerplant:* 1 Pratt & Whitney J75-P-19W, 17,200 lb thrust *Span:* 35 ft
*Length:* 64·2 ft *Weight:* 38,034 lb
*Armament:* 1 × 20-mm cannon; 8000 lb internal plus 6000 lb external ordnance load
*Speed:* Mach 2·1 at 36,000 ft

### Supermarine Scimitar
This naval fighter was a victim of the swing towards missiles in the late 1950s, when its original armament of four 30-mm cannon was replaced by four Sidewinder missiles
  *Crew:* 1 *Powerplant:* 2 Rolls-Royce Avon 202, 11,250 lb thrust each *Span:* 37·1 ft
*Length:* 55·3 ft *Weight:* 27,000 lb
*Armament:* 4 × 30-mm cannon
*Speed:* 710 mph at 10,000 ft

carrying airplane; the second stage is the guided weapon which delivers the warhead.

Fighter-fighter combat was and remains different. No matter how carefully planned are the tactics, and no matter if the first salvos are made on a lead-collision intercept basis, the contact soon swings into a swirling dog-fight, with the simple objective of getting on the enemy's tail. In the Korean war and subsequently, typical fighting ranges were on the order of 500 to 750 yards; beyond those ranges, the chances of kills were greatly reduced.

With guided missiles, that range could be increased greatly; ranges of several thousand yards were thought of as typical for the missiles. And so, fighter designs switched to an armament of all missiles, and the cannon was considered obsolete. Air-to-air missiles, either with radar or infrared homing, were further developed and became the standard weapons to the exclusion of anything else.

And then there was another war, in Southeast Asia, and there it became obvious that a gun was an absolute necessity for air-to-air combat. Missiles did not do so well at short ranges. And there were times when the enemy fighters did not stay off at long ranges, as they so often had in Korea – this time, the enemy fighters were eager to

Most modern jet fighters have some strike capability: here a Phantom drops napalm over Vietnam

close for combat. In such situations a fighter pilot calls for guns – and fast. The answer was supplied in the form of external podded guns that could be hung below the wings of fighters for either air-to-air combat or strafing ground positions.

Those external stores, of course, slow the fighter, and may even reduce its manoeuvrability considerably, so they were not the ideal solution. Further, the guns were distant from the pilot, robbing him of that aiming feeling that he normally has with guns on the fuselage centreline. Finally, unless the pods were unusually rigid, and rigidly mounted, they might deflect enough under the loads of firing to increase the scatter and degrade the accuracy of the burst. At best the pods were a temporary solution for fighter-fighter combat.

As one result of the Vietnam action, guns have been reintroduced into USAF fighters where they had been lacking. (The Navy had never given them up.) Nor did most designers succumb to the siren call of the missiles. Very few European and Russian fighters have been built during the last 30 years around a missile battery only.

As the weapons progressed from form to form, so did the rest of the armament system. Gunsights, which were simple optical types with a lead-angle computer in early jet fighters, soon added more and more semi-automatic features. Today, firing controls range from the nearly automatic to the automatic. In a modern interceptor, it is possible for the pilot to find, fix and destroy his target without ever seeing it, or initiat-

ing any action except arming the system and the weapons. The automatic systems on board have searched for the target, found it, locked on to it, computed its position, corrected the aircraft course, moved into firing position, released the weapon, and confirmed the destruction of the enemy target, and all without the touch of the pilot's hands.

For interception missions, this is the future, as well as the present, until the unmanned guided missiles take over completely. For air-to-air combat between fighters, the trend is back to the automatic cannon with a high rate of fire, like the Vulcan. For fighter against bomber, the future is the same as that of the interceptor.

For the common foray against ground targets, there are whole packages of weapons that can be loaded under the wings.

Future fighters will continue to be armed as in the past. They will carry a single cannon battery with a high rate of fire, and probably a pair of long-range dog-fight missiles. They will have strong points under the wing and fuselage for a variety of external weapons. The fire-control systems will be advanced and semi-automatic, with an automatic mode for intercept or all-weather missions. Sighting data will be shown on the windshield in a head-up display.

A Fiat BR 3 medium bomber. Capable of carrying a
1320-lb bombload, it served with the Italian air force
throughout the 1930s

Fiat

# BOMBERS 1914–1939

## *Bryan Cooper*

No aspect of modern warfare has stimulated as much controversy as the bombing offensive — particularly the 'strategic' raids against civilian and industrial targets. And while the effects of bombing raids during the First World War seem almost trivial compared with the destruction they caused less than thirty years later, and again in the saturation bombing of North Vietnam, all the basic concepts of the use of bombers in war had been evolved by 1918.

This chapter traces the development of the bomber from its earliest beginnings, when pilots dropped steel darts, artillery shells and cans of kerosene on enemy troops, through the brief but spectacular career of the Zeppelin and the development of the specialised heavy bomber. We follow this development up to 1939, when the foundations were being laid for the giant bombers of the Second World War.

*A Caproni Ca 41, one of the Ca 4 series of gigantic triplane bombers evolved by Caproni during the First World War*

# CONTENTS

Imperial War Museum

**CAPRONI Ca 42**

**Gross weight:** 16,535 lb **Span:** 98 ft
1 in **Length:** 49 ft 6 in **Engine:** 3 × 270 hp
Fiat, Isotta-Fraschini, or American Liberty
**Armament:** 4 machine-guns **Crew:** 5
**Speed:** 87 mph at ground level **Ceiling:**
9,840 ft **Range:** 7 hr **Bomb load:**
3,910 lb

# HOW THE BOMBER WAS BORN

Aviation is good sport, but for the army it is useless.' This comment, attributed to General Foch, one of the more enlightened of military leaders, typified the prevailing attitude to an innovation which was to play an increasingly important role in warfare . . .

German ground crew member loading a 25-kg bomb

Like so many concepts of modern warfare, aerial bombing on a large scale was first developed during the First World War. The two main ingredients – bombs and a means of delivering them – were already in existence. The bombs were simple metal shells containing explosives and a detonator, which could be set off by a time fuse, and with fins to give more stable and accurately predictable flight. Such weapons had been conceived of since the earliest days of flight, but use of aeroplanes for bombing came about only when it was found how vulnerable airships were to defensive fire from the ground, and to attacks in the air.

Although several types of aeroplane were available in 1914, and experimental bombing from aeroplanes had been carried out before the war, their military potential was not generally appreciated by the strategists of the combatant nations. Consider-

ing the flimsy nature of these craft, mostly two seaters, it was understandable. Most were unable to fly higher than a few thousand feet and their average speed of about 60 miles per hour could be reduced to a crawl by strong headwinds. They were not only vulnerable to ground fire but their performance was also seriously affected by carrying the added weight of bombs or machine-guns.

For the most part, the military leaders of all the major powers saw the main function of aircraft as reconnoitring for the infantry and spotting the fall of shot for the artillery – merely an extension of the observation balloon. It was the pilots and observers who showed the fighting potential of their machines by taking pot-shots at enemy planes with hand-guns. This eventually led to the development of specifically designed single-seat fighters armed with fixed forward-firing machine-guns.

In a similar way, the use of early reconnaissance aircraft for dropping bombs led to the evolution of the true bomber – that is, an aeroplane designed to carry and deliver bombs in level flight, against a specific target, over medium or long range distances.

### The true beginning

At the start, just as reconnaissance aircraft were armed only with pistols, many of the first bombs dropped were simply grenades or canisters filled with gasoline, dropped over the sides of open cockpits. The French even threw down steel darts (*fléchettes*) in the hope of disrupting German infantry and cavalry formations.

The first aircraft used for both fighting and bombing were the general purpose types which were already in existence when war broke out. In fact, a Voisin III, the type most widely used for bombing by the French in the early days of the war, was also the first to shoot down an enemy aircraft in aerial combat. And it was with an Avro 504, the first British bomber, and the type to make the first organised raid in history, that some of the earliest experiments were made in fitting a machine-gun, leading to the development of the true fighter.

Two basic configurations were predominant at that time: the tractor aeroplane, with the engine and propeller in front, and the pusher type, like the Voisin, in which the engine and propeller were mounted behind a tub-like nacelle seating the pilot

The main production model of Caproni's Ca 4 series of huge triplanes was brought into service in 1918. It was used by the British RNAS and air units of the Italian army and navy

and observer, the tail being carried on booms. Both were made in biplane and monoplane form, biplanes being the more common. The pusher type soon became obsolete, but it was the most successful in the early days since the observer in front had a much better field of vision. (In two-seater tractor aircraft the second crew member invariably sat in front but since his view and ability to fire were restricted by the propeller and wing struts, the positions were later reversed.)

Using the experience with these general purpose types, different classes of aircraft were developed to meet specific requirements. An account of bombers might well be limited to the type defined as 'an aircraft designed to drop bombs in level flight'. These have been built in great variety up to the present day and indeed are the ones mainly dealt with here. But there have been others such as dive-bombers, torpedo-bombers and fighter-bombers combining a dual role. These, with aircraft designed primarily for a ground attack role in support of infantry, might be grouped separately as strike aircraft. However, the most important examples are included in this book. In fact, as we have implied, almost every basic tactical and strategic use of bombers was first tried out during the First World War.

## Combined role

The fighter-bomber is a particular case in point. Like the Voisin, the earliest bombers combined both roles, but thereafter, the development of bombers took a separate course from that of the fighters designed to intercept them. Except for a period in the 1930s when a few notable bombers actually flew faster than the fighters of the day, they generally had to make up for their lack of speed by carrying heavier armament. This, together with the need to carry the largest possible bomb load, necessitated a continual compromise in design between weight, size and range on the one hand, and speed and altitude on the other. Although some remarkable engineering advances were made by bomber designers, such as the geodetic form of construction, for the most part they had to make use of the materials currently available. And most of the pioneering effort, in terms of speed at any rate, went into fighter design. For example, a number of exceptional engines were specifically designed for fighters whereas bombers usually had to make use of the same type of engines.

In more recent times however, with the development of small but powerful atomic missiles – as well as the atom bomb itself – there has not been the same need to consider weight of bomb load as the measure of a bomber's capability. The development of sophisticated defence systems, including supersonic fighters, has meant that bombers must be able to compete on an equal footing in both armament and speed. Largely because of the enormous costs involved the future of the true bomber was at one time in some doubt. This has been partly resolved by the American decision to go ahead with the B-1 strategic bomber, capable of three times the speed of sound. However, for reasons of cost and modern strategic planning, the fighter-bomber is one of the most common types now in service with the air forces of the world. The wheel has turned full circle back to the kind of aeroplane first used offensively sixty years ago.

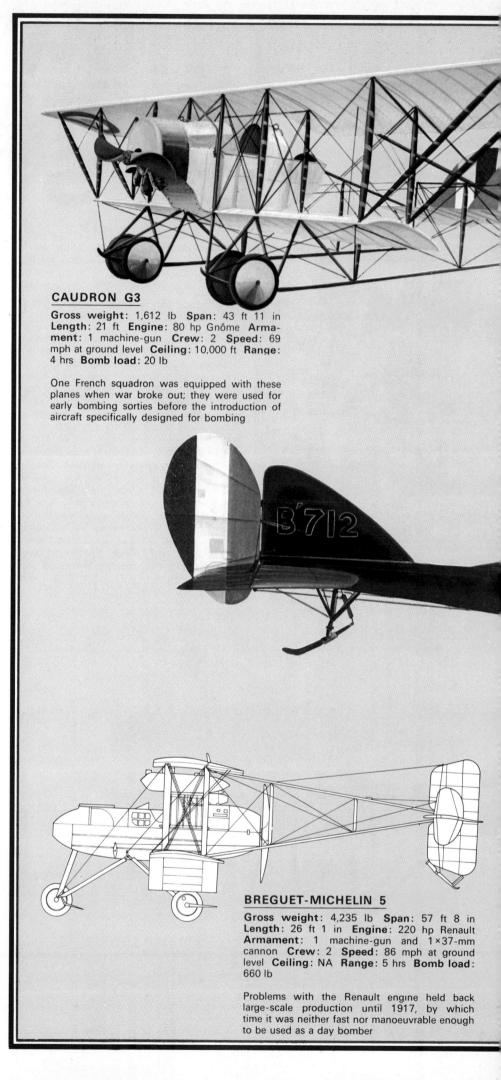

**CAUDRON G3**

**Gross weight:** 1,612 lb **Span:** 43 ft 11 in **Length:** 21 ft **Engine:** 80 hp Gnôme **Armament:** 1 machine-gun **Crew:** 2 **Speed:** 69 mph at ground level **Ceiling:** 10,000 ft **Range:** 4 hrs **Bomb load:** 20 lb

One French squadron was equipped with these planes when war broke out; they were used for early bombing sorties before the introduction of aircraft specifically designed for bombing

**BREGUET-MICHELIN 5**

**Gross weight:** 4,235 lb **Span:** 57 ft 8 in **Length:** 26 ft 1 in **Engine:** 220 hp Renault **Armament:** 1 machine-gun and 1×37-mm cannon **Crew:** 2 **Speed:** 86 mph at ground level **Ceiling:** NA **Range:** 5 hrs **Bomb load:** 660 lb

Problems with the Renault engine held back large-scale production until 1917, by which time it was neither fast nor manoeuvrable enough to be used as a day bomber

## VOISIN 5

**Gross weight**: 3,240 lb **Span**: 52 ft 4½ in
**Length**: 31 ft 6½ in **Engine**: 150 hp Canton-
Unné **Armament**: 1 machine-gun **Crew**: 2
**Speed**: 74 mph **Ceiling**: 1,500 ft **Range**: 3½ hrs
**Bomb load**: 130 lb

The Voisin pusher biplane, which equipped four
of the French squadrons at the outbreak of war,
was the most widely used bomber during the
first two years of operations. The Voisin 5 variant
was introduced at the end of 1915

## BE 2e

**Gross weight**: 2,142 lb **Span**: 37 ft **Length**:
27 ft 3 in **Engine**: 90 hp RAF 1a **Armament**:
1–4 Lewis machine-guns **Crew**: 2 **Speed**:
72 mph at 6,500 ft **Ceiling**: 10,000 ft **Range**:
NA **Bomb load**: 112–224 lb

Introduced into RFC service in the summer of
1916, in time to take part in the Battle of the
Somme, but showed little improvement on previous BE2 types; the observer had to be left at
home when a maximum bomb load was carried

## IL'YA MUROMETS TYPE V

**Gross weight**: 10,130 lb **Span**: 97 ft 9 in
**Length**: 56 ft 1 in **Engine**: 4×150 hp Sunbeam
**Armament**: 3–5 machine-guns **Crew**: 4
**Speed**: 75 mph **Ceiling**: 9,840 ft **Range**: 5 hrs
**Bomb load**: 650 lb

The bomber version of the world's first four-
engined passenger aircraft, designed by Igor
Sikorsky and built in 1913

# UP THERE, A NEW KIND OF WAR WAS BEGINNING

**Though several countries had shown interest in military aviation in the years before the war, this quickly lapsed, and most of the early experiments in bombing were carried out through individual enterprise rather than official policy. Many of these early experiments were to prove of value as the war progressed**

Even though bomber aircraft as such did not exist at the beginning of the First World War, the threat of bombing was well recognised. Visionaries like H. G. Wells had foreseen the destruction that might some day rain from the skies. As early as 1670 a Jesuit monk, Francesco de Lana, produced the first known design for a lighter-than-air craft. It did not work, but from his writings, it is clear that de Lana saw the possibilities of airborne invasion and bombing, and indeed referred to the destruction of cities and ships by fireballs hurled down from the sky.

The difficulties of bombing from balloons were self-evident for they could only travel where the wind took them. Nevertheless, such attempts were frequently made in the latter half of the 19th century, one of the first being by the Austrians in 1849 when unmanned hot air balloons carrying 30-lb time-fused bombs were launched against Venice.

It was the invention of the dirigible (steerable) airship at the turn of the century that made bombing a practical and a frightening possibility. Although such craft were also developed in Britain, France, the United States and Italy, it was in Germany, with the pioneering work of Count Ferdinand von Zeppelin, that the rigid airship made its greatest strides. At the beginning of the war the Germans possessed more and better airships than any other nation. These, rather than planes, were intended to provide Germany's primary means of bombing.

Until a few years before the war Britain

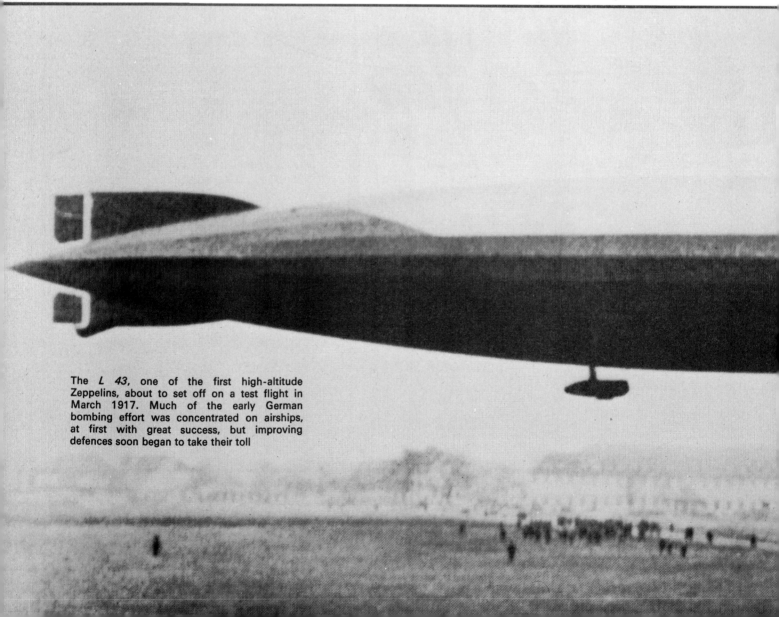

The *L 43*, one of the first high-altitude Zeppelins, about to set off on a test flight in March 1917. Much of the early German bombing effort was concentrated on airships, at first with great success, but improving defences soon began to take their toll

had largely neglected aviation development, but government indifference and indecision between lighter- and heavier-than-air machines gave way under the threat of Zeppelin attack to a belated attempt to develop defensive aeroplanes.

Meanwhile, the first experiments at dropping bombs from aircraft were carried out in the United States, by the great pioneering designer/constructor Glenn Curtiss in 1910, using dummy bombs. The following year, live bombs were dropped during a military exercise by Lt M. S. Crissy from a Wright biplane. Both the US Army and the US Navy showed some initial interest. In fact, in 1908 the US Army had been the first combat service in the world to buy an aeroplane for evaluation. But the authorities lost enthusiasm for military aviation and although the US Army and Navy both established aeronautical divisions, the US Army Air Arm had only 20 planes on its strength when war broke out in Europe, all of them obsolete, and the Naval Air Arm was almost non-existent. While the Americans later built thousands of French and British aircraft under licence, no aeroplane of American design fought in France during the First World War. The only American aircraft used in combat were Curtiss flying-boats, employed by the British Royal Naval Air Service for anti-submarine patrols.

However, one result of those early bombing experiments was the evident necessity for some kind of aiming device to enable bombs to be dropped accurately. This led Lt Riley Scott to invent the first bomb-sight in 1912. It was installed in a Wright biplane of the US Army, and was a simple device of wires and nails. Of particular interest was the fact that the bomb-aimer lay in a prone position in the nose of the aircraft and viewed the ground through a mica window, anticipating the bombing method of many years later. Also, the bombs were carried horizontally underneath the aeroplane, also common practice years later, instead of being dropped over the side or carried vertically in racks.

## Italy in the lead

Italy was the first country in the field of military aeronautics, having established an Army Aeronautical Section equipped with balloons as early as 1884, and it was also the first country to drop bombs on an enemy during war. This was on 1 November 1911, during the Italo-Turkish war, when Lt Gavotti dropped four grenades of 4.4 lb each on Turkish troops in Libya. Further raids followed, causing more consternation than damage. Then the Turks protested that an Italian aircraft had bombed a military hospital at Ain Zara and an immediate controversy arose in the Italian, Turkish and neutral Press. There was no independent way at that time of establishing whether a hospital had actually been bombed and the Italians reasonably pointed out that no similar protest had been made when their warships shelled Ain Zara a few days earlier. But many felt there was something particularly inhuman about aerial bombs, and the controversy about the subject has continued ever since.

When Italy entered the war in May 1915 against Austria and Hungary, her Army Air Service was perhaps more highly trained than any other and in an excellent state of readiness. This was offset to some extent by the fact that most of her 150 aircraft were French types such as Nieuport and Blériot monoplanes and Maurice-Farman pusher biplanes, already out-dated. An exception, however, was the Caproni Ca 30/33 series which, together with the Russian Il'ya Muromets, were the first very large planes built. These certainly gave Italy the lead in heavy bombing during the early part of the war.

The series began in 1913 with the Ca 30 powered by three Gnôme rotary engines, one of which was mounted as a pusher in the central nacelle while the other two drove, indirectly, two tractor propellers in twin fuselage booms. The indirect drive was not successful and was abandoned in the Ca 31, the outer pair of engines being re-located at the front of the booms. This became the standard configuration for the series. The first bomber to be built in quantity for the Italian Army Air Service was the Ca 32, in which the rotary engines were replaced by three Fiat inline water-cooled engines of 100 hp each. Deliveries began within three months of Italy's declaration of war and went into action almost immediately. They were easy to handle, had an excellent range, and could carry up to 1,000 lb of bombs.

## AVRO 504

**Gross weight:** 1,574 lb   **Span:** 36 ft
**Length:** 29 ft 5 in   **Engine:** 80 hp Gnome
**Armament:** Optional Lewis   **Crew:** 2
**Speed:** 62 mph at 6,500 ft   **Ceiling:**
13,000 ft   **Range:** 4½ hrs   **Bomb load:**
4x20 lb

Britain's first effective light day bomber,
notable for RNAS raids on Zeppelin sheds
in Germany. Also used for bombing airships
in flight.

## CAPRONI Ca 32

**Gross weight:** 7,280 lb   **Span:** 72
ft 10 in   **Length:** 35 ft 9 in   **Engine:**
3x100 hp Fiat A10   **Armament:**
2 machine-guns   **Crew:** 3–4   **Speed:**
72 mph at ground level   **Ceiling:** NA
**Range:** 340 miles   **Bomb load:**
200 lb

The first production version of the
Caproni to enter service with the
Italian Army Air Force, in August
1915 and the basis of Italy's strategic
bomber arm. An excellent plane to
handle with greater range than most
aircraft of that time.

The Ca 33 version was put into production the following year. It had a multi-wheel landing gear for operations from rough ground and included, for the first time, two gun positions – one in the front cockpit, ahead of the two pilots who sat side by side, and another at the rear. This type was so successful that it was also built under licence in France.

### France moves in
France had made greater headway in aviation than any other country before the war, with the result that the French Army possessed the widest range of flying machines – about 138 in all. An Army Air Arm had been established by 1914 and with the approval of General Joffre, one of the few military leaders to see a potential for aircraft in the war beyond mere reconnaissance, this force began to explore various offensive possibilities. The smaller single-seat tractor scouts like the Nieuport biplane and Morane-Saulnier monoplane were formed into fighter squadrons (*escadrilles de chasse*) which were attached to each army for the purpose of harrying German reconnaissance craft. And within weeks of the outbreak of war, the larger two-seater pusher biplanes were formed into a bomber group (*l'ère Groupe de Bombardement*) consisting of three six-machine squadrons, under Commandant Göys.

The French bomber group, quickly joined by another two, was equipped with various Henri and Maurice Farman types, the twin-engined Caudron, the Bréguet-Michelin and the Voisin 13.50 (so named from its 13.50 metre wingspan). These pusher biplanes were completely blind from the rear, and as the idea of aerial combat did not exist before the war and since the bomb-aimer/gunner in front enjoyed a wide range of vision, they had things very much their own way to begin with, being sturdy and dependable. The Voisin was the most widely used for bombing in the early days. It could carry a bomb load of about 200 lb for three hours at a maximum speed of 55 mph and could take off and land on the roughest ground. More than 2,000 IIIs were built and supplied not only to France but also to Britain, Belgium, Italy and Russia.

### The Zeppelins
In Germany, both the army and navy had established aviation corps before the war in which the pride of place was given to Zeppelins. Compared to the planes of that period, the dirigible airship was a formidable weapon. It had a speed of about 50 mph, barely less than that of most aircraft, and a much greater ceiling and range. It could cruise far behind the front lines and carry bomb loads of around 1,000 lb – impossible with aircraft until the development of multi-engined types like the Caproni. The Military Aviation Service, formed in 1912, had a fleet of six large airships and three smaller ones, to which were added three airships from a commercial airline service. The German Naval Air Service was primarily committed to the airship for the task of reconnaissance over the North Sea. But as a result of two airship disasters in 1913, it had only one Zeppelin actually in operation at the outbreak of war.

As well as airships, the Military Aviation Service was equipped with a mixed collection of 246 aircraft comprising single and two-seater Type A monoplanes like the Rumpler and Etrich Taube, and Type B

biplanes of which the Albatros, Aviatik and LVG were the most common.

These were all tractor types, since Germany did not then possess any pusher planes, and none of them was armed. In fact, the development of faster aircraft had been restricted in case speed should interfere with careful observation. The Service was completely subservient to the Army with its squadrons disposed between Army HQ and Army Corps.

Within two months of the outbreak of war, however, some of the Type Bs were equipped as bombers and formed into a bomber force known as the *Fliegerkorps des Obersten Heeresleitung* (Air Corps of GHQ) comprising thirty-six aircraft divided into two wings which were given the code-name 'carrier-pigeon units'. This force was led by Major Wilhelm Siegert, a pilot himself and an aviation enthusiast, who was to play a vital part in the development of the German Army Air Service.

### Britain – a late starter
The air units of all the nations mentioned came under the control of the army, the navy, or both. Britain was late in the field and it was not until 1911 that the Air Battalion of the Royal Engineers was formed, comprising a miscellany of planes, small airships, man-lifting kites and balloons. A year later the Royal Flying Corps was formed as a joint service with naval and military wings. Although the objective of military aviation in Britain was limited to reconnaissance over land and water, the RFC had one great advantage: it was a single organisation and not split up into army and navy groups as in most other countries. However, six weeks before war broke out, when Britain had 113 aircraft operational, the Admiralty decided to form its own Royal Naval Air Service. This became largely responsible for the defence of Britain, especially against the threatened Zeppelin raids. The RFC was relegated to the status of a corps attached to the army; in August 1914, the four RFC squadrons operational at the time were sent to France to join the British Expeditionary Force.

As far as equipment was concerned, a major difference in policy marked Britain's two services. The War Office had decided in 1911 that it would be more economical to build its own planes. Accordingly, it established the Army (later Royal) Aircraft Factory at Farnborough, employing Geoffrey de Havilland as chief designer and test pilot. The research effort at Farnborough concentrated on producing an aeroplane of inherent stability which would fly straight and level with little effort required on the part of the pilot so that he could devote most of his attention to observation. This was achieved with the BE2, which was given military trials in 1912. It was the first of a series of two-seater tractor biplanes, which reached its peak with the BE2c version. Well over 1,000 of this model were delivered during the first years of the war, making it the RFC's standard observation machine. However, very few were available at the outbreak of war and the War Office had to continue buying French aircraft to make up the deficiency. The result was that the RFC squadrons consisted of a mixed bag of Farman pusher biplanes, Bleriot and Morane-Saulnier monoplanes alongside a few BE2s and BE8 'Bloaters'. This last, named from its fish-like appearance, in the

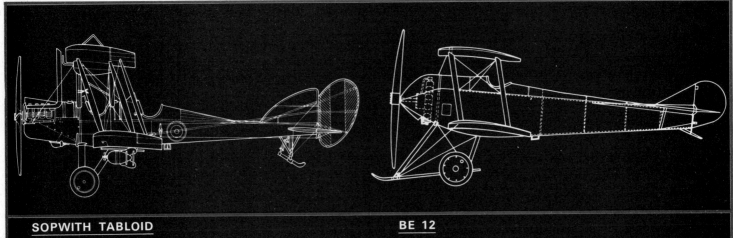

**SOPWITH TABLOID**

**Gross weight**: 1,120 lb **Span**: 25 ft 6 in **Length**: 20 ft 4 in
**Engine**: 80 hp Gnôme **Armament**: 1 Lewis **Crew**: 1 **Speed**:
92 mph **Ceiling**: NA **Range**: 3½ hrs **Bomb load**: 20–40 lb

The fastest biplane in the world when it first appeared in 1913.
Capable of climbing to 1,200 ft in one minute

**BE 12**

**Gross weight**: 2,352 lb **Span**: 37 ft **Length**: 27 ft 3 in **Engine**:
150 hp RAF 4a **Armament**: 1–2 Lewis machine-guns **Crew**: 1
**Speed**: 102 mph at ground level **Ceiling**: 12,500 ft **Range**: NA
**Bomb load**: 2×20 lb

A more powerful version of the BE 2c, introduced in an attempt to
combat the menace of the Fokker monoplane in 1916

early days was used as one of the RFC's
first bombers.

The Royal Navy, on the other hand, had
preferred to buy the products of the
embryonic British aircraft industry and
thus supported the development of some of
the best aircraft used during the war. In
addition to some French types, the RNAS
was equipped with the single-seat Sopwith
Tabloid, Bristol Scout, and Martinsyde SI,
the two-seater Avro 504 – the first British
bomber – and seaplanes like the two-seater
Short Folder and the single-seat Sopwith
Schneider. It was with a Folder, in July
1914, that the RNAS began to drop torpedoes,
although the first torpedo launch from a
plane had been made by the Italians several
months earlier.

**The Russian version**
Most of the 224 aircraft in service with
the Russian Army at the outbreak of war
were of French and German design, built
under licence in Russia, and as the war
progressed, further aircraft were imported
from France, Britain and the United States.
Only in one field did the Russians lead the
way with planes of their own design, and
that was with the heavy bomber. The *Grand*
biplane, designed by Igor Sikorsky (who was
later to achieve fame as the pioneer of
helicopters) was built in 1913. This was the
world's first four-engined aeroplane, intended to carry passengers in great comfort.
From it, Sikorsky developed a production
series, the Il'ya Muromets, but with the
threatened outbreak of hostilities ten of
these were purchased by the Russian Army
for military trials.

After experiments with various types of
armament and bomb racks it was found
that the huge planes with their roomy
cabins had too low a speed and limited an
altitude for offensive purposes. Accordingly,
Sikorsky designed a lighter version, the
Il'ya Muromets Type V, and deliveries of
these began early in 1915 to the special
'Squadron of Flying Ships' formed ·the
previous December. This was both the first
specialised bomber to be built and also the
first to be powered by four engines – initially
British Sunbeams of 150 hp each. The name
Il'ya Muromets became a class name for a
variety of wartime types, some of them
mounting machine-guns. These could fire

from the sides of the fuselage through doors
and windows, upwards from a crow's-nest
position in the centre of the top wing,
downwards from a platform under the
fuselage and rearwards from an installation
in the extreme tail, thus anticipating the
heavily armed large bombers of the Second
World War.

Over seventy of these remarkable aircraft
were produced, and in some 400 raids over
East Prussia only one was lost in air
combat. The need for long-range and consequently big aircraft was obvious in a
country as large as Russia, but the success
of the Il'ya Muromets was also responsible
for the Russians' preference for the heavy
bomber as a weapon. which has lasted to
the present day.

The needs of the other combatant nations
were supplied primarily by the major powers,
the Belgian Army for instance being
equipped with French planes, while the

even smaller Turkish air units used German
types, most of them passed on after they
had become out-dated on the Western Front.
The more extensive squadrons of Austria-
Hungary were equipped at the start with
a variety of Etrich Taube monoplanes and
Lohner arrow-wing (*Pfeil*) biplanes, together with Albatros B Is which had been
designed in Germany by Ernst Heinkel and
were made under licence in Austria by the
Phoenix company.

Throughout the war Austria continued
to rely on aircraft supplied by Germany or
built under licence, but in its own right
Austria was an important supplier of
engines, especially the Austro-Daimler and
the Hiero, on which Dr Ferdinand Porsche
worked. Since engines were of such vital
importance in aircraft design, it is worth
considering their development before
describing the first use of bombers in the
First World War.

Caproni Ca 41 triplanes used by the Royal Naval Air Service in Italy for a short period
towards the end of the war
Caproni, Milan

# THE FLYING MOTOR CARS

The major — and often insurmountable — problem confronting early bomber designers, was getting the machines to fly at all. The major choices open to them — air-cooled or water-cooled engines; monoplane, biplane or triplane construction; of wood, metal or a combination — were often made on the basis of availability rather than sound reasoning. Nevertheless, tremendous strides had been made by the end of the war

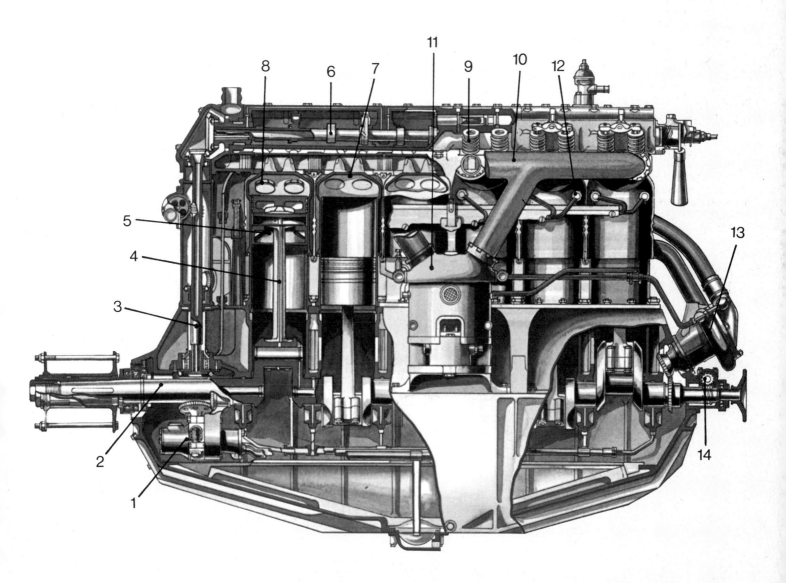

| | | | |
|---|---|---|---|
| **1** | RECIPROCATING PLUNGER OIL PUMP | **8** | VALVES |
| **2** | CRANK SHAFT | **9** | VALVE SPRINGS |
| **3** | CAM SHAFT DRIVE SHAFT | **10** | INDUCTION MANIFOLD |
| **4** | CONNECTING RODS | **11** | DUPLEX CARBURETTOR |
| **5** | ALUMINIUM PISTON (SECTIONED) | **12** | SPARK PLUGS |
| **6** | OVERHEAD CAM SHAFT | **13** | CENTRIFUGAL WATER PUMP |
| **7** | CYLINDER HEAD | **14** | INTERRUPTER GEAR CAM |

**AUSTRO-DAIMLER ENGINE**

**Power:** 200 hp at 1,400 rpm **Cylinders:** 6
**Weight/power ratio:** 3.64 lb per BHP

By August 1914 the designers of aircraft engines had already established the main families that were to last until the general introduction of the gas turbine more than thirty years later. Many of the earliest engines were based on those used in motor-cars and thus had cylinders cooled by water jackets and disposed in rows, driving on a multi-throw crankshaft. But there was one outstanding engine, the Gnôme rotary, invented by Laurent Séguin in France in 1907, that established a totally new configuration.

Although the American Farrand car had tried a cruder version of the same idea, the Gnôme set a new standard in engineering design, and by weighing only about two-and-a-half pounds for each horsepower it roughly doubled the ratio of power to weight generally available at that time. The first Gnôme had seven cylinders arranged radially, like the spokes of a wheel. A mixture of petrol and castor oil (lubricant) was admitted to the crankcase and then, via flap-valves in the pistons, to the combustion spaces in the cylinders. The crankshaft was fitted to the aircraft so that when the engine was running, the cylinders and the crankcase rotated, taking the propeller with them. The spinning cylinders had high gyroscopic inertia which damped out vibration and made for sweet running, but they tended to impart gyroscopic forces to single-engined aircraft. It was this factor that gave the fighting scouts much of their manoeuvrability – and also made them difficult to handle. The only drawback was that although they could turn very smartly in one direction, they were very sluggish in turning to the other and enemy aircraft could usually guess what they would do in combat. For this reason some engines were adapted to reverse the propeller's normal direction. The gyroscopic characteristic was less marked with multi-engined bombers that used the rotary engine, such as the twin-engined French Caudron G4, G6 and the Italian Ca 30 and 31 Capronis, but it was certainly evident in the RFC's single-engined BE8 'Bloater', especially when carrying a 100-lb bomb.

### New techniques
Unlike almost every other petrol engine of the time, the Gnôme was machined from high-tensile steel. This was costly even in mass-production, but it established a trend of fine precision engineering which gradually eliminated the crop of engines put together from cruder pieces of steel, cast iron, brass and copper. The spinning cylinders were easy to cool by the slipstream but introduced problems of carburisation and mixture distribution which were only partly overcome by the development of the *Monosoupape* Gnôme, with only one valve in each piston serving for both inlet and exhaust. Though the original 50 hp engine was developed to give 80, 150, and finally, in two-row form, over 200 hp, it was obsolescent by 1916. So too was the only other important family of engines with radially disposed cylinders in the First World War. These were the Swiss/British/French Canton-Unné or Salmson engines – static radials with fixed liquid-cooled cylinders and propellers mounted on the rotating crankshaft. They gave good service, especially in the Voisin pusher biplanes and the Short bomber seaplanes, but the experimental French SM I bomber of 1916 could hardly expect to be successful as its

Salmson engine, arranged sideways in the fuselage, drove two propellers on the wing struts by means of long shafts and bevel gears.

By 1916 almost all the engines pouring off the assembly lines in ever increasing numbers were of the in-line or Vee type. A few had air-cooled cylinders (the cooling scoop of the British RE8 reconnaissance biplane could be seen a mile away) but the vast majority had water cooling, even though this was clumsy, heavy and vulnerable. Practically every aircraft on the side of the Central Powers used a six-in-line water-cooled engine, the most common makes being Mercedes, Maybach (based on an early airship engine) and Austro-Daimler. The French developed advanced and powerful models of Renault and Hispano-Suiza engines, first used in the Letord series and the Spad S XI respectively. And the winning Mercedes Grand Prix racing car of 1914 served as a starting point for the British Rolls Royce aero engines. By 1917 Rolls Royce had produced versions of the Vee-12 Eagle giving 360–375 hp, with outstanding efficiency and reliability, for the giant Handley Pages and Vickers Vimy bombers.

Though by this time the air-cooled radial was suddenly coming into favour, the powerful and efficient Vee-12 was dominant at the end of the war. This was also the arrangement adopted in the United States for the Liberty engine, created in a matter of days by a committee of engineers (mainly from automobile companies) and swiftly pushed through a troublesome development into production on an unprecedented scale. The Liberty, ultimately rated at well over 400 hp, served the US Army for fifteen years, beginning with the DH4, built under licence in America.

The greatest scope for variation in design lay in choosing the location of the radiator made necessary by the water-cooled engine. One of the simplest and least vulnerable schemes was the frontal car-type arrangement with the radiator forming a bluff face immediately behind the propeller as in the Handley Pages and DH9a (or, in the case of a pusher, such as the Gotha, at the front of the nacelle). Other bombers had radiators in a box flush with the top wing (LVG and Halberstadt); inclined on each side of the engine cowl (Voisin and Bréguet); disposed vertically on the struts above the engine (Caproni Ca 40); arranged vertically up the sides of the fuselage and then slanting in to meet above the engine (Armstrong Whitworth FK8); or in a box in front of the upper centre-section (Lloyds and Lohners). Many bombers of the Central Powers such as Aviatiks and DFWs had Hazlet radiators, looking just like those used for domestic heating, fixed on the side of the fuselage in sections so that they could be shortened in cold weather and enlarged in summer.

### Wire and plywood
A major factor in design was the materials available at any given time. Early aircraft made use of obvious proven materials, the preferred choice being carefully selected hardwood. For example the widely used Rumpler C-1 had ash for the top (compression) booms of the wing spars and spruce for the lower (tension) booms. Ribs were generally built up from pieces of hardwood and plywood, glued and pinned. Only at major joints and places bearing especially heavy loads were metals used – the most common material being steel. On the other hand, a large number of aircraft were built up from thin-walled steel tube,

### GOTHA GV
**Gross weight:** 8,745 lb   **Span:** 77 ft 9 in
**Length:** 40 ft 7 in   **Engine:** 2×260 hp Mercedes
D IVa   **Armament:** 3–4 machine-guns   **Crew:** 3
**Speed:** 87 mph at ground level   **Ceiling:** 21,320
ft   **Range:** 520 miles   **Bomb load:** 1,300 lb

### BREGUET 14 B2

**Gross weight**: 3,892 lb **Span**: 49 ft **Length**: 29 ft 1 in **Engine**: 300 hp Renault 12 Fe **Armament**: 1 Vickers (fixed forward-firing) 2 Lewis (aft) **Crew**: 2 **Speed**: 112 mph at ground level **Ceiling**: 18,000 ft **Range**: 435 miles **Bomb load**: 32×22 lb

During the last year of the war this day-bomber version of the Bréguet 14 series established a formidable reputation with the bombardment squadrons of France's First Air Division, and was also widely used by the American Expeditionary Forces in France

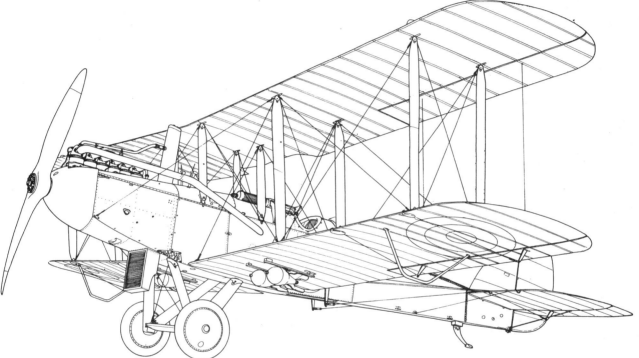

### DH 9

**Gross weight**: 3,669 lb **Span**: 42 ft 5 in **Length**: 30 ft 6 in **Engine**: 230 hp BHP **Armament**: 1 Vickers; 1–2 Lewis machine-guns **Crew**: 2 **Speed**: 118 mph at 10,000 ft **Ceiling**: 17,500 ft **Range**: NA **Bombs**: 2×230 lb or 4×100 lb

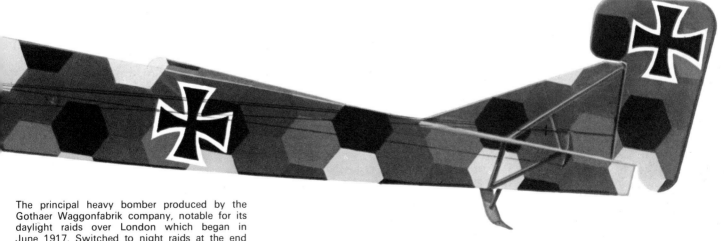

The principal heavy bomber produced by the Gothaer Waggonfabrik company, notable for its daylight raids over London which began in June 1917. Switched to night raids at the end of that year, when British fighters were introduced with the speed and operational altitude necessary for successful interception

Lamblin radiators on the DH 9

Flight International

either welded, or held by bolted or riveted joints. This was simple and quick to make and the weight factor was competitive. Even wooden aircraft often had steel-tube tails. Covering was invariably of doped fabric, though often the front fuselage was skinned in plywood. Undercarriage and wing struts were nearly always of steel tube, sometimes with light wood fairings to change the round tube to a streamline form. Wheels were thin steel rims held by multiple wire spokes, often faired by a disc of fabric. Very often there was no trailing-edge member in the wing except for a single wire, which was bent by the tension of the doped fabric to give a scalloped outline.

Such were the difficulties of making a strong, rigid biplane cellule (wing assembly) that the Paul Schmitt bombers had no fewer than twelve structural bays, with six pairs of interplane struts on each side of the fuselage. On such machines, streamlined rigging wire was a great advantage, especially the British Rafwire with a cross-section like that of a raindrop. Some of the fastest bombers, the Ansaldo SVA family for example, had no bracing wires at all. The Voisin bombers, though of primitive design and low performance, were of all-steel construction and could be left out in any weather in the certainty that they would not deteriorate or warp.

Deliberate warping of the wings was the common way of providing lateral control in 1914, but by 1915 nearly all new bombers had ailerons, often on all wings. Another change was that, whereas many 1914 bombers had rudders and elevators only, by 1915 it was usual to hinge these surfaces to fixed fins and tailplanes. Wheel brakes were never fitted but some aircraft, especially German ones, had a claw arrangement which would plough a furrow across the airfield to bring the machine more quickly to rest.

In November 1916 Bréguet produced the Type Br 14 in which extensive use was made of aluminium. This light metal was used for most of the wing structure and undercarriage, and even in sheet form for the fuselage decking around the cockpits. Though use was still made of wood, steel and fabric, aluminium helped to keep many thousands of Br 14 bombers in worldwide use until after 1930. In 1917, German Professor Hugo Junkers produced his J-1 in which almost the whole airframe was duralumin (aluminium/copper alloy), the unbraced wings and tail having a robust skin of corrugated dural.

## Design errors

In the earliest bombers the first objective was to design a machine that would fly at all, because many failed to achieve even that. Next came reliability, while such considerations as flight performance and bomb load were to some degree bonuses. Aircraft design at that time was anything but an exact science. It was not uncommon for a design team to follow a winner with a distinctly inferior machine. For example, the early Caproni Ca 30 bombers were followed late in 1917 by the giant triplane Ca 40 family. These, though robust and well engineered, had such a 'built-in headwind' that they were easy meat for defending scouts and AA artillery in the daytime and had to be used at night. This made bombing even more random than before. These impressive-looking aircraft illustrated the fact that adding more power did not necessarily give better performance. In contrast the small Italian SVA series, with only 220 hp, carried a useful bomb load and were possibly the fastest machines used in the First World War.

Before 1920, aircraft design was very much a matter of opinion, often vehemently expressed yet backed by the slenderest of evidence. Some designers contended that the structural integrity of the triplane more than made up for its other shortcomings; most favoured the biplane, and a very small number (none was a designer of large bombers) considered the monoplane in-

herently superior. In the same way there was no evidence to suggest that four 100 hp engines might be superior to two 200 hp, beyond the obvious fact that, first, the aircraft would fly better with one engine failed, and second, the chances of engine failure were twice as great.

With most multi-engined bombers it was possible for a courageous crew member to reach a faulty engine in flight and attempt to rectify the trouble. In some designs, notably the German Siemens-Schuckert R-types, the engines were installed in the fuselage purposely to render the whole propulsion system and cooling radiators readily accessible during a mission. The usual fate of such designs was a short period on operations, followed by a much longer period as a trainer in an environment where poor performance was less likely to prove lethal.

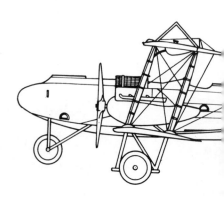

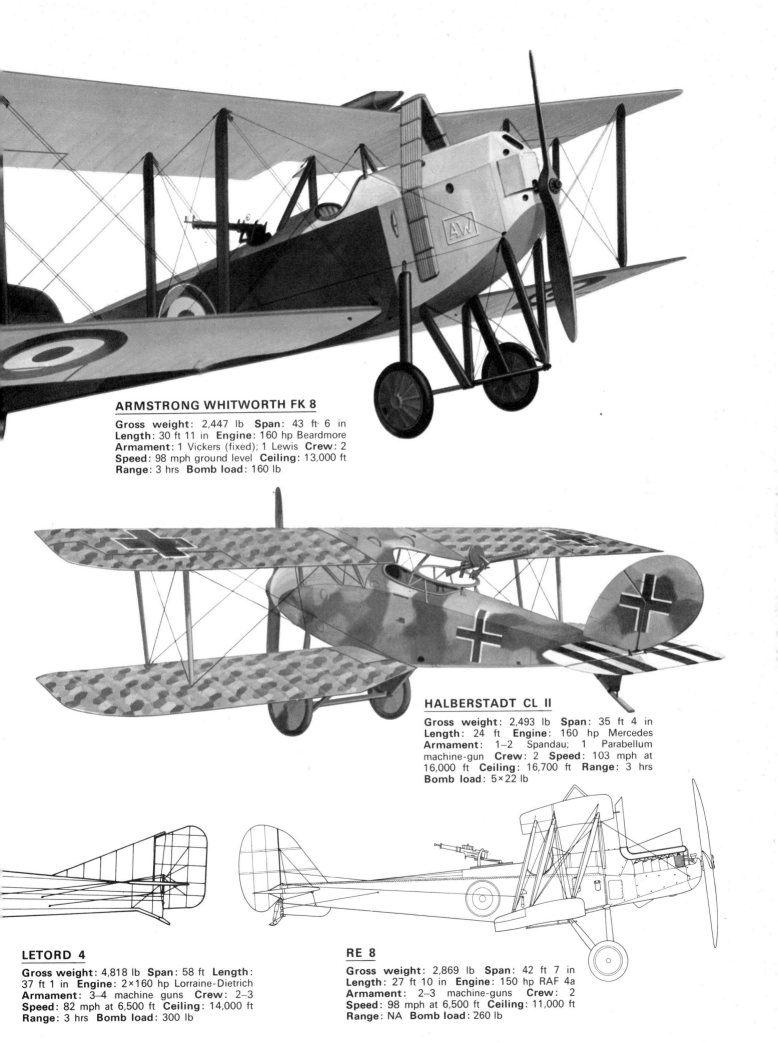

## ARMSTRONG WHITWORTH FK 8

**Gross weight**: 2,447 lb **Span**: 43 ft 6 in
**Length**: 30 ft 11 in **Engine**: 160 hp Beardmore
**Armament**: 1 Vickers (fixed); 1 Lewis **Crew**: 2
**Speed**: 98 mph ground level **Ceiling**: 13,000 ft
**Range**: 3 hrs **Bomb load**: 160 lb

## HALBERSTADT CL II

**Gross weight**: 2,493 lb **Span**: 35 ft 4 in
**Length**: 24 ft **Engine**: 160 hp Mercedes
**Armament**: 1–2 Spandau; 1 Parabellum
machine-gun **Crew**: 2 **Speed**: 103 mph at
16,000 ft **Ceiling**: 16,700 ft **Range**: 3 hrs
**Bomb load**: 5×22 lb

## LETORD 4

**Gross weight**: 4,818 lb **Span**: 58 ft **Length**:
37 ft 1 in **Engine**: 2×160 hp Lorraine-Dietrich
**Armament**: 3–4 machine guns **Crew**: 2–3
**Speed**: 82 mph at 6,500 ft **Ceiling**: 14,000 ft
**Range**: 3 hrs **Bomb load**: 300 lb

## RE 8

**Gross weight**: 2,869 lb **Span**: 42 ft 7 in
**Length**: 27 ft 10 in **Engine**: 150 hp RAF 4a
**Armament**: 2–3 machine-guns **Crew**: 2
**Speed**: 98 mph at 6,500 ft **Ceiling**: 11,000 ft
**Range**: NA **Bomb load**: 260 lb

# THE MENACE OF THE ZEPPELINS

The advantages of airships as bombers were enormous in the early days of the war. Almost as fast as the planes then in existence, their range and bomb-load were far superior. But their days were numbered: more effective defences and better fighters made their vulnerability painfully clear. By the end of their short career, however, the Zeppelins had firmly established the concept of strategic bombing

German airships were the first craft to be used for bombing in the First World War. Only two days after Germany's invasion of Belgium, the Zeppelin Z 6 was sent to attack Liège. It was damaged by gunfire from the ground however, and was wrecked during a forced landing near Cologne. Before the month of August was out, two more Zeppelins had been shot down while on their first operational missions. One of them was captured by the French. It was at such an early stage of the war that the Germans learned the bitter lesson of just how vulnerable were the mighty Zeppelins on which they had pinned such faith.

Their basic mistake had been to direct them over strongly defended battle areas during the hours of daylight when they made such large, slow-moving targets. Nevertheless the disasters shattered the enthusiasm of the German High Command for the airship as a weapon. Although a small number of army airships were later used on bombing raids over Britain, it was the German Naval Air Service which exploited the Zeppelin to its fullest extent, not only in strategic raids over Britain, but even more so in the reconnaissance over the North Sea which was its main task. Much of the success of the German navy against Allied shipping was due to the observation maintained by airship patrols, which during the course of the war, made 971 scouting flights over the North Sea, and 220 over the Baltic.

Meanwhile, the army learned to limit tactical Zeppelin raids over the front line to the hours of darkness, thus anticipating the dark-painted night bomber.

## The first attack

Without the more sophisticated bombing devices developed later, bombs dropped at night were even more inaccurate than those dropped by day. So the Military Aviation Service turned its main attention to aircraft which were beginning to show they had uses other than mere reconnaissance. As early as 13 August 1914 a Taube flown by Lt Franz von Hiddeson dropped two light bombs on the outskirts of Paris. Urgent orders were put through for some of the large bomber aircraft which German designers had on their drawing boards but which had previously aroused little interest. At the same time a special bomber force of existing aeroplanes, unsuitable though they were, was set up under GHQ command.

In taking over the main responsibility for Zeppelin bombing raids on Britain, the German Naval Air Service pioneered some of the techniques used later by aircraft. In

the same way, their efforts led to the development of anti-aircraft defences which stood Britain in good stead at the end of 1916 when the first German heavy bombers came into service. It was early in the January of 1915 that the Kaiser sanctioned airship raids on Britain but limited them to military establishments such as shipyards and arsenals. However, since the raids were made at night and the airships often drifted far off course, the commanders had little idea where their bombs might fall. Many fell on undefended villages in East Anglia, particularly after the black-out became widespread, making it even more difficult to differentiate between military and civilian targets.

The first airship raid took place on the night of 19 January 1915 when two Zeppelins dropped bombs on Kings Lynn, Yarmouth, and several villages in the area. The bombs themselves were not large, mostly 110-lb high explosives and 6½-lb incendiaries with a limited destructive capability. Less than

half-a-dozen people were killed, but this was the first time that a civilian population had ever been subjected to systematic bombing, and the terror and sense of outrage that it produced was out of all proportion to the weight of bombs dropped.

Reprisals were called for, leading eventually to raids on Berlin by British bombers. Meanwhile, the Zeppelin raids continued: in twenty raids during 1915, thirty-seven tons of bombs killed 181 people and injured 455. So there seemed little point in pretending *not* to bomb London when the airship crews had little idea where their bombs were falling anyway. In fact the capital was bombed by mistake on 31 May, when seven civilians were killed and thirty-five injured. Two months later, the Kaiser lifted his ban on bombing London, and raids on that city, as well as others such as Liverpool and the Tyneside area, began in earnest.

The Zeppelin crews, though, didn't have everything their own way. They had to contend with storms and gales that often

forced them to crash land, usually with fatal results. This was the fate, only one month later, of the two airships which had made the January raid. And the organisation of the British defences improved as time went on. To begin with, the Royal Navy was responsible for the defence of London and the major cities while the Army took care of ports and military installations. Guns such as 3 and 4-in quick firers and 1-pounder pom poms were converted for anti-aircraft use and even machine-guns were placed on high-angle mountings and fitted to motorcars for mobility. Searchlights were introduced, manned first by special constables.

## Counter measures

Until the crews gained experience, there were many false alarms when illuminated clouds were mistaken for Zeppelins. Nevertheless, just as the bombing brought a new terror not exactly justified by the damage to property and loss of life actually caused, so the airship crews learned what it was like to be in a huge, slow-moving target, flying at no more than 6,000 feet, held in the beam of a searchlight and being fired at from the ground. Such was the fear this aroused that when five Zeppelins tried to raid London in August, only one got within thirty miles and the returning crews told dramatic stories of searchlights and anti-aircraft fire which did not actually exist. They bombed the wrong targets and the firing was only from rifles.

The new series of giant Zeppelins which came into service in 1916 were able to fly as high as 12,300 ft, and a novel device was introduced to reduce their chance of being caught in searchlight beams. While the airship cruised safely above the clouds, a streamlined car with an observer inside was lowered nearly 3,000 feet on a steel cable so that the observer could direct the bombing by telephone. In the spring and summer of that year, the Zeppelin raids reached their peak, with a dozen or more craft taking part in a single attack. But the height at which they flew, although making it easier to elude the defences, imposed even greater strains on the crews who were often airborne for twenty-four hours or more at temperatures of 30° below zero in winter. And the ground defences were also improving. In February the Army took over the entire responsibility for home defences and several months later, the RFC's BE2c aircraft were fitted with fixed forward-firing machine-guns for the first time, following the 1915 invention by the French and Germans of interrupter gear which enabled these guns to be synchronised to fire through the propellers. It was this invention that turned the early reconnaissance aircraft into lethal fighters.

Aircraft had brought down Zeppelins before but by the unorthodox method of flying above and dropping bombs on them. The first to be destroyed in this way – and the first to be destroyed in aerial combat – was brought down on 7 June 1915 by Flt Sub-Lt R. A. J. Warneford, of the RNAS, flying a French Morane Parasol Type L. He set fire to the airship LZ 37 over Ghent by dropping six 20-lb bombs on it and was awarded the Victoria Cross. But by mid-1916, the British aircraft were armed with machine-guns which could not only be aimed more accurately but could fire incendiary bullets. These, along with phosphor shells from anti-aircraft guns spelled doom for the hydrogen-filled airships.

The first victory by a defending fighter at night came on 2 September 1916 when Lt William Leefe Robinson of 39 Squadron RFC shot down one of twelve naval and four military airships which had set out to raid London. Flames from the burning Schutte-Lanz airship SL 11 could be seen fifty miles away. Robinson too, was awarded the Victoria Cross for his success, which marked the beginning of the end of the airship menace.

The new class of airship which was coming into service could fly at 18,000 ft in an attempt to avoid the fighters, but at that height the crew suffered even more severely from the cold and had to use oxygen, while the craft themselves were subject to greater hazards from treacherous wind changes. When eleven airships set out to bomb London at the beginning of October, they became so scattered that only one person was killed in the city, while one of the Zeppelins was lost with its entire crew – shot down by a BE2c over Potters Bar. In the last raid of 1916, eight weeks later, two out of ten Zeppelins were lost. The total number of raids during the year was twenty-three, during which 125 tons of bombs were dropped, killing 293 people and wounding 691.

## End of the airship

The Germans were reluctant to admit defeat but the short day of the airship as a bomber was over. Only eleven more raids were made against England, seven in 1917 and four in 1918. The last was on 5 August 1918 when five Zeppelins launched a surprise attack which resulted in the loss of their latest and best airship, the L 70, together with *Fregattenkapitan* Peter Strasser, the officer who had led the naval airship force so courageously since the very beginning of the war. It was not an end to bombing, of course, as towards the end of 1916 aircraft were beginning to take over that role with growing effectiveness. But the Zeppelins had played their part, albeit at a tremendous cost. Of the eighty-eight airships built during the war (seventy-two for the German navy), over sixty were lost; thirty-four due to accidents and forced landings caused by bad weather and the remainder shot down by Allied aircraft and ground fire. In fifty-one raids on Britain they dropped 5,806 bombs (a total of 196½ tons), killing 557 people, injuring a further 1,358, and causing an estimated £1½ million damage. This was not too highly significant in itself, considering the thousands who were dying in the trenches in France every day, but the raids were of considerable military value as they hampered war production and diverted men and equipment from more vital theatres of war.

It is estimated that the airship and aircraft raids together reduced the total munitions output by one sixth. And by the end of 1916, the British home-based air defences included twelve RFC squadrons and a large force of anti-aircraft guns and searchlights, requiring 12,000 men.

The airship raids were the first ever attempt to defeat a country by the bombing of military and civilian targets in the homeland rather than attacks on military targets in the battle zone – strategic and psychological bombing as distinct from tactical bombing. Its success gave rise to a whole new school of military thought which was put into effect with far greater destruction during the Second World War. Much of the initial success was due to the fact that bombing was an unknown factor before the First War and its effect on an unprepared civilian population was therefore considerably greater.

The wreckage of Zeppelin *L 10* in the shallows off Cuxhaven. Returning from a scouting flight on 3 September 1915, she was struck by lightning and fell in flames, killing all on board

# HEAVIER BOMBS, BETTER GUNS

Early aeroplane bombing raids were limited to the battle-zone and its vicinity. The German Zeppelins had the range to attack more distant targets, but the first strategic raid by plane was carried out by the British – not by the RFC but by the navy

One of the great advantages which the Zeppelin of 1914 had over the aeroplane was, of course, its range – more than 2,000 miles at a time when most planes were hard pressed to fly 150 miles without having to land to refuel. It was range that made it possible for the Germans to pioneer the bombing of an enemy's homeland and in particular his capital city. Military leaders believed that the capture of an enemy's capital city somehow marked his defeat. London was chosen by the Germans because it could be reached by flying across the North Sea and thus avoiding the battle zones of the Western Front, whereas Paris could only be reached by battling through the Allied air patrols over France. Only two airship raids were made on the French capital throughout the war and one of two Zeppelins making one attempt was destroyed. But apart from the need to avoid the Western Front, range gave a wide choice in the selection of targets. In fact, airship range was always greater than that of aircraft. In 1918, the Germans were even able to contemplate raids on New York with their newest Zeppelins which had a range of 7,500 miles. Strasser, head of the German Naval Air Service, was planning such a raid when he was killed with the destruction of the L 70.

Meanwhile, during the early stages of the war when planes were first used for bombing, their missions generally had to be confined to the battle-zone or targets in the near vicinity. The French Voisin and Farman bombers were particularly effective in tactical missions against military targets, proving themselves to be virtually an extension of artillery, for which the French had been enthusiasts since the days of Napoleon. The earliest French bombs were

actually converted Canton-Unné 90-mm and 155-mm artillery shells with the addition of fins and impact fuses. The French might well have undertaken the first strategic aeroplane bombing raid of the war – a mass attack on the Kaiser's personal headquarters at Mézières in September 1914 – but for some reason the planned raid was cancelled.

The German 'carrier-pigeon' bomber units were also used during the first months of the war, although to a lesser extent because of the emphasis placed on airships. They might have been used for strategic raids as early as February 1915, when plans were made for them to bomb England. An experimental night bombing raid was carried out on Dunkirk at the end of January, but before the raid itself could be carried out, the force was transferred to the Eastern Front where it gave considerable assistance at the battle of Gorlice-Tarnow in March. As it was, the first ever strategic raid by a plane was carried out by the British – not by the RFC, as might have been expected, but by the RNAS.

### 'Samson's boys'

In August 1914, while the four RFC squadrons were sent to France to operate under army command from a base at Maubeuge, an RNAS squadron under Cdr Samson was despatched to Ostend to assist an attempted naval diversion by a brigade of marines.

Among the assorted aircraft were two Sopwith Tabloids, tiny single-seat tractor biplanes, which before the war had been a favourite sporting machine. The diversion was planned by Winston Churchill, then First Sea Lord, but it did not materialise and within three days the marines were

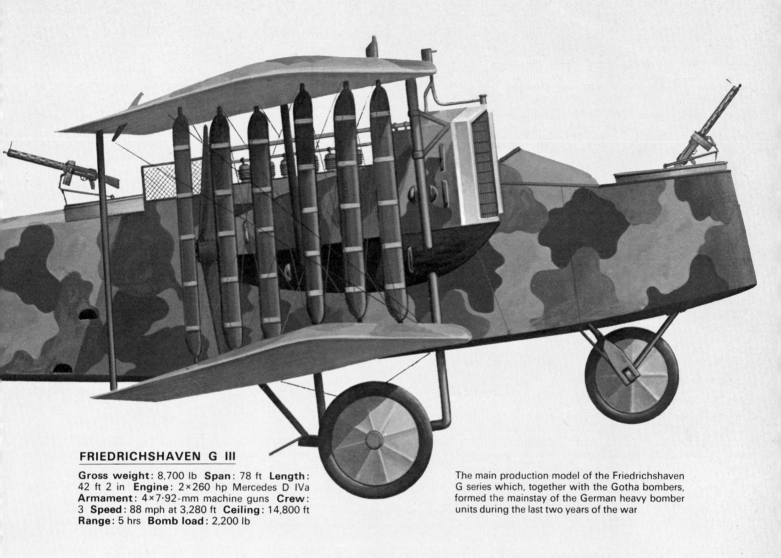

## FRIEDRICHSHAVEN G III

**Gross weight**: 8,700 lb **Span**: 78 ft **Length**: 42 ft 2 in **Engine**: 2×260 hp Mercedes D IVa **Armament**: 4×7·92-mm machine guns **Crew**: 3 **Speed**: 88 mph at 3,280 ft **Ceiling**: 14,800 ft **Range**: 5 hrs **Bomb load**: 2,200 lb

The main production model of the Friedrichshaven G series which, together with the Gotha bombers, formed the mainstay of the German heavy bomber units during the last two years of the war

## HANDLEY PAGE 0/100

**Gross weight**: 14,000 lb **Span**: 100 ft **Length**: 62 ft 10 in **Engine**: 2×250 hp Rolls Royce Eagle **Armament**: 4–5 machine guns **Crew**: 4 **Speed**: 95 mph at ground level **Ceiling**: 7,000 ft **Range**: 6 hrs **Bomb load**: 2,000 lb

The world's first effective heavy night bomber – the biggest British warplane of the war recalled. Samson's squadron was supposed to go with them, returning to one of the RNAS air stations that had been established along the east coast as part of Britain's defences. But this did not suit the fiery Commander who had pioneered flight from ships at sea. Using Channel fog as an excuse he landed his squadron at Dunkirk and enlisted the aid of the British Consul for him to stay on to support the French. Churchill agreed to the establishment of a naval air base on the French coast to help protect Britain from airship raids, and so 'Samson's boys' became a permanent feature at Dunkirk.

The RNAS squadron was intended to carry out reconnaissance and engage any Zeppelins or enemy aircraft sighted. But Samson preferred more vigorous action and decided to attack the Zeppelins at their actual bases. The first British air raid into German territory took place on 22 September 1914 but no damage was caused. Meanwhile, several of the aircraft had been sent to Antwerp to help in the battle to prevent the German Army reaching the sea. British forces were forced to withdraw on 7 October and a day later the last RNAS aeroplanes to leave the Belgian city were the two Sopwith Tabloids, flown by Squadron Commander Spenser Grey and Flt-Lt R. L. G. Marix.

Before returning to Dunkirk they were ordered by Samson to attack the Zeppelin sheds at Cologne and Düsseldorf. Spenser Grey was prevented by mist from seeing his target and dropped bombs on Cologne's railway station instead. But Marix was more successful and from 600 ft his 20-lb Hale high explosive bombs fell on the Zeppelin shed at Düsseldorf which exploded into flames, completely destroying the brand-new Zeppelin Z 9 inside. This was the first-ever strategic bombing raid, since it was designed to cripple the enemy's war effort by striking at a military target behind the front line rather than in the battle zone. A few weeks later, three Avro 504 biplanes made a similar raid on the Friedrichshafen Zeppelin works on Lake Constance, a mission requiring a flight of about 250 miles over hostile territory, from a take-off point at Belfort on the French-Swiss border. The Avros attacked by flying to within ten feet of the lake's surface – probably the first low-level bombing strike in history – destroying one Zeppelin under construction, as well as the gasworks used for filling the airships.

**Heavier bombers**
The success of these attacks led the Air Department of the Admiralty to issue a specification for a much larger twin-engined bomber able to carry much heavier bombs than the 20-pounders which were in general use at that time, and the biggest that most planes could carry. This resulted in the first of the famous Handley Page series of bombers, the 0/100.

At the same time, the Germans were developing large R-class bombers such as the Friedrichshafen and Gotha types. These also made their appearance in 1916, the latter being one of the finest bombers of the war. The French were slower in building heavy twin-engined bombers and it was not until 1917 that the first of the Letord series, designed by Colonel Dorand of the French Service Technique, came into operation. The Russians of course already possessed a number of Il'ya Muromets, but these were hampered by maintenance problems and

## HANDLEY PAGE 0/400

**Gross weight**: 14,022 lb  **Span**: 100 ft  **Length**:
62 ft 10 in  **Engine**: 2×250 hp Rolls Royce Eagle
**Armament**: 5×·303 machine guns  **Crew**: 4–5
**Speed**: 97 mph at ground level  **Ceiling**: 8,000 ft
**Range**: 4 hrs  **Bomb load**: 2,000 lb

Improved version of the Handley Page 0/100
heavy night bomber (see page 21) and the most
widely used by the British during 1917 and 1918
for raids on Germany

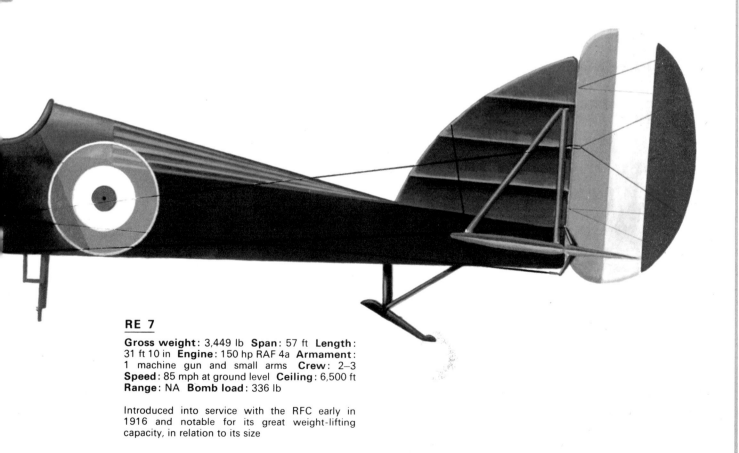

## RE 7

**Gross weight**: 3,449 lb  **Span**: 57 ft  **Length**:
31 ft 10 in  **Engine**: 150 hp RAF 4a  **Armament**:
1 machine gun and small arms  **Crew**: 2–3
**Speed**: 85 mph at ground level  **Ceiling**: 6,500 ft
**Range**: NA  **Bomb load**: 336 lb

Introduced into service with the RFC early in
1916 and notable for its great weight-lifting
capacity, in relation to its size

later operations were affected by the Russian revolution. The Italians, who only joined the war in May 1915, were also well advanced in the development of the heavy bomber, and the Caproni Ca 32 was in service by the end of the same year. It was evident that 1916 would see the introduction of specialised bombers in the air services of all the major combatant powers.

In the meantime, however, they had to use the planes already in existence. From early in 1915, the Allies increased their raids, while the French, in retaliation for attacks on Paris, began to include targets in Germany itself. Much of the French strategic bombing effort was, in fact, based on a doctrine of retaliation. On 26 May, for instance, following a German poison-gas attack the previous month, Commandant Göys led three squadrons of Voisin bombers on a raid against the poison-gas factories near Mannheim, a flight of five hours. Considerable damage was achieved and only one Voisin failed to return. Unfortunately this was the one piloted by Göys, who had been forced to land in Germany because of engine trouble. He was taken prisoner but eventually escaped and found his way back to France.

By June the French had begun to penetrate as far as Karlsruhe. Much of the early success of these bombing missions was due to the fact that there was virtually no opposition from German aircraft. This meant that raids could be carried out in daytime and at relatively low altitudes, so that even by the primitive method of dropping bombs over the side of the cockpit, a surprising degree of accuracy could be achieved. The British and French crews had a healthy respect for the Germans' improvised anti-aircraft guns. Many of these had been hastily converted from other uses: the 3.7-cm *maschinen flak*, for example, which could fire three shots a second with an accuracy up to 9,000 ft, was originally used by the navy on torpedo-boats. But there was little to fear from German aircraft, whose only armament was the small arms carried by the crews.

Even with such relatively ineffective weapons, the Allied pusher aircraft had an advantage over the German tractor types for their gunners had a much less restricted field of fire. This was even more marked when, in late 1914, the French began to arm their Voisin bombers with machine-guns carried by the gunner/observer seated in front of the pilot. These would have been more effective still had they not been the heavy Hotchkiss 8-mm type, which hampered the plane's performance and were difficult to operate in the air. Nevertheless they were more than a match for the German aircraft and during the early period of 1915, French bombers were virtually unmolested as they raided German targets.

### Wider fire power
Meanwhile, a better and lighter air-cooled machine-gun had been designed by Col Isaac Newton Lewis of the US Army, and experiments in its use were made as early as August 1914 by individual pilots such as Lt L. A. Strange of No. 5 Squadron RFC. Simple mountings were devised to enable the observer of a two-seater – or even the pilot of a single-seater – to fire over the side of the cockpit. This meant that an enemy plane had to be approached in a crab-like manner to make it a suitable target, but many victories were obtained in

this way, and aerial fighting became commonplace. An even more successful method in the case of single-seaters was to mount a machine-gun above the centre section of the top wing, so that it could be fired forwards above the propeller arc by means of a cable attached to the trigger.

This period of German docility, which gave the Allies supremacy in the air, came to an end in the spring of 1915 when the more powerful C-class planes of Rumpler, Albatros and Aviatik design came into service. These were armed with a Parabellum machine-gun, fired by the observer, who now sat behind the pilot so that he had a wider field of fire. It was now the turn of the slower pusher aeroplanes to suffer since they were blind from behind. The French followed the German example, but inexplicably, the British retained the old seating arrangement and the standard Farnborough-designed BE types of the RFC began to suffer disastrous losses. With the Germans fighting back, the Allied bombers no longer always got through.

An even greater blow to the Allies came in the summer of 1915 when the German single-seat Fokker monoplane (*Eindekker*) appeared. It was armed with a machine-gun placed directly in front of the pilot, and synchronised to fire between the blades of the revolving propeller. The aircraft itself could now be aimed at a target. making it much more effective than the free-swinging weapon that the observers on two-seater aircraft had to use. In spite of the lead that might have been achieved by the French, who had been the first to fit deflectors to propeller blades, and similar ideas that had been put forward and rejected in Britain, Allied fighter aircraft were not fitted with gun synchronising gear until 1916. The winter of 1915–1916 saw the Fokker achieving such supremacy in the skies above the Western Front that British pilots were angrily labelled 'Fokker fodder'.

Increasingly, the Allied bomber squadrons had to call on the faster and better armed single-seat scouts to provide fighter escorts, which severely restricted their range. To give themselves better protection the French bomber units developed the idea of formation flying, first realised by the remarkable Capitaine Happe of *Escadrille* MF29 and later adopted by the British, while more and more raids had to be carried out at night, reducing the effectiveness of the bombing. By the autumn of 1915 it was apparent to the military leaders of all the combatant powers that air power was an important new element in warfare. Reconnaissance was still its most useful function, especially in view of the growing use of photography. But it was equally important to deny the enemy similar facilities for reconnaissance, and this involved not only fighters to patrol the battle zone, but also bombers to raid the enemy's airfields and bomb his planes on the ground.

The fighters had advanced considerably with the development of fast and manoeuvrable single-seat aircraft armed with machine-guns. But the bombers, by and large, were still the general purpose types which had been in existence when the war began. It was now necessary to design bombers especially adapted to the task, with the defensive armament to take on fighters. From this point onwards, the story of air warfare became a continual struggle, both in the air and on the drawing board, between bomber and fighter aircraft.

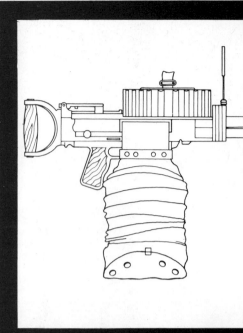

### LEWIS ·303-in MACHINE-GUN
Invented by Col Isaac Newton Lewis of the US Army and adopted early in the war as their standard light machine gun by both the British Army and the RFC – and later by the French and Belgian forces. Gas-operated and fed at first by 47-round revolving drums – later by 97-round drums. For aerial use, a cartridge-case deflector and receptacle was provided to prevent ejected cartridges damaging the aeroplane, and electric heaters prevented freezing-up at height. Rate of fire was increased by 1918 to 850 rounds per minute

### PARABELLUM 7.92-mm MACHINE-GUN

**Weight:** 22 lb **Cyclic rate:** 700 rpm

The most widely-used observers' machine gun in German aircraft from early 1915 until the end of the war. A lightweight form of Maxim with a firing rate of 700 rounds per minute, and a water-jacket slotted for air cooling in later models

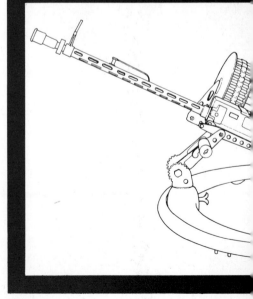

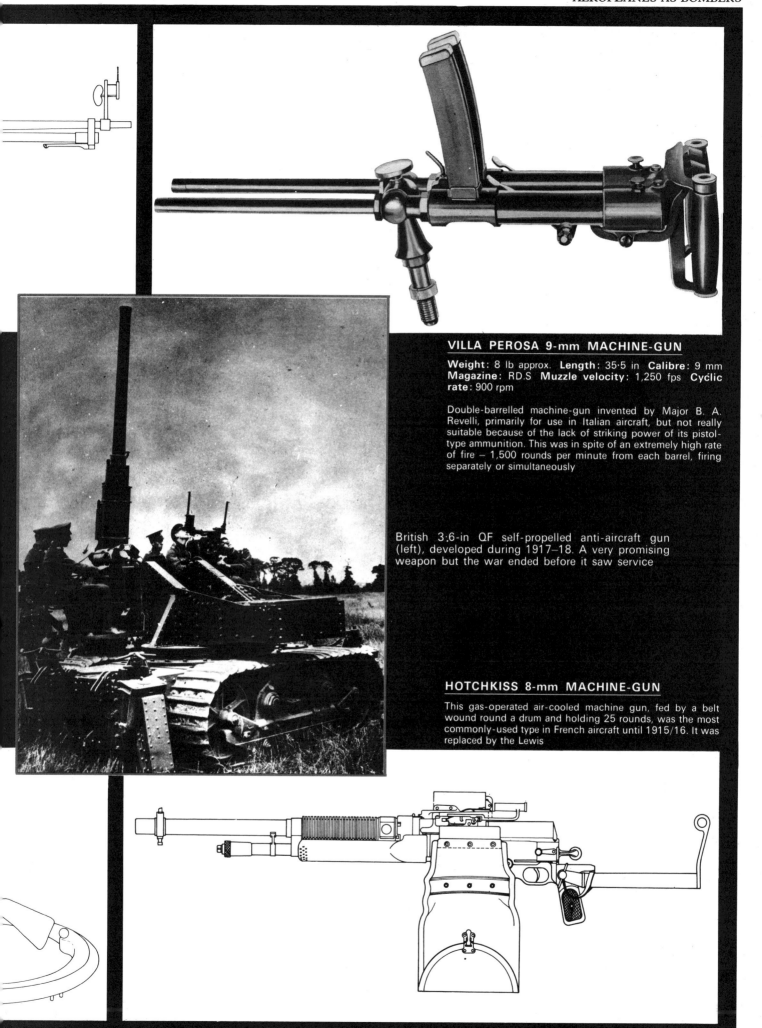

## VILLA PEROSA 9-mm MACHINE-GUN

**Weight:** 8 lb approx. **Length:** 35·5 in **Calibre:** 9 mm
**Magazine:** RD.S **Muzzle velocity:** 1,250 fps **Cyclic
rate:** 900 rpm

Double-barrelled machine-gun invented by Major B. A.
Revelli, primarily for use in Italian aircraft, but not really
suitable because of the lack of striking power of its pistol-
type ammunition. This was in spite of an extremely high rate
of fire — 1,500 rounds per minute from each barrel, firing
separately or simultaneously

British 3·6-in QF self-propelled anti-aircraft gun
(left), developed during 1917–18. A very promising
weapon but the war ended before it saw service

## HOTCHKISS 8-mm MACHINE-GUN

This gas-operated air-cooled machine gun, fed by a belt
wound round a drum and holding 25 rounds, was the most
commonly-used type in French aircraft until 1915/16. It was
replaced by the Lewis

# THE BOMBS

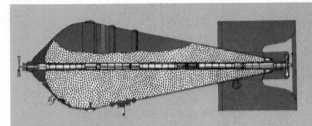

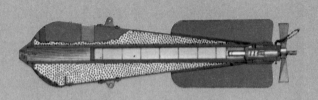

### H.E.R.L. 520-lb

**Actual weight:** 525 lb **Case:** 180 lb **Explosive:** 340 lb **Case material:** steel **Dimensions:** 5 ft 1 in long × 1 ft 7½ in wide **Fuse:** nose and tail fuses

### H.E.R.L. 16-lb

**Actual weight:** 16 lb **Case:** 9 lb **Explosive:** 7 lb **Case material:** M S steel **Dimensions:** 25·15 in long×5 in dia. **Fuse:** tail fuse

### H.E. 'HALE' 20-lb

**Actual weight:** 18·5 lb **Case:** 14 lb **Explosive:** 4·5 lb Amatol **Case material:** steel **Dimensions:** 23¼ in long × 5 in wide **Fuse:** tail fuse

Hardly less important than the planes themselves were the bombs they carried and the techniques employed to drop them with some degree of accuracy. Both made very great advances during the war, to the point where high-explosive and incendiary bombs were beginning to resemble the weapons used to such devastating effect in the Second World War, and bomb-sights were taking into account not merely height and airspeed but also wind velocity and drift when flying across wind. Special telescopic sights were also developed for night raids.

The idea of launching a bomb from the air dated back to the early days of lighter-than-air craft and with the arrival of the Zeppelin as a practical means of bombing, the Germans in 1912 developed the first bomb intended specifically for aircraft use. This was the APK (Artillery Test Commission), a spherical cast-iron shell filled with high explosive and detonated by an impact fuse. It was produced in 5-kg and 10-kg sizes but was too light for effective aerial use and was not used in the war. From experiments with the APK, the more advanced Carbonit series of bombs was developed, ranging in

size from 4·5-kg to 20-kg and widely used by the Germans until late 1915.

A similar type of bomb was designed by F. Martin Hale in 1913 for the RNAS which pioneered British bombing; the RFC possessed no bombs when the war began, since its function had been limited to reconnaissance. The first Hale bomb was a 20-pounder but by early 1915, 10 and 100-lb versions were being manufactured by the Cotton Powder Company and used by both the RNAS and RFC.

The Carbonit and Hale bombs represented the first attempts to modify the external shape to give stable and predictable flight. Both were pear-shaped and had tail-fins, although the Carbonit merely had a tin cylinder attached by stays which failed to provide sufficient accuracy in aiming at a target. The Carbonit did however have a steel-tipped nose for better penetration; this was one of the aims of later bombs which had a more streamlined shape, based on a study of ballistics.

From the beginning, one of the main problems was to ensure that bombs could be carried safely and jettisoned if necessary without exploding, but that when aimed at

a target, they would arm themselves when falling and detonate on impact. This led to a great deal of work on fuses, carried out all through the war. There was an analogy with artillery shells, which also had to be safe when transported and handled and had to withstand the shock of discharge from a gun without premature detonation, and yet still remain in a condition to function on impact. The advantage with the artillery shell was that the large translational and rotational forces arising on discharge could be used to arm the fuse. These forces were so great that fuses could be designed without any detachable safety devices. Aircraft bombs on the other hand were not subject to such forces on release and other means had to be found to arm their fuses.

After the initial use of steel darts and small Aasen bombs of Danish origin, the French solved this problem in 1915 by equipping their daybombing units with artillery shells fitted with tail-fins, a logical development since they regarded bombing basically as an extension of artillery. The most commonly used shells were 75, 90 and 155 mm calibre. Later in the war, more conventional bombs were manufactured by the

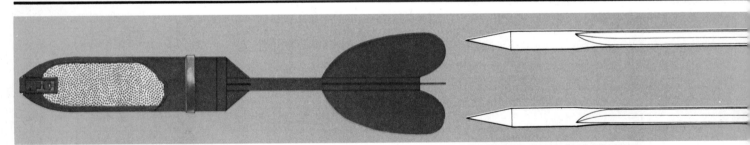

### FRENCH BOMB MADE FROM 75-mm SHELL AND TIN FIN

The French, early exponents and innovators of artillery, regarded bombing as essentially an extension of artillery, and it was not surprising that most of their bombs were modified artillery shells. This was one of the best and most widely used throughout the war

### RANKEN ANTI-PERSONNEL DARTS (flèchettes)

Approx 5 in long and made of steel, they were dropped from canisters each containing 500 darts. On reaching ground from 5,000 ft, they attained the speed of a rifle bullet

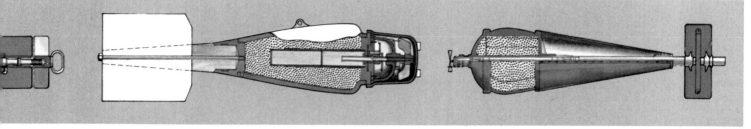

**H.E. COOPER 20-lb**

**Actual weight:** 24 lb **Case:**
20 lb **Explosive:** 4 lb Amatol
**Case material:** steel **Dimensions:** 24·4 in long × 5·1 in wide
**Fuse:** nose fuse

**H.E. RAF 336-lb**

**Actual weight:** 336 lb **Explosive:** 70 lb compressed TNT
**Case material:** cast steel **Dimensions:** 6 ft 9$\frac{1}{8}$ in long × 14 in
dia. **Fuse:** nose fuse

French, such as the Gros-Andreau type made in 22, 55 and 110-lb sizes. But they were much less interested than the Germans or the British in the concept of strategic bombing with heavy bombers and had no need of larger bombs. In consequence, the modified 75-mm artillery shell became the most widely used and successful of the French bombs, used right until the end of the war.

Before considering the development of bomb fuses, carried on mainly in Germany and Britain, it is first necessary to describe the nature and components of the high-explosive bomb of the First World War. All such bombs contained three different types of explosive. The main bursting charge was some safe and insensitive explosive such as TNT or amatol. Amatol was the most commonly used in the early stages of the war but thereafter, bomb fillings followed the development of explosives in general. The bursting charge could only be detonated by a violent explosion and therefore an exploder was required, usually tetryl which resembled TNT but was more sensitive. The exploder was set off by a detonator containing fulminate of mercury. This ex-

plosive was highly sensitive and had to be handled with extreme care. The exploder and detonator were fitted into the bomb before loading it into an aircraft. When the bomb hit the ground, the detonator was hit by a striking pin, igniting the exploder which, in turn, exploded the bursting charge.

In the case of the early Carbonit and Hale bombs, the fusing mechanism was armed by means of a small propeller mounted in the tail which unwound a spindle inside the bomb during flight and allowed the striker to come into contact with the detonator on impact with the ground. Thus the bomb was not fused until the airscrew had been turned a certain number of times. In later, more sophisticated bombs, this meant that the distance they had to fall before becoming armed could be predetermined. A clip was attached as a safety device to prevent the propeller turning while in the aircraft, and this had to be removed before release. This was done by hand with the early bombs, which were carried loose on the floor of the cockpit and dropped over the side. But they rolled about on the cockpit floor and tended to get under the pilot's feet, inter-

fering with the controls. A crude method of overcoming this was to carry two bombs, one attached to either end of a length of rope slung over the fuselage in front of the cockpit; the pilot released them by simply cutting the rope with a knife.

When special bomb release gear was developed for carrying bombs in racks, at first externally and later in bomb bays inside the aircraft, safety devices could either be removed by the automatic trigger mechanism on release, or retained if the bombs had to be jettisoned without exploding.

The propeller type of fuse in various forms was used in all the British bombs manufactured in wartime. From late 1914, the Royal Laboratory at Woolwich began designing a series of bombs, ranging in size from 16-lb to 550-lb, while during 1915–1916, the Royal Aircraft Factory also developed a range up to 585-lb for the larger aircraft then coming

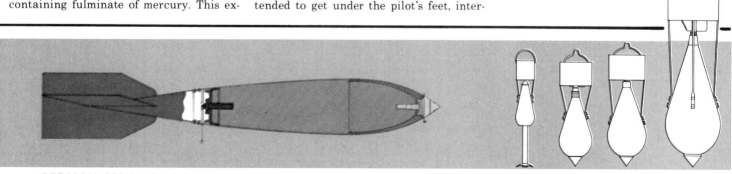

**GERMAN 300-kg P.u.W.**

**Actual weight:** 300 kg **Case:** steel **Dimensions:** 2,750 mm long × 365 mm diameter. **Fuse:** nose fuse

Its offset fins gave this bomb high speed rotation for penetration

**CARBONIT H.E. BOMBS**

A group of German bombs, left to right: 4·5 kg; 10 kg; 20 kg; 50 kg into use. The 20-lb Hale became obsolete towards the end of the war and was replaced by the 20-lb Cooper bomb. Two main classes of fuse were developed and used in accordance with the type of target being attacked. The nose fuse had the arming propeller in the nose of the bomb and, since the striker spindle was the first portion to hit the ground, the explosion was instantaneous on impact so that the bomb did not enter the ground. Only a shallow crater was formed while fragments of the bomb were scattered over a wide area. In consequence, it was generally desirable to use nose fuses with thick case bombs to attack personnel and light targets such as road transport, aeroplanes on the ground, and military equipment. The weight of the case was sometimes three or four times the weight of the explosive it carried.

The tail fuse, on the other hand, involved a delayed action since the fuse was some distance from the exploder; the delay was at least 1/20th of a second but could be set to as long as 15 seconds. Even the smallest delay caused tail fuse bombs to bury themselves in the ground before exploding, producing a considerable crater but scattering no fragments. This type of bomb therefore was used with a light case where violent local destruction was needed, in attacking railways or buildings for instance. In this example the weight of the explosive carried could be up to 50 per cent more than the weight of the case.

## Faulty fuses

Although each type of fuse had its particular advantages, it was not uncommon for bombs to fail to explode because of a faulty fuse mechanism. Another problem was the general shortage of steel which meant that some bombs had to be made of cast iron which was liable to break up on impact with hard surfaces before the tail fuse could operate. For this reason, many of the British bombs were fitted with both nose and tail fuses so that at least some result would be obtained if one failed to operate. Heavy and light case bombs were manufactured in varying sizes, the given weight including both the casing and the explosive material.

By the end of the war, four main types had become standard. These were the small 20-lb heavy case Cooper bomb, the medium 50-lb heavy case RL Mk IV, the large 112-lb heavy case RL Mks VI and VII, and the large 230-lb light case RFC. Heavier bombs such as the 550-lb heavy case RL Mk I/N were occasionally used during the last year of the war and a special series of heavy SN bombs was developed for the large Handley-Page bombers, ranging from 1,600-lb to 1,800-lb and finally to 3,360-lb. This last could only be carried by the V/1500 Handley-Page, although it was not, in fact, used.

The early German Carbonit bombs also used the propeller-actuated firing pistol but in 1915 the *Prüfanstalt und Werft der Fliegertruppe* (Test Establishment and Workshop of the Air Service) co-operated with the Görz company at Friedenau, already manufacturing bomb sights for use in Zeppelins, to produce a new type of bomb. The PuW bomb, as it became known, had a streamlined torpedo shape to give it better aerodynamic performance and was in many respects the true prototype of the modern aircraft bomb. It was made of high-grade steel rather than cast iron, which gave it superior penetration power. But of more

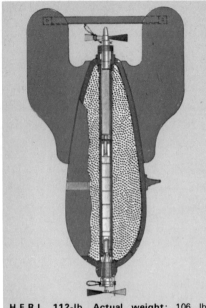

H.E.R.L. 112-lb  Actual weight: 106 lb
Case: 79 lb  Explosive: 27 lb  Case material:
Cast steel  Dimensions: 29·1 in long × 9 in dia.
Fuse: nose or tail fuse

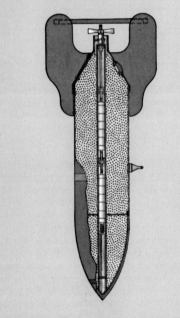

H.E. RFC 230 lb  Actual weight: 230 lb
Explosive: 140 lb  Case: 90 lb  Case material:
Steel  Dimensions: 50½ in long × 10 in dia.
Fuse: tail fuse

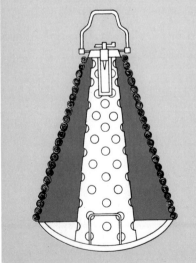

ROPE BOUND INCENDIARY with inertia fuse. The detonating material (in the perforated brass centre tube) and incendiary material around it only had to burn through rope – not a metal case

significance were the slanting fins fitted to the tail which gave a rotary motion to the bomb as it fell. In this way the fuses could be armed by centrifugal force instead of propeller devices, in a similar way to artillery fuses.

Unlike most Allied bombs at that time – which were carried vertically in racks either under the wing or fuselage in the case of British aircraft, or on the fuselage sides as favoured by the French – the PuW bombs could be suspended horizontally in racks, under the wing or fuselage at first and later in bomb bays inside the aircraft. This lessened wind resistance. A PuW type of bomb was also used in the latter part of the war by the Austro-Hungarian air services while the Italians used a light torpedo with thin casing for anti-personnel attacks and a heavy mine type with thick casing for raids on buildings.

## The first incendiary

Incendiary bombs were also developed during the war, particularly by the Germans. These bombs, generally around 20 to 40 lb in weight, were in an entirely different category from high explosives and could be said to date back to the 'Greek fire' of ancient times, with aircraft outclassing all previous methods of launching them. The early British incendiaries were filled with petrol, carcass or black powder, with a propeller-actuated fuse firing a Very Cartridge. They were of little practical use, but a 40-lb phosphorous incendiary introduced in the last year of the war was more successful, often employed against kite balloons as the time fuse could be set to ignite at any height above ground and spread a shower of burning phosphorus over a circle of about 250 yards diameter.

The first French incendiaries were also phosphorus-filled but other fillings such as a mixture of celluloid, paraffin and resin, or cotton, potassium and resin were later developed. German incendiaries were first of all filled with petrol, kerosene and liquid tar, a highly inflammable mixture which was only loaded into the bomb shortly before take-off. This was later succeeded by a mixture of benzol, tar and thermite which burned at temperatures as high as 3,000°C. And by 1918, Germany had developed an even more advanced bomb of which the body itself, made of magnesium, formed the main incendiary material. This magnesium bomb weighed about 2 kg and was used mainly for anti-personnel attacks. A bomb rack containing twenty such bombs could be fitted to any two-seater aircraft.

As the war progressed, more powerful aircraft were introduced which could carry increasingly heavy bomb loads, from the 50 lb of the 1914 Maurice-Farman and RE8 to the 4,500 lb load which could be carried by the Zeppelin-Staaken of 1918. It was, of course, with the later heavy bombers in mind that the larger sizes of high explosive bombs were designed, primarily by the British and Germans for strategic raids on each other's cities. But in practice, the big bombers usually carried a larger number of small size bombs rather than one giant bomb. One reason was the difficulty of loading and carrying such a weapon, but there was also the problem of being able to aim it with sufficient accuracy to make it worth risking an aircraft for a single bomb. The more bombs that could be dropped, even small ones, the better the chance that at least some would hit the target.

# THE BOMB-SIGHTS

In the early days of the war, when bombs were mostly dropped by hand, the pilots had two advantages which partly made up for the lack of any mechanical sighting mechanism. Because of little opposition, except gunfire from the ground and even that was not well organised to begin with, they could fly by day and at sufficiently low altitudes to give a good visual appreciation of the target. They learned to make allowance for the effect of the speed of their aircraft on the trajectory of the bombs and from this experience, the first elementary system of sighting used by the Allies was simply three nails driven into the side of the aircraft, to be lined-up on a target.

Another simple method was to dive the aircraft at a target which gave the pilot a better view and increased the stability of the plane itself, making for a more accurate drop. This should not be confused with the specialised technique of dive-bombing which was developed after the war, but it was a step in that direction.

As fighter aircraft were developed, and defences improved, more sophisticated methods of sighting were essential. German bomb-sights were initially the most advanced because of the thought that had been given to the Zeppelin as a bomber. Even before the war, the Zeiss optical firm had produced a telescopic device of this kind, intended for operation at altitudes higher than most aircraft could then fly. This sight was adopted for use in aircraft by the German Air Service and by 1916 it had been improved to take account of the fact that aircraft flew faster than airships. The observer noted with a stop-watch the time it took for a landmark on the ground to pass across a certain part of the sight and could set the sight at the correct aim by relating this to his altitude, based on a predetermined table of scales. A special Görz telescopic sight was devised for night use, protruding through the cockpit floor in some aircraft. Towards the end of the war, the Görz-Borjkow sight was introduced. By means of a built-in clockwork mechanism, this automatically determined the correct moment at which to release a bomb.

The first British bomb-sight was simply an arrangement of nails and wire, not dissimilar from that devised by Lt Scott of the US Navy in 1912, and evolved soon after the outbreak of war by Lt R. B. Bourdillon of No. 6 Squadron RFC. As a result, he was sent to the Central Flying School at Upavon and co-operated in designing the CFS bomb-sight which became standardised in the RFC and RNAS and remained in service until the end of 1917. Like the German Görz sight, it ascertained the speed of an aircraft by means of a stop-watch to measure the time taken for a landmark to pass two sightings, and the foresight was moved along a time-scale to obtain the correct angle for bomb-dropping.

A similar system was adopted by the French in 1915 with the Dorand and Lafay sights. By the following year it was general practice for sights to be mounted internally and viewed through glass panels in the floor, either in the nacelle of pusher or twin-engined aircraft, or in the rear cockpit of two-seaters for use by the observer or gunner.

## More scientific

During the last eighteen months of the war, bomb-sights became vastly more complicated as more was understood about the theory of bombing. Sights fitted to the large bombers could take account of the height of an aircraft above the target, its airspeed, the wind velocity, and the amount of drift when flying across the wind. First spirit levels and then gyrostats were used to level the sight. In the case of single-seater aircraft, more simple sights were devised which the pilot could view through the floor of the cockpit but these did not give a high degree of accuracy. Usually they were calibrated for use only at three heights – about 6,000 ft, 10,000 ft and 15,000 ft – and the pilot had to guess the wind direction and velocity. This could either be done from a previous knowledge of the weather forecast – not reliable when flying above cloud where the wind could be blowing in a reverse direction from that on the ground – or by watching the smoke bursts of shells fired from the ground.

Left: A negative-lens bomb-sight calibrated for use at three heights. Aircraft must fly at a predetermined height for each of these altitude readings: 6,000 ft = 90 mph; 10,000 ft = 80 mph; 15,000 ft = 70 mph. Above: A detail of the calibration plate. Below: A bomb-sight installed in a DH 4 fuselage

## HANDLEY PAGE V 1500

**Gross weight:** 24,700 lb **Span:** 126 ft **Length:** 62 ft **Engine:** 4×375 hp Rolls Royce Eagle VIII **Armament:** 4 Lewis machine-guns **Crew:** 4–5 **Speed:** 90 mph at 6,500 ft **Ceiling:** 11,000 ft **Range:** 6 hrs **Bomb load:** 7,500 lb

Built secretly in Belfast towards the end of the war for the purpose of carrying a large bomb-load from British bases to Berlin. It was the biggest aeroplane to be built in Britain up to that time but it appeared too late for operational service

# THE BOMBER SPREADS ITS WINGS

The first bombers were little more than general purpose aircraft adapted to carry indifferent bomb-loads. But in order to carry out the growing number of strategic raids against effective fighter opposition, armament, speed, range and ceiling all had to be increased. The first country to develop large, multi-engined bombers was Russia, but by 1916 most of the major powers were working on their own 'heavies'

The first multi-engined heavy bombers to be used in the war were the Russian Il'ya Muromets with which the Squadron of Flying Ships of the Russian Imperial Air Service was equipped at the end of 1914. Operations started in February 1915 with raids on towns and military targets in East Prussia, well inside the Muromets' combat range of about 300 miles. Among the successful attacks was the destruction of the German seaplane base on Angern Lake and the virtual obliteration of the HQ of the German Commander-in-Chief, General von Bulow, at Shavli in Lithuania.

These bombers were tough enough to take considerable punishment and still keep flying, helped by specially designed fire-proof fuel tanks and some metal-plate protection under the pilot's cabin. The heavy armament they carried, normally three machine-guns, with up to seven on some models, was usually more than a match for the German fighters of the period. The Muromets were also equipped with excellent bomb-sights which gave them an average of 75 per cent successful hits on target. If it had not been for constant maintenance problems and difficulties in obtaining engines and spare parts, the Russian bombing offensive would have been even more effective. As it was, the morale of the German aircrews fell to a very low ebb when it seemed that nothing could prevent the Muromets getting through. It was in an effort to inspire morale that Oswald Boelcke, the German fighter ace who had become a national hero after scoring eighteen victories in France, was sent on a tour of the Eastern Front in the spring of 1916. But not until 12 September of that year was the first and only Muromets brought down in air combat – and only after it had shot down three of its opponents and damaged a fourth.

### Enclosed cockpits

The first Muromets type to enter operational service was the IM-B, which carried a crew of four – pilot, co-pilot, bombing officer and air mechanic – in completely enclosed accommodation with glazed window panels in the nose and fuselage sides. Its bomb load was 1,120 lb and three machine-guns were mounted in the fuselage sides for shooting at ground targets. It was powered by two 200 hp and two 135 hp Salmson radial engines. The next and most widely used type was the IM-V which had a smaller wing span and 150 hp Sunbeam liquid-cooled engines, originally bought from England and later made under licence. The bomb load was increased by 600 lb.

As a result of the introduction of armed fighters by the Germans, the defensive armament of the later types, the IM-G1, G2 and G3, was increased to include machine-guns mounted on the upper wing section, under the fuselage, and in the nose and tail, for which the crew was increased to seven: two pilots, one navigator, one mechanic and three gunners. The wing chord was also increased and the 2,000 lb bomb load was stored internally and the bombs dropped either vertically or horizontally.

Largest of the Muromets was the IM-YeA, with a wing span of 113 ft 2¼ in and a gross weight of 15,432 lb, but it was not put into production. A few of the slightly smaller IM-YeI were built, armed with seven machine-guns and powered by four 220 hp Renault engines. Although the Muromets crews were the élite of the Russian Air

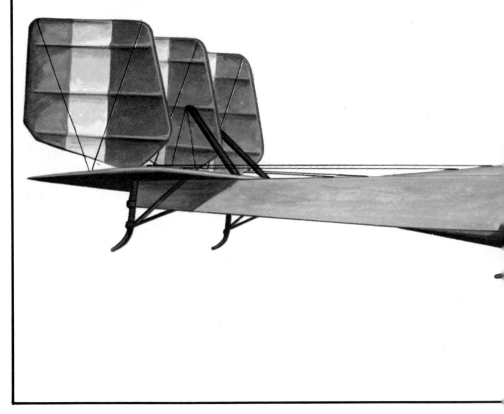

**CAPRONI Ca 5**

**Gross weight:** 11,700 lb **Span:** 77 ft **Length:** 41 ft 4 in **Engine:** 3×300 hp Fiat **Armament:** 2 machine-guns **Crew:** 3 **Speed:** 95 mph at ground level **Ceiling:** 15,000 ft **Range:** 4 hrs **Bomb load:** 1,188 lb

With the Ca 5 series in 1917, Caproni returned to a biplane configuration after the Ca 4 triplane. Widely used as a day and night bomber in 1918

Service, the plane itself was not easy to fly and considerable training was required before crews could be sent to the Front. About half of the eighty IMs built were used for training purposes. Most of the machines were destroyed by their crews after the Revolution and the German invasion to prevent them falling into enemy hands, but a few remained to inaugurate passenger services in Russia after the Civil War.

The next country to introduce large bombers into service was Italy, following its declaration of war against Austria-Hungary on 24 May 1915. These were the Caproni series, based on the original pre-war design of the Ca 31, but modified considerably as the war progressed. The first to enter operational service, in August 1915, was the Ca 32 biplane, powered by three 100 hp Fiat liquid-cooled engines. Its range of 340 miles enabled it to be used for strategic bombing attacks, but as the Austrian fighters improved in performance, the Ca 32 later had to be relegated to night bombing.

Meanwhile the most widely used type, the Ca 33, came into service towards the end of 1916 with a higher maximum speed of 94 mph. The two pilots sat side-by-side in the centre fuselage, a gunner operated a

Revelli machine-gun from the nose position, and a fourth crew member in the rear cockpit operated up to three similar guns, fixed to fire in different directions.

The Ca 3 series, as they were designated, were all biplanes but with the next (Ca 4) series, starting with the Ca 40, a triplane design was used with huge 100 ft wings. The main production model was the Ca 42, powered by three 270 hp Fiat, Isotta-Fraschini or American Liberty engines. It was a sturdy and reliable plane with a combat range of seven hours, and could carry a bomb load of 3,910 lb, made up of small bombs mounted externally under the bottom plane. But its size and relatively low speed made it an easy target for enemy fighters, and it was reserved for night-bombing operations. For the next (Ca 5) generation Caproni returned to a biplane layout and these were beginning to replace the Ca 3 series when the war ended.

Semi-rigid airships and smaller bombers were also used by the Italians throughout the war but by far the greatest bombing effort was made by the Capronis. The first raids, beginning at the end of August 1915, were against military targets to the rear of the Austrian armies, such as railways, supply depots, troop concentrations and

engines. So impressed was Murray Sueter with the design that he asked Handley Page to improve on it in order to produce a 'bloody paralyser' of an aeroplane. The result was the Handley Page 0/100 biplane, one of the most famous bombers of the war, which went into service in November 1916. It was intended to have two 150 hp Sunbeam engines but as construction progressed so did work by Rolls Royce on two engines which were to become the renowned Eagle and Falcon. Both were water-cooled twelve-cylinder Vee types, the Eagle being the largest. This was chosen for the 0/100, its horsepower de-rated from 300 to 250 to increase reliability. They were enclosed in armoured nacelles, each with a separate armoured fuel tank. Unusual features included the facility for folding the huge 100-foot wings, so that the aircraft could be stored in canvas field hangars, and a biplane tail unit.

The crew of three sat in a cabin which was originally intended to be enclosed with bullet-proof glass and armour plate, but in the production models, the enclosure and much of the armour plate was removed. The maximum bomb load was 2,000 lb, consisting either of a single 1,650-lb bomb or eight 250-lb, sixteen 112-lb, or three 520 or 550-lb bombs. Defensive armament consisted of one or two Lewis guns on a ring-mounting in the nose cockpit, one or two Lewis guns in the upper rear cockpit, and a single Lewis gun mounted to fire rearward and downward through a hole in the fuselage floor just behind the wings.

## More stability
While work on the Handley Page 0/100 continued, in the spring of 1916 the RNAS went ahead with the formation of two strategic bombing wings at Luxeuil in Belgium and Dunkirk in France, in preparation for the Sopwith 1½ Strutter and the Short Bomber which had been ordered the previous year. By that time it was apparent that bombers would have to fight their way past defending fighters, in particular the Fokker *Eindekker*. Coincidentally the British had at last developed interrupter gear to enable the fixture of fixed forward-firing machine-guns, either Lewis or the more successful belt-fed Vickers which was later generally adopted.

This gear, originally designed by Lt-Cdr V. V. Dibovsky of the Russian Imperial Navy, and developed by Warrant Officer F. W. Scarff of the Admiralty Air Department (who had also been responsible for designing one of the best ring-mountings of the war for use by observer/gunners) was first installed in the RFCs single-seat Bristol Scout. But the first two-seater to use this synchronised gun was the 1½ Strutter, in addition to the usual free-mounted Lewis gun in the rear cockpit. The Strutter had been designed as a bomber, capable of carrying a 130 lb bomb load with the emphasis laid on stability rather than manoeuvrability. However, its armament made it an excellent fighter as well, especially when German pilots attacked from the front to avoid the observer's gun, only to be met by a deadly hail of fire from the pilot's synchronised Vickers.

With the Battle of the Somme about to begin the RFC was desperately short of fighter aircraft, and after appealing to the RNAS for help, the Admiralty handed over a large number of Strutters for army use, thus delaying its own plans for a strategic bombing offensive against Germany.

The other plane which had been ordered

especially airfields. Tactical operations remained the prime objective of the Italian Air Service until the end of the war but early in 1916, mainly at the instigation of Gabriele d'Annunzio, long-range strategic raids were undertaken.

## Daylight attacks
The Capronis did not have the range to reach Vienna, 260 miles from their main base at Pordenone, but targets were available some 100 miles away across the Adriatic, such as the Austro-Hungarian naval base at Pola and the industrial city and seaport of Trieste. Daylight attacks were made on these and other strategic targets, at first by Capronis flying in formation but later escorted by Italian-built Nieuports. One of the biggest raids was made on 2 October 1917 by nearly 150 Capronis and eleven flying-boats on the naval base at Pola. As a result of the Caporetto disaster later in the month, however, when the Pordenone base was captured by the Austrians after the Italian retreat, strategic bombing became too much of a luxury and for the rest of the war the Caproni squadrons concentrated on tactical operations, mainly against Austrian airfields. A number of American pilots were

attached to Caproni units to gain experience in bombing operations.

Just as the first British efforts at bombing were pioneered by the RNAS, so the first British heavy bomber was developed by the Admiralty for use by the navy's Air Arm. Following the success of the RNAS raids on Zeppelin sheds towards the end of 1914, the Air Department of the Admiralty was convinced of the value of air bombardment and laid plans for the creation of a strategic bombing force that could operate at long range from bases in France and England. There were, of course, no planes then available for such a purpose and engines were too scarce in 1914 and 1915 for a twin-engined layout to be attempted. As an interim measure a single-engined bomber was ordered from Sopwith, resulting in the introduction of the 1½ Strutter, while the Short company developed a single-engined landplane from their successful Short 184 seaplane. In the meantime, Captain (later Rear Admiral) Murray Sueter, Director of the Air Department, issued a specification for a large twin-engined aircraft capable of extended patrols over the sea.

The challenge appealed to Sir Frederick Handley Page who set about designing such a bomber, powered by two 120 hp Beardmore

233

## VICKERS VIMY IV

**Gross weight**: 12,500 lb  **Span**: 67 ft 2 in
**Length**: 43 ft 6 in  **Engine**: 2×360 hp Rolls
Royce Eagle VIII  **Armament**: 6 Lewis machine-
guns  **Crew**: 5–6  **Speed**: 103 mph at ground
level  **Ceiling**: 7,000 ft  **Range**: 985 miles
**Bomb load**: 2,476 lb

Initiated in 1917 as a heavy bomber to take part in
Britain's plan for the strategic bombing of Germany, but it arrived too late to see operational
service before the Armistice. After the war, the
Vimy distinguished itself for long-distance flights,
including the first non-stop crossing of the Atlantic
by Captain John Alcock and Arthur Whitten Brown
on 14/15 June 1919

was the Short Bomber, powered by a 250 hp
Rolls Royce engine and with a maximum
bomb load of 920 lb, twice that of the
Caudron G4 which the RNAS also used at
that time. With the introduction of the
Short Bomber in the autumn of 1916,
together with the Caudrons and a few
Strutters, the RNAS began a bombing
offensive against military and industrial
targets in Germany and on German naval
forces at Ostend and Zeebrugge, although
not on the scale that had been planned.

The Short Bomber had a range of over
400 miles, which enabled it to raid targets
far behind the front line, but its maximum
speed was only 77 mph and it was inade-
quately armed, with only a single Lewis gun
on top of the centre-section. Even to reach
that the observer had to climb out of his
cockpit and stand up on the fuselage. In
company with the Caudron bombers, the
Short had to be relegated to night bombing.

With the arrival of the Handley Page
0/100 in November 1916 however, the RNAS
was able to increase its operations. The first
two bombers were delivered to the 5th Wing
at Dunkirk; a third landed by accident
behind the enemy lines and was studied in
great detail by the Germans. The first raids
by the 0/100 were carried out in daylight,
and it proved to be a formidable bomber,
able to carry three times the load of the
Short Bomber and six times that of the DH 4
day-bomber which had also just been
brought into service. In the meantime the
capability of fighter aircraft had also ad-

vanced, and when one of the valuable
Handley Page bombers was brought down
into the sea (admittedly after sinking an
enemy destroyer), the type was used only
for night bombing. Raids were directed
mainly against U-boat bases, railway
centres and airfields, while four aircraft
were brought back to England for anti-
submarine patrols over the North Sea. One
machine made a remarkable flight from
England to the island of Lemnos in the
Aegean Sea, with stops in Paris, Rome and
the Balkans. In June 1917 it bombed the
Turkish capital of Constantinople, but on
a second attempt two months later it came
down into the sea with engine failure and
the crew were taken prisoner, including
the pilot, Flt-Lt J. Alcock (later one of the
first two men to fly the Atlantic non-stop).

### New fuel system

In the spring of 1917 the Germans began a
systematic campaign to bomb London and
other targets in England, using large
bomber aircraft far more frightening than
the Zeppelins. In retaliation, the British War
Cabinet decided to launch a strategic
bombing offensive against German cities
and industrial targets. The task was given
to Major General Hugh Trenchard, then
commanding the RFC in France. In October
1917 he formed the 41st Wing at Ochey
from where it was possible to reach German
towns in the Saar, and such cities as Karlsruhe, Mainz, Koblenz, Cologne, Frankfurt
and Stuttgart. In addition to the RFC day-

bombers (FE 2bs and DH 4s), the Wing
included a RNAS squadron of Handley
Page 0/100s for night bombing. These were
more effective than the day-bombers and
led to the development of an improved
version, the 0/400, which became the most
widely used of the Handley Page types;
some 400 were delivered before the end of
the war, compared with less than fifty of
the 0/100.

One of the features of the 0/400 was a
completely redesigned fuel system, in which
the fuel tanks were moved from the engine
nacelles to a position above the bomb-bay,
thus enabling the nacelles to be considerably shortened. The crew consisted of
a pilot who sat in the main cockpit, a bomb-
aimer who doubled as front gunner and
observer from a cockpit in the extreme nose
of the fuselage, and one or two rear gunners
who occupied a cockpit in the mid-section
just behind the wings. The bombs were
suspended nose-upwards in separate honeycomb cells in the bomb-bay, each covered
by a door which was pushed open by the
weight of the falling bomb and closed by a
spring. The bombs were released by cables
from the nose cockpit where a bomb sight
was mounted externally. One or two Lewis
guns were mounted in both nose and rear
cockpits; in the latter position, one gun was
fired sideways or backwards from a raised
platform while the other, when carried,
could be fired downwards and backwards
through a trapdoor in the floor.

Although starting the war with very

different functions, the RNAS and RFC had come to co-operate with each other for many operations, including Trenchard's strategic bombing wing which was later re-designated the VIII Brigade. During the last two years of the war there was a complete reorganisation of British military aircraft development in which the army followed the navy's policy of leaving aircraft design and manufacture to private industry. The Royal Aircraft Factory at Farnborough (which, in spite of some failures, produced in the SE 5a one of the best Allied fighters of the war) was directed to concentrate solely on research.

On 1 April 1918 the RNAS and RFC were merged into one service, the Royal Air Force, with the prime intention of pooling resources for a sustained strategic bombing campaign against Germany. Under the command of General (later Marshal of the RAF, Viscount) Trenchard, who was directly responsible to the newly formed Air Ministry, the RAF was the world's first major independent air service. On 6 June the strategic bombing wing he had previously commanded became the famous Independent Force of the RAF.

The Handley Page night bombers contributed a large proportion of the 665 tons of bombs dropped by the Independent Force and its predecessors the 41st Wing and VIII Brigade. One daring attack made was on the Badische Anilin factory at Ludwigshafen during the night of 25 August when the two Handley Pages came down to

200 and 500 feet respectively to place their bombs with the greatest accuracy, in spite of the searchlights and heavy anti-aircraft fire. This was no mean hazard by that stage of the war, considering the great advances made in the development of anti-aircraft defences. In raids during a single night the following month, six of the large bombers were brought down by anti-aircraft fire while attacking Saarbrücken and Trier. The five night bombing squadrons which were in operation during the last months of the war suffered eighty-seven crew members killed or missing and 148 aircraft destroyed. In the latter stages, the bombers were often employed against tactical targets to support the Allied offensives, and 220 tons of the bombs dropped were on enemy airfields, destroying many aircraft. From September onwards the Handley Pages often employed 1,650-lb bombs; one dropped on Kaiserlautern wiped out an entire factory.

### Great secrecy

In order to carry an even bigger bomb load of 7,500 lb from bases in Britain to Berlin, including a single 3,300-pounder, the Handley-Page V/1500 was built in great secrecy in Belfast. Six had been delivered by the time the Armistice was signed, although none were used operationally. Nevertheless the V/1500 marked a triumph for British aeronautical development, with 126-foot wings which could still be folded, four 375 hp Rolls Royce Eagle VIII engines and a tail gun position which could be reached

by climbing along a cat-walk. With a combat range of 1,200 miles and a maximum speed of just under 100 mph, it showed how far the design of bomber aircraft had advanced since the days of the Avro 504 and symbolised the policy of long-range strategic bombing on which Trenchard set such importance. It was a policy which was continued after the war and culminated in the RAFs mighty Bomber Command striking force of the Second World War. Two other heavy bombers were also developed for use by the Independent Force but, like the V/1500, arrived too late to see operational service. These were the twin-engined three-seater DH 10 Amiens and Vickers Vimy. The Amiens operated as a mail carrier after the war while the Vimy achieved fame with a series of record long distance flights, including the first non-stop crossing of the Atlantic by Alcock and Brown and the first flight from Britain to Australia.

One of the myths of early aviation history was that the Handley Page 0/100 which had been captured after accidentally landing behind German lines in November 1916 was used as a model for Germany's own heavy bombers, the famous Gotha biplanes, which appeared shortly afterwards. This was certainly not true for even before the war German designers including Count von Zeppelin had drawn up plans for multi-engined aircraft. But the main effort was concentrated on building airships, and it was only with the realisation that lighter-than-air craft were too vulnerable that development was pushed ahead on heavy bombers.

Even earlier, however, the Germans had produced twin-engined aircraft which were used as bombers. Late in 1914 the Friedrichshafen company, though mainly concerned with the design and construction of naval seaplanes, built the GI bomber, a twin-engined pusher biplane with a biplane tail unit. This did not enter production, but it did lead to the G II, small numbers of which entered service in late 1916. Powered by two 200 hp Benz engines and able to carry a bomb-load of about 1,000 lb, it had two Parabellum machine-guns for defensive armament, one in the nose and the other in a dorsal position. A larger and more powerful development was the G III, with 260 hp six-cylinder water-cooled Mercedes engines, a monoplane tail unit, and the ability to carry over twice the bomb-load. This was the main production model of the Friedrichshafen G series which entered service in early 1917. It carried a crew of three in the central fuselage area and an interesting feature of its design was the steel-tube frame of the square-section fuselage, covered with wood at the nose and tail and fabric over the central part. Modifications to later models included a return to the biplane tail unit on the G IIIa and tractor propellers on the G IV.

The Friedrichshafen types were generally similar in design to the series of twin-engined bombers produced by the Gothaer company, and together they formed the mainstay of the German bomber units (*Bombengeschwadern*) from 1916 until the end of the war. The principle Gotha types were the G IV and G V and it was their ability to fly at a high altitude – over 20,000 feet – carrying a bomb-load of 1,000 lb in external racks that made them ideal for taking over long-range bombing duties from Zeppelin airships. Another asset was that in addition to the usual machine-gun in the front cockpit, they had a second machine-gun mounted behind the wings, which could fire not only upwards, but also downwards and rearwards beneath the tail, for defence against fighters attacking from behind and below.

### Terrifying raids

Although the first aeroplane raid on London took place on 28 November 1916, when a single LVG CII dropped six 22-lb bombs near Victoria, it was the Gotha which introduced a new and terrifying form of warfare by a succession of day and night raids on the city from May 1917 to May 1918. The first attempt, by a formation of twenty-one bombers on 25 May 1917, ended when they had to turn back because of heavy cloud and dropped their bombs on towns in Kent instead, killing nearly one hundred civilians. There was an outcry from the British public who, congratulating themselves that the Zeppelin menace had been overcome, now had to face daylight raids with no warning of the enemy's approach.

Worse was to follow. On 13 June a formation of fourteen Gothas led by Hauptmann Brandenburg circled over London with contemptuous ease, in full view of people watching from below. They dropped nearly one hundred bombs, mostly in the region of Liverpool Street station, killing 162 people and wounding 438 – higher casualties than all the Zeppelin raids had caused, and more

than any single bombing attack on Britain during the entire war. Although ninety-two fighter aircraft took off from various parts of England, the bombers were able to fly at an even greater altitude on the way home, lightened of their bombs, and had disappeared by the time the fighters could climb to that height. Not a single bomber was lost in the raid.

In spite of the serious situation in France, where every aircraft was needed, two squadrons of the latest fighters – SE5as and Sopwith Pups – had to be withdrawn to help protect London from bombing. It was this dissipation of the Allied war effort on the Western Front, together with the effect on morale at home, which was the greatest success of the Gotha raids. Gradually, with strengthened fighter units and a complete reorganisation of anti-aircraft gun defences, the British began to take a heavy toll of the Gotha bombers, and in September they were forced to turn to night bombing. By this time they were being supported by the R-class bombers, produced by several different companies, but generally known as Zeppelin Staaken 'Giants'. They were the

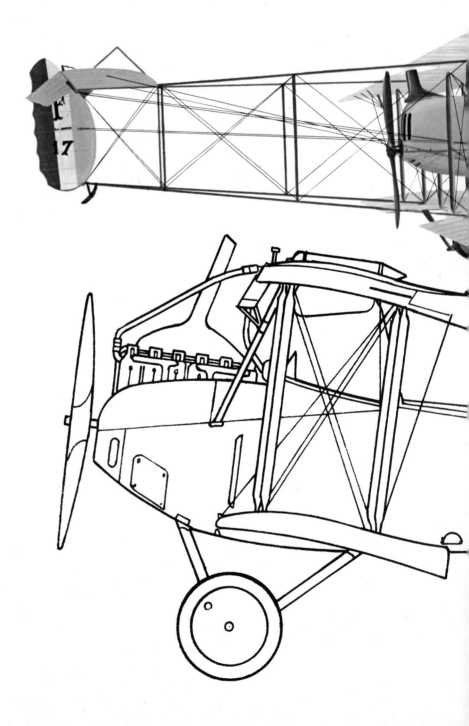

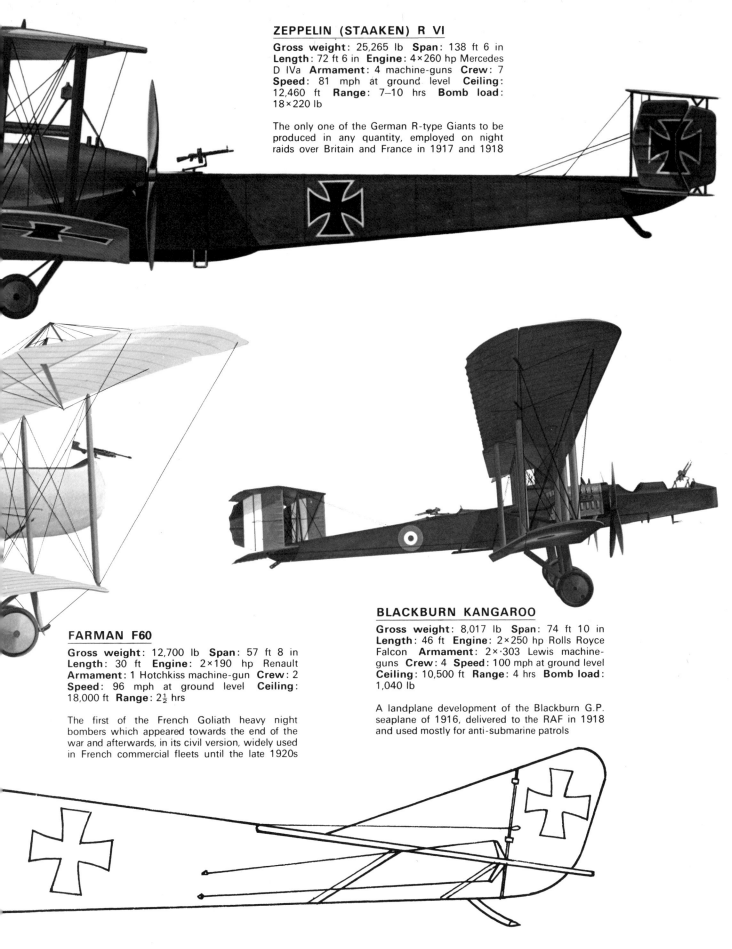

## ZEPPELIN (STAAKEN) R VI

**Gross weight**: 25,265 lb  **Span**: 138 ft 6 in
**Length**: 72 ft 6 in  **Engine**: 4×260 hp Mercedes
D IVa  **Armament**: 4 machine-guns  **Crew**: 7
**Speed**: 81 mph at ground level  **Ceiling**:
12,460 ft  **Range**: 7–10 hrs  **Bomb load**:
18×220 lb

The only one of the German R-type Giants to be
produced in any quantity, employed on night
raids over Britain and France in 1917 and 1918

## FARMAN F60

**Gross weight**: 12,700 lb  **Span**: 57 ft 8 in
**Length**: 30 ft  **Engine**: 2×190 hp Renault
**Armament**: 1 Hotchkiss machine-gun  **Crew**: 2
**Speed**: 96 mph at ground level  **Ceiling**:
18,000 ft  **Range**: 2½ hrs

The first of the French Goliath heavy night
bombers which appeared towards the end of the
war and afterwards, in its civil version, widely used
in French commercial fleets until the late 1920s

## BLACKBURN KANGAROO

**Gross weight**: 8,017 lb  **Span**: 74 ft 10 in
**Length**: 46 ft  **Engine**: 2×250 hp Rolls Royce
Falcon  **Armament**: 2×·303 Lewis machine-
guns  **Crew**: 4  **Speed**: 100 mph at ground level
**Ceiling**: 10,500 ft  **Range**: 4 hrs  **Bomb load**:
1,040 lb

A landplane development of the Blackburn G.P.
seaplane of 1916, delivered to the RAF in 1918
and used mostly for anti-submarine patrols

## LVG CII

**Gross weight**: 3,091 lb  **Span**: 42 ft 2 in
**Length**: 26 ft 7 in  **Engine**: 160 hp Mercedes
D III  **Armament**: 1 or 2×7·92-mm machine-
guns  **Crew**: 2  **Speed**: 81 mph at ground level
**Ceiling**: 10,000 ft  **Range**: 4 hrs  **Bomb load**:
150 lb

A light bomber, as well as one of the newly-
established C class of armed two-seater reconnaissance aircraft introduced by the Germans at
the end of 1915, this plane is credited with the
first daylight raid on London, which took place in
November 1916

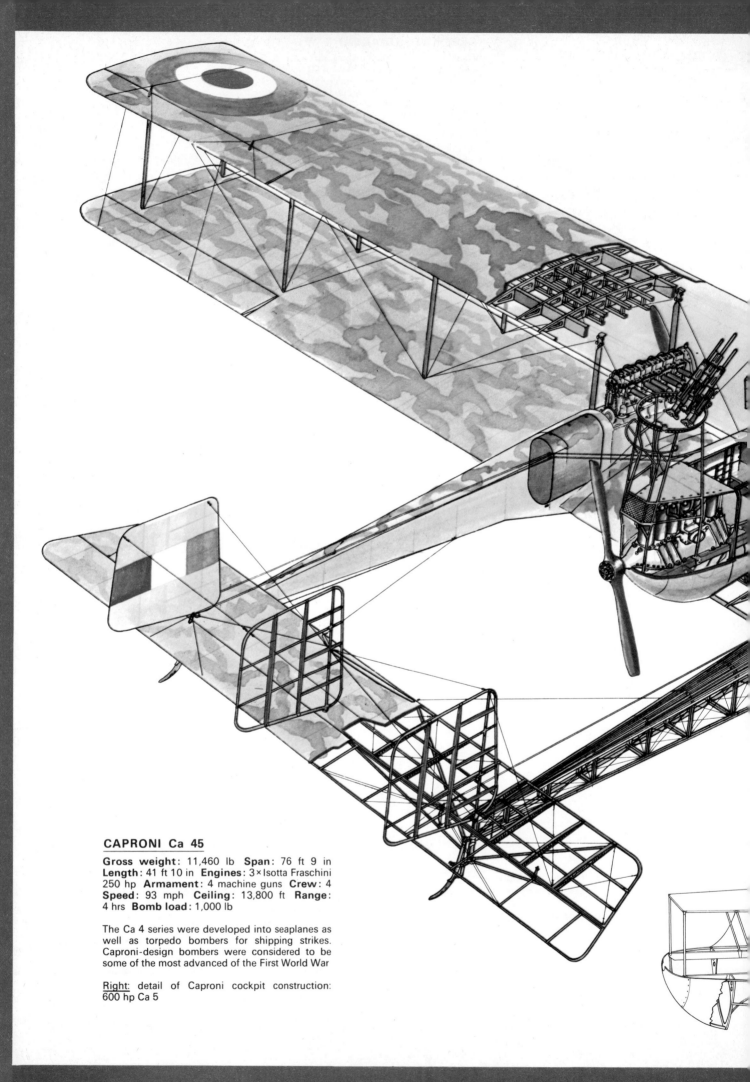

## CAPRONI Ca 45

**Gross weight**: 11,460 lb **Span**: 76 ft 9 in
**Length**: 41 ft 10 in **Engines**: 3×Isotta Fraschini
250 hp **Armament**: 4 machine guns **Crew**: 4
**Speed**: 93 mph **Ceiling**: 13,800 ft **Range**:
4 hrs **Bomb load**: 1,000 lb

The Ca 4 series were developed into seaplanes as
well as torpedo bombers for shipping strikes.
Caproni-design bombers were considered to be
some of the most advanced of the First World War

Right: detail of Caproni cockpit construction:
600 hp Ca 5

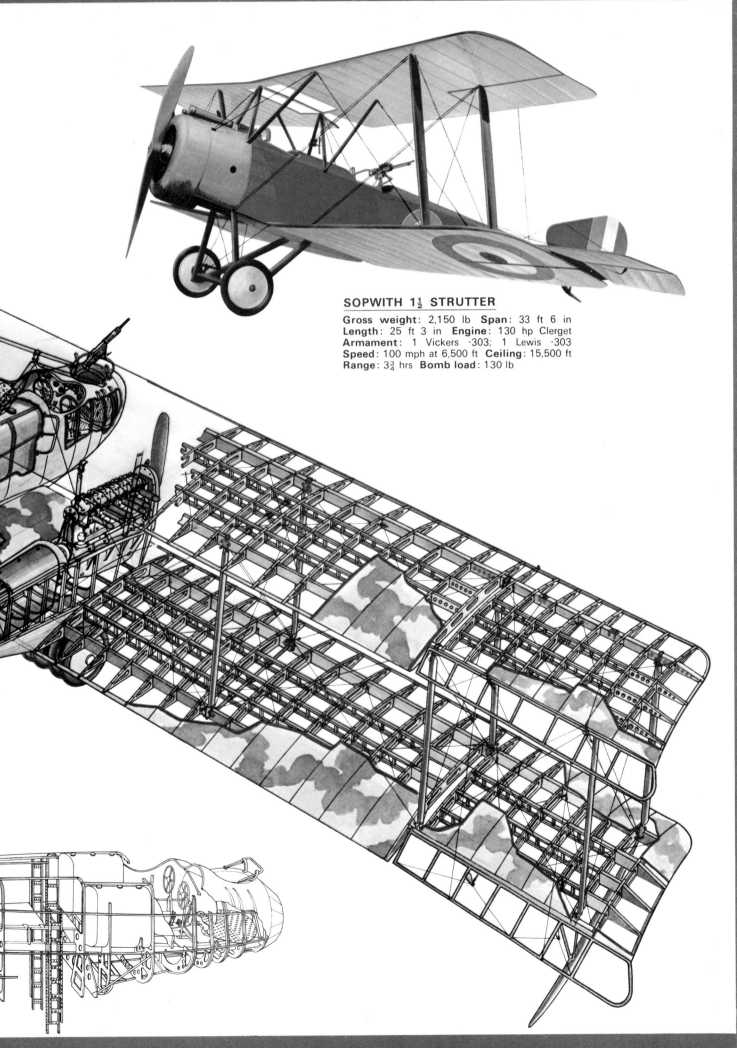

## SOPWITH 1½ STRUTTER

**Gross weight:** 2,150 lb  **Span:** 33 ft 6 in
**Length:** 25 ft 3 in  **Engine:** 130 hp Clerget
**Armament:** 1 Vickers ·303; 1 Lewis ·303
**Speed:** 100 mph at 6,500 ft  **Ceiling:** 15,500 ft
**Range:** 3¾ hrs  **Bomb load:** 130 lb

biggest and in some respects the most remarkable aircraft of the war, and were produced only as single examples or in very small numbers.

The first of these leviathans to appear was the VGO I, with a wing span of over 138 ft, powered by three 240 hp Maybach engines, which flew for the first time in April 1915. The engine power was not sufficient for the huge, nine-ton aircraft and the later VGO III had six 160 hp Mercedes engines instead. Two of these were mounted in tandem in each of the port and starboard nacelles, driving pusher propellers, with the other pair side-by-side in the nose of the fuselage driving a single tractor propeller. After various modifications the R VI appeared in mid-1917 (the 'R' designation standing for *Riesenflugzeug* – giant aeroplane). This was the only type to be produced in any quantity.

The nose-mounted engines were abandoned and instead, four Maybach or Mercedes 260 hp engines were mounted in tandem pairs between the wings, driving tractor and pusher propellers. The nose cockpit was fitted with a machine-gun and the release mechanism for the eighteen 220-lb bombs carried internally. Two guns were located in a dorsal position and another two ventrally. The main cockpit was enclosed and the usual crew was seven. So great was the weight (nearly twelve tons fully loaded) that the undercarriage consisted of no less than eighteen wheels, two of which were under the nose.

In July 1917 the giant R-class bombers were first flown operationally on the Eastern Front, and gained some successes in attacks on railways and military installations. Between August and February the following year they were transferred to the Western Front and used initially for night raids on Britain, usually escorted by Gothas. In eleven such raids, during which single 2,200-lb bombs (the largest used in the war) were occasionally dropped, not one Giant was lost in action. The Gothas were less fortunate however, mainly because pilots of such improved fighters as the Sopwith Camel had learned to fly and intercept at night. After a raid on the night of 19 May 1918, when seven Gothas out of forty-three aircraft sent to attack London and Dover were shot down by fighters and ground fire, the German High Command ordered a stop to the bombing of Britain.

In fifty-two raids on Britain, one more than were carried out by airships, German aircraft achieved a statistically better result, killing 857 people and injuring a further 2,058. Just under £1,500,000 worth of damage was caused for the expenditure of only 2,772 bombs totalling about 196 tons.

Although the loss of life and amount of damage caused was hardly significant compared with what was happening in France, the psychological effect was considerable. This, however, operated to the detriment of Germany, for the anger aroused was directly responsible for retaliatory raids by the British, and was one of the factors which led to the merger of the RNAS and RFC to form the RAF.

The Giants continued to make bombing raids over Paris until those too were stopped. Thereafter they were used singly for tactical attacks on military targets a few miles behind the Allied lines, but it was a serious mishandling of machines that had been developed for long-range strategic bombing. They were too slow and made too big a target to operate successfully in the battle zone, and casualties were very heavy.

A better aircraft for tactical purposes was the other main type of twin-engined bomber built by the Germans, the AEG series.

This was a development of the K I three-seat general purpose biplane of 1915, later re-designated the G I when the Germans decided on the classification *Grossflugzeug* (big aeroplane) for all such types, irrespective of manufacturer. This model was used mainly on the Eastern Front during 1915 and, like the G II which followed, was only built in small numbers. The G III which began to appear in December 1915 carried a 660-lb bomb load and two machine-guns for defence. The major production model however was the G IV which came into service towards the end of 1916, powered by two 260 hp Mercedes engines driving four-blade opposite-rotating propellers. Neither its bomb load of 770 lb, nor its range – about 350 miles against the Gotha's 550 – could compare with those of the Gotha and other big bombers, but its speed of over 100 mph made it very suitable for attacking short-

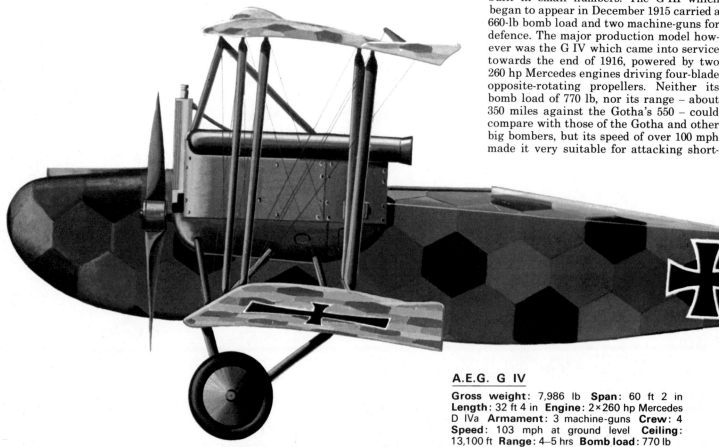

**A.E.G. G IV**
**Gross weight:** 7,986 lb **Span:** 60 ft 2 in
**Length:** 32 ft 4 in **Engine:** 2×260 hp Mercedes
D IVa **Armament:** 3 machine-guns **Crew:** 4
**Speed:** 103 mph at ground level **Ceiling:**
13,100 ft **Range:** 4–5 hrs **Bomb load:** 770 lb

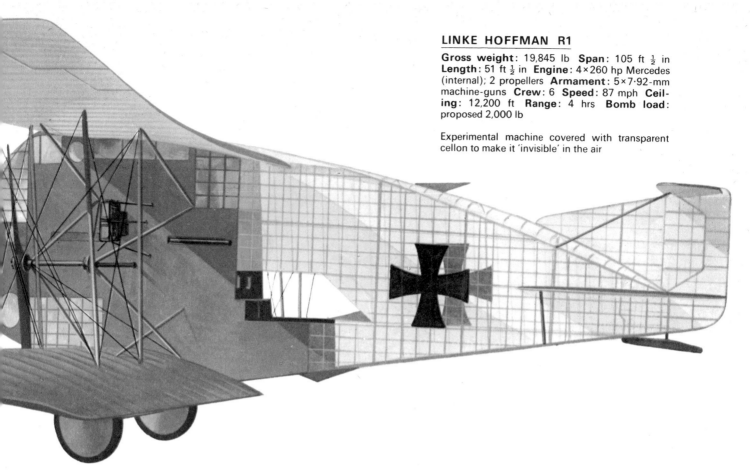

## LINKE HOFFMAN R1

**Gross weight:** 19,845 lb **Span:** 105 ft ½ in
**Length:** 51 ft ½ in **Engine:** 4×260 hp Mercedes
(internal); 2 propellers **Armament:** 5×7·92-mm
machine-guns **Crew:** 6 **Speed:** 87 mph **Ceiling:** 12,200 ft **Range:** 4 hrs **Bomb load:**
proposed 2,000 lb

Experimental machine covered with transparent
cellon to make it 'invisible' in the air

range tactical targets and for photographic reconnaissance. A larger version, the G V, with a biplane tail unit and a bomb load of 1,320-lb was beginning to appear at the end of the war.

Although the French made a number of spectacular day and night bombing raids in the early months of the war, mainly through the efforts of enthusiasts like Laurens, Happe and Göys, the French High Command was not very impressed with the results.

Towards the end of 1915 the bomber groups which had been built up under the command of Commandant de Göys were split up and the aircraft dispersed among various army commanders. In fact, the French showed little interest in any long-range strategic bombing during the war, largely due to Marshal Foch's obsessive desire to win the war through ground attack alone. The Independent Force of the RAF was originally intended to include French bomber squadrons but these never materialised. Consequently, France was much slower than the other major powers in developing large twin-engined specialised bombers.

The first to appear, in mid-1917, was the Letord 3, one of a series of three-seat aircraft designed for reconnaissance, fighter and night bombing roles by Colonel Dorand of the French Service Technique, and built in relatively small numbers in the last two years of the war. The Letord 3 had the characteristic back-staggered wings of other types in the series, and was powered by two 200 hp Hispano engines. A better version was the Letord 5 which was of sesquiplane configuration (the lower wing being half the size of the upper) with 220 hp Lorraine engines giving a speed of nearly 100 mph. It could carry a bomb load of 440 lb and two machine-guns. The Letord 7 which appeared in 1918 had the same wing arrangement as the Letord 3 only larger, and was notable for the cannon mounted in the nose of the fuselage in place of a machine-gun. Only the Letord 9, with a wing span of 85 feet and two 400 hp Liberty engines, began to match the Handley Page bombers, and that only flew in prototype before the war ended.

The only big French bomber to see any service during the last months of the war was the two-seat Farman F50, powered by two 265 hp Lorraine engines mounted on the lower wing on either side of the fuselage and resembling the general lines of the Gotha. Eight 165-lb bombs could be carried under the fuselage between the legs of the

undercarriage, and the bomb aimer/gunner sat in the nose of the fuselage, in front of the pilot's cockpit. Only two Voisin units were re-equipped with F50s at the time of the Armistice, while a few were flown by the American Expeditionary Forces in France. The prototype of a larger bomber, the excellent Farman F60 Goliath, first flew in 1918 and, although too late for the war, it later became the standard French night bomber. Another type which was too late to see war service was the Caudron C23 powered by two 260 hp Salmson engines. In fact, a smaller Caudron twin-engined aeroplane, the G4, had been used for bombing since the spring of 1915, but it was too slow and poorly armed for daylight raids. Even in night bombing it suffered many casualties, and it was withdrawn the following year to be used mostly for training.

In general, all the big bombers that were built and brought into operation during the last two years of the war – the Capronis, Handley Pages, Gothas and Zeppelin-Staakens – were quickly transferred from day to night bombing as the capabilities of fighter aircraft improved and ground anti-aircraft defences became better organised. Such big aircraft, which after the war were to be of vital importance in the development of passenger airliners, were too expensive and in too short supply to be risked in hazardous daytime operations. Even at night they sometimes suffered heavy casualties – for instance, over one-third of the Handley Pages of the Independent Force were lost.

Night bombing inevitably was much less accurate than day operations but there were some targets, especially military installations, which required a high degree of accuracy if any success was to be achieved. There was an obvious need for smaller and faster bombers which could elude – or battle their way through – fighter opposition by day. These proved to be some of the best aircraft types produced during the war.

# THE FIGHTERS MEET THEIR MATCH

Although large-scale strategic raids have always been the most dramatic form of bombing, it is debatable whether they have proved as valuable as local tactical raids in support of ground troops or against specific military objectives. This was certainly the case in 1914–18, and it was the small fast planes used in the latter role that made the most significant contribution

Bombing did not make a very great contribution to the First War and reconnaissance remained the most important use of aircraft. In spite of great technological advances, neither the bombers themselves, nor the bombs they carried were sufficiently powerful, or available in large enough numbers, to determine the outcome of any particular battle. Some of the bombing attacks were indeed spectacular, but it was the psychological fear they caused that created the spectre of terror bombing in the years after the war. This was reinforced by air force leaders who were to claim more for the power of strategic bombing than they could actually fulfil when the time came. But during the First World War, the most important contribution of both fighters and bombers was to control the air space over the battle zones by destroying the enemy's aircraft in the air and on the ground. Complete control could never be achieved, of course, and the balance swung to and fro depending on the quality and number of aircraft possessed by one side at any particular time.

In later wars, such as Germany's attack on Poland in 1939 and Israel's pre-emptive strike against the Arabs in the six-day war of 1967, complete control of air space was a decisive factor, but only in relation to the effort of ground forces and the strategic aims of the war. The much greater air superiority enjoyed by the Americans in Vietnam failed to achieve a decisive result because of the vagueness of the strategic aims. As the history of bombing has proved time and time again, attacks on military targets, especially airfields, are perhaps the most useful contribution that bombers can make in an overall war effort. The Luftwaffe's failure to maintain such attacks during the Battle of Britain, because the temptation to bomb civilian targets was too irresistible, altered the course of the

Second World War. As the Germans, the British and – in another war – the Americans discovered, bombing civilians diverted an effort which might more usefully have been directed elsewhere. Not only did it fail to produce the expected psychological results, but it invariably strengthened a country's morale.

## Shorter landing

During the First World War, therefore, in spite of the dramatic raids of heavy night bombers and their subsequent effect on the thinking of air force planners, it was the smaller and faster bombers, used for tactical raids in the battle areas, that made the significant contribution. Some remarkable and highly successful aircraft were produced for that purpose. Although the British, through the efforts of the RNAS, were the first to carry out strategic raids, it was the French who employed bombing most widely during the early part of the war. The Voisin was the most predominant type of bomber aircraft at that time and it continued in service in various forms right up to the end of the war. It was very sturdy, due to the extensive use of steel in its construction, and was able to operate from small, rough fields, making for a short landing run. It was the first aircraft to be equipped with wheel brakes. But the early advantage of the pusher type of aeroplane was lost as soon as machine-guns were mounted in the rear cockpits of the faster tractor-driven types, and even more so with the development of fixed, forward-firing machine-guns.

The pusher was inevitably slower because of the drag created by the booms which carried the tail, the three-bay wing structure, and the complicated system of bracing struts and wires associated with such a configuration. By the autumn of 1915, the Voisin and the other pushers, with maximum speeds of little more than 70 mph,

had to be relegated to night bombing. The resulting decrease in their effectiveness was largely responsible for the lack of interest shown by the French at this time in strategic bombing.

The first of a new family of Voisins, the Voisin VIII, with longer wings and distinctive streamlined fuel tanks mounted between them, appeared early in 1917. It was powered by a converted lorry engine, the 220 hp in-line Peugeot 8a, which gave it a speed of over 80 mph and a bomb load of 400 lb. Another feature was the replacement of the normal machine-gun with a 37-mm Hotchkiss quick-firing cannon. But the engine did not prove reliable, and the speed was still too low for it to be used for anything other than night bombing.

The Voisin X which appeared at the front early in 1918 had a much better performance and range, able to deliver 660 lb of bombs against a target 150 miles distant and return, but it too had to be confined to night bombing, for which purpose it was painted entirely in black.

Another pusher type of bomber which remained in service until the end of the war was the Bréguet-Michelin. This should have appeared in 1915, when its performance, superior to that of the Voisin, would have made it a useful day-bomber. In a competition with one of the early tractor biplanes produced by Paul Schmitt, the Bréguet-Michelin won hands down and promised to meet the specification for a powerful aeroplane capable of destroying enemy munition factories. But delays were caused by the initial failure of the 220 hp Renault engine, and by the time it appeared on the Western Front, in 1916, the development of fast fighter interceptors meant that it, too, had to be confined to night bombing. Meanwhile, Paul Schmitt had overcome his early difficulties and much was expected of the PS 7 single-engined tractor type. Again, how-

## MARTINSYDE G 100 'ELEPHANT'

**Gross weight:** 2,458 lb **Span:** 38 ft **Length:** 26 ft 6 in **Engine:** 120 hp Beardmore **Armament:** 2 Lewis machine-guns **Crew:** 1 **Speed:** 108 mph at ground level **Ceiling:** 16,000 ft **Range:** 4½ hrs **Bomb load:** 1x230 lb or 4x65 lb

Based on a design for a long-range fighting scout, the Martinsyde 'Elephant' was introduced in 1915 as a bomber because of its ability to carry one of the large British bombs then being developed. It served in the interim before the specially designed bombers became available and was particularly effective in low-level attacks

ever, delays in production held back deliveries until 1917, and although it was used as a day-bomber, it was by then virtually obsolete.

The most successful of the single-engined French bombers, the Bréguet 14, was a remarkable aircraft in its own right, remaining in general service until 1930. It was a tractor type which first flew in November 1916 and was brought into service late in 1917. With a maximum speed of 112 mph at sea level, a service ceiling of 18,000 feet, a range of 435 miles and capable of carrying thirty-two 22-lb bombs, it at last provided the French with a first class day-bomber which could take on fighters on something approaching equal terms. It was built largely of light alloy and was powered by a 300 hp Renault engine which enabled it to climb to 16,500 feet in 39 minutes.

Of the seventeen versions produced, three were used as bombers – the B2 two-seater day-bomber, the BN2 two-seater night-bomber, and the B1 single-seater bomber. The Bréguet 14 enabled the French bombardment squadrons to resume daylight operations on a scale never before possible, and under the command of such men as Vuillemin and de Göys, the bomber crews took over from the fighter pilots as the élite of the French aviation groups. From early 1918 until the end of the war, the Bréguet formations of the First Air Division made daily raids on German military targets. A striking example of the power of mass-bombing was given on 4 June 1918, when a concentration of German troops in a ravine near the forest of Villers-Cotterets was virtually obliterated.

Also in June 1918, the first day-bomber squadron of the American Expeditionary Forces, the 96th Aero, commenced operations with Bréguet 14 B2s supplied three months earlier. The rest of the US bombing units, which flew day-bombers only, were equipped with American-built DH 4s with Liberty engines, one of the most successful British planes of the war. In August 1918 all the American air squadrons at the front were grouped into the Air Service of the First Army, under the command of General W. Mitchell. In 150 American bombing raids before the end of the war, about 140 tons of bombs were dropped.

Early in 1915 the German C-class of two-seater tractor biplanes began to enter service with engines giving up to 180 hp double that of the unarmed B-class machines. The new types were notable for the Parabellum machine-gun mounted in the rear cockpit giving them a 'sting in the tail', and which, for a period, turned the tables against the Allies. The British BE2c observation machine suffered particularly heavy casualties, because it retained the original layout with the pilot in the rear seat, leaving the observer/gunner hemmed in by wings, wires and struts. During the following year a number of the German C-class planes were developed as light bombers and formed into bombing groups (*kampfgeschwader*), but because of the shortage of aircraft they couldn't often be spared for strategic bombing. A raid on the night of 20 July 1916 showed what might have been achieved, when four machines bombed a British ammunition dump near St. Omer, destroying over twenty sheds and some 8,000 tons of ammunition.

### Staff re-organisation
The Battle of the Somme in the summer of 1916 left the Germans even more desperately short of aircraft. Most of them were required to protect and escort the vital reconnaissance machines, leaving none to spare for bombing sorties. The aviation units at this time were divided among the various army groups with little co-ordination between them and this also detracted from their operations. Accordingly, in October 1916, all the units were combined under one command, directly responsible to Army GHQ. General von Hoeppner was appointed commander of the new German Army Air Service with Major Thomsen as his chief of staff. The only aviation force not controlled by this new organisation was the Naval Air Service which continued to come under navy command. While the heavy multi-engined bombers began night bombing raids on both the Eastern and Western fronts, von Hoeppner built up C-class units for daylight sorties against military targets on the Western Front. In November 1917 the designation of the bombing groups was again changed to *Bombengeschwader*.

First of the C-class bombers to enter service was the Aviatik C III, with a maximum speed of 100 mph, a combat range of about 250 miles, and the ability to carry a bomb load of some 200 lb. The Albatros types used were the CIII, CVII and CX, the latter appearing in 1917 with a 260 hp Mercedes engine. All these were notable for their rounded tail units which had a 'fishtail' appearance. From the Rumpler firm came the CI and CIa, both of which were capable of carrying up to 220 lb of small bombs. The LVG C II, first appearing at the end of 1915, was credited with making the first daylight raid on London in November of the following year. During 1917, a CL category of plane was brought into service as a small, lightweight two-seater to undertake fighter escort duties. Two of these types, however, found additional employment in a ground-attack role in support of the infantry, when four or five 22-lb bombs could be dropped and enemy trenches machine-gunned. The Halberstadt CL II made its mark in attacking British troops during the Battle of Cambrai in November 1917. The other type, the Hannover CL IIIa, was unique among single-

**SALMSON II**

**Gross weight:** 2,798 lb **Span:** 38 ft 7 in
**Length:** 27 ft 11 in **Engine:** 260 hp Salmson
**Armament:** 2–3 machine-guns **Crew:** 2
**Speed:** 116 mph at 6,500 ft **Ceiling:** 20,500 ft
**Range:** 3 hrs **Bomb load:** 200 lb

engined aircraft of the period for its biplane tail unit. It was so small and compact in design that Allied pilots often mistook it for a single-seater until, coming up to attack from behind in order to avoid the pilot's forward-firing gun, they were met with an unexpected hail of fire from the observer's machine-gun.

As well as the Caproni heavy bombers, the Italians also produced one of the fastest light bombers of the war, the Ansaldo SVA 5, in the summer of 1917. This single-seater biplane, powered by a 220 hp SPA in-line engine and with twin synchronised Vickers machine-guns mounted on top of the cowling, was originally intended to be a fighter. Although its maximum speed of 136 mph compared favourably with most fighters of the time, its lack of manoeuvrability made it unsuitable for fighting duties. However, it possessed two attributes which made it an excellent day-bomber; a range of over 600 miles, equal to that of the German Giants and almost double that of the Capronis, and the ability to carry a 200-lb bomb load. The SVA 5 entered service as a bomber in February 1918 and quickly established a reputation for itself, enabling the Italians for the first time to carry out long-range strategic raids on cities as far away as Innsbruck, Zagreb, Ljubljana and Friederichshafen. A two-seater version, the SVA9, was brought into service during the last months of the war, and one of these machines, with Gabrielle d'Annunzio as observer, led six SVA 5s on a 625-mile round journey from San Pelagio to Vienna to drop leaflets on the Austrian capital.

Many different types of British aircraft brought into service in the early days of the war were used as bombers, even those originally intended as fighters such as the FE2b two-seat pusher biplane produced by the Royal Aircraft Factory. Although not fitted with interrupter gear, the FE2b achieved considerable success as a fighter, with its two Lewis machine-guns and unobstructed view forward. One of its successes was claimed to be the shooting down of the German Fokker ace, Max Immelmann.

During daylight fighter-reconnaissance missions, eight 20-lb bombs could be carried, and a few aircraft of the type, fitted with

Vickers one-pounder pom-pom guns, were found to be ideal for attacking trains. This kind of low-level ground-attack was primarily a British idea and was to lead later to the development of strike aircraft specifically designed for that purpose. By the end of 1916, however, the FE2b had been outclassed by the new Albatros and Halberstadt single-seat fighters and was used mainly for night bombing when up to three 112-pounders were carried.

British fighters continued to carry small bomb loads until the end of the war, primarily for ground-attack duties in support of the infantry. The two-seat Bristol Fighter, one of the best general purpose aircraft of the war, could carry up to twelve 20-lb bombs in racks under the bottom wing, and even the fast and highly manoeuvrable

SE 5a and Sopwith Camel single-seat fighters could carry four 20-lb bombs. It was as a result of this experience that an experimental version of the Camel was built early in 1918 as an armoured trench fighter, fitted with two downward-firing machine-guns in addition to the one firing forwards, and with a sheet of armoured plate to protect the pilot from ground fire. The first British aircraft to be built specifically for this purpose was the heavily armour-plated Sopwith Salamander which could carry a remarkably heavy bomb load of nearly 650 lb in addition to its two machine-guns. It was built in some numbers but did not see active service before the war ended.

Meanwhile by 1915, the concept of bombing had been accepted, and plans put in hand for multi-engined heavy bombers.

**DH 10 AMIENS Mk III**

**Gross weight:** 9,000 lb **Span:** 65 ft 6 in
**Length:** 39 ft 7 in **Engine:** 2×400 hp Liberty
**Armament:** 4 Lewis machine-guns **Crew:** 3
**Speed:** 116 mph at 6,500 ft **Ceiling:** 16,500 ft
**Range:** 5 hrs **Bomb load:** 900 lb

One of several promising bombers being developed for the RAF, which arrived too late to see operational service before the end of the war. Later used for long-distance air mail services

**DH 4**
**Gross weight**: 3,312 lb **Span**: 42 ft 5 in **Length**: 30 ft 8 in **Engine**: 250 hp Rolls Royce Eagle **Armament**: 1 × ·303 Vickers; 1 × ·303 Lewis **Crew**: 2 **Speed**: 117 mph at 6,500 ft **Ceiling**: 16,000 ft **Range**: 3½ hrs **Bomb load**: 4 × 112 lb

But there was also a need for fast day-bombers which could drop larger bombs than the four 20-pounders carried by most of the adapted fighters. The first answer was the Martinsyde G 102 Elephant which appeared late in 1915, so named because it was a relatively large aeroplane for a single-seater, and designed to have an endurance of 5½ hours. It could carry loads of up to a single 336-lb bomb. While it had a high speed for the time of 104 mph, it was not very manoeuvrable, its ceiling was limited to 16,000 feet and it was used most successfully as a low-level bomber. Another type which could carry the 336-lb bomb was the Farnborough-designed RE7 which came into service early in 1916. This was originally intended as a two or three-seater reconnaissance machine, and the crew were normally

armed only with rifles and pistols. The observer in the front cockpit had a very restricted field of fire, surrounded as he was by bracing struts and wires. The RE7 had a short-lived career, due to the development of fast and better-armed German fighters, but for a while its great weight-lifting capacity made it a useful bomber.

Early in 1917, there appeared in service with the RFC on the Western Front one of the best combat aircraft of the war and certainly the most outstanding day-bomber. This was the DH 4, designed by Geoffrey de Havilland and produced by the Aircraft Manufacturing Company (Airco). It was a straightforward two-seater tractor biplane, but unlike many aircraft coming into service at that time, part of whose construction was of metal, the DH 4 was

built almost entirely of wood. It was fabric-covered, except for the front half of the fuselage, which was covered with plywood, and this improved both its appearance and its strength. There was more than the usual distance between the two cockpits, giving the pilot an excellent view for bombing, while the observer/gunner was far enough back to have a wide field of fire for his Lewis machine-gun. The only drawback was the difficulty in communication between the two crew members during combat. A speaking tube connecting the cockpits on some machines was of little use in view of the noise of the engines and slipstream, and most crews worked out a satisfactory system of hand signals. The pilot was provided with a Vickers machine-gun synchronised by the Constantinesco system. Later models had two forward-firing Vickers and some, built for the RNAS, had two Lewis guns on pillar mountings in the rear cockpit. The normal bomb load was two 230-lb or four 112-lb bombs, carried on racks under the fuselage and lower wings.

The prototype DH 4, first tested in the autumn of 1916, was powered by a 230 hp Beardmore-Halford-Pullinger engine, but this proved troublesome and was soon changed for the excellent 250 hp liquid-cooled Vee-twelve Eagle produced by Rolls-Royce. This gave it a remarkable speed of about 130 mph at 10,000 feet, and it could climb to this height in nine minutes. When the 375 hp Eagle was fitted later, the performance of the DH 4, both in speed (143 mph at sea level) and ceiling (22,000 ft), outclassed all but a very few of the opposing German fighters. In addition to the Western Front, it saw service in Italy, the Aegean, Macedonia and Palestine. It was the only British aircraft to be built in any number by the Americans, who produced nearly 5,000 with Liberty engines and twin forward-firing Marlin guns. Some 600 were in service with the American bomber units in France at the end of the war.

From the moment of its introduction, the DH 4 was successfully used by the RFC for daylight attacks on military targets, while the RNAS used the type mostly for anti-Zeppelin patrols – it was a DH 4 which shot down the Zeppelin L 70 in August 1918. But the real impetus given to British bomber

production came after the Gotha raid on London in June 1917 when more damage was caused than during all the previous Zeppelin raids. It was decided to increase the strength of the RFC from 108 squadrons to 200, most of the new ones to be equipped with bombers to undertake a retaliatory strategic bombing campaign against German cities and industrial targets. Large numbers of the DH 4 were ordered, together with a new version, the DH 9, which had a longer range. This was basically similar to the DH 4, except that the pilot's cockpit was moved aft so that he could communicate more easily with his observer.

The DH 9 seemed to offer all the advantages of its predecessor and more, for it was expected to carry a heavier bomb load as well. But there were development problems with the BHP engine which had to be de-rated to 230 hp and further modified to facilitate production, at which point it was re-named the Siddeley Puma. The resulting loss of performance – the DH 9 could barely reach 13,000 feet with a full bomb and petrol load – rendered it considerably inferior to the older Rolls-Royce powered DH 4. By the time this was fully appreciated it was already being produced in large numbers and brought into service with the Independent Force. A marked improvement was achieved with the DH 9A, powered mostly by the 400 hp Liberty engine, although a few were fitted with the Rolls-Royce Eagle. This version could carry a maximum bomb load of 660 lb, and a normal load of two 230-lb bombs, at 17,000 feet without loss of height. It had a good enough performance to carry out daylight raids without escort, but unfortunately there were difficulties in obtaining the Liberty engine. Only four units had been re-equipped with 9As by the end of the war and they were only in active service for about two months.

The ill-fated DH 9 therefore had to bear the brunt of daytime operations with the Independent Force, with some pilots making as many as six sorties a day. Losses were high because of the reduced speed and ceiling, as well as the unreliability of the engines. During one raid against Mainz in July 1918, seven out of twelve DH 9s were shot down by German fighters and three had to turn back with engine failure. The day-bombers paid the highest price for the strategic bombing offensive against Germany; casualties among the four de Havilland squadrons were 25 killed, 178 missing and 58 wounded, with over 100 aircraft brought down over enemy territory and 201 wrecked in crashes on the Allied side of the lines.

This was a very different story from the early months of the war. It is fair to say that the bomber came of age during the First World War, proving all fears of its destructiveness to be well-founded. But defences against the bomber – fighters and anti-aircraft guns – had also developed. Some of the highest losses in men and aircraft were sustained by the bomber squadrons, especially those whose task it was to undertake precision bombing by day.

Light bombers on the production line of a French factory during 1918

## SHORT BOMBER

**Gross weight**: 6,800 lb **Span**: 85 ft **Length**: 45 ft **Engine**: 250 hp Rolls Royce Eagle **Armament**: 1 Lewis machine-gun **Crew**: 2 **Speed**: 77·5 mph **Ceiling**: 9,500 ft **Range**: 6 hrs **Bomb load**: 920 lb

A landplane development of the Short 184 seaplane, brought into service with the RNAS in 1916 to initiate the concept of strategic bombing behind the front line

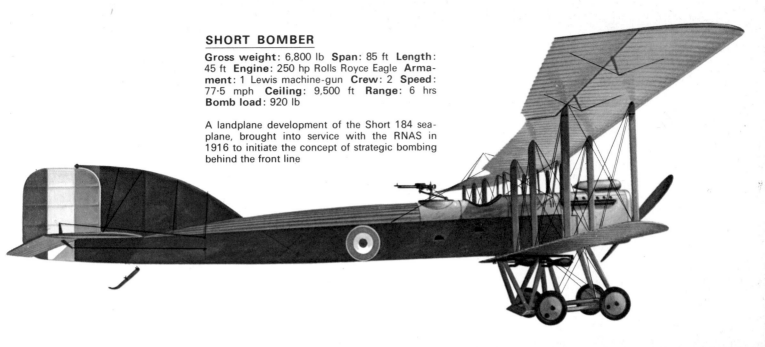

# BETWEEN THE WARS

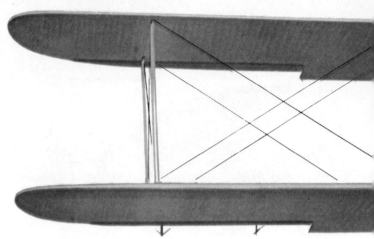

In the post-war years, de-militarisation and financial stringency put many obstacles in the way of the bomber's development. Private designs proliferated, but in the absence of any official policy, progress was haphazard and unco-ordinated. Some remarkable aircraft were produced nonetheless, and all the bombers used operationally during the Second World War were being developed by the mid-1930s

In many ways the most advanced plane of the First World War was the Junkers D 1 which embodied many of the features that were to become generally accepted in the 1930s. It was a cantilever monoplane with an all-metal airframe covered with thin sheet iron, resulting in such a strong structure that no struts or bracing wires were required to support the wings. Although the D1 was first flown in 1916, only a few were built during the war because of production problems. Under wartime conditions, aircraft were needed in such large numbers – France produced some 60,000, Britain 53,000, Germany 48,000, Italy 20,000 and the USA 12,000 – that preference often had to be given to simplicity of construction over ingenuity of design.

The large number of aircraft produced meant that there was a considerable surplus when the war ended, except in Germany. This ultimately turned out to Germany's advantage, since she had to build up a new air force from scratch, without the encumbrance of older types of aircraft. Meanwhile, a number of the large Allied bombers such as the DH 10 Amiens and the Il'ya Muromets, were used to pioneer civil transportation, opening up passenger, freight and mail routes around the world. Expenditure on military aircraft was severely cut back after the 'war to end wars', and it was not long before aircraft specifically designed as civil airliners had taken a technical lead over the bomber types from which they were originally developed.

Almost all the bombers used during the Second World War were the result of a technological revolution which took place in the early 1930s. This was led by the development of airliners in the USA and saw the increasing use of all-metal cantilever monoplanes with stressed alloy skins, retractable undercarriages, flaps, constant-speed propellers with variable pitch, and with radios among more efficient navigational aids. But for most of the period between the wars however, the biplane had remained the predominant type, and the first bombers to

be produced after the First World War continued to be constructed mainly of wood.

In Britain, for instance, the twin-engined Vickers Virginia, the main RAF heavy night bomber from 1924 to 1937, was of conventional wood and fabric construction, as was the Martin MB-2, in service from 1919 to 1927. This, the first American-designed bomber, was intended to improve on the performance of the Handley Page 0/400 then being built under licence in the USA. A similar French type was the Farman F 60 series, which appeared in the closing stages of the First World War, and remained in service until 1928, while a civil version developed from the Goliath was used with considerable success by French commercial fleets during the same period. In Italy, the large multi-engined Caproni bombers were replaced by the single-engined Caproni Ca 73 series, notable for their unusual inverted sesquiplane arrangement. Another single-engined bomber of conventional wood and fabric construction was the de Havilland R1, a Soviet version of the remarkable DH 9A – the first Russian aircraft to be mass-produced – which served from 1923 to 1935.

## Limited service

The gradual change from wood to metal construction for military aircraft took place almost universally in the mid-1920s. Although the Avro Aldershot, one of the first new bombers to be designed for the RAF after the First World War, had only a single engine, it was intended as a heavy night bomber and could in fact carry a bomb load of 2,000 lb, equal to that of most twin-engined types of the period. But the Air Staff decided against the idea of single-engined heavy bombers and the Aldershot saw only limited service.

A similar policy decision was taken in the USA where the Huff-Daland bomber (the name was later changed to Keystone), brought into service to replace the Martin MB-2 in 1927, was changed from a single

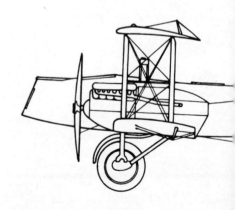

## AVRO ALDERSHOT

**Gross weight**: 10,950 lb **Span**: 68 ft **Length**: 45 ft **Engine**: 650 hp Rolls Royce Condor III **Armament**: 1 Lewis machine-gun **Crew**: 3 **Speed**: 110 mph at ground level **Ceiling**: 11,500 ft **Range**: 652 miles **Bomb load**: 2,000 lb

Entered RAF service in 1924 with a bomb-load equal to that of many twin-engined bombers

## VICKERS VIRGINIA

**Gross weight**: 12,467 lb **Span**: 86 ft 6 in **Length**: 50 ft 7 in **Engine**: 2×450 hp Napier Lion **Armament**: 2–4×·303 machine-guns **Crew**: 4 **Speed**: 104 mph at ground level **Ceiling**: 15,530 ft **Range** 985 miles **Bomb load**: 3,000 lb

The main heavy night bomber in service with the RAF from 1924 to 1937

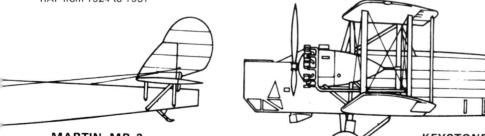

## MARTIN MB-2

**Gross weight**: 12,064 lb **Span**: 74 ft 2 in **Length**: 42 ft 8 in **Engine**: 2×420 hp Liberty **Armament**: 5×0·3-in machine-guns **Crew**: 4 **Speed**: 99 mph at ground level **Ceiling**: 8,500 ft **Range**: 558 miles **Bomb load**: 2,000 lb

Designed by Glenn L. Martin, one of America's leading air pioneers, the MB-2 formed the bulk of the US Army's bomber force in the early 1920s

## KEYSTONE B-4A

**Gross weight**: 13,209 lb **Span**: 74 ft 8 in **Length**: 48 ft 10 in **Engine**: 2×575 hp R-1860-7 **Armament**: 3×0·3 machine-guns **Crew**: 5 **Speed**: 121 mph at ground level **Ceiling**: 14,000 ft **Range**: 855 miles **Bomb load**: 2,500 lb

One of the final production series, for which orders were placed in 1931, of the long line of Huff-Daland/Keystone bombers

A postcard of the period showing the Curtiss Army biplane winning the Schneider Trophy in 1925. This was the last occasion on which a biplane won the trophy. The engine was similar to that adapted by Richard Fairey for use in the Fairey Fox, the fastest bomber of its day (see below)

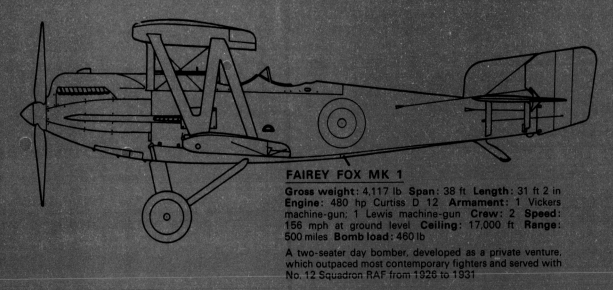

### FAIREY FOX MK 1

**Gross weight:** 4,117 lb **Span:** 38 ft **Length:** 31 ft 2 in **Engine:** 480 hp Curtiss D 12 **Armament:** 1 Vickers machine-gun; 1 Lewis machine-gun **Crew:** 2 **Speed:** 156 mph at ground level **Ceiling:** 17,000 ft **Range:** 500 miles **Bomb load:** 460 lb

A two-seater day bomber, developed as a private venture, which outpaced most contemporary fighters and served with No. 12 Squadron RAF from 1926 to 1931

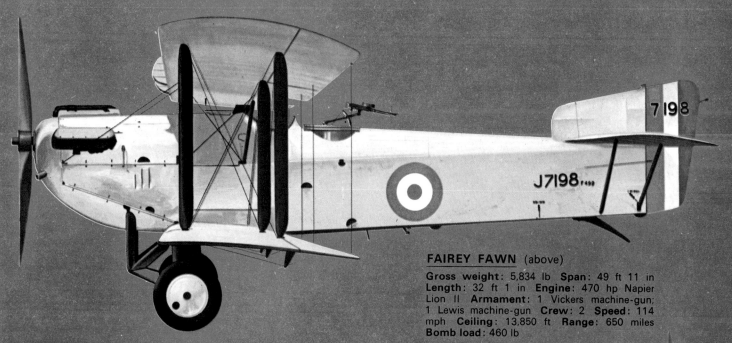

## FAIREY FAWN (above)

**Gross weight:** 5,834 lb **Span:** 49 ft 11 in
**Length:** 32 ft 1 in **Engine:** 470 hp Napier
Lion II **Armament:** 1 Vickers machine-gun;
1 Lewis machine-gun **Crew:** 2 **Speed:** 114
mph **Ceiling:** 13,850 ft **Range:** 650 miles
**Bomb load:** 460 lb

## MITSUBISHI B2M (right)

**Gross weight:** 7,937 lb **Span:** 49 ft 11 in
**Length:** 33 ft 8 in **Engine:** 600 hp Mitsubishi-
built Hispano-Suiza **Armament:** 2×7·7-mm
machine-guns **Crew:** 3 **Speed:** 132 mph
**Ceiling:** 14,300 ft **Range:** 1,100 miles **Bomb
load:** 1×1,764 lb Torpedo

Built for the Japanese Navy Air Force to a British
Blackburn design, and in service from 1932 to 1937

## SOPWITH SALAMANDER

**Gross weight:** 2,512 lb **Span:** 31 ft 3 in
**Length:** 19 ft 6 in **Engine:** 230 hp B.R. **Arma-
ment:** 2×·303 Vickers machine-guns **Crew:** 1
**Speed:** 125 mph at 500 ft **Ceiling:** 13,000 ft
**Range:** 3 hrs **Bomb load:** NA

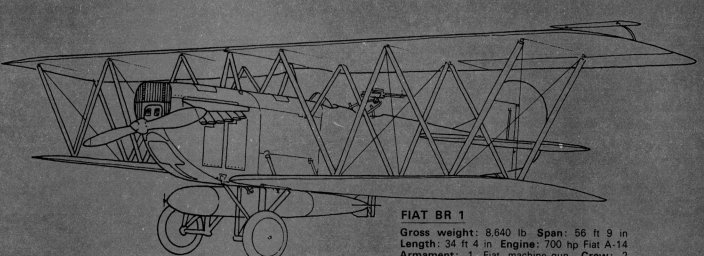

## FIAT BR 1

**Gross weight:** 8,640 lb **Span:** 56 ft 9 in
**Length:** 34 ft 4 in **Engine:** 700 hp Fiat A-14
**Armament:** 1 Fiat machine-gun **Crew:** 2
**Speed:** 153 mph **Ceiling:** 18,000 ft **Range:**
404 miles **Bomb load:** 1,000 lb

First of the BR (*Bombardamento Rosatelli*) series
produced in the early 1920s for the Italian Air
Force

engine to a twin-engined type. One reason was safety, in that a multi-engined plane could keep flying even with one engine out of action; another was that by placing the engines on or between the wings, the nose was left clear, giving the gunner and/or bomb aimer a better field of vision.

There was still a need for light two-seater day-bombers powered by a single engine, although such a classification became obsolete in the 1930s, when many day-bombers were built with two engines. The first new day-bomber to be received by the RAF after the war was the Fairey Fawn, a wood and fabric biplane, in service from 1923 to 1926, whose performance was actually inferior to the wartime DH 9A. This was replaced by the Hawker Horsley which remained in service until 1934. The Horsley, of mixed wood and metal construction, had an excellent load-carrying ability – either 600 lb of bombs or a single 2,150-lb torpedo in the torpedo-carrier version – and was almost as manoeuvrable as the fighters of those days.

But undoubtedly the best of the light bombers of wood construction was the Fairey Fox. Part of the reason for the poor performance of the Fairey Fawn had been the stringent official specifications to which it had been built. Richard Fairey decided to build a much faster bomber as a private venture in the hope of obtaining a production order. In the early 1920s he had seen an American Curtiss seaplane win the Schneider Trophy at a speed of over 177 mph and he was so impressed that he obtained the right to use certain features of its design, particularly the slim Curtiss D 12 in-line engine which he built under licence as the Fairey Felix. The Fairey Fox, which appeared in 1926, was one of the most beautiful biplanes ever built and not only the fastest bomber of its day, with a speed of 156 mph, but faster and more manoeuvrable than most fighters. In fact, only one RAF squadron was equipped with this outstanding aircraft, and in the event, production models were powered by the Rolls-Royce F Kestrel. But it established a line of bombers which saw much use in Belgium, where both Kestrel and Hispano-Suiza engines were installed.

### Japan's first
The first light bomber to be built in Japan for the Japanese Army Air Force was the Mitsubishi 2MB1, introduced in 1927, while two years later the Kawasaki Type 88 reconnaissance biplane was adapted to perform light bombing duties.

The transition in bomber design from wood to metal, though still fabric-covered to begin with, took place in the mid-1920s, with the final change to all-metal stressed-skin airframes beginning by the early 1930s. Before describing the introduction of these types, however, consideration has to be given to the other vital aspect of design, namely engine development.

In 1917 it had been decreed that the dominant British engine should be the ABC Dragonfly radial, a nine-cylinder air-cooled unit of modern design which was supposed to give over 400 hp. Had the war continued, this would have led to a crisis in British aviation, because by the time it was realised that it was a complete technical failure the Dragonfly was being produced on a large scale. It was not until after 1920 that the fine Rolls-Royce Eagle was backed up by two other outstanding engines, the Napier Lion and Bristol Jupiter.

The Lion, designed by Rowledge, had three banks each containing four water-cooled cylinders, the so-called 'W' or broad-arrow arrangement. It was rigid and refined, and though used at 450 hp in bombers such as the Handley Page Hyderabad (the last twin-engined bomber of wooden construction to be used by the RAF except for the Mosquito of the Second World War), it gave 1,320 hp in racing form. Roy Fedden's Jupiter, fitted originally in place of the Eagle in the Vickers Vimy, was the world-dominant engine of the 1920s, and was built under licence in no less than sixteen countries. Starting in 1918 as a 375–400 hp nine-cylinder radial, it grew to more than 500 hp and was then developed into the Pegasus, rated at more than 1,000 hp by 1938. Its history was one of strenuous effort to improve a basically sound mechanical design whilst introducing geared drive, supercharging, and better forms of low-drag aircraft installation.

### Britain's prejudice
The Jupiter/Pegasus family took full advantage of the fact that there is always a great difference in temperature between an air-cooled cylinder head and the slipstream, whereas in the hottest countries there is less difference between the temperature of the air and the water passing through a radiator. Conversely, in the coldest climates a water-cooled engine could freeze. Thus many British bombers designed for water-cooled engines were sold in very hot or very cold countries with air-cooled radials instead. But in Britain there was a prejudice in favour of the liquid-cooled engine, partly because of its success when used in Schneider Trophy racing. The belief grew that the supposedly 'streamlined' vee-12, exemplified by the 450–640 hp Rolls-Royce Kestrel and later by the 1,000 hp Merlin, was more efficient than the bluff-looking radial. No such belief was harboured in the USA, even though there was no lack of American in-line engines.

The vee-12 Hispano family, of 600–1,200 hp, was made in vast numbers, not only in France and other European countries but also in the Soviet Union where it was practically the standard engine for the most powerful military aircraft until after 1941. France's Lorraine Dietrich and Italy's Fiat and Isotta-Fraschini engines served to underline the European reliance on liquid-cooled vee types. In Germany the new *Luftwaffe* was born around the BMW VI (used in the prototypes of such bombers as the Dornier Do 17 and Heinkel He 111), some radials, and the unique opposed-piston diesel two-stroke, developed painstakingly by Junkers in the mistaken belief that it would give greater efficiency and longer range.

In the USA, in spite of extreme federal parsimony, the Curtiss D 12 was developed into the 575 hp Conqueror by 1926 and powered bombers well into the 1930s. Packard had a share of the business with a series of big vee-12 engines, some of which gave no less than 800 hp. But the Huff-Daland bomber, fitted with these big engines, was replaced by a model equipped with twin 400 hp Liberty engines, mounted upside down to raise the thrust axes to the correct levels. Increasingly, the air-cooled radial became dominant in the USA, especially after the emergence of the superb 425 hp Wasp, made by the newly-formed Pratt & Whitney company, late in 1925.

By 1930 nearly all American bombers were powered by various single-row and twin-row radials built by Pratt & Whitney and the Wright company. By 1937 the major engines of the future were seen to be the P & W R-1830 (radial, 1,830 cubic inches capacity) Twin Wasp, giving 1,000 hp, and the similarly powered Wright R-1820 Cyclone which, unlike the 14-cylinder Twin Wasp, had a single row of nine big cylinders. In 1937 the Cyclone entered combat service in the Boeing B-17 Fortress, the first really successful heavy bomber, and the Douglas B-18 derived from the DC 3. All the big American radials had one inlet and one exhaust valve in each cylinder head. They had geared drives and General Electric was nearing final success in its 20-year effort to perfect a turbo-supercharger spun by the white-hot exhaust gas. This was to prove of vital importance in the Second World War.

Meanwhile, Fedden's team at Bristol was looking keenly at the American radials which in turn owed so much to their own Jupiter engine. Fedden was daunted by the mechanical complexity of trying to devise valve gear for a two-row radial engine with four valves in each head. He had long used two inlet and two exhaust valves, giving better 'breathing' than the American engines, and was reluctant to halve the number. At the same time, though his Mercury (used in the Blenheim) and Pegasus engines were by 1930 showing great promise for a wide range of future aircraft, he wanted to look much further ahead. After extensive experiments with several schemes, he made the bold decision to try to develop a successful sleeve-valve engine. This type had been used in various forms since the turn of the century, but had never been completely successful. It nearly eluded his own team too, and vast effort and expense was needed. The problem lay not so much in perfecting the engine, but in making the engine a standard production type with interchangeable sleeves, instead of a hand-built engine

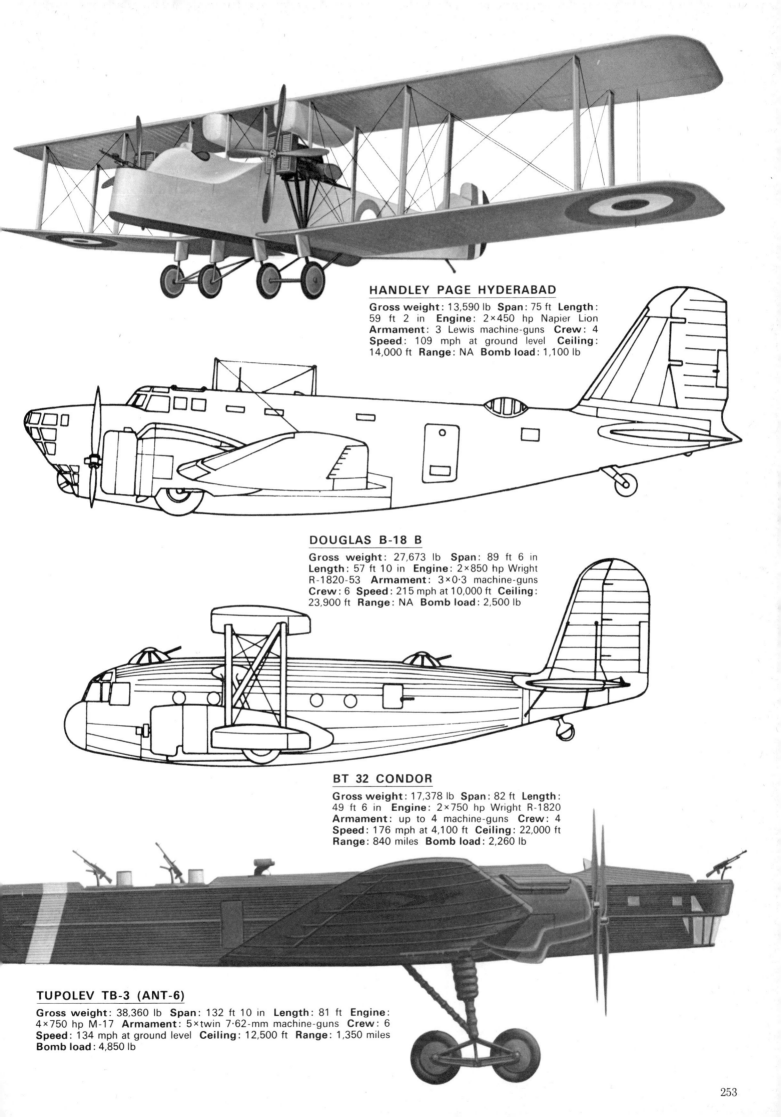

## HANDLEY PAGE HYDERABAD
**Gross weight**: 13,590 lb **Span**: 75 ft **Length**: 59 ft 2 in **Engine**: 2×450 hp Napier Lion **Armament**: 3 Lewis machine-guns **Crew**: 4 **Speed**: 109 mph at ground level **Ceiling**: 14,000 ft **Range**: NA **Bomb load**: 1,100 lb

## DOUGLAS B-18 B
**Gross weight**: 27,673 lb **Span**: 89 ft 6 in **Length**: 57 ft 10 in **Engine**: 2×850 hp Wright R-1820-53 **Armament**: 3×0·3 machine-guns **Crew**: 6 **Speed**: 215 mph at 10,000 ft **Ceiling**: 23,900 ft **Range**: NA **Bomb load**: 2,500 lb

## BT 32 CONDOR
**Gross weight**: 17,378 lb **Span**: 82 ft **Length**: 49 ft 6 in **Engine**: 2×750 hp Wright R-1820 **Armament**: up to 4 machine-guns **Crew**: 4 **Speed**: 176 mph at 4,100 ft **Ceiling**: 22,000 ft **Range**: 840 miles **Bomb load**: 2,260 lb

## TUPOLEV TB-3 (ANT-6)
**Gross weight**: 38,360 lb **Span**: 132 ft 10 in **Length**: 81 ft **Engine**: 4×750 hp M-17 **Armament**: 5×twin 7·62-mm machine-guns **Crew**: 6 **Speed**: 134 mph at ground level **Ceiling**: 12,500 ft **Range**: 1,350 miles **Bomb load**: 4,850 lb

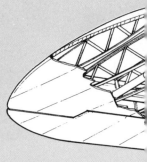

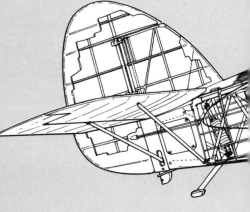

## HAWKER HART (above right)

**Gross weight:** 4,554 lb  **Span:** 37 ft 3 in
**Length:** 29 ft 4 in  **Engine:** 525 hp Rolls Royce
Kestrel  **Armament:** 1 Vickers machine-gun;
1 Lewis machine-gun  **Crew:** 2  **Speed:** 184
mph at 5,000 ft  **Ceiling:** 21,350 ft  **Range:**
470 miles  **Bomb load:** 520 lb

Section drawing of the RAF's standard light day
bomber from 1930 to 1939. Its performance was
better than most contemporary single-seater
fighters. Used especially for colonial policing
duties, for which many variants were produced

<u>Above</u>: armourers working on Browning and
Lewis guns, aligning the gun sights

<u>Centre</u>: aligning gun sight

<u>Left</u>: apprentices receiving instruction on bomb
preparation

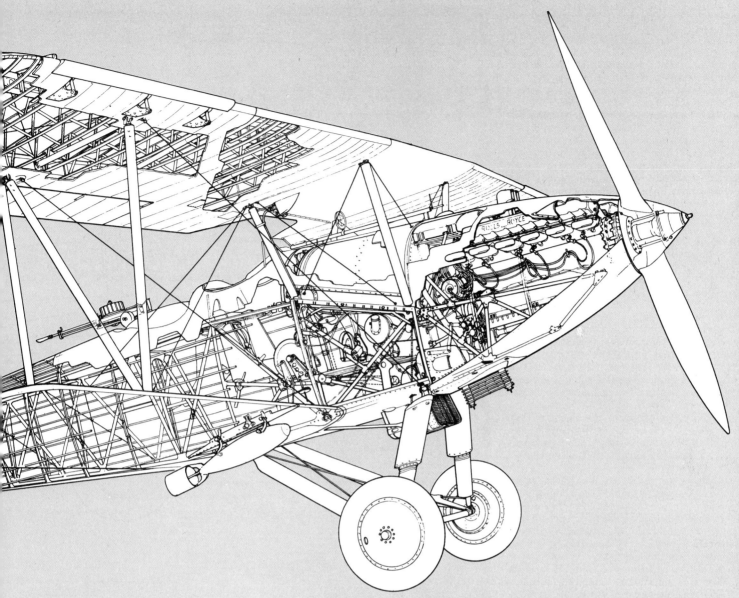

with each set of sleeves individually matched to its own set of cylinders and pistons.

In most of the aircraft-building countries, there were numerous companies making aero engines. Their efforts to make a living resulted in a profusion of engine types, when it would probably have been more cost/effective to have concentrated on one or two designs in each broad power class. In the first half of the 1930s British bomber design would not have been notably handicapped if the Bristol Pegasus had been specified universally and all other engines in the 500–900 hp range had been cancelled. But no such decision was taken, and a diverse profusion of engines resulted, despite the fact that until 1935, when Hitler's announcement revealed the *Luftwaffe* to be among the foremost of the world's air forces, most governments cut bomber procurement to a minimum.

The atmosphere of economy largely explains why the big bomber with three, four, or even more engines, common in 1916–18, was almost non-existent in 1919–1938. Only in the latter year was a production order placed for an American four-engined bomber (the first B-17 Fortresses), and the RAF four-engined 'heavies' did not come into use until more than two years after the war began. The only nation to use fleets of four-engined bombers was the Soviet Union, whose Tupolev TB-3 monoplane came into service in 1931. This massive aircraft, powered by four Hispano-Suiza vee-12 engines, had a greater loaded weight than any contemporary landplane. Experiments were even conducted into carrying two parasite fighters on the wings. The TB-3 and its derivatives were sufficiently reliable to play a major role in early Russian polar exploration in the era before the Second World War.

## More metal structures

It was the use of metal in place of wood that gave large aircraft sufficient strength to enable them to be constructed in monoplane form. In France, the metal preferred by most designers was light alloy duralumin. This was used in the construction of the twin-engined Lioré and Olivier LeO 12, first exhibited at the Salon de l'Aéronautique in 1924 and later taken into squadron service on an experimental basis. Few machines of the type were actually built but a widely used development was the three-seat LeO 20, also built of duralumin, which equipped many French night bomber squadrons from 1927–1937, rivalling the Farman F 160 to 168 series. One of the most successful metal biplanes however was the single-engined Bréguet 19 two-seat light bomber which remained in service from 1925 for some fifteen years. It was also used by many foreign air forces, and built under licence in Belgium, Greece, Japan, Spain and Yugoslavia.

Another single-engined biplane of light alloy construction, which appeared in 1928 was the Amiot 122 BP3, the latter designation showing it to be a bomber-escort three-seater. The Amiot 122 had the handling characteristics of a light single-engined aircraft with the load-carrying ability of larger twin-engined types, and remained in service in various forms until 1935. Two other excellent single-engined biplane light bombers were the Czech Aero A11 and the Dutch Fokker CV-C, both introduced into service in 1923. The CV-C, one of the best combat aircraft ever designed by Anthony Fokker, had a welded steel-tube fuselage and wooden wings, and remained operational until the late 1930s.

The first British heavy bomber of metal construction was the Handley Page Hinaidi, a 1929 development of the Handley Page Hyderabad. The Hinaidi was replaced in 1933 by the Handley Page Heyford whose outstanding feature was the attachment of the fuselage to the upper instead of the lower wing. It also had a rotatable ventral turret which could be drawn up into the fuselage when not in use, its cylindrical shape quickly giving rise to the nickname 'dustbin'. These Handley Page twin-engined biplanes were built of a steel-tube structure with internal wire bracing. Another metal twin-engined biplane which came into service in 1928 was the Boulton Paul Sidestrand, but instead of being a night bomber, it was sufficiently fast and manoeuvrable to be used for daytime duties and was designated as the RAF's first medium bomber. It was replaced in 1934 by an improved version, the Boulton Paul Overstrand, which could carry a heavier bomb-load and was the first British bomber to have a power-operated enclosed gun turret in the nose. Apart from these two, British day-bombers were all single-engined two-seaters until the arrival of monoplanes in the late 1930s.

The usual metal in the early composite or metal-framed bombers was high-tensile steel, and there was a considerable reluctance on the part of designers to use light alloys. The steel was used both as tube and as strip, often welded at the joints. When aluminium alloys were brought in there was a great difference of opinion as to how they should best be employed. Sydney Camm, at the Hawker company, devised a patented form of 'bulb flange' in the shape of a tube assembled from sections of strip rolled to particular profiles. These flanges were then riveted to aluminium-alloy sheets to serve as spar booms – the strong top and bottom of the wing spars – in the way that later aircraft used booms of much thicker angle and T-sections.

Camm's fuselages were typical in having either circular-section tube with the sides flattened at the joints or else square-section tube throughout. The flat sides could then fit snugly against heavy bolted or riveted plates which were added to reinforce the main joints. Ways were also found of making strong streamlined struts from hollow light-alloy sections, though high-tensile steel wires were invariably still needed for bracing and for 'rigging' the structure (adjusting the tensions in different wires to obtain exactly the desired shapes, wing angles and tail incidence). It was because this new form of metal construction was easy to maintain in operational conditions that Hawker won a competition in the late 1920s when the Air Ministry decided to have a high-performance all-metal aircraft.

The Hawker Hart remained the standard RAF light day-bomber from 1930 to 1937, with a speed of 175 mph and a better all-round performance than most single-seat fighters of the period. The last RAF biplane bomber was the Hawker Hind. This type was in service from 1935 to 1938 as an interim replacement for the Hart, until the monoplane Fairey Battles and Bristol Blenheims began to enter service in RAF Bomber Command in 1937.

In Japan, the last biplane bomber to be produced for the Japanese Army Air Force was the Kawasaki Ki-3, a single-engined two-seater which entered service in 1933 as the Type 93 light bomber. By then, however, advances were being made in the production of bombers in monoplane form.

### The big question

It was during the 1920s and 1930s that aircraft design became something of an exact science, though questions remained which caused endless arguments: should a bomber have air-cooled radial or liquid-cooled vee engines? Should it be a fabric-covered biplane or a fabric-covered monoplane – or, boldest of all, an all-metal stressed-skin monoplane? Fabric was used as the covering for the RAF's first twin-engined cantilever monoplane, the low-wing Fairey Hendon which entered service in 1936, and the Caproni Ca 101 and Ca 111 high-wing monoplanes, used extensively during the Italian campaign in Ethiopia. Some, such as the Handley Page Harrow and the Dornier Do 11, 13 and 23 bombers,

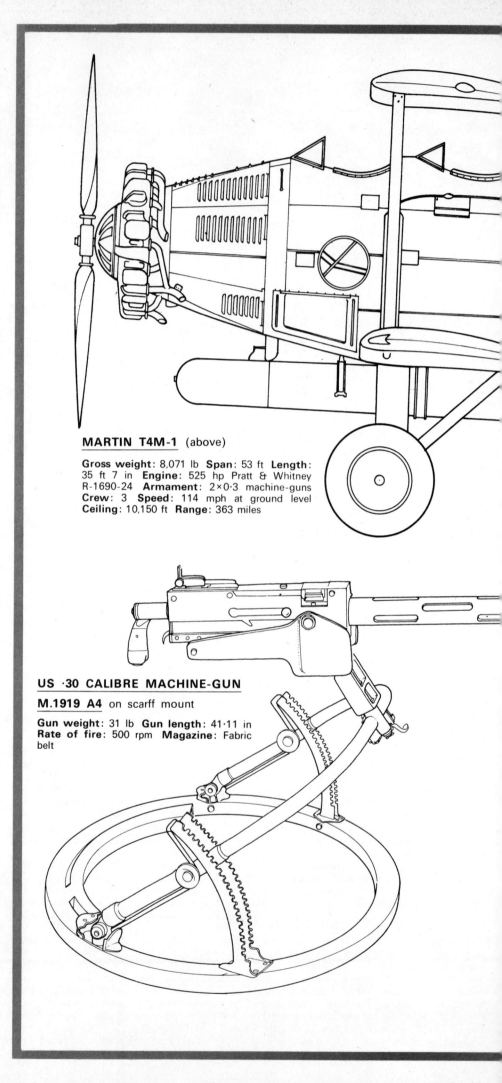

**MARTIN T4M-1** (above)

Gross weight: 8,071 lb  Span: 53 ft  Length: 35 ft 7 in  Engine: 525 hp Pratt & Whitney R-1690-24  Armament: 2×0·3 machine-guns  Crew: 3  Speed: 114 mph at ground level  Ceiling: 10,150 ft  Range: 363 miles

**US ·30 CALIBRE MACHINE-GUN**

**M.1919 A4** on scarff mount

Gun weight: 31 lb  Gun length: 41·11 in  Rate of fire: 500 rpm  Magazine: Fabric belt

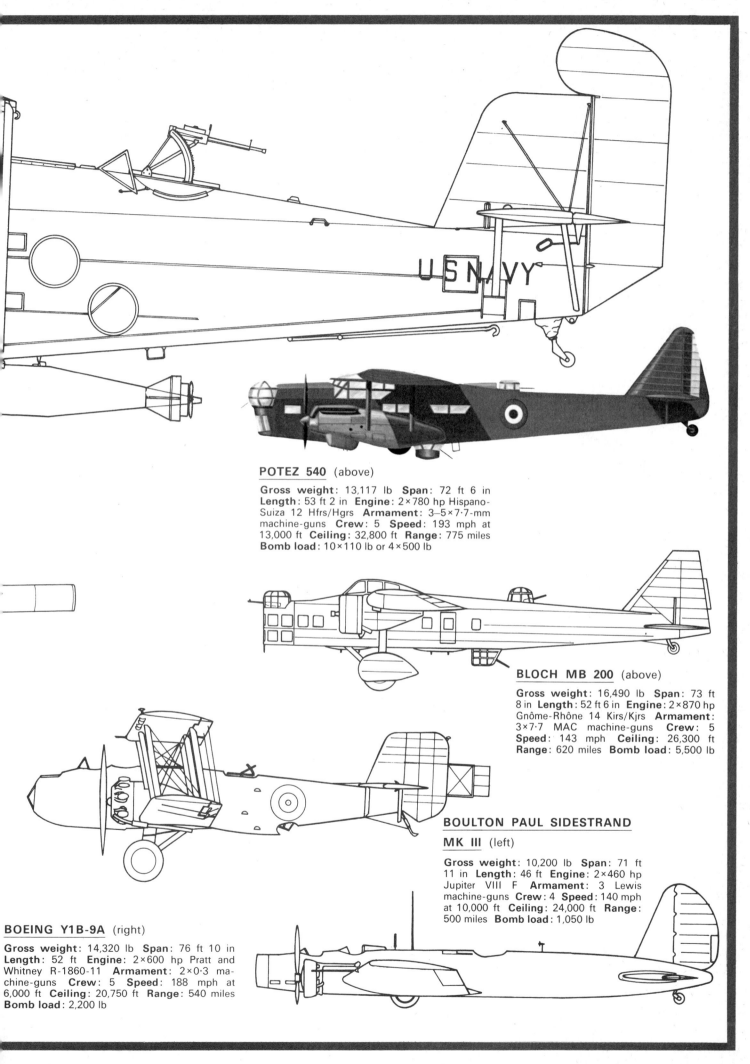

**POTEZ 540** (above)

**Gross weight**: 13,117 lb  **Span**: 72 ft 6 in
**Length**: 53 ft 2 in  **Engine**: 2×780 hp Hispano-
Suiza 12 Hfrs/Hgrs  **Armament**: 3–5×7·7-mm
machine-guns  **Crew**: 5  **Speed**: 193 mph at
13,000 ft  **Ceiling**: 32,800 ft  **Range**: 775 miles
**Bomb load**: 10×110 lb or 4×500 lb

**BLOCH MB 200** (above)

**Gross weight**: 16,490 lb  **Span**: 73 ft
8 in  **Length**: 52 ft 6 in  **Engine**: 2×870 hp
Gnôme-Rhône 14 Kirs/Kjrs  **Armament**:
3×7·7 MAC machine-guns  **Crew**: 5
**Speed**: 143 mph  **Ceiling**: 26,300 ft
**Range**: 620 miles  **Bomb load**: 5,500 lb

**BOULTON PAUL SIDESTRAND**

**MK III** (left)

**Gross weight**: 10,200 lb  **Span**: 71 ft
11 in  **Length**: 46 ft  **Engine**: 2×460 hp
Jupiter VIII F  **Armament**: 3 Lewis
machine-guns  **Crew**: 4  **Speed**: 140 mph
at 10,000 ft  **Ceiling**: 24,000 ft  **Range**:
500 miles  **Bomb load**: 1,050 lb

**BOEING Y1B-9A** (right)

**Gross weight**: 14,320 lb  **Span**: 76 ft 10 in
**Length**: 52 ft  **Engine**: 2×600 hp Pratt and
Whitney R-1860-11  **Armament**: 2×0·3 ma-
chine-guns  **Crew**: 5  **Speed**: 188 mph at
6,000 ft  **Ceiling**: 20,750 ft  **Range**: 540 miles
**Bomb load**: 2,200 lb

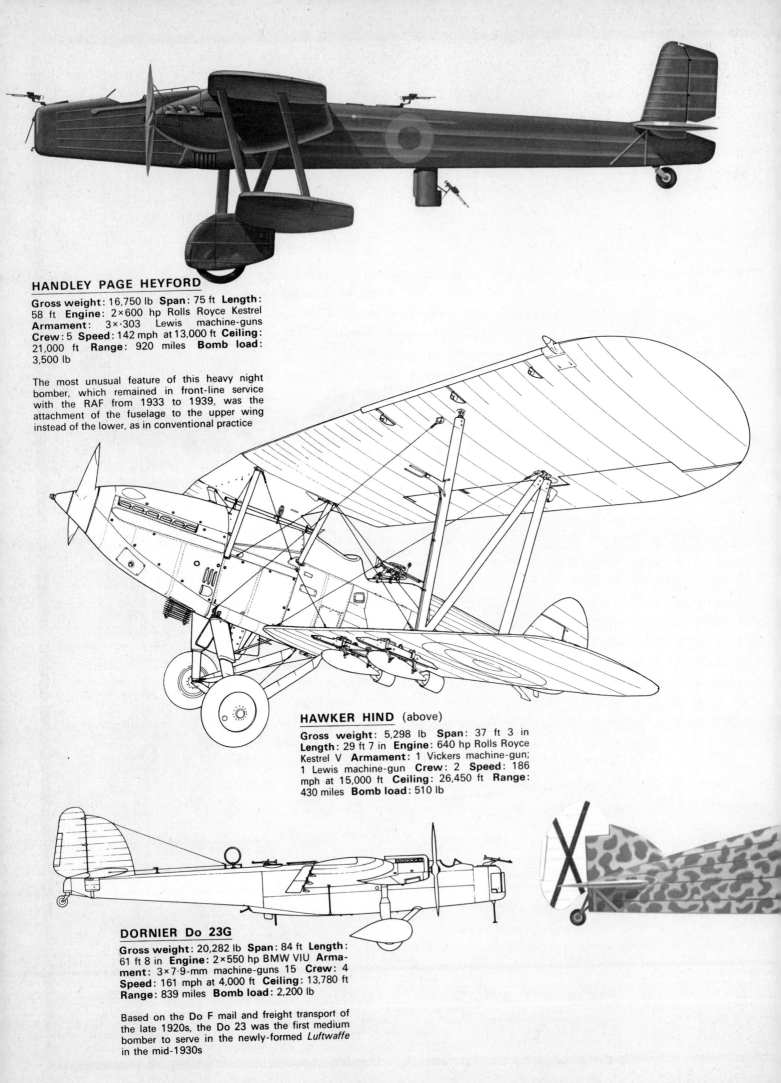

## HANDLEY PAGE HEYFORD

**Gross weight:** 16,750 lb **Span:** 75 ft **Length:** 58 ft **Engine:** 2×600 hp Rolls Royce Kestrel **Armament:** 3×·303 Lewis machine-guns **Crew:** 5 **Speed:** 142 mph at 13,000 ft **Ceiling:** 21,000 ft **Range:** 920 miles **Bomb load:** 3,500 lb

The most unusual feature of this heavy night bomber, which remained in front-line service with the RAF from 1933 to 1939, was the attachment of the fuselage to the upper wing instead of the lower, as in conventional practice

## HAWKER HIND (above)

**Gross weight:** 5,298 lb **Span:** 37 ft 3 in **Length:** 29 ft 7 in **Engine:** 640 hp Rolls Royce Kestrel V **Armament:** 1 Vickers machine-gun; 1 Lewis machine-gun **Crew:** 2 **Speed:** 186 mph at 15,000 ft **Ceiling:** 26,450 ft **Range:** 430 miles **Bomb load:** 510 lb

## DORNIER Do 23G

**Gross weight:** 20,282 lb **Span:** 84 ft **Length:** 61 ft 8 in **Engine:** 2×550 hp BMW VIU **Armament:** 3×7·9-mm machine-guns 15 **Crew:** 4 **Speed:** 161 mph at 4,000 ft **Ceiling:** 13,780 ft **Range:** 839 miles **Bomb load:** 2,200 lb

Based on the Do F mail and freight transport of the late 1920s, the Do 23 was the first medium bomber to serve in the newly-formed *Luftwaffe* in the mid-1930s

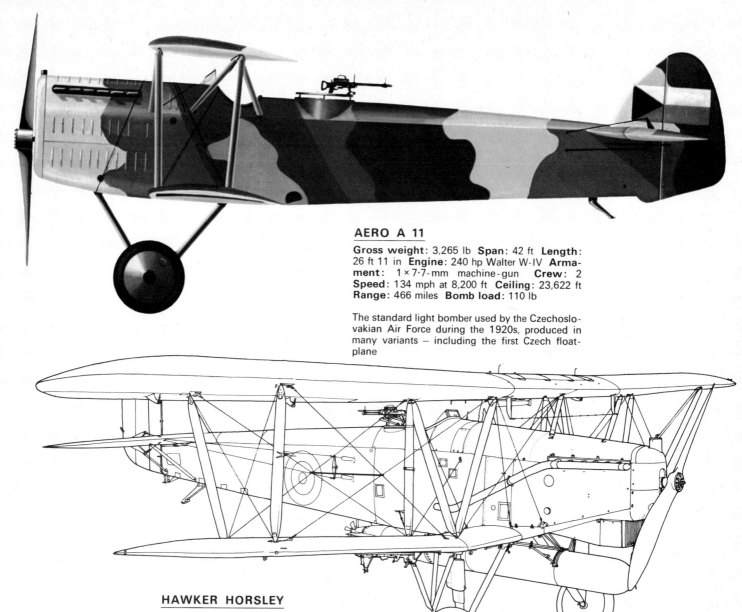

## AERO A 11

**Gross weight:** 3,265 lb **Span:** 42 ft **Length:** 26 ft 11 in **Engine:** 240 hp Walter W-IV **Armament:** 1 × 7·7-mm machine-gun **Crew:** 2 **Speed:** 134 mph at 8,200 ft **Ceiling:** 23,622 ft **Range:** 466 miles **Bomb load:** 110 lb

The standard light bomber used by the Czechoslovakian Air Force during the 1920s, produced in many variants – including the first Czech floatplane

## HAWKER HORSLEY

**Gross weight:** 7,800 lb **Span:** 56 ft 9 in **Length:** 38 ft 2 in **Engine:** 670 hp Rolls Royce Condor IIIA **Armament:** 1 Vickers machine-gun; 1 Lewis machine-gun **Crew:** 2 **Speed:** 126 mph **Ceiling:** 14,000 ft **Range:** 10 hrs **Bomb load:** 1,500 lb

This plane replaced the Fairey Fawn (see page 51) in RAF service in 1927 and remained the standard British day bomber until 1934

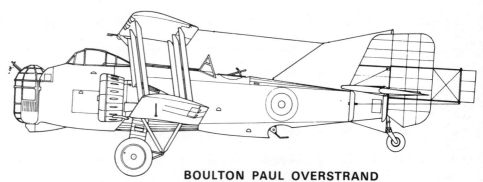

## CAPRONI Ca 101

**Gross weight:** 10,968 lb **Span:** 64 ft 7 in **Length:** 47 ft 2 in **Engine:** 3×240 hp Alfa Romeo D 2 **Armament:** 3×7·7-mm machine-guns **Crew:** 5 **Speed:** 103 mph at 3,500 ft **Ceiling:** 20,000 ft **Range:** 1,240 miles **Bomb load:** 1,102 lb

## BOULTON PAUL OVERSTRAND

**Gross weight:** 12,000 lb **Span:** 72 ft **Length:** 46 ft **Engine:** 2×580 hp Bristol Pegasus 2M3 **Armament:** 3 Lewis machine-guns **Crew:** 5 **Speed:** 153 mph at 6,500 ft **Ceiling:** 22,500 ft **Range:** 545 miles **Bomb load:** 1,600 lb

developed from the Do F mail and freight transport, at a time when the *Luftwaffe* was being secretly built up under the guise of civil aircraft, used a mixture of fabric and metal covering for different parts of the fuselage and wings. But structurally, military aircraft were conservative and all-metal stressed-skin airframes were rare until well after 1935.

This was surprising because, during the First World War, Junkers and other designers had shown that light-alloy skin could be used to bear part of the structural loads. A major advance was made in 1920 when at the London Aero Show, Oswald Short displayed the Silver Streak biplane which had a monocoque metal-covered fuselage, not only stronger and lighter than wood but easy to mass-produce.

The first Russian-designed metal bomber – the two-seat single-engined Tupolev R-3 biplane brought into service in 1926 – was also covered with corrugated Kolchug aluminium sheeting. This alloy was named after the Russian town where it was originally produced and claimed to be stronger than normal duralumin. A similar all-metal construction and covering was used in the Tupolev TB-1 twin-engined cantilever low-wing monoplane which was brought into service the following year. The TB-1 was a very large machine with an especially thick wing, capable of carrying a crew of six and a maximum bomb-load of 6,600 lb, and set the pattern for Soviet bomber design until the end of the 1930s.

## 'Iron Annie'

Germany had a long tradition of metal built and covered aircraft, and in 1926 the German designer Rohrbach perfected a complete system for making stressed-skin monoplanes with no bracing anywhere. Most German 'civil' aircraft of the middle and late 1920s were designed with military uses in mind, and such types as the three-engined Junkers Ju 52 'Iron Annie', which made its operational debut during the Spanish Civil War, served as interim bombers until the arrival of aircraft developed from airliners and specifically designed for bombing.

Some of the earliest mass-produced stressed-skin machines were the ugly, unstreamlined bombers of the French Armée de l'Air, such as the Amiot 143 multi-purpose battleplane and the Farman F 222, both twin engined high-wing monoplanes. Stressed-skin construction did not appear to do much either for the Bristol Bombay and Armstrong Whitworth Whitley, the two earliest large RAF aircraft of this type, though it was essential for faster machines like the Bristol Blenheim and Fairey Battle.

It was the USA in the early 1930s that led the switch to modern monoplane design. Stressed-skin construction made possible the rapid development of fast cantilever monoplane civil transports, notably the Boeing 247, Douglas DC-2 and Lockheed Electra, and stimulated the use of this form of construction for combat aircraft. The first of the new bomber designs was the twin-engined Boeing YB-9 low-wing monoplane in 1931, which incorporated such advanced features as semi-retractable undercarriage and variable-pitch propellers. Although not built in quantity and used only on a trial basis by the US Army Air Corps, the YB-9 offered, without any reduction of bomb-load, a dramatic improve-

## JUNKERS Ju 52

**Gross weight:** 24,200 lb  **Span:** 95 ft 10 in  **Length:** 62 ft  **Engine:** 3×830 hp BMW 132T  **Armament:** 2×7·92-mm machine-guns 15; 1×13-mm machine-gun;  **Crew:** 4  **Speed:** 165 mph at 3,000 ft  **Ceiling:** 16,600 ft  **Range:** 800 miles  **Bomb load:** 3,300 lb or 10–15 paratroops

Although this, the famous 'Iron Annie' of the *Luftwaffe* made its operational debut as a bomber during the Spanish Civil War, it became better known as a military transport plane in the Second World War. It saw service in every major German invasion campaign. It is shown here in its Second World War colours

ment over the Keystone biplane bombers which were its predecessors.

An even greater advance was made with the Martin B-10, brought into service in 1934, with a fully-retractable undercarriage and enclosed cockpit. Its maximum speed of 212 mph at 6,500 ft made it faster than

most fighters, and it was the first American monoplane bomber to be built in quantity. At the end of the 1930s, the MB-10 was being replaced by the Douglas B-18, developed from the DC-2 civil transport. Although an excellent aircraft, this was overshadowed by the four-engined Boeing B-17 Flying Fortress, produced from the Boeing 299 prototype of 1934.

Several of the American machines, especially the Northrop and Douglas types, had multi-spar wings with the advantage, not fully appreciated at the time, of being able to log large numbers of flying hours without suffering from any form of fatigue. Even if any structural member happened to crack, there were always alternative load-paths to bear the stresses.

It was partly in a search for a safer and

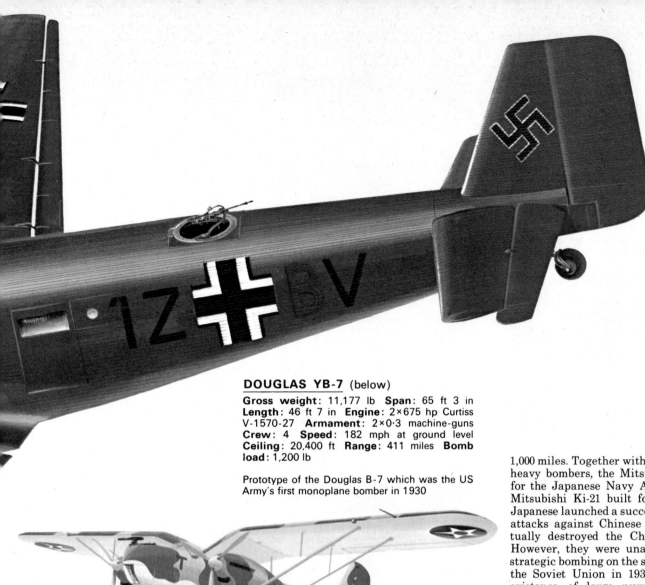

**DOUGLAS YB-7** (below)
**Gross weight**: 11,177 lb  **Span**: 65 ft 3 in
**Length**: 46 ft 7 in  **Engine**: 2×675 hp Curtiss
V-1570-27  **Armament**: 2×0·3 machine-guns
**Crew**: 4  **Speed**: 182 mph at ground level
**Ceiling**: 20,400 ft  **Range**: 411 miles  **Bomb
load**: 1,200 lb

Prototype of the Douglas B-7 which was the US
Army's first monoplane bomber in 1930

more efficient form of metal construction that Barnes Wallis devised his 'geodetic' system in the early 1930s. Wallis had been chief designer of the Vickers airship R 100 of 1929, and he developed its structure into a completely new form of metal basketwork, assembled from large numbers of standard metal sections. Riveted together by small tabs and connectors, they formed a complete wing or fuselage. All the members had the shape of intersecting curves and each carried either tension or compression but no bending.

The first geodetic aircraft, the private venture single-engined Vickers Wellesley which ultimately went into service in 1937, showed its great efficiency compared with the rival types built to the official G 4/31 specification. Its high-aspect ratio wing helped the Wellesley set the world distance record in 1938, by which time Vickers were in production with the geodetic Wellington. One of the big advantages of geodetic construction was that battle damage could easily be repaired by cutting out and replacing the small pieces of basketwork. A feature of this type of structure was that the skin should be unstressed, so that all the geodetic machines had fabric covering.

Most of the bombers used operationally

during the Second World War were being developed in the mid-1930s. Meanwhile, several modern all-metal cantilever monoplane types entered service towards the end of the 1930s but had been largely replaced by the time war broke out. One was the Bloch 200 which met a French specification for a five-seat night bomber in 1932. Unfortunately its top speed of 143 mph was some 30 mph slower than expected and by the time of the German offensive none of this type was in squadron service. The Bloch 210 was an improved version, with a speed of 186 mph at 13,000 feet, and saw operational service with the Republican forces during the Spanish Civil War. It was during that war of course that the Germans took the opportunity of testing their new aircraft in combat conditions, discovering for instance that the performance of the early Ju 86s was inadequate against contemporary fighter opposition.

Another war in which bombers were given their first important test in combat was the Sino-Japanese war which broke out in 1937. Two single-engined monoplane bombers had by that time been brought into service with the Japanese Army Air Force, the Kawasaki Ki-32 and the Mitsubishi Ki-30 which had a combat range of over

1,000 miles. Together with the twin-engined heavy bombers, the Mitsubishi G3M built for the Japanese Navy Air Force and the Mitsubishi Ki-21 built for the Army, the Japanese launched a succession of strategic attacks against Chinese airfields and virtually destroyed the Chinese Air Force. However, they were unable to undertake strategic bombing on the same scale against the Soviet Union in 1939 because of the existence of large numbers of Russian fighters.

### The new generation
The disarmament policy of most of the Allied powers prevented the construction of aircraft to meet such policies as strategic bombing, until after 1935 when the threat of war spurred rearmament programmes. Design was dominated by the immediate needs of the customer. The wars of the 1920s were mostly colonial skirmishes in which European colonial powers needed little more than light single-engined bombers for their policing operations – hence the emphasis, especially in Britain, on building fighter-bombers. It was partly the success of bombing operations against ill-armed tribesmen that gave rise to exaggerated claims for the importance of strategic bombing.

The Second World War was to see the introduction of a completely new generation of bomber aircraft, flying at three times the speed of their First World War counterparts and delivering bombs ten times more powerful than the largest used in 1918, while the development of radar added a new dimension to aerial warfare. But the actual ways in which bombers were used differed little from the first war, except that they were on a vastly greater scale. The difference in the Second World War was that air power had become of such vital importance that without at least some degree of balance in the control of air space, battles would be lost and entire countries fall to an enemy who had mastery of the skies.

**MITSUBISHI Ki-2I**
**Gross weight**: 10,031 lb **Span**: 65 ft 6 in
**Length**: 41 ft 4 in **Engine**: 2×570 hp Nakajima
**Armament**: 2×7·7-mm machine-guns **Crew**: 5
**Speed**: 158 mph at 9,500 ft **Ceiling**: 22,890 ft
**Range**: 560 miles **Bomb load**: 660 lb

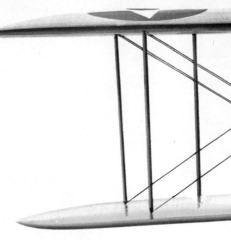

**AMIOT 143** (left)
**Gross weight**: 19,568 lb **Span**: 80 ft 2 in
**Length**: 58 ft 11 in **Engine**: 2×870 hp Gnôme-
Rhône **Armament**: 4×7·5-mm MAC machine-
guns **Crew**: 5 **Speed**: 190 mph at 13,000 ft
**Ceiling**: 21,200 ft **Range**: 800 miles **Bomb
load**: 1,984 lb

**MARTIN B-10**
**Gross weight**: 14,600 lb **Span**: 70 ft 6 in
**Length**: 44 ft 9 in **Engine**: 2×775 hp Wright
R-1820-25 **Armament**: 5×·303 Browning
machine-guns **Crew**: 4 **Speed**: 213 mph at
10,000 ft **Ceiling**: 24,200 ft **Range**: 1,240
miles **Bomb load**: 2,260 lb

**DH 9A**
**Gross weight**: 4,645 lb **Span**: 45 ft 11 in
**Length**: 30 ft 3 in **Engine**: 400 hp Liberty
**Armament**: 1 Vickers machine-gun; 2 Lewis
machine-guns **Crew**: 2 **Speed**: 116 mph at
10,000 ft **Ceiling**: 17,000 ft **Range**: 3 hrs
**Bomb load**: 660 lb

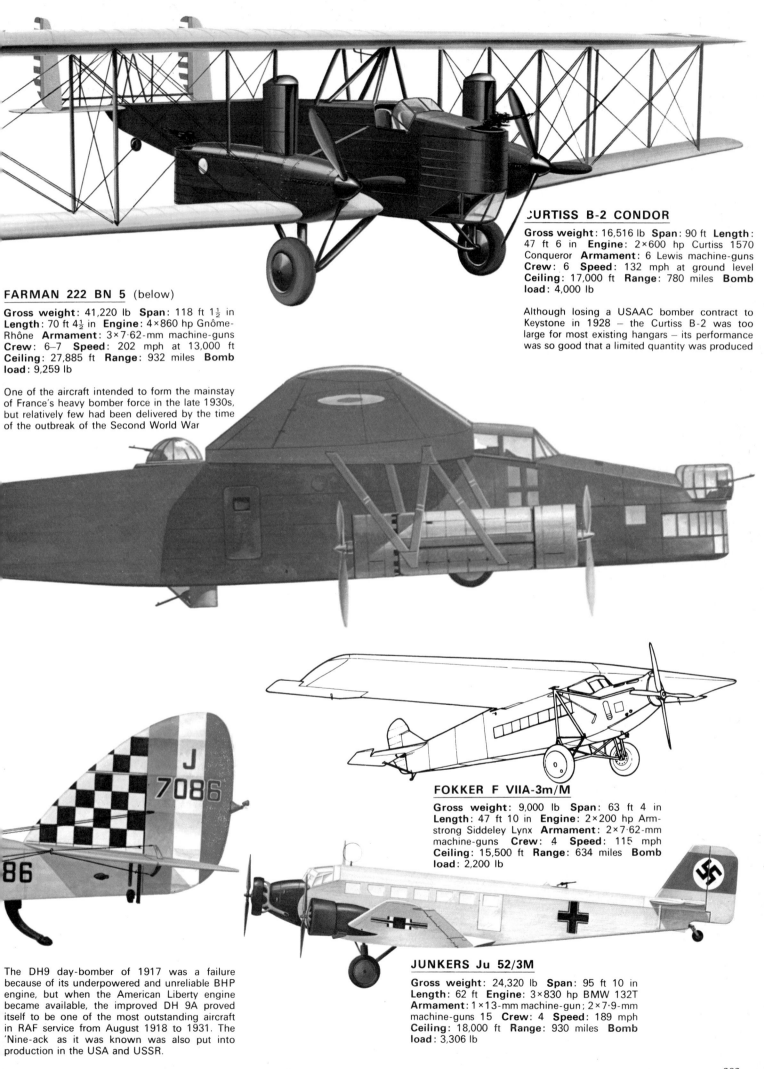

## CURTISS B-2 CONDOR

**Gross weight:** 16,516 lb **Span:** 90 ft **Length:** 47 ft 6 in **Engine:** 2×600 hp Curtiss 1570 Conqueror **Armament:** 6 Lewis machine-guns **Crew:** 6 **Speed:** 132 mph at ground level **Ceiling:** 17,000 ft **Range:** 780 miles **Bomb load:** 4,000 lb

Although losing a USAAC bomber contract to Keystone in 1928 – the Curtiss B-2 was too large for most existing hangars – its performance was so good that a limited quantity was produced

## FARMAN 222 BN 5 (below)

**Gross weight:** 41,220 lb **Span:** 118 ft $1\frac{1}{2}$ in **Length:** 70 ft $4\frac{1}{2}$ in **Engine:** 4×860 hp Gnôme-Rhône **Armament:** 3×7·62-mm machine-guns **Crew:** 6–7 **Speed:** 202 mph at 13,000 ft **Ceiling:** 27,885 ft **Range:** 932 miles **Bomb load:** 9,259 lb

One of the aircraft intended to form the mainstay of France's heavy bomber force in the late 1930s, but relatively few had been delivered by the time of the outbreak of the Second World War

## FOKKER F VIIA-3m/M

**Gross weight:** 9,000 lb **Span:** 63 ft 4 in **Length:** 47 ft 10 in **Engine:** 2×200 hp Armstrong Siddeley Lynx **Armament:** 2×7·62-mm machine-guns **Crew:** 4 **Speed:** 115 mph **Ceiling:** 15,500 ft **Range:** 634 miles **Bomb load:** 2,200 lb

## JUNKERS Ju 52/3M

**Gross weight:** 24,320 lb **Span:** 95 ft 10 in **Length:** 62 ft **Engine:** 3×830 hp BMW 132T **Armament:** 1×13-mm machine-gun; 2×7·9-mm machine-guns 15 **Crew:** 4 **Speed:** 189 mph **Ceiling:** 18,000 ft **Range:** 930 miles **Bomb load:** 3,306 lb

The DH9 day-bomber of 1917 was a failure because of its underpowered and unreliable BHP engine, but when the American Liberty engine became available, the improved DH 9A proved itself to be one of the most outstanding aircraft in RAF service from August 1918 to 1931. The 'Nine-ack as it was known was also put into production in the USA and USSR.

The Hawker Hart was the RAF's standard light day-bomber from 1930 to 1937, and was the basis for a number of specialized derivatives

# BOMBERS 1939–1945

*Bryan Cooper*

While the basic tactical and strategic uses of the bomber were all evolved during the First World War, the planes of 1939 were far from realising their full potential. But under the stimulus of war, the phenomenal ingenuity of the Germans and the vast resources of the US had produced, by 1945, the world's first jet bomber, the gigantic Superfortress – and the Atom Bomb.

Of course, there were many other fascinating bombers, and in this chapter we have combined an authoritative text with John Batchelor's magnificent drawings to tell the full story of the bomber's development during those years when it reached an unprecedented level of destructive power.

# CONTENTS

# WAR!

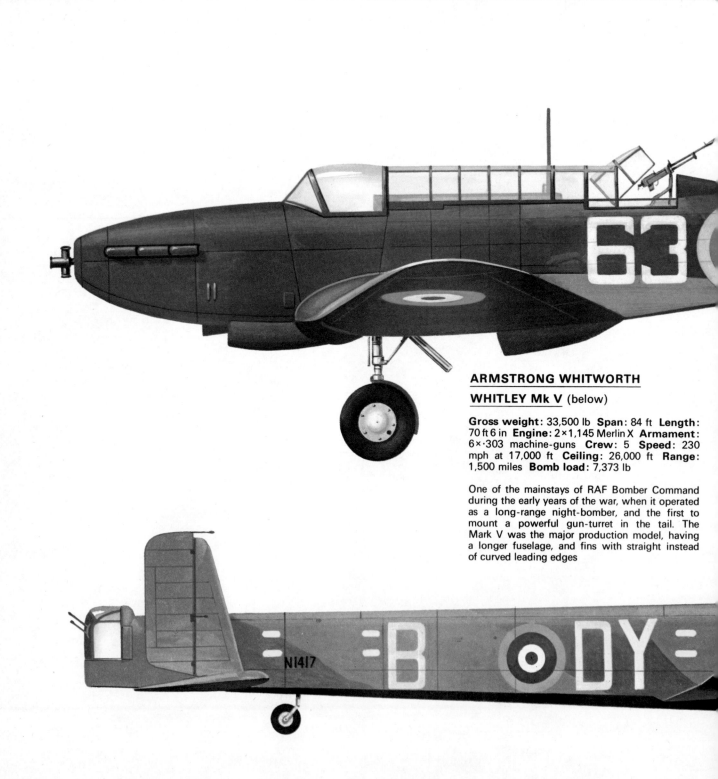

**ARMSTRONG WHITWORTH**

**WHITLEY Mk V** (below)

**Gross weight:** 33,500 lb  **Span:** 84 ft  **Length:** 70 ft 6 in  **Engine:** 2×1,145 Merlin X  **Armament:** 6×·303 machine-guns  **Crew:** 5  **Speed:** 230 mph at 17,000 ft  **Ceiling:** 26,000 ft  **Range:** 1,500 miles  **Bomb load:** 7,373 lb

One of the mainstays of RAF Bomber Command during the early years of the war, when it operated as a long-range night-bomber, and the first to mount a powerful gun-turret in the tail. The Mark V was the major production model, having a longer fuselage, and fins with straight instead of curved leading edges

# How the world's bombers lined up...

Most of the bombers used in the Second World War were either flying before the start of the war or resulted from programmes initiated before 1939. Unlike fighters, which could be designed and put into production relatively quickly, heavy bombers were so complicated even in those days that it was a daunting task to develop a completely new one. Very few new types were designed and brought into service within the period of the war, and those initiated before were largely based on earlier developments

**FAIREY BATTLE** (above)

**Gross weight:** 10,792 lb  **Span:** 54 ft  **Length:** 52 ft 2 in  **Engine:** 1,030 hp Rolls Royce Merlin  **Armament:** 2×·303 machine-guns  **Crew:** 3  **Speed:** 250 mph at 20,000 ft  **Ceiling:** 25,000 ft  **Range:** 1,050 miles  **Bomb load:** 1,000 lb

Entered RAF service in 1937 as a fast day bomber. Already outdated when war broke out, it suffered heavy losses in France in 1939/40 when it formed the vanguard of the British Advanced Air Striking Force. Later relegated for training purposes

# BRITAIN

## Battle against time

The main bomber types in service with the RAF in 1939 were the Vickers Wellesley and Fairey Battle – the last single-engined bombers to be built in Britain – and the twin-engined Armstrong Whitworth Whitley, Bristol Blenheim, Handley Page Hampden and Vickers Wellington. These bore the brunt of operations in the early part of the war.

The Wellesley, first to use the geodetic form of construction, was brought into service in 1937 and was used operationally for the first two years of the war, especially from overseas bases against the Italians. An unusual feature was the method of carrying its bomb-load in containers under the wings. The Fairey Battle, powered by the same Merlin engine used by Spitfire and Hurricane fighters, though obsolete when the war started, was the only bomber available to form the vanguard of the British Advanced Air Striking Force in France during the winter of 1939–40. Heavy losses were suffered during daytime operations; during an attack on German pontoon bridges at Sedan on 20 May 1940, forty out of seventy-one were lost. The Battle was later withdrawn from bomber squadrons and used for training, especially in Canada under the Commonwealth Air Training programme.

Whitleys were the first British bombers to fly over Germany on the first night of the war, when they dropped some six million leaflets in an attempt to persuade the German people that the war could be avoided. Accompanied by Hampdens, they made the first bombing raid on German soil, on 19 March 1940, and were the first RAF

## AVRO ANSON Mk I

**Gross weight:** 8,000 lb **Span:** 56 ft 6 in **Length:** 42 ft 3 in **Engine:** 2×350 hp Armstrong Siddeley Cheetah IX **Armament:** 2×·303 machine-guns **Crew:** 3 **Speed:** 188 mph at 7,000 ft **Ceiling:** 16,000 ft **Range:** 660 miles **Bombs:** 360 lb

Although Ansons first entered service in 1936 and were becoming obsolete by the outbreak of war, they continued in production until 1952, used for training and transport after 1941, and served with many other air forces

bombers to attack Italy. Early in 1942 they were withdrawn from front-line service, but they continued to be used for parachute dropping and glider-towing duties.

A smaller and much faster bomber was the Blenheim, which created a sensation when it entered service in 1937 with a speed of 280 mph, outpacing some of the best fighters. Many changes were made as the war progressed, particularly by fitting armour-plate and heavier armament, but these considerably reduced its speed and by 1942 the type was outdated. Blenheims had the distinction of being the only aircraft to serve in all the RAF wartime Commands – Bomber, Fighter, Coastal, Army Co-operation and Training. As a night-fighter, the Blenheim pioneered the use of the highly secret Airborne Interception radar in 1940, though it was not really fast enough.

The Hampden was the last medium bomber with only two engines to enter service with the RAF, in 1938. It had a serious deficiency in defensive armament and suffered very heavy losses in early daytime raids. Known as the 'Flying Suit-case' because of its deep forebody and slender tail, it was reserved for night operations from 1940–42, taking part in the first raid on Berlin and the famous 1,000-bomber raid on Cologne. At the end of 1942 it was withdrawn from bombing operations but continued as a torpedo-bomber.

### Backbone of Bomber Command
Last and best of the pre-war bombers was the long-range Wellington. This plane formed the backbone of Bomber Command's offensive against Germany in the early years of the war, and was popular among its crews for its ability to return to base even after sustaining severe battle damage. Affectionately known as the 'Wimpey', it was generally used for night operations after early daylight bombing raids disproved the theory that the combined firepower of bombers flying in formation could beat off fighters. In April 1941, Wellingtons were the first to drop the new 4,000-lb 'block-buster' bomb during a raid on Emden, and

they were used to help start the Pathfinder tactics for indicating targets. Nearly 11,500 Wellingtons were built in many different versions, including reconnaissance, troop-carrying and transport as well as bombing. They remained in front-line service until the end of the war, by which time, because of the strength and lightness of the geodetic structure, the 21,000 lb all-up weight of the first models had been increased to 36,500 lb, with a maximum bomb-load of 6,000 lb.

A two-engined medium-heavy bomber, the Avro Manchester, was brought into service in November 1940, but constant trouble with the Rolls-Royce Vulture engine led to its being withdrawn from service in 1942, when only 200 machines had been built.

In 1936, when Germany's aggressive intentions were becoming obvious even to the pacifist British, the British Air Staff initiated a programme for three heavy, four-engined bombers. They were not ready when the war started, but later played the major role in RAF Bomber Command's strategic offensive against Germany. In spite of the

valiant efforts of the lighter bombers during the early years of the war, the use of air power built up very gradually; 83 per cent of all the bombs dropped on Germany by the Allies were delivered from 1944 onwards.

The first of the heavy bombers to enter service was the Short Stirling in 1940, with a maximum bomb-load of 14,000 lb at a range of 590 miles. But its usefulness was limited by the fact that the heaviest bomb it could carry was 4,000 lb. The first raid carried out by Stirlings was on 10 February 1941, against oil storage tanks at Rotterdam.

The Stirling was followed late in 1940 by the Handley Page Halifax medium-heavy bomber. Over 6,000 were built in several different versions and the Halifax proved itself a worthy successor of Handley Page's four-engined 0/400 'Bloody Paralyser' of the First World War. But it was inevitably overshadowed by the other heavy bomber with which it shared the major part of the RAF's night-bombing offensive against Germany – the Avro Lancaster. Whereas the Halifax flew 75,532 bombing sorties,

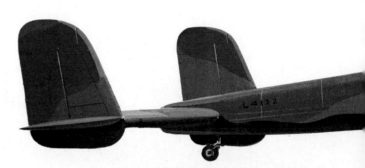

### HANDLEY PAGE HAMPDEN Mk I

**Gross weight:** 18,750 lb **Span:** 69 ft 2 in **Length:** 53 ft 7 in **Engine:** 2×1,000 hp Bristol Pegasus XVIII **Armament:** 4×·303 Vickers machine-guns **Crew:** 4 **Speed:** 265 mph at 15,500 ft **Ceiling:** 22,700 ft **Range:** 1,990 miles **Bomb load:** 4,000 lb

## BRISTOL BLENHEIM Mk I

**Gross weight**: 12,500 lb **Span**: 56 ft 4 in
**Length**: 39 ft 9 in **Engine**: 2×840 hp Bristol
Mercury VIII **Armament**: 2×·303 machine-guns
**Crew**: 4 **Speed**: 285 mph at 15,000 ft **Ceiling**:
27,280 ft **Range**: 1,125 miles **Bomb load**:
1,000 lb

Faster than the best fighters when it entered
RAF service in 1937, the Blenheim light bomber
later served in many wartime roles, including
night-fighting, ground attack, and anti-shipping

dropping 227,610 tons of bombs, the
Lancaster flew no less than 156,000 sorties,
and delivered 608,612 tons.

Although the most famous and successful
heavy night bomber used in Europe during
the war, the Lancaster did not have the
adaptability of the Halifax and was used
almost exclusively for bombing operations.
It went into service early in 1942 and by
mid-1943 had established its excellent all-
round performance – only one Lancaster
was lost for every 132 tons of bombs
delivered, compared with 56 tons for each
Halifax lost and 41 tons for each Stirling.

A major feature of the Lancaster –
described by the chief of RAF Bomber
Command as the greatest single factor in
winning the Second World War – was its
cavernous bomb-bay. This was initially
designed to take bombs of up to 4,000 lb,
with a maximum bomb-load of 14,000 lb, but
it was progressively modified to carry 8,000,

12,000, and eventually the 22,000-lb 'Earthquake' or Grand Slam bombs. The Lancaster
was also chosen to carry the remarkable
'spinning drum' bomb designed by Dr Barnes
Wallis for the famous raid on the Mohne
and Eder dams on 17 May 1943. Lancasters
were also responsible for sinking the *Tirpitz*
on 12 November 1944, using Barnes Wallis's
12,000-lb 'Tallboy' deep-penetration bomb.
Well over 7,000 Lancasters served in various
air forces, many until long after the end of
the war. A development of the Lancaster –
the Avro Lincoln, produced just after the
war – was the mainstay of RAF Bomber
Command as the last piston-engined heavy
bomber before the introduction of the jets.

The second most successful bomber used
by the RAF was the extremely versatile

twin-engined de Havilland Mosquito,
initiated as a private-venture light bomber
in 1938. One of its claims to distinction was
that it was built almost entirely of wood;
another was that it was so fast – 400 mph
plus – that it could generally avoid fighter
interception and therefore had no need to
be armed.

Mosquitoes were used for many purposes,
including night-fighting and photographic
reconnaissance. As a bomber it could
initially carry four 500-lb bombs, but more
than fifty Mark IVs were adapted to carry
4,000-pounders, previously taken only by
heavy bombers. The Mosquito first went into
service in 1941 and was the fastest type in
Bomber Command for nearly ten years until
the introduction of the Canberra jet.

Entered RAF service in 1938 as the last of the
twin-engined medium bombers ordered by Bomber Command, but suffered heavy losses because
of inadequate armament. Later models were considerably improved and took part in night raids on
Germany

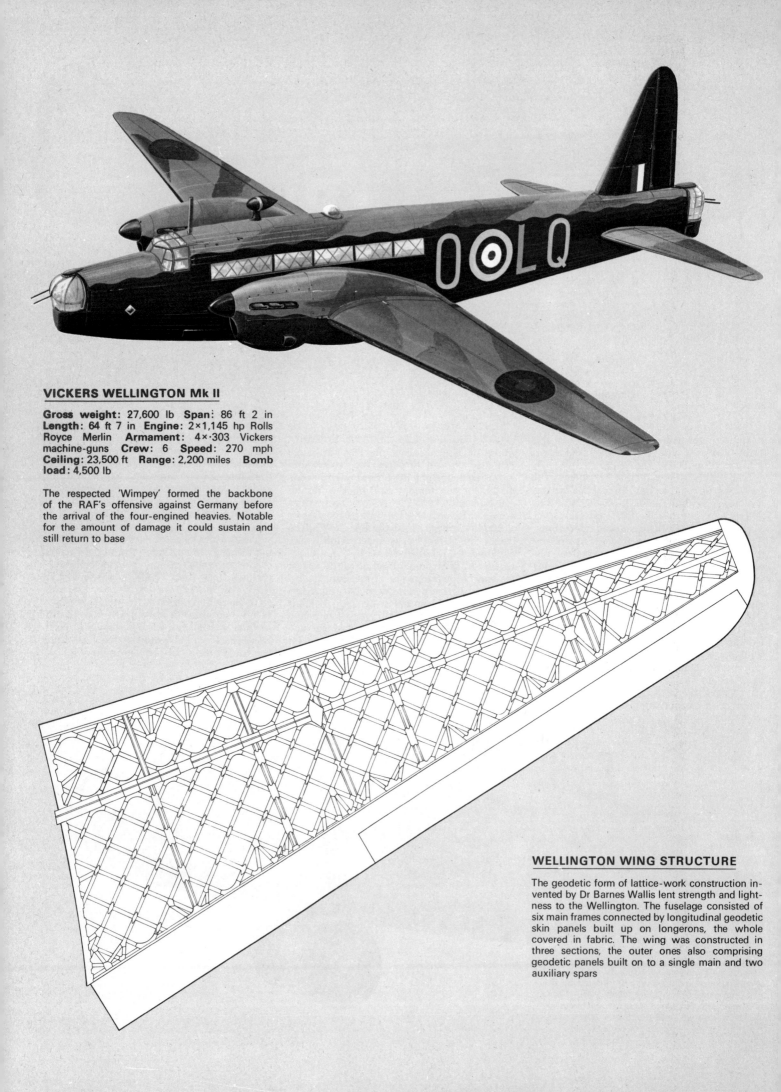

## VICKERS WELLINGTON Mk II

**Gross weight:** 27,600 lb **Span:** 86 ft 2 in
**Length:** 64 ft 7 in **Engine:** 2×1,145 hp Rolls
Royce Merlin **Armament:** 4×·303 Vickers
machine-guns **Crew:** 6 **Speed:** 270 mph
**Ceiling:** 23,500 ft **Range:** 2,200 miles **Bomb
load:** 4,500 lb

The respected 'Wimpey' formed the backbone
of the RAF's offensive against Germany before
the arrival of the four-engined heavies. Notable
for the amount of damage it could sustain and
still return to base

## WELLINGTON WING STRUCTURE

The geodetic form of lattice-work construction invented by Dr Barnes Wallis lent strength and lightness to the Wellington. The fuselage consisted of
six main frames connected by longitudinal geodetic
skin panels built up on longerons, the whole
covered in fabric. The wing was constructed in
three sections, the outer ones also comprising
geodetic panels built on to a single main and two
auxiliary spars

# FRANCE

## Too little, too late

Military aviation in France in the mid-1930s was in an even worse state than in Britain. At the start of the war, many French squadrons were equipped with such obsolete bombers as the Amiot 143, which undertook a few bombing missions over enemy-occupied territory, and the Bloch 210, used by the Republican forces in the Spanish Civil War but which, by 1940, was 100 mph slower than German bombers. A re-armament programme was under way in France but it was not due for completion until 1942 and France capitulated before very many of the new types had been delivered to the French Air Force. Even these suffered from a lack of equipment and serious engine problems, the result of government incompetence and refusal to take the German threat seriously during the mid-1930s. The development of French bombers naturally ceased after the Armistice, but some of the new twin-engined types were used in the early months

of the war, while others were taken over and used operationally by the Axis Powers.

One of these was the Bréguet 690 series which entered production in 1938 as a two-seater light assault bomber. The 693 version was used in May 1940 for low-level attacks on German trenches, and many were later used for the same purpose by the Italian Air Force. Another type which would have made an excellent bomber but which was developed too late was the Lioré et Olivier 451. Again there were problems with unreliable and underpowered engines, although the prototype achieved a speed of 310 mph in level flight as early as 1937. The few that had been delivered by the time of the German invasion of France in May 1940 were later used by the *Luftwaffe* as transports.

With the Potez 630 series, it was not only engine troubles but also a shortage of propellers which kept them grounded at the

### BLOCH 174 A3

**Gross weight:** 15,748 lb  **Span:** 58 ft 9½ in  **Length:** 40 ft 1½ in  **Engine:** 2×1,140 hp Gnôme-Rhône  **Armament:** 7×7·5-mm machine-guns  **Crew:** 3  **Speed:** 329 mph at 17,060 ft  **Ceiling:** 30,090 ft  **Range:** 1,025 miles  **Bomb load:** 1,500 lb

This 1938 French design was switched to a reconnaissance role because of its limited bomb load, but from it was developed the Bloch 175. None saw action before the French surrender but the Germans continued production until 1942. French naval units used a torpedo-carrying version until 1953

time of the German attack. A greater degree of success was achieved by other countries which purchased machines of this type, including Greece, Rumania and Japan. One bomber which was kept in production by the Germans until the end of 1942 was the Bloch 175, a development of the 174, which had been given a reconnaissance role because of its limited bomb-load. Later, it was decided to use the self-contained engine-propeller-cowling units of the 175 in the huge Messerschmitt Me 323 transports. Finally, there was the Amiot 350 series which was being delivered to the French Air Force early in 1940. Lack of armament prevented these machines being used operationally, and many were destroyed to prevent them falling into German hands.

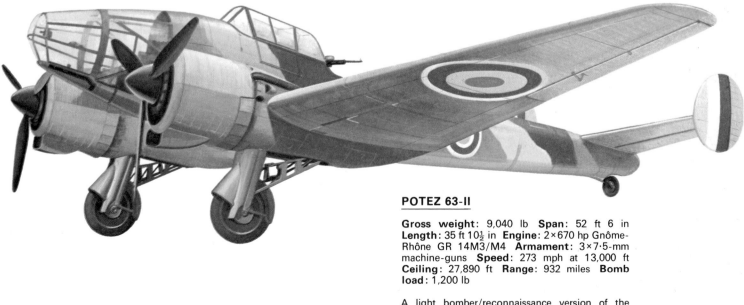

### POTEZ 63-II

**Gross weight:** 9,040 lb  **Span:** 52 ft 6 in  **Length:** 35 ft 10½ in  **Engine:** 2×670 hp Gnôme-Rhône GR 14M3/M4  **Armament:** 3×7·5-mm machine-guns  **Speed:** 273 mph at 13,000 ft  **Ceiling:** 27,890 ft  **Range:** 932 miles  **Bomb load:** 1,200 lb

A light bomber/reconnaissance version of the Potez 630 day and night fighter

## JUNKERS Ju 86K

**Gross weight:** 18,070 lb **Span:** 73 ft 10 in **Length** 57 ft 9 in **Engine:** 2×Swedish-built Bristol Mercury **Armament:** 3×7·9-mm machine-guns **Crew:** 5 **Speed:** 202 mph at 9,800 ft **Ceiling:** 22,300 ft **Range:** 1,240 miles **Bomb load:** 2,200 lb

Shown here in Swedish markings, the Ju 86 revealed deficiencies as a bomber during the Spanish Civil War and despite improvements it was used primarily as a reconnaissance aircraft by the *Luftwaffe* in the early years of the war

## LeO 451

**Gross weight:** 26,000 lb **Span:** 73 ft 11 in **Length:** 56 ft 4 in **Engine:** 2×1,000 hp Gnôme-Rhône 14 N20/21 **Armament:** 1×20-mm cannon; 4×7·5-mm machine-guns **Crew:** 4 **Speed:** 310 mph at 18,000 ft **Ceiling:** 27,000 ft **Range:** 1,040 miles **Bomb load:** 3,080 lb

The only really modern bombers in the French Air Force in September 1939 were LeO 451s taken from an experimental squadron but they arrived too late to be of much value. A number were later used by the *Luftwaffe* and by the Vichy Government

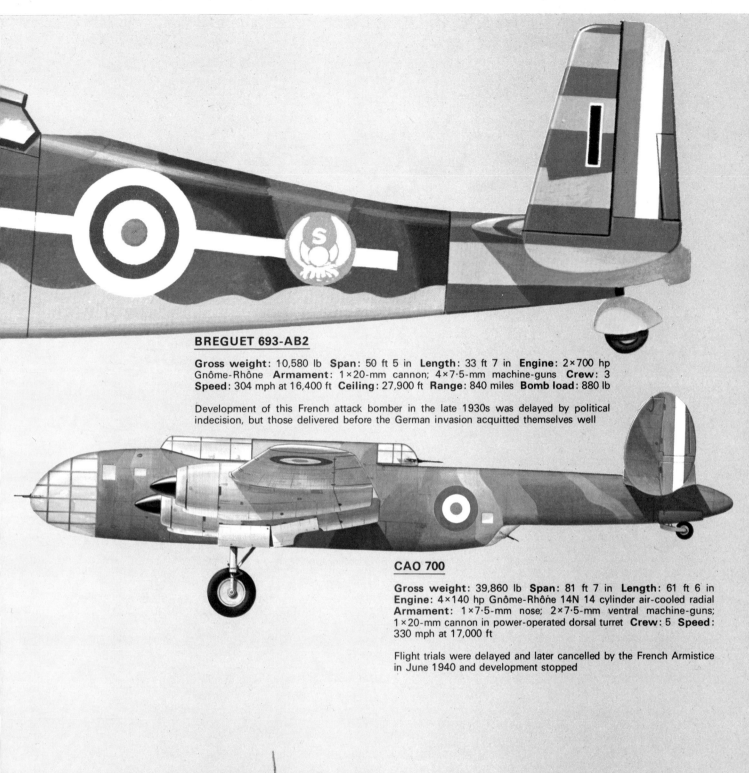

## BREGUET 693-AB2

**Gross weight:** 10,580 lb  **Span:** 50 ft 5 in  **Length:** 33 ft 7 in  **Engine:** 2×700 hp Gnôme-Rhône  **Armament:** 1×20-mm cannon; 4×7·5-mm machine-guns  **Crew:** 3 **Speed:** 304 mph at 16,400 ft  **Ceiling:** 27,900 ft  **Range:** 840 miles  **Bomb load:** 880 lb

Development of this French attack bomber in the late 1930s was delayed by political indecision, but those delivered before the German invasion acquitted themselves well

## CAO 700

**Gross weight:** 39,860 lb  **Span:** 81 ft 7 in  **Length:** 61 ft 6 in **Engine:** 4×140 hp Gnôme-Rhône 14N 14 cylinder air-cooled radial **Armament:** 1×7·5-mm nose; 2×7·5-mm ventral machine-guns; 1×20-mm cannon in power-operated dorsal turret  **Crew:** 5  **Speed:** 330 mph at 17,000 ft

Flight trials were delayed and later cancelled by the French Armistice in June 1940 and development stopped

## HEINKEL He 111H-6

**Gross weight:** 27,400 lb **Span:** 74 ft 1½ in **Length:** 54 ft 5½ in **Engine:** 2×1,340 hp Junkers Jumo 211 F-2 **Armament:** 6×7·9-mm MG 15 machine-guns; 1×20-mm cannon **Crew:** 6 **Speed:** 258 mph at 16,400 ft **Range:** 1,760 miles **Bomb load:** 5,510 lb or 2 torpedoes

The He 111 first proved itself during the Spanish Civil War, and gave excellent service with the *Luftwaffe* throughout the Second World War. The H-6 was a first-rate torpedo-bomber

# The LUFTWAFFE -born in secret

The three great German bomber manufacturers – Junkers, Heinkel and Dornier – were all gaining experience in the mid-1930s with aircraft either originally designed for civil use or disguised as such, a necessary deception in view of the Versailles Treaty which prohibited Germany from building military aircraft. Bombing operations during the Spanish Civil War provided the newly-formed *Luftwaffe* with valuable experience. Germany never showed much interest in long-range heavy bombers and concentrated from first to last on relatively small tactical machines for use within a European theatre of operations.

Towards the end of the Second World War, prototypes and even a few production models of larger aircraft appeared. The Junkers Ju 390, a very large six-engined machine which on one occasion flew to within ten miles of New York, could have formed the basis of a powerful night-bomber force. The Messerschmitt Me 264 was specifically planned as a bomber capable of striking the eastern seaboard of the USA, its four engines mounted on a wing of very high aspect ratio with a span of 141 feet. But little effort was put behind these developments and they eventually petered out.

With the exception of the four-engined Ju 290 maritime patrol and reconnaissance bomber, produced in small numbers from 1943 onwards, and the Heinkel He 177 which went into service at the same time, all the German bombers were limited to either one or two engines. In fact, the He 177 was the only long-range strategic bomber put into production by Germany during the war, but persistent problems with the coupled engines – they were mounted in pairs in two nacelles, each pair driving a single propeller – and the fact that they caught fire easily made it far from efficient.

The main bombers in service with the *Luftwaffe* when war broke out were the twin-engined Do 17, Ju 88 and He 111, together with the single-engined Ju 87 dive-bomber. In addition, there was the Ju 86, whose poor performance in the

## JUNKERS Ju 87 B2 'STUKA' COCKPIT

1   Visual dive indicator
2   Gun sight
3   Artificial horizon
4   Compass repeater
5   Speedometer
6   Boost pressure
7   Altimeter
8   Rev counter
9   Flap indicator
10   Intercom connection
11   Crash pad
12   Manual engine pump
13   Engine priming pump
14   Electrics panel (radio)
15   Oil cooler flap control
16   Rudder bar pedal
17   Target view window
18   Control column
19   Target view window flap control
20   Fuel metering hand priming pump
21   Throttle
22   Starter switch
23   Main electrics switch
24   Coolant temperature
25   Fuel contents
26   Oil temperature
27   Oil contents
28   Compass
29   Oil pressure gauge
30   Clock
31   Dive pre-set indicator
32   Fuel pressure gauge
33   Radio altimeter
34   Rate of climb indicator
35   Water cooler flap indicator

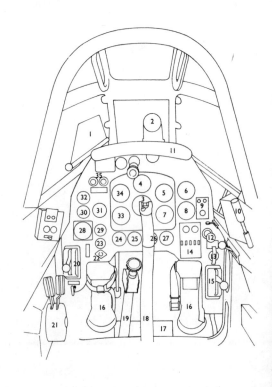

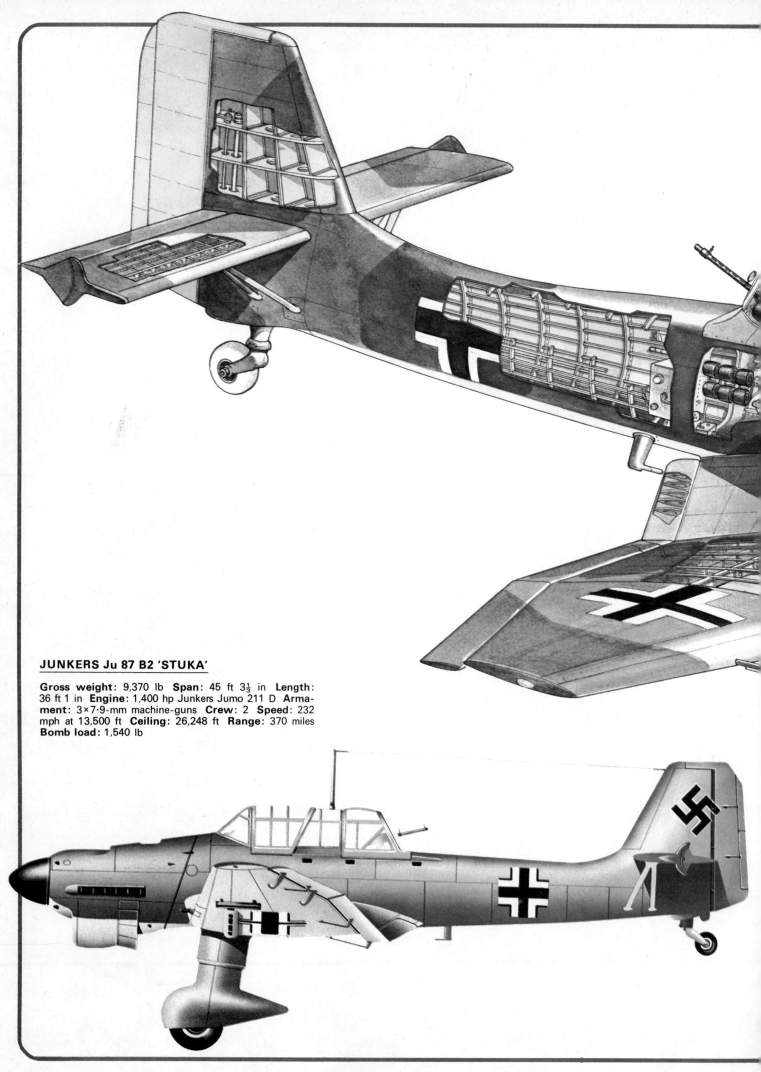

## JUNKERS Ju 87 B2 'STUKA'

**Gross weight:** 9,370 lb  **Span:** 45 ft 3⅓ in  **Length:** 36 ft 1 in  **Engine:** 1,400 hp Junkers Jumo 211 D  **Armament:** 3×7·9-mm machine-guns  **Crew:** 2  **Speed:** 232 mph at 13,500 ft  **Ceiling:** 26,248 ft  **Range:** 370 miles  **Bomb load:** 1,540 lb

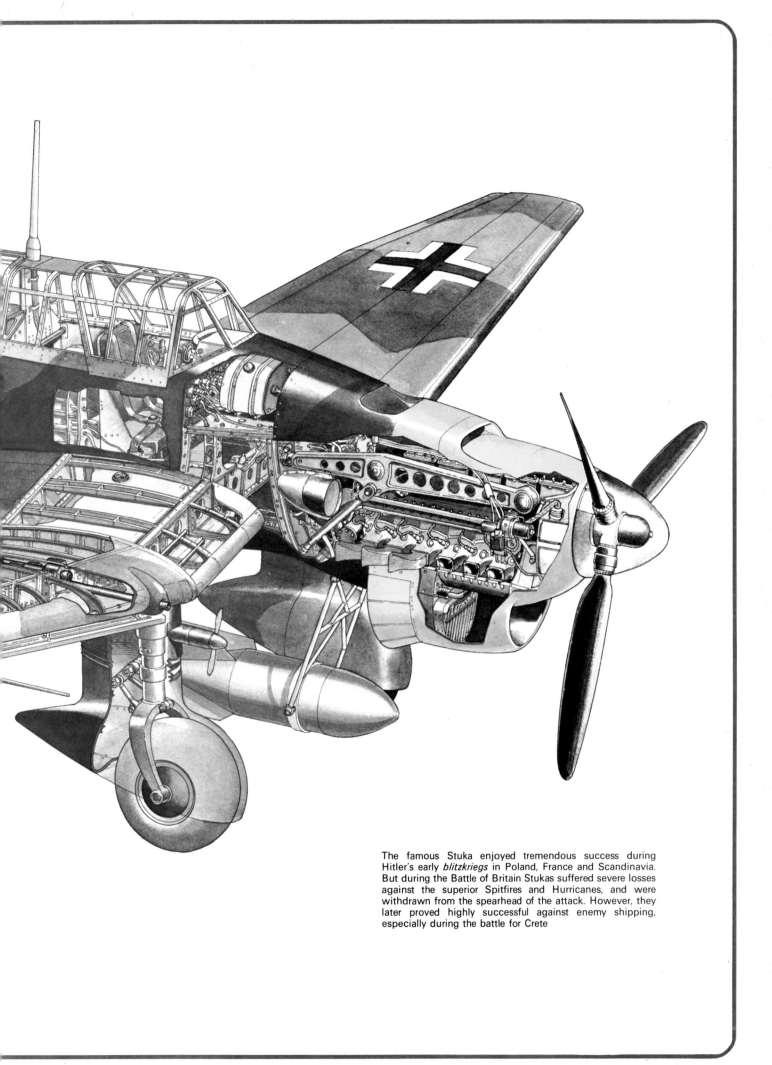

The famous Stuka enjoyed tremendous success during Hitler's early *blitzkriegs* in Poland, France and Scandinavia. But during the Battle of Britain Stukas suffered severe losses against the superior Spitfires and Hurricanes, and were withdrawn from the spearhead of the attack. However, they later proved highly successful against enemy shipping, especially during the battle for Crete

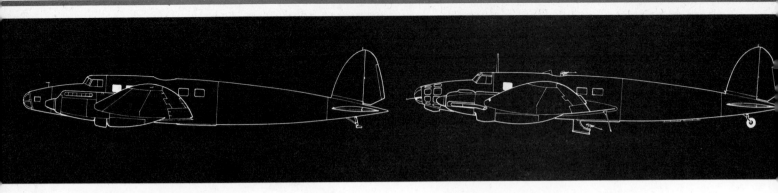

### HEINKEL He 111/VI

The first prototype of the He 111 appeared at the end of 1934, fitted out as a bomber armed with three 7·9-mm machine-guns in nose, dorsal and ventral positions, powered by two 660 hp BMW VI liquid-cooled engines, and able to carry a 2,200-lb bomb load. Later prototypes were developed into commercial transports, and deliveries of the first military version, with more powerful DB 600A engines, began in 1936

### HEINKEL He 111 E3

The E-series of the He 111, produced in 1937, was powered by Junkers Jumo 211A-3 engines which gave a maximum speed of 267 mph and a service ceiling of 22,900 ft. Semi-retractable radiators were adopted to reduce drag. Up until the beginning of the Second World War the He 111 was probably the best medium bomber flying, but improvements could not keep pace with changing requirements

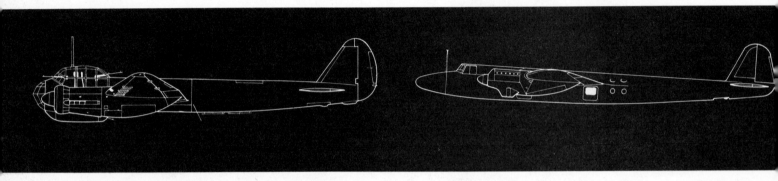

### JUNKERS Ju 88 A-1

**Gross weight:** 27,500 lb **Span:** 59 ft 11 in **Length:** 47 ft 1½ in **Engine:** 2×1,200 hp Jumo 211B **Armament:** 3×7·9-mm machine-guns **Crew:** 4 **Speed:** 286 mph at 18,000 ft **Ceiling:** 30,675 ft **Range:** 1,550 miles **Bomb load:** 5,500 lb

### DORNIER Do 17 V1

The Do 17 V1 first flew in 1934, and was quickly nicknamed the 'Flying Pencil' for its clean, slim lines. Intended as a high-speed commercial aircraft (a six-passenger mailplane was developed from it, and three built for *Lufthansa*), it was powered by two 660 hp BMW V1 liquid-cooled engines, and had two cabins for the passengers and a crew of two

Spanish Civil War had underlined its ineffectiveness against fighter opposition. Considerable improvements were made to later models, the Ju 86P and Ju 86R, which were still only really suitable for reconnaissance. Their main asset was an ability to fly at high altitudes, achieved by greatly increased wing span, supercharged diesel engines, and the installation of pressure cabins for the two-man crew. The service ceiling of the Ju 86R for instance, with its wing span increased from 74 to 105 ft, was over 49,000 ft.

Towards the end of the war, apart from the Ju 290 and He 177, the only other notable bomber to be brought into service was the Arado Ar 234 Blitz (Lightning), the world's first operational jet bomber, although only a few saw combat. The Messerschmitt Me 262 played a greater part. Though originally designed as a jet fighter, on Hitler's insistence it was pressed into service as a bomber, which delayed its introduction from early in 1944 as had been originally

planned to the end of that year.

Most of the German bomber effort during the war went into developing and improving the four major types already in service. One of the most successful early on was the Ju 87 'Stuka', the result of a German vogue in the late 1930s for dive-bombing as an integral part of the *blitzkrieg* tactics. It first appeared in 1938, and, during the early months of the war, achieved all that had been expected of it in helping the German armies blast their way across Poland, France, Belgium and Holland. The high-pitched scream made by the Ju 87's dive brakes (required to hold the speed steady enough to aim the bombs) brought a new terror to the confused armies and civilian populations of Europe. But when the Germans employed the Stuka to attack British airfields in the Battle of Britain, it proved highly vulnerable to the much faster Spitfires and Hurricane fighters and suffered heavy losses. Attempts to provide Me 109 fighter escorts were unsuccessful, because

in a dive they could not keep down to the speed of the Stukas and had to leave them behind to the mercy of the British fighters.

Later improvements included a doubling of engine power, an increase in bomb-load from 1,000 lb to 3,960 lb, and additional armour protection for the crew. Various Ju 87 types were used until the end of the war, with particular success on the Russian front and as a 'tank-buster' ground-attack machine during and after the Normandy invasion. Nearly 5,000 of all versions were completed during the war period.

### The 'Flying Pencil'

The Do 17, originally designed as a fast six-passenger mail plane when the first prototype flew in 1934, earned itself the nickname 'Flying Pencil' for its slender appearance from the side. In 1937 it outstripped the best fighters then being produced in Europe, and was put into service with the *Luftwaffe* as a medium bomber. It was among the types selected to equip the

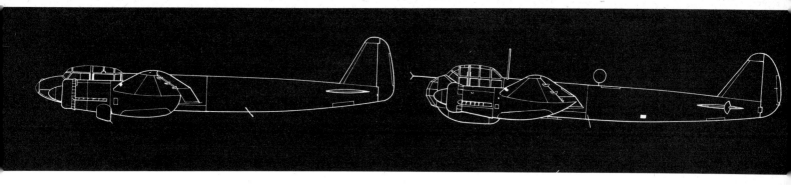

## JUNKERS Ju 88 V1

Following the 1935 German specification for a fast twin-engined medium bomber, the Ju 88 V1 development prototype made its maiden flight in December 1936, powered by two 900 hp DB 600 liquid-cooled engines

## JUNKERS Ju 88 V4

Following the V1, six months later, the Ju 88 V4 had a redesigned cockpit to take a fourth crew member, and used 950 hp Jumo 211A engines. This set the pattern for the first models to be taken into service, the production Ju 88 A-1s arriving in time to take part in attacks on Britain in September 1939

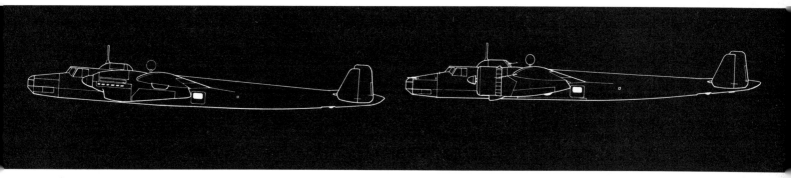

## DORNIER Do 17 V8

Of a further series of prototypes, the V4 (1935) replaced the single fin-and-rudder with twin fins and rudders – a feature of all later models. The Do 17's military potential was revealed by the V8 model entered for the Military Aircraft Competition at Zurich in 1937

## DORNIER Do 17K

One of the first orders for the Do 17 was received from the Yugoslav Government for an export model, the Do 17K, with two 986 hp Gnôme-Rhône 14N radial engines instead of the BMW engines in the earlier models

---

famous Condor Legion during the Spanish Civil War and, of the improved versions introduced by 1940, the Do 17Z was most commonly used in raids against England. This type had a completely new, more bulbous forward section for the crew of four or five, with extensive glazing. As further improvements were made, the original beautiful shape was gradually obscured by more pieces of equipment – the fate of many military aircraft. The Do 17Z series was followed by the Do 215, a number of which were used as night-fighters.

Last of the line was the Do 217 which first came into service in 1940, proving itself one of the *Luftwaffe's* most valuable bombers and used operationally until the end of the European war. It was powered at various times by both in-line and radial engines and carried out a wide range of operations, including dive bombing, mine-laying and torpedo bombing, as well as being used as a night-fighter with Lichtenstein airborne interception radar. The E-5 type was developed with radio equipment to launch and guide the Henschel Hs 293 glider bomb.

### The most versatile bomber

Although originally designed as a fast medium bomber when brought into service in 1939, the Ju 88 became the most versatile aircraft of the Second World War, even more so than the Mosquito. The standard bomber type of the early years of the war was the Ju 88A, with a maximum bomb-load of 5,500 lb and air brakes fitted for dive-bombing operations, although this was not its primary function. Later developments included increased armour protection, a longer wing-span, and increased power and defensive armament. With the last bomber version, the Ju 88S, speed was increased to 370 mph at 20,340 ft in an attempt to elude Allied fighters during daytime operations. Total production of the Ju 88, including day and night fighter versions, amounted to some 15,000 machines.

From this highly efficient basic design was produced the more advanced Ju 188 which first appeared in 1942, but relatively few were used purely as bombers. Development continued with the Ju 388 but the only version to go into service was for photographic reconnaissance. The final stage in development, the Ju 488, was still under construction at the end of the war.

The Heinkel He 111, which first flew in 1935, unconvincingly disguised as a commercial transport, was another German plane used by the Condor Legion in the Spanish Civil War. The success achieved then and during the bombing of Poland in 1939 was shattered when it came up against Spitfires and Hurricanes during daylight raids on Britain in 1940, and it had to be assigned to night raids. Nearly 1,000 He 111s had been built by that time and the production of numerous versions, although brought to a halt in Germany at the end of 1944 when over 5,200 had been built, continued for some years afterwards in Spain.

## HEINKEL He 111 B-2

**Gross weight:** 22,046 lb  **Span:** 74 ft 2 in  **Length:** 57 ft 5 in  **Engine:** 2×950 hp Daimler-Benz 600  **Armament:** 3×7·9-mm machine-guns  **Crew:** 4  **Speed:** 186 mph at ground level  **Ceiling:** 22,966 ft  **Range:** 1,030 miles  **Bomb load:** 3,307 lb

The B-2 version of the He 111 served with the Condor legion in the Spanish Civil War, outpacing all opposing fighters and carrying out unescorted raids at will. But the resulting over-confidence was shattered by opposition from Spitfires and Hurricanes in 1940 and the He 111 was soon relegated to night operations

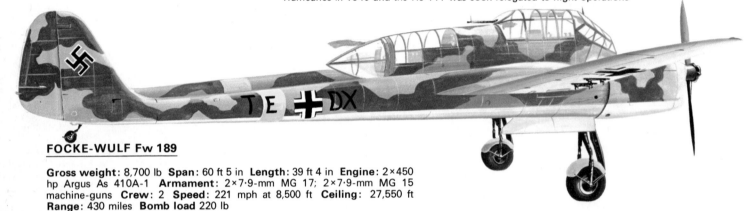

## FOCKE-WULF Fw 189

**Gross weight:** 8,700 lb  **Span:** 60 ft 5 in  **Length:** 39 ft 4 in  **Engine:** 2×450 hp Argus As 410A-1  **Armament:** 2×7·9-mm MG 17; 2×7·9-mm MG 15 machine-guns  **Crew:** 2  **Speed:** 221 mph at 8,500 ft  **Ceiling:** 27,550 ft  **Range:** 430 miles  **Bomb load** 220 lb

A light bomber/ground attack machine which entered service with the *Luftwaffe* at the end of 1940 and was later used primarily against the Russians

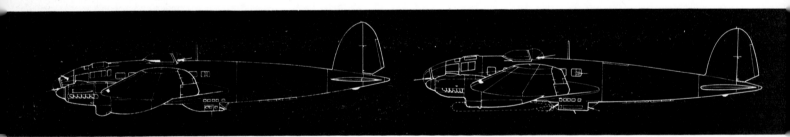

## HEINKEL He 111 P-6

The He 111 P-series entered production in 1938. The P-6 was powered by 1,200 hp DB 601N engines, and had an oval nose section which gave excellent visibility. Three 7·9-mm machine-guns were mounted, and the maximum bomb load was 4,410 lb

## HEINKEL He 111 H-6

The more effective H-series, with Jumo engines, was the most important bomber variant. The H-6 was used both as a bomber and a torpedo-carrier, some of which were fitted with a remotely-controlled 7·9-mm MG 17 in the extreme tail of the fuselage

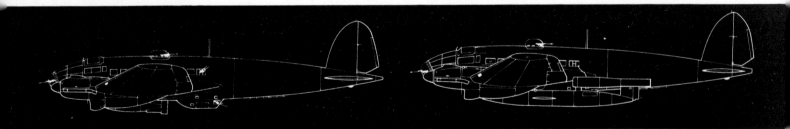

## HEINKEL He 111 H-21

## HEINKEL He 111 H-22

The H-21 and H-22 were basically similar, except in defensive armament. The H-21, the last bomber variant of the He 111, was produced in 1944 and powered by two 1,600 hp Jumo 213 E1 engines which gave a maximum speed of 295 mph and a service ceiling of 32,800 ft. Its defensive armament consisted of one 13-mm hand-held MG 131 in the nose, an electrically-operated dorsal turret with an MG 131, a similar gun in a heavily-armoured ventral gondola and twin 7·9-mm MG 81s in each beam position. Intended exclusively for night-bombing, it had large flame dampers on the exhaust pipes

# ITALY

# One of the war's best bombers

## SAVOIA-MARCHETTI SM 79

**Gross weight**: 23,100 lb **Span**: 69 ft 7 in **Length**: 51 ft 10 in **Engine**: 3×750 hp Alfa Romeo **Armament**: 3× 7·7-mm; 1×12·7-mm machine-guns **Crew**: 5 **Speed**: 267 mph at 13,100 ft **Ceiling**: 21,325 ft **Range**: 1,180 miles **Bomb load**: 2,750 lb

Not only the best Italian but one of the best of any land-based bombers used during the Second World War. Developed in 1936 from a commercial airliner design and later active on virtually every battlefront involving the *Regia Aeronaútica*

When Italy entered the war in 1940, a large number of Caproni Ca 133 high-wing medium bombers were still in service with the *Regia Aeronautica*, and although these were too outdated for bombing duties, they gave excellent service as transports throughout the war. The much faster Ca 135 mid-wing medium bomber, first flown in 1935, was a great improvement but it was not taken into service as better machines were available from other Italian manufacturers. In fact, it would have been superior to the Ca 309 to 314 series of bombers which were used in the early years of the war, and for which Caproni adopted a low-wing arrangement after trying all the other positions.

The closest Italian bomber to the German and British ones of the time was the Fiat BR20 Cicogna (Stork) which was used both in Ethiopia and Spain. It was the only Italian bomber to operate against Britain (from bases in Belgium) and was later used in the Italian invasion of Greece in 1941. A number of improvements were made, including increased defensive armament, and there

were plans to produce it in large numbers before the Italians surrendered. The Caproni and Fiat bombers were twin-engined types, but several three-engined bombers were also produced and used with even greater success. The CRDA Cant Z 1007 *bis* Alcione (Kingfisher), brought into service in 1939, was constructed mostly of wood, but even so it stood up well to the extreme climatic conditions experienced during operations in North Africa and Russia, as well as serving throughout the Mediterranean and Aegean theatres.

## Three-engined Hawk

But undoubtedly the best of the Italian bombers, and one of the best of all land-based bomber aircraft of the Second World War, was the three-engined Savoia-Marchetti SM 79 Sparviero (Hawk). This was also built largely of wood and came into service in 1937, two years after the SM 81 Pipistrello (Bat). The SM 81 was used in Ethiopia and in a wide range of areas after 1940, but because of its relatively

slow speed it had to be relegated to more mundane duties towards the end of the war. The SM 79 was much faster, although the clean lines of the eight-passenger commercial airliner from which it was developed in the late 1930s were somewhat marred by the addition of a dorsal hump in the bomber version, housing two 12·7-mm Breda-SAFAT machine-guns. It was active throughout the war on almost every front, including the Mediterranean, North Africa and the Balkans, and a number served in a transport or training role until 1952.

The only heavy four-engined bomber produced by Italy during the war was the Piaggio P.108B, first used in action in 1942 over Gibraltar. It could carry a maximum bomb-load of 7,700 lb and was notable for its defensive armament of eight 12·7-mm machine-guns. In addition to single guns in nose and ventral turrets and in sideways-firing barbettes amidships, two remote-controlled pairs were installed in the rear upper cowling of each outboard engine nacelle.

## FIAT BR 20

**Gross weight:** 22,266 lb **Span:** 70 ft 9 in
**Length:** 52 ft 9 in **Engine:** 2×1,000 hp
Fiat A80 RC41 **Armament:** 2×7·7-mm;
1×12·7-mm machine-guns **Crew:** 4
**Speed:** 267 mph at 13,120 ft **Ceiling:**
24,935 ft **Range:** 1,700 miles **Bomb
load:** 3,500 lb

A fast, well-armed light bomber which
came into service with the *Regia Aeronautica*
in time to see combat during the Spanish
Civil War. Used for night raids until Italy's
surrender in the Second World War

## SAVOIA-MARCHETTI SM 81

**Gross weight:** 23,000 lb **Span:** 78 ft 9 in
**Length:** 58 ft 5 in **Engine:** 3×680 hp
Piaggio P IX **Armament:** 6×7·7-mm
machine-guns **Crew:** 5 **Speed:** 200 mph
at 3,280 ft **Ceiling:** 22,965 ft **Range:**
930 miles **Bomb load:** 2,200 lb

A 1935 military version of the SM 73
commercial airliner which followed the
current Italian vogue for three engines. It
took part in the Ethiopian campaign and the
Spanish Civil War; outdated by the outbreak of the Second World War, it was
quickly relegated to night operations

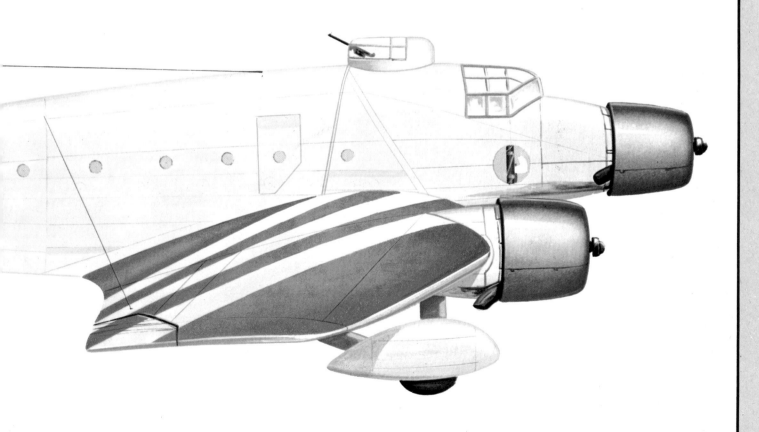

# Fighters lose to bombers

In addition to the bombers produced by the major powers, several types were designed and built by other European nations and used in the early stages of the Second World War, in spite of the extra cost of bombers as against fighters. The most modern aircraft type in the Czechoslovak Air Force in 1938 was the Aero A 304 bomber, based on the prototype for a civil airliner. The A 304 was a three-seat reconnaissance/light bomber, able to carry up to 660 lb of bombs. After the German occupation, the few machines that had been built were handed over to the Bulgarian Air Force for training purposes. In the case of the Fokker TV twin-engined fighter-bomber, which came into service in small numbers with the Netherlands Air Force in 1938, none remained to be taken over by the Germans after their invasion on 10 May 1940, the nine then operational having all been destroyed in attacks on enemy installations. Another Fokker aircraft used for bombing, although it was designed primarily as a bomber-interceptor, was the twin-engined GIA.

The most outstanding bomber produced by the small European nations, the Polish twin-engined PZL P 37 Lós, first flew in 1936, and in spite of being one of the smallest and fastest aircraft of its kind, it could carry up to 5,685 lb of bombs – equal to its own unladen weight. About forty of the type were operational with the Polish Air Force in 1939 and they inflicted heavy casualties on the Germans before Poland finally succumbed. The remaining machines were taken over by the Rumanian Air Force and used throughout the war against Russia.

**PZL P37 LOS B** (above)

**Gross weight:** 19,577 lb  **Span:** 58 ft 10 in  **Length:** 42 ft 5 in  **Engine:** 2×918 hp Pegasus XX  **Armament:** 3×7·7-mm machine-guns  **Crew:** 5  **Speed:** 276 mph at 11,154 ft  **Ceiling:** 19,685 ft  **Range:** 932 miles  **Bomb load:** 5,688 lb

This all-metal stressed-skin medium bomber was one of the best machines produced by the Polish aircraft industry before the war, and inflicted heavy casualties on the invading German armies before Poland's capitulation

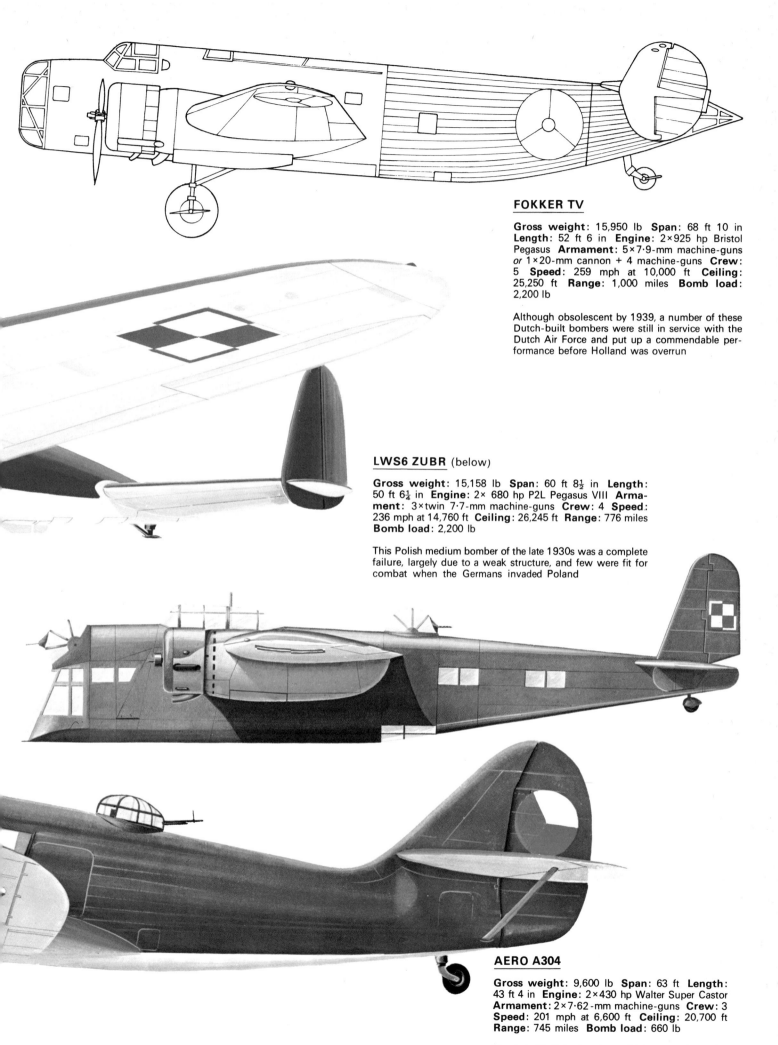

## FOKKER TV

**Gross weight**: 15,950 lb **Span**: 68 ft 10 in
**Length**: 52 ft 6 in **Engine**: 2×925 hp Bristol
Pegasus **Armament**: 5×7·9-mm machine-guns
*or* 1×20-mm cannon + 4 machine-guns **Crew**:
5 **Speed**: 259 mph at 10,000 ft **Ceiling**:
25,250 ft **Range**: 1,000 miles **Bomb load**:
2,200 lb

Although obsolescent by 1939, a number of these
Dutch-built bombers were still in service with the
Dutch Air Force and put up a commendable per-
formance before Holland was overrun

## LWS6 ZUBR (below)

**Gross weight**: 15,158 lb **Span**: 60 ft 8½ in **Length**:
50 ft 6¼ in **Engine**: 2× 680 hp P2L Pegasus VIII **Armament**: 3×twin 7·7-mm machine-guns **Crew**: 4 **Speed**:
236 mph at 14,760 ft **Ceiling**: 26,245 ft **Range**: 776 miles
**Bomb load**: 2,200 lb

This Polish medium bomber of the late 1930s was a complete
failure, largely due to a weak structure, and few were fit for
combat when the Germans invaded Poland

## AERO A304

**Gross weight**: 9,600 lb **Span**: 63 ft **Length**:
43 ft 4 in **Engine**: 2×430 hp Walter Super Castor
**Armament**: 2×7·62-mm machine-guns **Crew**: 3
**Speed**: 201 mph at 6,600 ft **Ceiling**: 20,700 ft
**Range**: 745 miles **Bomb load**: 660 lb

Czech-built light bomber which entered service
with the Czechoslovak Air Force in 1938

# SOVIET UNION

# Obsolescent air force

## PETLYAKOV Pe-8

**Gross weight:** 63,052 lb **Span:** 131 ft 1 in **Length:** 73 ft 9 in **Engine:** 4×AM-35A **Armament:** 2×20-mm; 2×12·7-mm; 2×7·62-mm machine-guns **Crew:** 5 **Speed:** 234 mph at ground level **Ceiling:** 32,972 ft **Range:** 2,038 miles **Bomb load:** 4,400 lb

Designed by A. N. Tupolev to meet a 1934 specification for a fast long-range bomber able to carry a heavy bomb-load, the Pe-8 was overshadowed by the very successful Ilyushin Il-4 and the Yer-2, and relatively few were built

During the early months of the war, until production of more advanced designs got under way, Russian bomber squadrons were equipped largely with obsolete aircraft. The most out-dated was the Tupolev TB-3, first brought into service in 1931, which for some years constituted the only really effective four-engined bomber force in the world. Later models took part in operations against the Japanese in 1938 and in Poland and Finland in 1939. After Germany's invasion of the Soviet Union in 1941 they were used for a short period as night-bombers. Thereafter they continued to see service as transports throughout the war.

A more advanced type was the twin-engined Tupolev SB-2 which first flew in 1934 and served on the Republican side in the Spanish Civil War. Like the Bristol Blenheim, with which it was comparable in performance, the SB-2 was too slow and too poorly armed for daytime operations and in the early stages of the war with Germany it was relegated to night bombing duties.

Apart from the TB-3, the only other Russian four-engined bomber to be introduced during the war was the Petlyakov Pe-8. It was actually designed by A. N. Tupolev as a fast long-range bomber, capable of carrying a heavy bomb-load, but V.M. Petlyakov was responsible for preparing it for series production in 1939. The top speed of the Pe-8 at altitudes between 26,000 and 29,000 ft was faster than the Messerschmitt

Bf 109 fighter. With a range of over 4,800 miles and carrying a bomb-load of 4,400 lb, it undertook many raids deep into Germany. It remained in production until 1944, but relatively few were built as the Russians found the performance of their twin-engined bombers satisfactory.

## Backbone of Russia's air force

The two which were produced in the largest numbers and formed the backbone of the Russian air force during the war were the long-range Ilyushin Il-4 and the Petlyakov Pe-2 tactical bomber, both introduced in 1940. The Il-4 was a development of the all-metal Ilyushin DB-3 of the early 1930s, with a completely redesigned forward fuselage and an elongated nose section to accommodate the navigator/bomb-aimer. Due to a metal shortage in Russia after the German attack, later models were built largely of wood, but this had only a slight effect on their high performance. The Il-4 saw service on every front on which the Russians were engaged against the Axis Powers, but was best known as a long-range bomber, taking part in continuous raids on Germany – including the first Russian attack on Berlin in August 1941.

The Petlyakov Pe-2 was based on an original design for a fast twin-engined fighter and a number were used for that purpose. Some versions were used for dive-bombing, with dive-brakes fitted under the

wings, but the Pe-2 also carried out level bombing and as such has to be considered a true bomber. Bombs could be carried not only internally in the fuselage and rear of the engine nacelles but also externally under the wings.

Another tactical bomber which was mass-produced from 1942 onwards, although not in the quantity of the Pe-2, was the Tupolev Tu-2, which had a maximum speed close to that of a fighter and could be adapted for either level- or dive-bombing. It continued in production after the war and served with the Soviet, Polish and Chinese air forces until the mid-1950s, seeing combat in Korea. In 1941, the Yermolayev Yer-2 long-range bomber was introduced as an intended replacement for the Ilyushin Il-4, but relatively few were built. This also applied to the only other Russian bomber of note in the Second World War, the Archangelskii Ar-2, which saw limited use during the German invasion in 1941.

Mention should be made, however, of the Ilyushin Il-2 Shturmovik, a ground-attack aircraft capable of carrying a bomb-load of up to 1,325 lb, built in single and two-seater versions, and in a class by itself. With its strong armour protection and remarkable handling characteristics at low levels, it was one of the Soviet Union's most effective weapons and was certainly superior to any other ground-attack machine used during the war.

# Armoury of the allies

The American aircraft industry made a tremendous contribution to the war effort, both in terms of numbers and different types of aircraft produced. Total production rose from about 18,000 machines in 1941, when the re-armament programme began to get under way, to 48,000 in 1942, over 90,000 in 1943 and a peak of more than 100,000 in 1944. Not only were the needs of the US armed forces satisfied but thousands of aircraft were supplied to Britain and Russia under lend-lease. Only in the USA was a sustained effort made throughout the war to develop more powerful long-range strike aircraft, and the design work on some of the later types was actually started after the war began.

This achievement was all the more remarkable in view of the fact that when war broke out in Europe, the US Army Air Corps (which became the US Army Air Force in 1942) was equipped largely with obsolete machines owing to the unwillingness of Congress to allocate adequate funds for military expenditure in the 1930s. The US Navy Air Force was in a better position, although even so it was not until 1942 that it received the Curtiss SB2C Helldiver, its most successful dive-bomber, which was also supplied to the Army as the A-25. But as far as USAAC was concerned, the only modern bomber being developed in the mid-1930s was the Boeing B-17.

One of the most remarkable four-engined bombers of the war, the B-17 ultimately bore the brunt of the massive high-altitude daylight raids on Occupied Europe by the USAAF, operating from bases in England. When design work on the B-17 started in 1934, however, it was intended as a purely defensive weapon for offshore anti-shipping operations. It began to enter service in 1939 and was flown operationally for the first time by the RAF in 1940. Although its bomb-load was less than that of the Lancaster, it had the advantage of being able to operate at much greater heights.

## The first Flying Fortress

Armament was less than adequate to begin with and it was only with the introduction of the more heavily armed E series that the type merited the name 'Flying Fortress'. With eight or more guns to each aircraft and flying in close formation to give covering fire, it was hoped that B-17 squadrons would have sufficient protection to fight their way through German air defences and carry out precision bombing by day. Some excellent results were achieved, but at a very heavy cost; as the British had learned in 1939 and the Germans in 1940, heavy bombers were vulnerable to determined fighter opposition, no matter how well armed they were.

Support for the American aircraft industry in the late 1930s was given by the British Purchasing Commission, which was concerned to meet those needs of the RAF which could not be provided by Britain's own industry. In 1938 an order was placed for the twin-engined Lockheed Hudson coastal reconnaissance bomber which was coming into RAF service at the outbreak of war. A Hudson from No. 224 Squadron had the distinction of being the first RAF aircraft to destroy a German aircraft in the war, during a coastal patrol on 8 October 1939. Under the American designation A-28, Hudsons also served with the USAAC. Another Lockheed bomber purchased for the RAF was the Ventura, of which deliveries began in 1942.

Meanwhile, from 1938 onwards, the US Army began holding competitions for aircraft designs to replace the mainly obsolete types then in service. It was in that year that Douglas won the award for a light bomber with its DB-7 Boston. This was ordered in large numbers by the French and British, entered service with the RAF in 1942 and turned out to be one of the best light bombers of the war. It was also built as a fighter under the name Havoc.

In 1939 a competition for medium bomber designs resulted in contracts being awarded for the North American B-25 Mitchell and the Martin B-26 Marauder. The Mitchell became an outstanding aircraft, flying for the first time only one year after its design had been accepted. It was taken into service in 1941, operating mainly in the Pacific to begin with and scoring its first success on

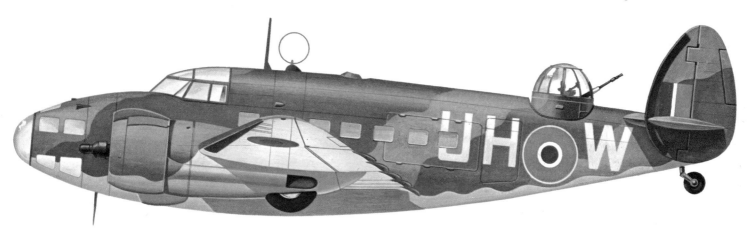

## LOCKHEED HUDSON

**Gross weight:** 20,500 lb  **Span:** 65 ft 6 in  **Length:** 44 ft 4 in
**Engine:** 2×1,200 hp Wright Cyclone GR-1820  **Armament:** 7× ·303 machine-guns  **Crew:** 4  **Speed:** 253 mph  **Ceiling:** 26,500 ft
**Range:** 2,160 miles  **Bomb load:** 1,400 lb

Purchased by the British in 1938 and used extensively by RAF Coastal Command from 1939 to 1943, especially carrying four 325-lb depth charges against submarines. A Hudson with No 224 Squadron was the first RAF machine to shoot down a German aircraft in the Second World War, on 8 October 1939

24 December 1941 against a Japanese submarine. The epic raid on Tokyo in April 1942, led by Lt-Col 'Jimmy' Doolittle, was made by sixteen Mitchells flying from the deck of the aircraft carrier *Hornet*, although they were never intended for such a purpose. The Martin Marauder was also a fine aircraft although its very high wing loading caused several early accidents. It came into service in 1941, mainly in the Pacific, and in 1943 became operational in Europe with the US 8th Air Force. It was also flown by the RAF under the lend-lease programme, as were the two other twin-engined Martin bombers produced during the war, the Maryland and the Baltimore.

While deliveries of the B-17 were starting in 1939, it was decided to design a new four-engined bomber with a higher performance. The resulting Consolidated B-24 Liberator showed little actual improvement on the Boeing, but it served with great distinction in many roles, particularly in the Pacific where its long range – nearly 3,000 miles in later versions – proved of immense value. This was achieved mainly because of a new type of high aspect ratio wing, named Davis after its inventor. The Liberator was also used by RAF Coastal Command as a patrol bomber over the Atlantic, in areas previously out of range to land-based bombers, and joined the B-17 in the USAAF's historic raids over Occupied Europe.

Early in 1940, an official requirement was issued for a 'Hemisphere Defense Weapon', having a very long range and the ability to carry a much greater bomb-load than previous American aircraft. Of the four designs submitted, that of the Boeing B-29 Superfortress – conceived as early as 1938 – was judged to be the best.

### Pressurised cabins
From the outset, the B-29 was planned to have pressurised cabins for its crew to allow it to operate at very great heights. The main factor which made the B-29 possible was the development of the large (3,350 cu in displacement) Wright R-3350 Duplex Cyclone 18-cylinder two-row radial engine, which had been launched as a bold private-venture programme by Wright Aeronautical in 1936. By 1940, prototypes of this engine were being provisionally rated at 2,200 hp, enabling Douglas to build the enormous but unsuccessful B-19, flown in 1941, and

opening the way for more capable and useful four-engined aircraft such as the B-29 (first flown on 21 September 1942) and the Lockheed Constellation.

The B-29 engines were each fitted with two General Electric turbo-superchargers, very similar to the single units fitted to the engines of the B-17 and B-24, to give high power at altitudes up to 35,000 ft. The propellers were four-blade Curtiss electrically controlled units, bigger than any previously used in a production aircraft. Operations of the B-29 by the USAAF began in the summer of 1944, initially from bases in India and China against the Japanese, and it was responsible for most of the devastating attacks on the Japanese mainland, including the two atomic bombs dropped on Hiroshima and Nagasaki in August 1945.

One of the companies which submitted designs for a Hemisphere Defense Weapon was Consolidated. Although it did not win, its XB-32 was felt to be promising enough to be given an order, resulting in the Consolidated B-32 Dominator which was first flown two weeks before the B-29. Smaller than the Boeing, the Dominator was adapted for a lower-altitude role in the Pacific, but relatively few were built.

Meanwhile, in 1941, as a result of the 'Atlantic Charter' meeting between the leaders of the USA and Britain, the USAAF planned a bomber far larger even than the B-29. This was intended for the almost incredible task of bombing Germany from bases in North America, should Britain capitulate. The required range of more than 7,500 miles, combined with a bomb-load of 10,000 lb – and a maximum of 72,000 lb over shorter distances – was almost beyond the limit of what was possible in the Second World War. Eventually, Convair won the design competition with its XB-36, designed around six as-yet-undeveloped Pratt & Whitney R-4360 engines in the 3,500 hp class, arranged inside a huge wing to drive pusher propellers. These propellers had to have a diameter of 19 ft, even larger than those of the B-29. So enormous was the task of creating the B-36 that it was not completed until after the war was over, first flying on 8 August 1946. Deliveries to the US Strategic Air Command began in 1947; the last version was retired from service with SAC in 1959.

One bomber which was brought into

combat service before the end of the war, even though the prototypes were not ordered until mid-1941, was the twin-engined Douglas A-26 Invader, intended to replace the Boston attack bomber. Operational use began at the end of 1944, first in the European theatre and later in the Pacific. The Invader lasted in operational service much longer than originally expected; it was used not only in Korea but also in Vietnam in the 1960s.

Finally, another Douglas aircraft which deserves mention is the single-engined SBD Dauntless, the most successful American dive-bomber of the Second World War. Although this was built primarily for the US Navy, beginning operational service at the end of 1941, it was also used by the USAAF.

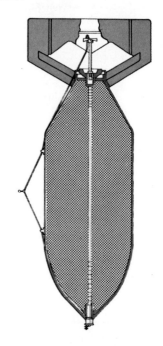

### US LIGHT CASE BOMB

**Total weight**: 4,000 lb **Length**: 117·25 in **Fin width**: 47·62 in **Charge/weight ratio**: 80%

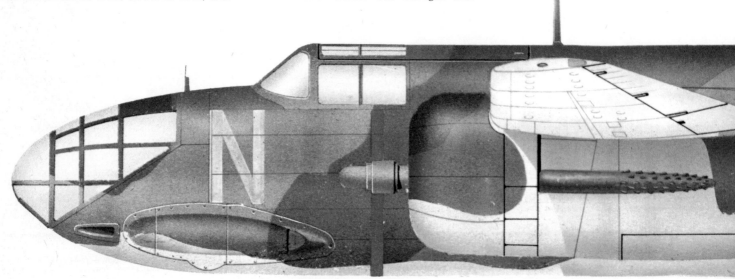

## US ARMOUR PIERCING BOMB

Total weight: 1,000 lb  Length: 73 in  Fin width: 16·6 in  Charge/weight ratio: 14·5%

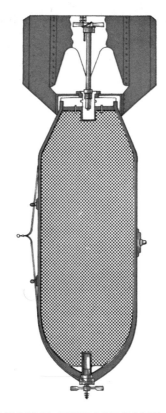

## US GENERAL PURPOSE BOMB

Total weight: 500 lb  Length: 59·16 in  Fin width: 18·94 in  Charge/weight ratio: 51%

## US SEMI-ARMOUR PIERCING BOMB

Total weight: 1,000 lb  Length: 70·4 in  Fin width: 20·72 in  Charge/weight ratio: 31%

## US FRAGMENTATION BOMB

(dropped in clusters of 6)

Total weight: 19·8 lb  Length: 21·8 in  Fin width: 5·13 in  Charge/weight ratio: 13%

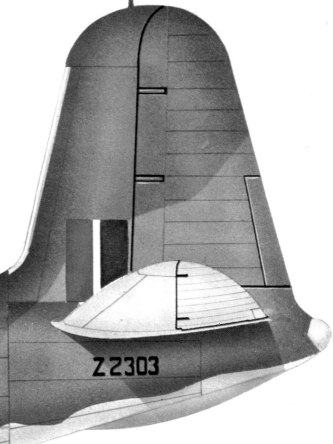

## DOUGLAS DB-7 BOSTON

Gross weight: 15,150 lb  Span: 61 ft 4 in  Length: 46 ft  Engine: 2×1,100 hp Pratt & Whitney R-1830  Armament: 7× ·303 machine-guns  Crew: 3  Speed: 322 mph at 12,500 ft  Ceiling: 33,800 ft  Range: 462 miles  Bomb load: 2,080 lb

One of the most widely used light bombers of the Second World War, the Boston was also developed as a fighter under the name Havoc

# JAPAN

# Less than the best

## MITSUBISHI Ki-30

**Gross weight:** 7,324 lb **Span:** 47 ft 9 in
**Length:** 33 ft 11 in **Engine:** 1×850 hp 14 cyc.
Nakajima **Armament:** 2×7·7-mm machine-guns
**Crew:** 2 **Speed:** 263 mph at 13,125 ft **Ceiling:**
28,120 ft **Range:** 1,056 miles **Bomb load:**
660 lb

The Ki-30 light bomber made its debut in the Sino-Japanese War and was still in JAAF service in the Pacific in 1942, although it was seldom used after the Philippines campaign

While Japanese carrier-based aircraft, such as the Mitsubishi A6M Zero-Sen fighter and the Nakajima B5N torpedo-bomber, were among the best of their kind used during the war, the same could hardly be said of Japanese bombers. Although the Japanese might have been expected to gain a great deal from a force of heavy long-range bombers, only one four-engined type ever entered service (the Nakajima G5N1, code-named 'Liz' by the Allies) and though much larger than any other of its generation, with a 138 ft wing-span, it was never used as a bomber at all. All the Japanese bombers brought into service just before and during the war had two air-cooled radial engines, the major constructor being Mitsubishi.

Mainstay of the Japanese Navy Air Force's land-based bomber force was the Mitsubishi G3M, first flown in 1935 as a major step forward in Japanese aviation and making its combat debut in the war against China in 1937. It was code-named 'Nell' under the Allied identification system and together with the later Mitsubishi G4M 'Betty' which was brought into service in 1941, took part in the attack and sinking of the Royal Navy battleships *Prince of Wales* and *Repulse* on 10 December 1941. The G4M had an impressive range of more than 3,000 miles but its necessarily large load of gasoline, carried in unprotected tanks, made it even more vulnerable than most Japanese aircraft, and it became known as the 'one-shot lighter' among

American pilots. Heavy losses were suffered, especially in the battle for the Solomon Islands in 1942, but only towards the end of 1943 was an attempt made to provide smaller, fully-protected fuel tanks. A large number of G4Ms were equipped to carry and launch the Ohka piloted suicide bomb which caused considerable damage in the later stages of the war.

The Japanese Army Air Force's standard heavy bomber at the time of the attack on Pearl Harbour on 7 December 1941 was the Mitsubishi Ki-21 ('Sally'). This was too slow and poorly armed to be very effective, but its intended replacement, the Nakajima Ki-49 ('Helen'), which began to enter service in 1941, suffered from poor range and inadequate bomb-load in trying to achieve higher speed with heavier armament. It was not until 1944 that the JAAF received a really first-class bomber, the Mitsubishi Ki-67 Hiryu (Flying Dragon), code-named 'Peggy', which first saw combat in the Battle of the Philippine Sea. Unlike most Japanese bombers, which had too light an armament, the Ki-67 was formidably armed with four heavy machine-guns and a 20-mm cannon. Its speed and manoeuvrability also showed a marked improvement on previous types. Towards the end of the year several of these aircraft were converted into three-seater suicide bombers.

A smaller and lighter machine was the Kawasaki Ki-48 ('Lily') which entered service with the JAAF in 1940 and was used both for level and dive-bombing. It

had the range to bomb Port Darwin, Australia, from bases in New Guinea, but its bomb-load of 660 lb (1,765 lb in later models) was too modest.

The last and best of the Japanese bombers to see combat before the war ended, although limited by a shortage of skilled pilots and high-octane fuel, was the Yokosuka P1Y1 Ginga (Milky Way), code-named 'Frances'. It had a speed in excess of 345 mph at 19,300 ft, was well armed with a 20-mm nose cannon and one or two dorsal heavy machine-guns, had a high rate of climb, and a combat range of nearly 3,000 miles. It was used for both level and dive bombing by the JNAF, while torpedo-bomber and night-fighter versions were also developed. During 1945, one of the machines was used to flight-test a turbo-jet engine, but the war ended before further advances in jet-propulsion could be made.

The only long-range strategic bomber with a pressurised cabin to be produced in Japan during the war was the Tachikawa Ki-74 ('Patsy'), originated in 1939 for the JAAF. Little interest was shown in its possibilities until after Pearl Harbour, and its maiden flight was not made until May 1944. With a range of 4,350 miles, a maximum speed of 360 mph at 28,000 ft, and an internal bomb-load of 2,200 lb, the Ki-74 could have been of considerable value to the Japanese, but although a small number had been delivered by the summer of 1945, the war ended before combat operations could be carried out.

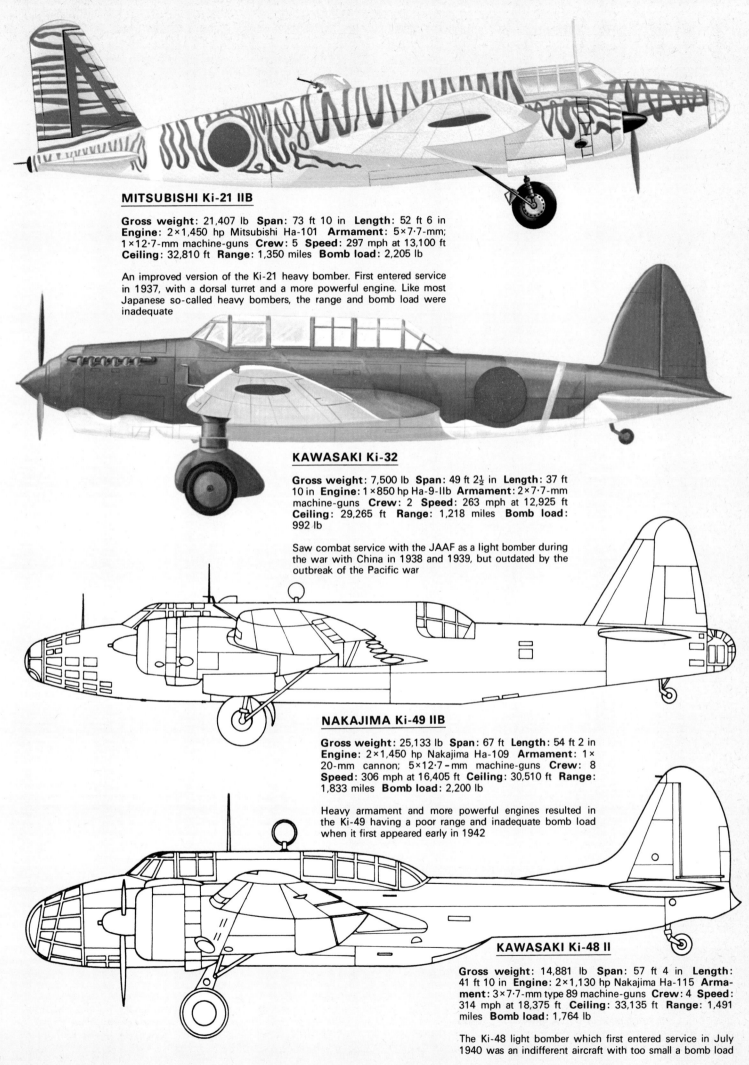

## MITSUBISHI Ki-21 IIB

**Gross weight:** 21,407 lb **Span:** 73 ft 10 in **Length:** 52 ft 6 in
**Engine:** 2×1,450 hp Mitsubishi Ha-101 **Armament:** 5×7·7-mm;
1×12·7-mm machine-guns **Crew:** 5 **Speed:** 297 mph at 13,100 ft
**Ceiling:** 32,810 ft **Range:** 1,350 miles **Bomb load:** 2,205 lb

An improved version of the Ki-21 heavy bomber. First entered service
in 1937, with a dorsal turret and a more powerful engine. Like most
Japanese so-called heavy bombers, the range and bomb load were
inadequate

## KAWASAKI Ki-32

**Gross weight:** 7,500 lb **Span:** 49 ft 2½ in **Length:** 37 ft
10 in **Engine:** 1×850 hp Ha-9-IIb **Armament:** 2×7·7-mm
machine-guns **Crew:** 2 **Speed:** 263 mph at 12,925 ft
**Ceiling:** 29,265 ft **Range:** 1,218 miles **Bomb load:**
992 lb

Saw combat service with the JAAF as a light bomber during
the war with China in 1938 and 1939, but outdated by the
outbreak of the Pacific war

## NAKAJIMA Ki-49 IIB

**Gross weight:** 25,133 lb **Span:** 67 ft **Length:** 54 ft 2 in
**Engine:** 2×1,450 hp Nakajima Ha-109 **Armament:** 1×
20-mm cannon; 5×12·7-mm machine-guns **Crew:** 8
**Speed:** 306 mph at 16,405 ft **Ceiling:** 30,510 ft **Range:**
1,833 miles **Bomb load:** 2,200 lb

Heavy armament and more powerful engines resulted in
the Ki-49 having a poor range and inadequate bomb load
when it first appeared early in 1942

## KAWASAKI Ki-48 II

**Gross weight:** 14,881 lb **Span:** 57 ft 4 in **Length:**
41 ft 10 in **Engine:** 2×1,130 hp Nakajima Ha-115 **Armament:** 3×7·7-mm type 89 machine-guns **Crew:** 4 **Speed:**
314 mph at 18,375 ft **Ceiling:** 33,135 ft **Range:** 1,491
miles **Bomb load:** 1,764 lb

The Ki-48 light bomber which first entered service in July
1940 was an indifferent aircraft with too small a bomb load

# STRUCTURE AND DESIGN
## Swordfish to Superfortress

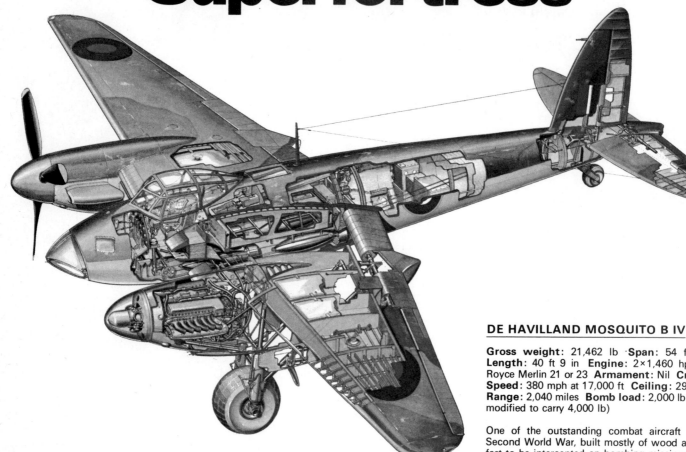

**DE HAVILLAND MOSQUITO B IV**

**Gross weight:** 21,462 lb ·**Span:** 54 ft 2 in **Length:** 40 ft 9 in **Engine:** 2×1,460 hp Rolls Royce Merlin 21 or 23 **Armament:** Nil **Crew:** 2 **Speed:** 380 mph at 17,000 ft **Ceiling:** 29,000 ft **Range:** 2,040 miles **Bomb load:** 2,000 lb (some modified to carry 4,000 lb)

One of the outstanding combat aircraft of the Second World War, built mostly of wood and too fast to be intercepted on bombing missions; guns were only carried on fighter variants. The B IV was the first bomber version to enter RAF service

The Second World War was delayed just long enough for almost all of the old aircraft with fabric-covered airframes of wood, mixed wood and metal or steel-tube, to have been withdrawn from service and replaced by modern types. The redoubtable Fairey Swordfish carrier-based torpedo-bomber was the only aircraft of the older type to serve throughout the war in the European theatre, its low performance being balanced by robust construction, ease of repair, good weight-lifting ability from small carrier decks and general utility except in the face of lethal defensive fire. Among other fabric-covered machines to survive were the Vickers Wellington and Warwick twin-engined bombers which remained in production until 1945.

But as a rule, bombers of the Second World War had light-alloy stressed-skin construction, built in cantilever monoplane form with monocoque fuselages. Undercarriages were made to retract into the wing, engine nacelles, or other compartments, at first by laborious winding of a hand-wheel and later by directing hydraulic power to a ram or electric power to a motor. The original hydraulic systems were crude and heavy, filled with oil at 1,000 lb/ sq in.

Later aircraft had more complex systems operating at 2,500 lb/sq in and serving undercarriage retraction, wheel brakes, landing flaps, bomb doors and sometimes propeller pitch or gun-turret drive. American aircraft favoured all-electric systems, with as many as one hundred and eighty motors to drive various items of equipment.

### More powerful flaps

With wing loadings rising from around 10 lb/sq ft, typical in 1930, to 40/65 lb/sq ft, the wings needed powerful flaps to increase lift at take-off (when they were depressed to about ten degrees) and at landing (depressed to the maximum of perhaps forty-five degrees to give both extra lift and extra drag). The Armstrong Whitworth Whitley, designed as a troop carrier around 1932, was redesigned as a bomber in 1934. The original design had no wing flaps and the wing was accordingly set at a large angle of incidence. The flapped Whitley always flew in a characteristic nose-down attitude, and would probably have been faster with the wing made thinner and set at a shallower angle.

Junkers devised a 'double wing' type of flap comprising a completely separate

auxiliary surface hinged well behind the main wing trailing edge. This was fitted to the Ju 86 and Ju 87, but the far more effective Ju 88 had powerful conventional slotted flaps. Both the Ju 87 and Ju 88 had special dive-brakes hinged under the wings and extended broadside-on to the airflow to permit very steep attacks (almost ninety degrees) without reaching excessive speeds.

In the USA, Lockheed used the advanced Fowler flap, still in evidence today. This extended backwards on wheeled carriages running on fixed tracks to give extra area on take-off. When fully extended, the tracks pulled the carriages sharply down to give very high drag as well. This was a useful feature of the Lockheed Hudson and Ventura, designed specifically for British requirements, which would otherwise have been difficult to land. Even with full flap, the Martin B-26 Marauder was not easy to land, and this hastened the development of the modern type of aircraft which is driven on to the runway at high speed and stopped with powerful brakes.

Special mention should be made of the four-engined Consolidated B-24 Liberator, possibly the first of the modern bombers. This was based on the Davis wing, a design of very high aspect ratio giving great efficiency and capable of heavy loading to well over 60 lb/sq ft. Such a wing was a radically new innovation when the Liberator first flew, as the Model 32, in 1939. It was only possible because of its advanced stressed-skin structure, which allowed a large quantity of fuel to be carried inside the slender wing as well as the main units of the landing gear and the Fowler flaps. Compared with British bombers such as the Avro Lancaster, which had thin skin and heavy highly-stressed spar booms, the wing of the Liberator was uniformly loaded throughout with slimmer spars and thicker skin, reinforced by many long stringers.

There was little that was unconventional about the structures of most of the main Second World War bombers. Nearly all had stressed-skin airframes, with quite thin skin, held by flush riveting and with large, high-tensile steel pins securing the fork fittings where wings joined the fuselage or where outer wings joined the centre section. Compared with earlier aircraft the smooth, regular skin made performance higher and more precisely predictable, and the time-consuming task of airframe rigging receded into the background. Yet occasionally, there were still maverick aircraft which did not conform to the general pattern. Either they consistently flew slower than the others, or suffered from disconcerting and

sometimes dangerous flight characteristics – or they proved themselves to be among the most successful aircraft of the war.

In the latter category, apart from the geodetic Vickers bombers, an outstanding example was the de Havilland Mosquito. This cantilever mid-wing monoplane, initiated as a private venture in 1938 and nearly cancelled before it first flew in November 1940, ultimately became one of the most versatile combat aircraft of the war. It was planned purely as a high-speed unarmed bomber fast enough to avoid interception, and for various reasons, including the metal shortage, it was built of wood. The de Havilland company had devised a method of making the main sections from a sandwich

## CONSOLIDATED B-24 LIBERATOR

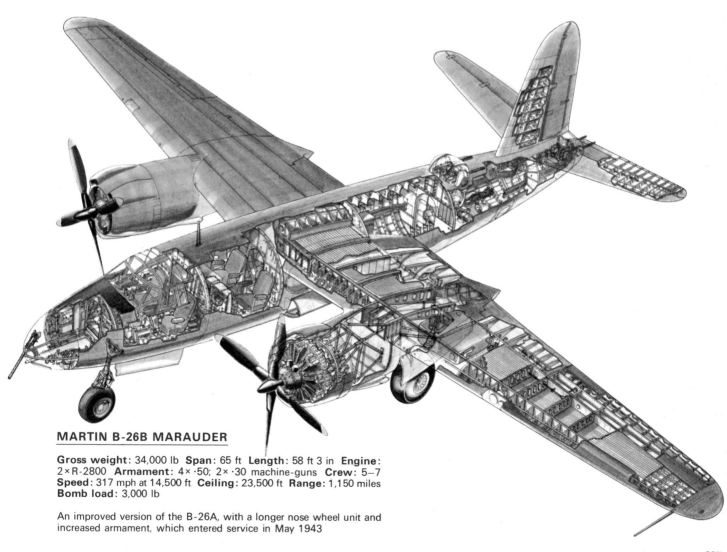

## MARTIN B-26B MARAUDER

**Gross weight**: 34,000 lb  **Span**: 65 ft  **Length**: 58 ft 3 in  **Engine**: 2 × R-2800  **Armament**: 4 × ·50; 2 × ·30 machine-guns  **Crew**: 5–7  **Speed**: 317 mph at 14,500 ft  **Ceiling**: 23,500 ft  **Range**: 1,150 miles  **Bomb load**: 3,000 lb

An improved version of the B-26A, with a longer nose wheel unit and increased armament, which entered service in May 1943

of hardwood ply veneers with a thick core of light balsa wood, which gave a robust, rigid and smooth-surfaced airframe. Light alloy and steel fittings were used at the main stress areas and joints, and fabric formed a top skin over the entire airframe, including the control surfaces which were generally of light alloy.

Since a wooden structural member has a larger section than a metal one, the Mosquito had an outstanding ability to absorb battle damage. Shell splinters and bullets which would have severed a metal structure merely left a hole in the wooden one. This ability, combined with its speed of over 400 mph, made Mosquito losses towards the end of the war by far the lowest of any RAF bomber – only one per 2,000 sorties. As wood structures are inherently heavier than metal, the efficiency and weight-lifting power of the Mosquito was remarkable, and it was the only light bomber capable of carrying a 4,000 lb block-buster bomb. It was particularly effective against pin-point targets, but its manoeuvrability and performance at altitudes of up to 40,000 ft also made it an outstanding night fighter.

Apart from the Fi 103 Flying Bomb, produced by the Germans towards the end of the war with a predictably simple structure assembled from welded steel sheet, the only other bomber to make a major break with tradition and still be put into large-scale production was the Boeing B-29 Superfortress, brought into operation in 1943. Planned as the ultimate strategic bomber that could be created with the technology of 1940, it brought together every available new advance that appeared worthwhile. The great range (3,250 miles) and payload (16,000 lb of bombs) demanded led to an unprecedented size and weight, and this in turn led to the use of light-alloy sheet several times thicker (up to 0·188 in) than anything used in such machines as the Heinkel He 111 or Lancaster.

### Absolute precision
This at once precluded the kind of workshop practice that had been common since the start of stressed-skin construction. No longer could sections of skin be filed to fit, or ill-fitting rivets be put into re-drilled holes. Each hole had to be a precision job,

in exactly the correct place to mate with the underlying structure, and countersunk to take a large rivet or screw fitting flush with the surface. It was no longer sufficient merely to bend the skins around the wings or fuselage as they were riveted on. They had to be formed in three dimensions by large press-tools. Though far stronger and more robust than any earlier airframes, the B-29 wing structure did have the drawback of being difficult to repair without special equipment.

The B-29 also marked a great increase in complexity. For example, it had five electrically-driven gun-turrets, four of them remotely controlled from sighting stations, with an override system to give a gunner control of all four turrets if he should have a good target. A further notable advance was that the crew were accommodated in a pressure cabin, linked by a pressurised tunnel to the smaller pressurised compartment of the two rear-fuselage gunners and the pressurised tail turret. All in all, the B-29 was a tremendous stride forward towards the even more complicated jet bombers of today.

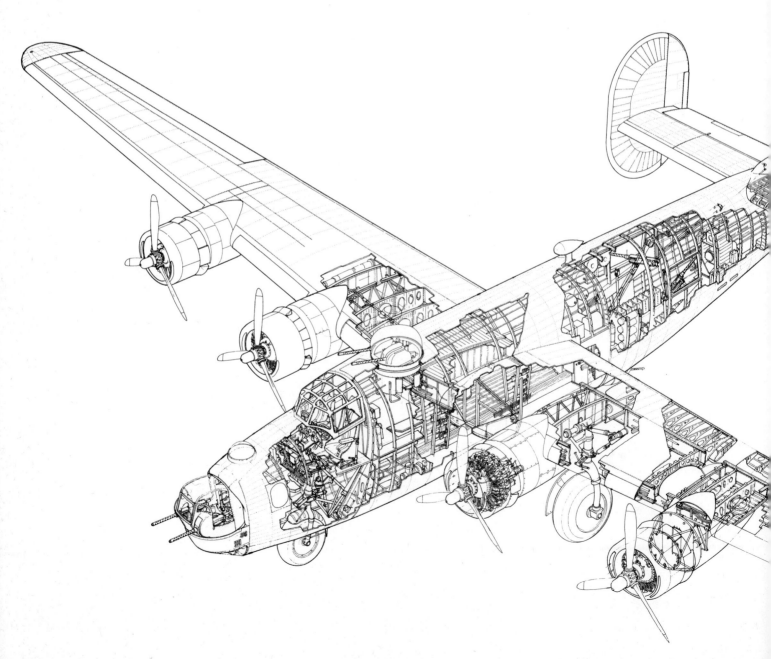

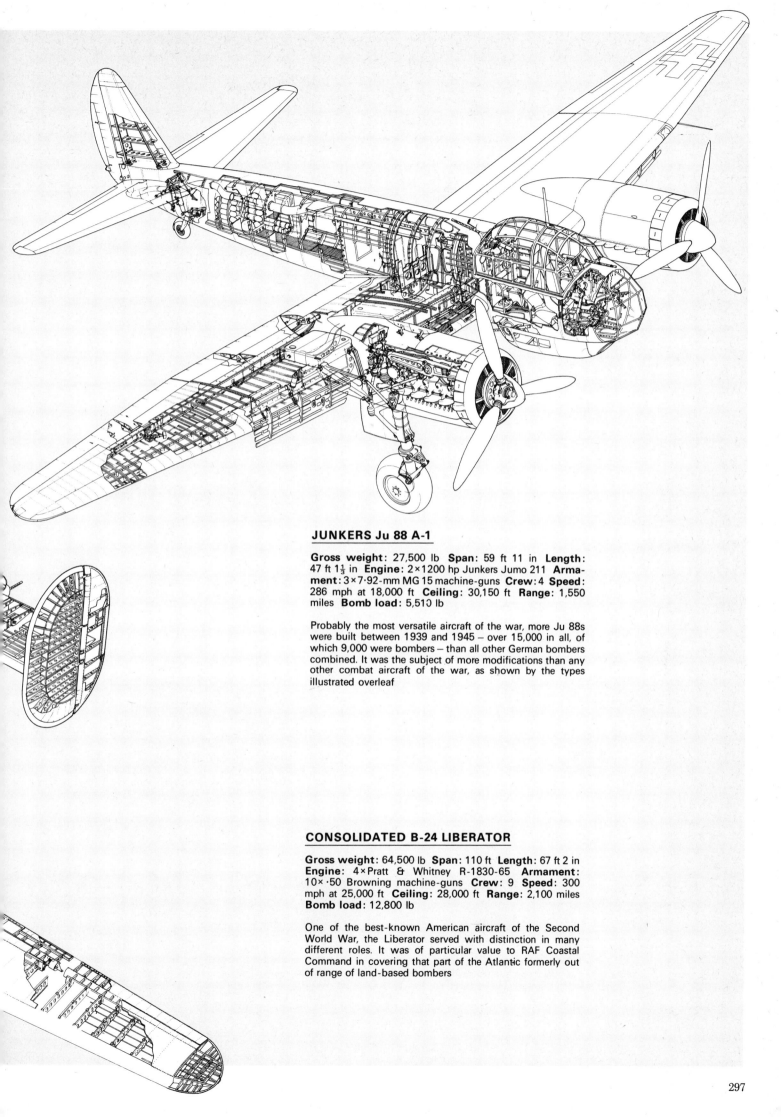

## JUNKERS Ju 88 A-1

**Gross weight:** 27,500 lb **Span:** 59 ft 11 in **Length:** 47 ft 1⅓ in **Engine:** 2×1200 hp Junkers Jumo 211 **Armament:** 3×7·92-mm MG 15 machine-guns **Crew:** 4 **Speed:** 286 mph at 18,000 ft **Ceiling:** 30,150 ft **Range:** 1,550 miles **Bomb load:** 5,510 lb

Probably the most versatile aircraft of the war, more Ju 88s were built between 1939 and 1945 – over 15,000 in all, of which 9,000 were bombers – than all other German bombers combined. It was the subject of more modifications than any other combat aircraft of the war, as shown by the types illustrated overleaf

## CONSOLIDATED B-24 LIBERATOR

**Gross weight:** 64,500 lb **Span:** 110 ft **Length:** 67 ft 2 in **Engine:** 4×Pratt & Whitney R-1830-65 **Armament:** 10×·50 Browning machine-guns **Crew:** 9 **Speed:** 300 mph at 25,000 ft **Ceiling:** 28,000 ft **Range:** 2,100 miles **Bomb load:** 12,800 lb

One of the best-known American aircraft of the Second World War, the Liberator served with distinction in many different roles. It was of particular value to RAF Coastal Command in covering that part of the Atlantic formerly out of range of land-based bombers

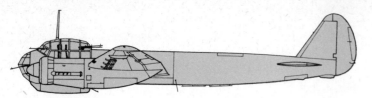

## JUNKERS Ju 88 A-5

The A-5 was similar to the A-1, which made its debut in September 1939, but had its wing-span increased to 65 ft 10½ in, and a larger bomb load (6,614 lb) which reduced its maximum speed to 273 mph

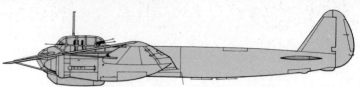

## JUNKERS Ju 88 A-6

Generally similar to the A-5, the A-6 carried as standard a balloon-cable fender and destroying gear. This was so unwieldly, however, that it was soon withdrawn from service

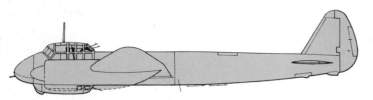

## JUNKERS Ju 88 C-4

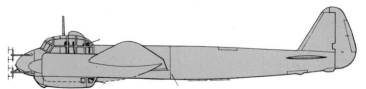

## JUNKERS Ju 88 C-6b

The C-series were all primarily fighters, developed parallel with the A-series of bombers

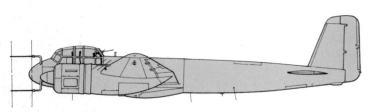

## JUNKERS Ju 88 G-4

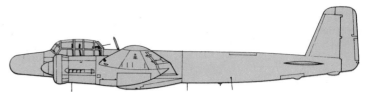

## JUNKERS Ju 88 G-7c

The G-series of Ju 88s were specialised night-fighters

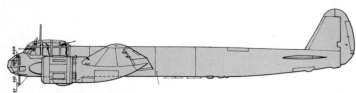

## JUNKERS Ju 88 H-1

The H-1, characterised by an elongated fuselage to house extra fuel tanks, was a long-range photo reconnaissance aircraft

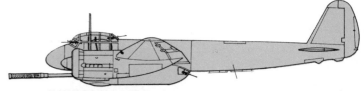

## JUNKERS Ju 88 P-1

The P-1 was an anti-tank ground attack machine, used mainly on the Russian front and armed with a single forward-firing 75-mm BK 7·5 long-barrel cannon, in addition to rearward-firing twin MG 81 machine-guns fitted in a gun fairing which could be jettisoned if necessary

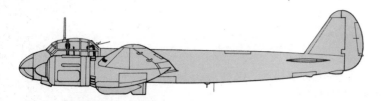

## JUNKERS Ju 88 S-1

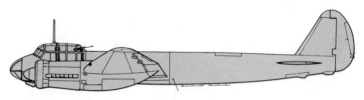

## JUNKERS Ju 88 S-3

By 1943 the A-series were too slow for unescorted daylight operations, so development went ahead on the S-series, in which the bomb-load was reduced to 1,760 lb, and the upper nose guns and ventral gondola removed in order to increase speed. The glazed nose reverted to a shape similar to the early prototypes. The S-1 was the initial production model of the series, with 1,700 hp BMW 801G engines with GM-1 power boost, and only one rearward-firing 13-mm MG 131. The S-3 was powered by 1,750 hp Jumo 213E-1 engines

# ARMAMENT

# The sting in the tail

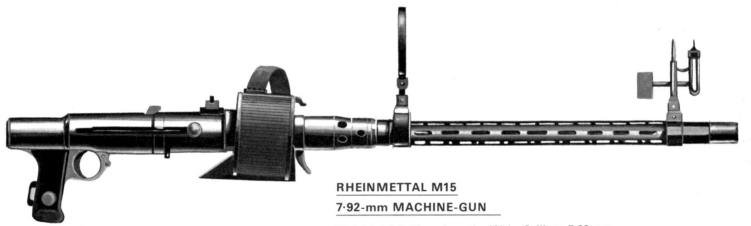

**RHEINMETTAL M15**

**7·92-mm MACHINE-GUN**

**Weight:** 15 lb 12 oz  **Length:** 42½ in  **Calibre:** 7·92-mm
**Magazine:** saddle type of 75 rds  **Rate of fire:** 1100 rpm

The atmosphere of extreme financial stringency that surrounded armaments in most countries during the 1920s tended to restrict the pace of development and efforts were concentrated on making the best and most economical use of available equipment. This was invariably similar to that used in the First World War, if not actually ex-wartime stock. Thus the dominant aircraft guns in the RAF were the manually trained Lewis and the fixed Vickers, while the Marlin remained the chief American gun, despite the fact that both the rifle-calibre Browning and the 20-mm Hispano had completed their initial development before November 1918 and were available for use.

In the inter-war period, the Scarff ring, usually carrying single or twin Lewis guns, was almost universal for RAF bomber defence and was widely used elsewhere. Much heavier guns were carried in some aircraft, but only in a somewhat ineffectual and experimental way. For example, several British machines such as the Vickers and Westland COW-armed designs, the Bristol Bagshot and the Westland Westbury, had fitted the big 37-mm (1·46-in) Coventry Ordnance Works cannon dating from 1918. Later, in 1934, the Blackburn Perth three-engined flying-boat, then the RAF's biggest and heaviest aircraft, went into service with one of these guns in the bows, in addition to three Lewis guns on the usual Scarff rings at bows, amidships and stern. The 1½ lb shells could have been dangerous to surfaced submarines, but there is no record that they were regarded as useful defence weapons.

Brief trials were also conducted in the USA, France and the Soviet Union with 20-mm and 37-mm guns on large aircraft, but with inconclusive results.

A far more important trend was the realization that, as aircraft speeds were rising steadily from the 100 mph maximum of the typical bomber of the mid-1920s to at least twice that speed, some form of shelter would be needed by the gunners. Indeed, there was evidence that the aerodynamic drag of the guns, when firing to either side, would become so great as to make manual aiming impossible. The answer appeared to be some form of power-driven turret, such as had long been used on warships; the main difficulties were weight and the provisions of drive power.

**Windshields to turrets**
The first companies to produce workable turrets were Martin in the USA, with a nose turret for the Type 130 B-10 bomber mounting a 0·30-in Browning gun, and Boulton Paul in Britain with a pneumatic-motor traversed nose turret for the Overstrand, fitted with a Lewis. Turrets were also produced in France by 1934, but these were at first manually operated and served merely as windshields. The Amiot 143 for instance used un-powered turrets for the Lewis gun in the nose and for two similar guns in the ventral position amidships; together with another gun in the rear cabin section under the wing, the Amiot 143 could defend itself against attack from any direction, foreshadowing the B-17 Flying Fortress. Later, the French MAC 1934 rifle-calibre gun supplemented and then replaced the Lewis.

Crude turrets also appeared on small numbers of Italian and Russian heavy bombers and on such flying-boats as the Short Sunderland, Consolidated PB2Y Coronado, Martin PBM Mariner, Blohm und Voss 138, Kawanishi H6K and H8K, and several Russian Beriev designs. There were many instances in the Second World War of

flying-boats fending off repeated attacks by fighters. A Sunderland, attacked by six Ju 88s, shot down one and crippled a second; another Sunderland shot down three Ju 88s out of eight – simply because of fire-power. Large aircraft were no longer defended merely by three Lewis guns on Scarff rings but by multiple belt-fed guns accurately trained by power-driven turrets.

By 1935, it was widely accepted that to have any hope of survival in enemy airspace, a large bomber would have to have power-driven gun-turrets. The turret was no longer a mere windshield but a precision aiming device, complete with a sighting system and operator controls so that the gunner could train his guns effortlessly and accurately, even with a 200-knot slipstream or when being pulled in a tight turn. By the start of the Second World War, there were almost twenty companies making power-driven bomber defence turrets. Every large bomber in production in either Britain or the USA had at least one turret.

But in most other countries, the power-driven turret was ignored, or accorded low priority. Germany clung to the belief that bombers could be adequately defended by two or four hand-held machine-guns, and similar thinking persisted in Italy, Japan and the Soviet Union. In France, however, every conceivable kind of bomber defence was tried – including catapults lobbing aerial mines timed to explode close to attacking fighters. Several French bombers had powered turrets, in some cases mounting 20-mm cannon, such as the extremely graceful twin-engined LeO 451, an excellent aircraft but which appeared too late to be of value to the French during the Second World War; a number were later used as transports by the *Luftwaffe*. The earlier Farman 223, first produced in 1937, had

299

20-mm guns in electrically-powered dorsal and ventral turrets which, in a hectic month of combat in June 1940, proved to be very effective even at long ranges, although handicapped by the fact that the gunners had only lately seen the equipment for the first time.

A typical example of how the turret evolved during the period before the war was the case of the Vickers Wellington. The prototype B 9/32, flown in June 1936, had transparent domes in the nose and tail which in production aircraft were expected to carry a single Lewis gun fired by hand through a slot, sealed on each side of the gun by a sliding wind-shield. By 1937, the Mark I production machine had power-driven turrets of a rudimentary kind. Each mounted twin belt-fed Browning guns which were elevated and traversed hydraulically by a Frazer-Nash system, while firing through an aperture in a broad flexible belt, arranged to slide freely in runners between the fixed upper (transparent) and lower parts of the installation. By 1939 the Mark 1A had gone into production with powered turrets of modern design, each with twin Brownings, and the Mark III of 1941 mounted the four-gun rear turret introduced to the RAF with the Whitley Mark IV and the production Sunderland of 1938.

This rapid advance in bomber armament had been made despite the fact that in the mid-1930s, the RAF had no modern gun of any kind, nor any under development. The position had been rectified in July 1935 by the conclusion of a licence agreement with the Colt Automatic Weapon Corporation for the conversion of their Browning machine-gun to take British 0·303-in rimmed ammunition and for the manufacture of the resulting gun in Britain. Later, in 1939, a licence was obtained for the 20-mm (0·787-in) Hispano cannon which went into large-scale production at the British MARC factory at Grantham and the BSA plant at Sparkbrook, though the first deliveries of this much bigger gun, in July 1940, were all for fighters.

With very few exceptions, such as the French bombers, the only defence of large combat aircraft in 1939 comprised rifle-calibre guns fired either by hand and trained manually or, in the case of British heavy bombers, aimed by a power-driven turret. It was to be a matter of profound importance that the *Luftwaffe*, misled by the ease with which the He 111 and Do 17 operated over Spain in 1937–39, standardised bomber-defence armament with three or four hand-held MG-15 machine-guns. Although this 7·92-mm gun was a good modern design, it was fed by hand-loaded magazines and in any case lacked the punch needed to deter the fighters the German bombers were soon to meet over England.

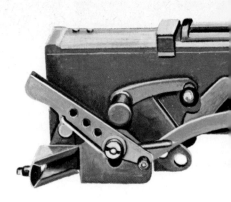

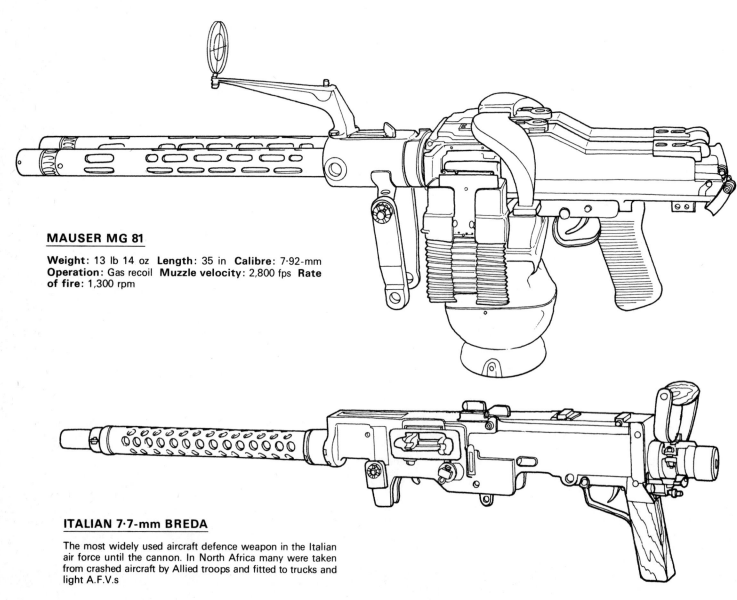

## MAUSER MG 81

**Weight:** 13 lb 14 oz **Length:** 35 in **Calibre:** 7·92-mm
**Operation:** Gas recoil **Muzzle velocity:** 2,800 fps **Rate of fire:** 1,300 rpm

## ITALIAN 7·7-mm BREDA

The most widely used aircraft defence weapon in the Italian air force until the cannon. In North Africa many were taken from crashed aircraft by Allied troops and fitted to trucks and light A.F.V.s

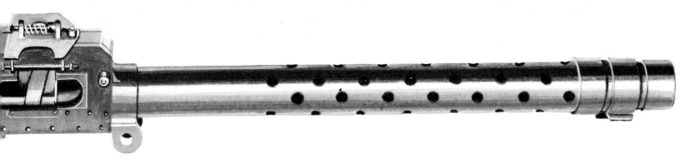

## JAPANESE 7·7-mm TYPE 89
## AIRCRAFT GUN

**Calibre**: 7·7-mm  **Magazine**: belt feed  **Muzzle
velocity**: 2,070 fps  **Rate of fire**: 550 rpm

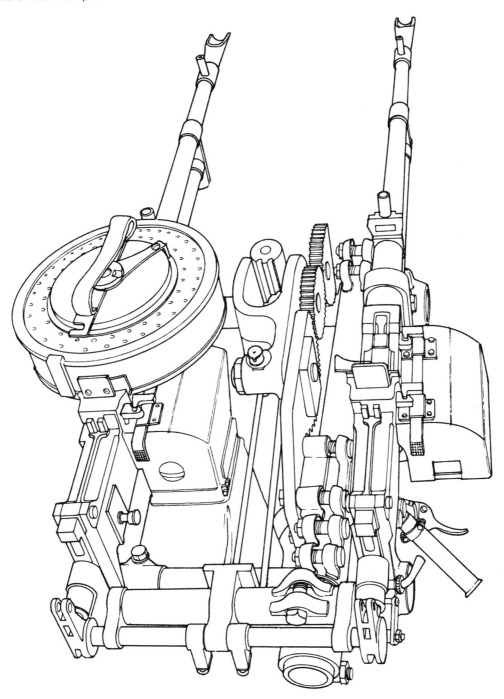

### TWIN ·303 LEWIS A.A. MOUNTING

A typical British A.A. mount used against low flying aircraft.
This pair of Lewis machine-guns could send up a total firepower of 1,150 rpm

## B-17D 'FLYING FORTRESS'

**Gross weight:** 50,000 lb **Span:** 103 ft 9 in **Length:** 67 ft 11 in
**Engine:** 4×Wright Cyclone R-1820 **Armament:** 6×·50; 1×·30
machine-guns **Crew:** 9 **Speed:** 323 mph at 25,000 ft **Ceiling:**
37,000 ft **Range:** 2,100 miles **Bomb load:** 10,500 lb

Spearhead of the USAAF's daylight raids on Occupied Europe. The
D model carried less armament than later versions, one of which – the
B-40 fighter – carried up to 30 machine-guns and cannon in an
unsuccessful attempt to provide an escort for B-17 formations

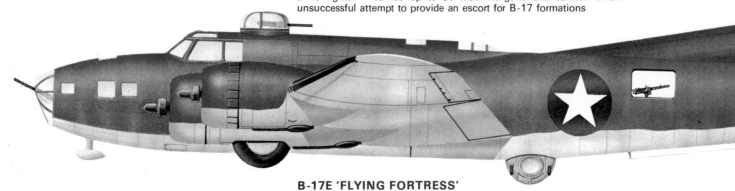

## B-17E 'FLYING FORTRESS'

**Gross weight:** 53,000 lb **Span:** 103 ft 9 in **Length:** 73 ft 10 in
**Engine:** 4×Wright R-1920-65 **Armament:** 15 machine-guns **Crew:**
9 **Speed:** 317 mph at 25,000 ft **Ceiling:** 36,600 ft **Range:** 3,300
miles **Bomb load:** 4,000 lb

This version of the famous B-17 was the first to live up to the name
'Flying Fortress', with the addition of tail, ventral and front upper gun
turrets, the last power-operated

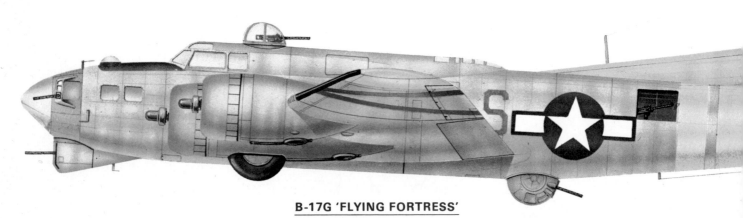

## B-17G 'FLYING FORTRESS'

**Gross weight:** 65,500 lb **Span:** 103 ft 9 in **Length:** 74 ft 4 in
**Engine:** 4×Wright Cyclone R-1820 **Armament:** 11 machine-guns
**Crew:** 10 **Speed:** 287 mph at 25,000 ft **Ceiling:** 35,600 ft **Range:**
2,000 miles **Bomb load:** 6,000 lb

Among other modifications, the G model of the B-17 introduced a
two-gun 'chin' turret to help repel attacking fighters

## BENDIX UPPER GUN TURRET
### TYPE A9B

**Elevation and operation power:** $\frac{1}{2}$ hp electric motor **Rotation power:** $\frac{1}{2}$ hp motor **Guns:** 2 × Browning M2 ·50-in **Radius:** 360° **Max. rotation:** 33° per second **Elevation:** horizontal – 92° **Foot charging:** left and right

A profile cam stops guns firing when any part of the aircraft comes in line of fire – to prevent the gunner shooting off his own tail, for example

Except for the special case of the de Havilland Mosquito, which was able to rely on its remarkable flight performance, British bombers of the Second World War relied for their defence mainly on power-driven turrets equipped with rifle-calibre Browning guns. Operational experience resulted in a gradual shift of the main weight of firepower towards the rear. For example, the Handley Page Halifax began life in November 1940 with a two-gun nose turret, two-gun dorsal turret and four-gun tail turret, but from 1943 onwards the standard armament was merely a single hand-operated gun in the nose and four-gun dorsal and tail turrets.

### Extended fire

Together with the Lancaster, the Halifax formed the backbone of RAF Bomber Command's night attacks on Occupied Europe between 1941 and 1945, and in both aircraft the turrets could fire for extended periods. In 1939, the typical ammunition capacity of a turret was 500 rounds per gun, but for the last two years of the war British heavy bombers frequently carried 8,000 rounds, weighing almost 800 lb, stored in long boxes in the rear fuselage. There was no powered feed system, each Browning merely pulling its long belt by tension from the gun itself. The gunner, seated in the turret behind local small areas of armour-plate, had a reflector sight and either a joystick or handlebar-type controls for elevation and traverse. Flash eliminators

prevented the gunner from losing his night-adapted vision when he fired at attacking fighters.

Although the penalty in weight and drag of these turrets was considerable, there was never any doubt they were effective even though the gunner often had to fire just as the aircraft entered a violent evasive manoeuvre. In sharp contrast, the policy of the day-bombers of the US Army Air Force (which became the US Air Force in 1946) was to maintain straight and level flight in formation and to defend themselves against fighter attack by massed firepower. By 1941, American bombers had abandoned most rifle-calibre guns in favour of the Browning 0·5-in (12·7-mm), used in both hand-held and turret mountings. By 1943 the heavies of the US 8th Air Force, which daily penetrated German airspace, were in greater need of defensive armament than bombers had ever been. Though each aircraft mounted ten or eleven 0·5-in guns, eight of them in four powered turrets, the opposition was so intense that losses were very severe. Increasingly, the enemy fighters devised methods of attack that reduced their exposure to the bombers' fire, eventually standing-off and firing at long range with 30-mm cannon, rockets and, ultimately, guided missiles. Though the American bombers carried more than 2,000 lb of armour, and the crews wore new flak jackets which protected the torso by overlapping squares of manganese steel, the slow day-bomber was sorely pressed.

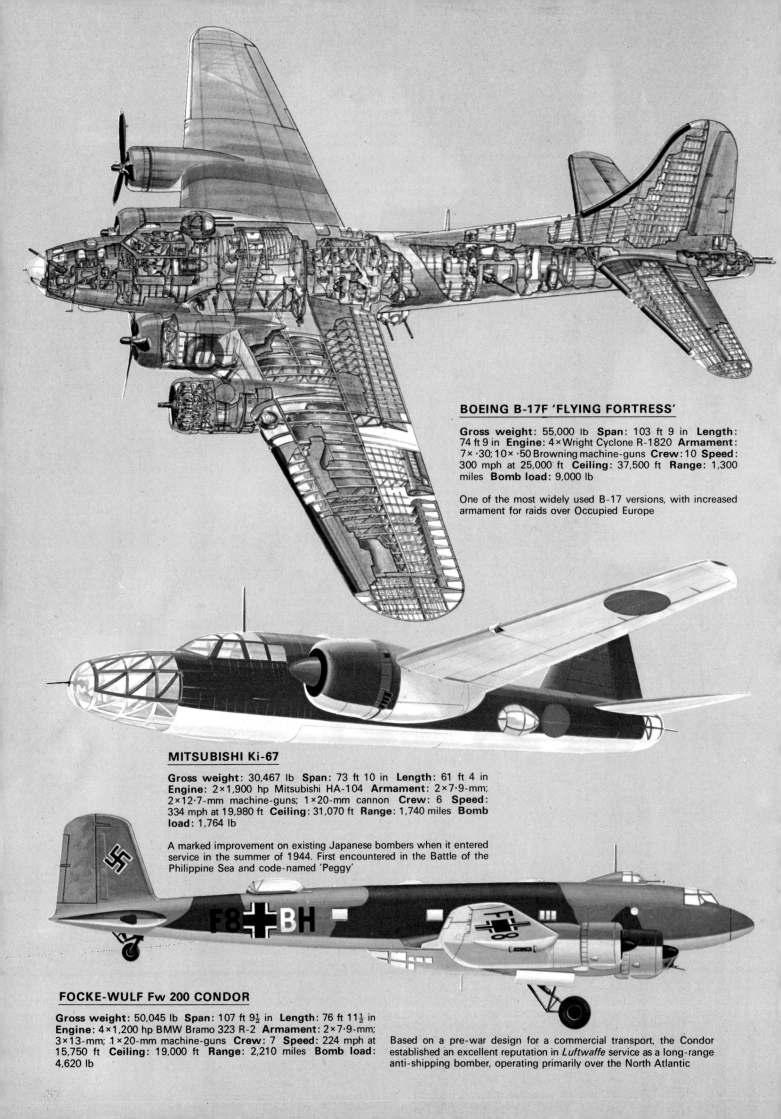

## BOEING B-17F 'FLYING FORTRESS'

**Gross weight:** 55,000 lb **Span:** 103 ft 9 in **Length:** 74 ft 9 in **Engine:** 4×Wright Cyclone R-1820 **Armament:** 7×·30; 10×·50 Browning machine-guns **Crew:** 10 **Speed:** 300 mph at 25,000 ft **Ceiling:** 37,500 ft **Range:** 1,300 miles **Bomb load:** 9,000 lb

One of the most widely used B-17 versions, with increased armament for raids over Occupied Europe

## MITSUBISHI Ki-67

**Gross weight:** 30,467 lb **Span:** 73 ft 10 in **Length:** 61 ft 4 in **Engine:** 2×1,900 hp Mitsubishi HA-104 **Armament:** 2×7·9-mm; 2×12·7-mm machine-guns; 1×20-mm cannon **Crew:** 6 **Speed:** 334 mph at 19,980 ft **Ceiling:** 31,070 ft **Range:** 1,740 miles **Bomb load:** 1,764 lb

A marked improvement on existing Japanese bombers when it entered service in the summer of 1944. First encountered in the Battle of the Philippine Sea and code-named 'Peggy'

## FOCKE-WULF Fw 200 CONDOR

**Gross weight:** 50,045 lb **Span:** 107 ft 9½ in **Length:** 76 ft 11½ in **Engine:** 4×1,200 hp BMW Bramo 323 R-2 **Armament:** 2×7·9-mm; 3×13-mm; 1×20-mm machine-guns **Crew:** 7 **Speed:** 224 mph at 15,750 ft **Ceiling:** 19,000 ft **Range:** 2,210 miles **Bomb load:** 4,620 lb

Based on a pre-war design for a commercial transport, the Condor established an excellent reputation in *Luftwaffe* service as a long-range anti-shipping bomber, operating primarily over the North Atlantic

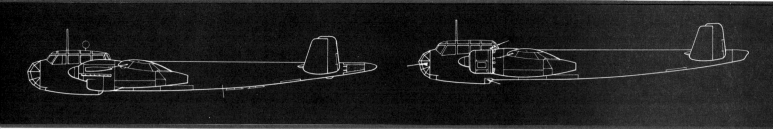

### DORNIER Do 217 VI

The Do 217 was a development of the basic Do 17 design, following the Do 215. The VI, the first prototype, flew in August 1938. It was stressed for bigger loads and had increased wing span (62 ft 4 in) and length (60 ft 10 in). A new air-brake in the form of a ribbed umbrella which opened out from the tail extension was also incorporated

### DORNIER Do 217 V7

The air-brake fitted to the VI prototype proved troublesome and was rarely used operationally. It was abandoned on the V7 prototype, which was the first to be fitted with radial engines – two 1,550 hp BMW 139s

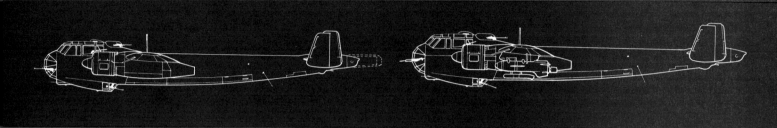

### DORNIER Do 217 E-2

The E-series were the first production models of the Do 217. The E-2, initially intended as a dive-bomber, was fitted with the umbrella air-brake, though this was soon removed. It carried a 2,200-lb bomb load, and was armed with one 13-mm MG 131 in a power-operated dorsal turret and another in the ventral position

### DORNIER Do 217 E-5

The final production model of the E-series, the E-5 was designed to carry the Henschel Hs 293 rocket-propelled radio-controlled glider bomb on carriers outboard of the engine nacelles

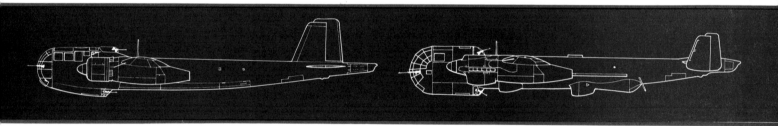

### DORNIER Do 217K VI

The K-series of which the K VI was a prototype, was developed from the E-series, with a redesigned, more rounded nose, and powered by two BMW 801D radials

### DORNIER Do 217P-O

A high-altitude reconnaissance-bomber developed in 1942. The P-series had increased wing-span (80 ft 4½ in) and were powered by two DB 603A engines, with a DB 605T in the fuselage driving a two-stage compressor. 2,200 lb of bombs and six MG 81 machine-guns were carried, while speed (388 mph) and ceiling (53,000 ft) were both increased

German bombers even earlier had shown themselves to be anything but invincible, particularly during the Battle of Britain. Their design was such that all that could be done, short of an extensive redesign for which the German aircraft industry was not equipped, was a succession of modifications. During the Battle of Britain, the Ju 88 was urgently modified until, in its A-4 version, the original three MG-15 machine-guns had been augmented by two to three additional MG-15s, a 13-mm MG-131 cannon and usually one or two fixed MG-81 machine-guns. In many aircraft the upper rear gunner had to manage four separate hand-held MG-15s projecting at different angles on ball-and-socket mountings. Turrets were rare, although MG-131 guns in electrically-powered turrets were fitted to the Do 217 E-2 and many other Do 217 sub-types, and various Fw 200 C Condors had turrets with either the MG-131 or the far more powerful MG-151/15, a high-velocity 15-mm weapon.

But German armament was at best a botch-up born of urgent necessity. Late models of the Do 217 had four fixed rearward-firing MG-81s in the extreme tail and some had a pair of these fixed in the rear of each

engine nacelle, all eight guns being 'sighted' and fired by the pilot – using nothing more accurate than a rear-view periscopic sight.

The one area in which the Germans did considerable pioneering work was in remotely controlled 'barbettes' (unmanned gun positions). These were lighter than the conventional form of turret and offered less drag. Moreover, in theory at any rate, it was possible for a gunner at a single sighting station to control several barbettes and bring many heavy-calibre guns to bear on a single target. The first aircraft to be equipped with this type of armament was the Messerschmitt Me 210 two-seater night fighter, with a 13-mm MG-131 on each side of the rear fuselage, aimed by the navigator in the rear cockpit. A refined version of the same scheme was used on the much more numerous Me 410. The He 177 heavy bomber usually had a 13-mm MG-131 barbette on the upper forward fuselage and all the candidates for the new Bomber-B specification – the Do 317, Fw 191 and Ju 288 – would have had three to five remotely controlled cannon barbettes offering very great firepower and low drag. But the war ended before these could be put into production.

The armament of Italian, Japanese and Russian bombers was less advanced and chiefly consisted of 7·7-mm, 12·7-mm or 20-mm guns in individual hand-held mountings. However, the Japanese did use a crude form of electrically-driven 20-mm turret on the Mitsubishi Ki-67 and G4M2 bombers, produced respectively for the Army Air Force and Navy Air Force, while the Soviet Union adopted a manned unpowered turret in the rear of both inner engine nacelles on the big four-engined Petlyakov Pe-8.

The greatest and most advanced form of firepower was used in the American heavy bombers. The Douglas A-26 Invader of 1944 was fitted with remotely-controlled upper and lower twin 0·5-in turrets and usually had very heavy fixed nose armament fired by the pilot, who could also use the upper turret when it was locked facing forward. But the biggest advance of the whole war was seen in the Boeing B-29 Superfortress strategic bomber which, though fully pressurised, had five sighting stations and five powered turrets (four of them remotely controlled) with ten or twelve 0·5-in and one 20-mm guns, giving it a remarkable concentration of firepower.

# ENGINE DEVELOPMENT

# Forcing the pace

**HANDLEY PAGE HALIFAX Mk II**

**Gross weight:** 60,000 lb **Span:** 98 ft 10 in **Length:** 71 ft 7 in **Engine:** 4×1,390 hp Rolls Royce Merlin XX **Armament:** 8×·303 Browning machine-guns **Crew:** 7 **Speed:** 285 mph at 17,500 ft **Ceiling:** 23,000 ft **Range:** 1,860 miles **Bomb load:** 13,000 lb

The performance of the Mk II Halifax was improved by higher powered engines. The nose gun turret of the Mk I, which experience had shown was seldom used, was replaced with a streamlined fairing

Most of the engines used in the bombers of 1939 were sound and reliable and capable of considerable further refinement. There was no need to increase the number of engines merely to provide against failure of one engine. For example one of the longest-ranged bombers of the 1930s, the Vickers Wellesley – which for many years held the world absolute distance record at 7,162 miles – had only a single Pegasus engine. The slightly later Vickers bomber, the Wellington, had two Pegasus engines simply because it was bigger and heavier. But during the war there was soon abundant evidence that attacks against well-defended targets were best made by four-engined aircraft (excepting the unique Mosquito) which could make it home even with one or two engines knocked out by defensive fire.

Whereas the reliable engines of the late 1930s had made the twin-engined airliner as safe as one with four engines, there seemed a good case for fitting a bomber with four engines or even more. It is strange that the British Air Ministry, among those of other nations, should have resisted this trend. In 1936 it grudgingly allowed four engines for the Short Stirling, yet planned the proposed new heavy bombers from Handley Page and A.V. Roe around two very powerful engines of a totally untried type. Fortunately for the Halifax these untried engines, the Rolls-Royce Vulture, were expected at first to be in short supply so its design was changed to specify four Merlins. The rival Avro Manchester stayed with the Vulture, with the result that it was plagued by engine failures and was taken out of service in 1942.

### Ignoring the obvious

In the *Luftwaffe* the position was even worse. The He 177, and later the Ju 288, Do 317 and Fw 191, were all planned to use two engines of new design. The result was that the He 177 was notorious for engine failure, while the others never completed their development. All the time it was glaringly obvious that with four DB 605, Jumo 213 or BMW 801 engines distributed conventionally across the wing span, the Germans could have fielded a bomber of excellent range and striking power.

It is paradoxical that, apart from the rather unsuccessful Stirling, the only four-engined bombers planned in the 1930s to make their mark during the war were the US Army's B-17 and B-24. Other four-engined heavy bombers were put into limited service by the Soviet Union, Italy and Japan, but played only a minor role.

At the outbreak of the Second World War, the Bristol air-cooled radial engines provided 52 per cent of the horsepower of the RAF – and for bombers the percentage was even higher. The pinnacles of development of the poppet-valve engine family were the 1,520 cu in Mercury, with nine $6\frac{1}{2}$-in-stroke cylinders, and the 1,753 cu in Pegasus with a one inch longer stroke.

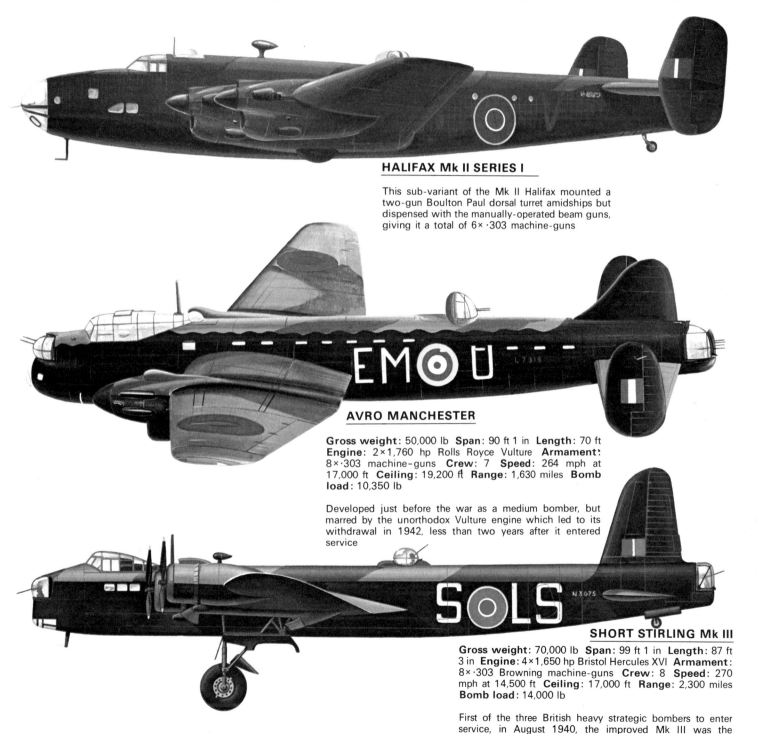

**HALIFAX Mk II SERIES I**

This sub-variant of the Mk II Halifax mounted a two-gun Boulton Paul dorsal turret amidships but dispensed with the manually-operated beam guns, giving it a total of 6×·303 machine-guns

**AVRO MANCHESTER**

**Gross weight:** 50,000 lb **Span:** 90 ft 1 in **Length:** 70 ft **Engine:** 2×1,760 hp Rolls Royce Vulture **Armament:** 8×·303 machine-guns **Crew:** 7 **Speed:** 264 mph at 17,000 ft **Ceiling:** 19,200 ft **Range:** 1,630 miles **Bomb load:** 10,350 lb

Developed just before the war as a medium bomber, but marred by the unorthodox Vulture engine which led to its withdrawal in 1942, less than two years after it entered service

**SHORT STIRLING Mk III**

**Gross weight:** 70,000 lb **Span:** 99 ft 1 in **Length:** 87 ft 3 in **Engine:** 4×1,650 hp Bristol Hercules XVI **Armament:** 8×·303 Browning machine-guns **Crew:** 8 **Speed:** 270 mph at 14,500 ft **Ceiling:** 17,000 ft **Range:** 2,300 miles **Bomb load:** 14,000 lb

First of the three British heavy strategic bombers to enter service, in August 1940, the improved Mk III was the standard version in 1943 and 1944

1 W/T visual indicator
2 Fuel jettison valve controls
3 Undercarriage hydraulic instruments
4 Fuel instruments
5 Bomb doors and bomb release lights
6 RPM indicator instruments
7 Throttles
8 Airscrew pitch controls
9 Mixture controls
10 Way to nose turret and bomb aimer's position
11 Compass
12 Pilot's seat
13 Control column
14 Trim controls
15 Ground steering controls
16 Oxygen connection
17 Bomb sight steering indicator
18 Blind landing indicator
19 Radio compass
20 Flap indicator
21 Main electrics switches
22 Undercarriage indicator lights
23 Boost instruments
24 Turn and bank indicator
25 Compass repeater
26 Altimeter
27 Air speed indicator
28 Artificial horizon
29 Climb and descent indicator

## HALIFAX COCKPIT

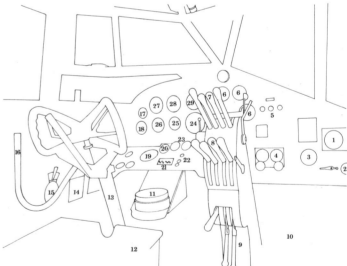

Both were exceedingly refined engines, with geared drive, single or two-speed supercharger, automatic boost control and provision for a variable-pitch or constant-speed propeller. Both gave over 900 hp, well over twice the power of the original 1,753 cu in Jupiter. In the second half of the 1930s both were put into enormous volume production by a scheme of 'shadow factories' set up by Bristol, the Air Ministry and several car firms. These provided engines for such aircraft as the Harrow, Bombay, Hampden, Blenheim, Sunderland, Swordfish, Wellesley, Wellington and the Gladiator fighter. The Pegasus, in particular, was also widely built in other countries.

By 1937 Roy Fedden's design team at Bristol had finally solved the problem of how to mass-produce standardized and interchangeable sleeves, and the first of the sleeve-valve engines, the 905 hp Perseus, was coming into production. The first unit to use them was an RAF squadron of Vildebeeste torpedo-bombers, which found justification for the claim that the new kind of engine offered smoother running, greater efficiency (with lower consumption of both fuel and oil) and much better potential for future growth in power. These factors, combined with lower cost, faster manufacture and greater reliability, all stemmed from a dramatic reduction in the number of parts. The new engines were cleared for production just in time to play a major role in the Second World War.

The Perseus was soon followed by the 1,050/1,200 hp Taurus, a two-row engine only 46 inches in diameter, used in the Beaufort and Albacore torpedo-bombers, and then by the bigger two-row Hercules. This began life in 1939 at 1,375 hp and by the end of the war was rated at up to 1,800 hp. The Hercules was vital to the RAF programme of large aircraft, though in 1939 it was flying only in test and experimental aircraft. Bristol had an even bigger engine, the Centaurus, planned for a power range from 2,000 to 3,500 hp, but this was still on the bench when war was declared.

Armstrong Siddeley Motors had never been able to rival the success of Bristol, but their 350 hp Cheetah radial was made in very large numbers and before the war was in service with the Avro Anson coastal reconnaissance bomber. The much bigger two-row Tiger, of 795/910 hp, powered the Blackburn Shark torpedo-bomber and the first models of the Armstrong Whitworth Whitley, but all the later Whitleys used Merlins.

The Rolls-Royce Kestrel, used in the Heyford and Hendon heavies of the RAF in 1933–39, hardly figured at all during the war although many hundreds of these 460/745 hp engines had been fitted into Hawker Hart and Hind day-bombers only a few years previously.

Around 1936, it looked as if future heavy bombers were going to use large and very complicated engines such as the Rolls-Royce Vulture, the Rolls-Royce Exe, the Fairey P24 Prince, and the Armstrong Siddeley Deerhound. In fact none of these was ever to see operational service apart from the short and unhappy period with the Vulture already mentioned.

In Germany there were three main classes of engine. The most conventional were the BMW 132 (derived from the American Pratt & Whitney Hornet) and Bramo Fafnir air-cooled radials, the former giving about 850 hp and the latter about 900/1,000 hp – but at the cost of higher weight and fuel consumption. The second German category consisted of the vee petrol engines. The BMW VI had its output raised from 620/660 hp to 725/750 hp by increasing the compression ratio, but in 1934, Daimler Benz and Junkers had been developing much more powerful inverted-vee engines which, by 1937, had been cleared for service. These were the DB 600 and the Jumo 210, both giving over 700 hp – the former soon reaching 900 hp.

### Fuel injection

By 1937 development was fast proceeding on two later engines, the DB 601 and Jumo 211, both in the 1,000 hp class. These were destined to play an all-important role in the coming conflict, and were unusual in several respects. One was their inverted arrangement, which was used to advantage by bomber manufacturers to achieve a clean installation, with the engine hung from two large, forged, light-alloy beams. A second was that their superchargers were mounted on one side at the rear, the impeller shaft running transversely. But the most important new feature was that by 1938 both engines used direct fuel injection. Instead of having a carburettor, they had a system for measuring the engine's fuel requirement

and arranging for a 12-plunger metering pump to deliver a precise amount of petrol to each cylinder in turn. This system could not ice up, gave rapid starting in all climates and continued to function perfectly even with the aircraft inverted or in any other manoeuvre. Its only possible drawback was that it was complex, though this only affected production man-hours, not reliability.

The third German category comprised the Jumo two-stroke diesels. The first of these unusually thin, flat engines was flown in 1929, and by 1934 the Jumo 205 was in production at about 570 hp for the Ju 86, the *Luftwaffe's* first really effective bomber. The heavy-oil diesels were more efficient than petrol engines (so that even allowing for the greater weight of the engine, the weight of engines and fuel for a long bombing mission was less), while their less-volatile fuel reduced the risk of catching fire after suffering battle damage. Yet the Jumo diesels faded from the scene and played only a small role in a most unexpected type of mission, bombing from the strato-sphere.

While the French and the Italians rather inefficiently produced a wide range of different families of engines of every conceivable type, the USA had by 1939 abandoned practically everything in favour of the excellent and highly refined air-cooled radials from Pratt & Whitney and Wright. The main P & W engines were the R-1830 Twin Wasp, which had reached the 1,200 hp level, and the much bigger R-2800 Double Wasp which in 1939 was running on the test-bed.

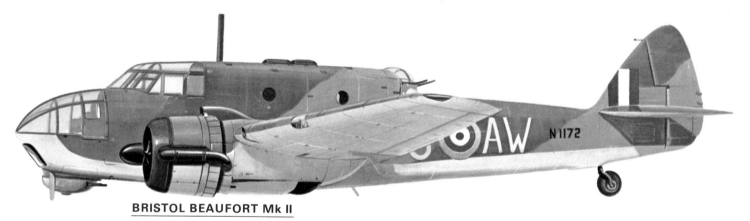

### BRISTOL BEAUFORT Mk II

**Gross weight:** 21,050 lb **Span:** 57 ft 10 in **Length:** 44 ft 7 in **Engine:** 2×1,200 hp Twin Wasp **Armament:** 4×·303 machine-guns **Crew:** 4 **Speed:** 268 mph **Ceiling:** 25,000 ft **Range:** 1,054 miles **Bomb load:** 1×1,605 lb torpedo or 1,500 lb bomb load

RAF Coastal Command's standard torpedo-bomber from 1940 to 1943, when it saw service in the North Sea, English Channel, the Atlantic, Mediterranean and North Africa

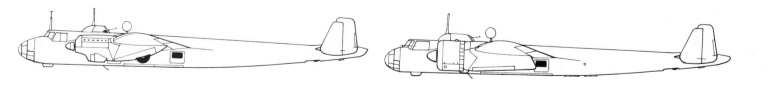

### DORNIER Do 17 E-1

The first production model of the Do 17, the E-1, produced in 1936, had two 750 hp BMW V1-7·3 engines, giving it a top speed of 220 mph, ceiling of 18,000 ft and 990 miles range with a 1,760-lb bomb load

### DORNIER Do 17 M-1

Powered by the Bramo 323A engine, the M-1 entered production in 1938, and had a downward firing MG 15 fitted as standard

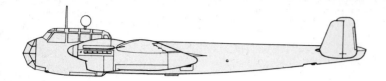

## DORNIER Do 17 S

Similar to the M-series, but with the deepened forward fuselage and fully glazed nose which were to become characteristic of the type. Powered by DB 600A or G liquid-cooled engines, the Do 17 S was used primarily as a reconnaissance-bomber

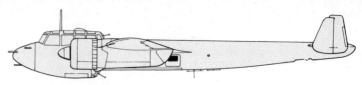

## DORNIER Do 17 Z-10

The last Do 17s before the designation was changed to Do 215, the Z-series reverted to the Bramo radials used on the M-series. A night fighter carrying an early form of Lichtenstein radar, the Z-10 had no less than four 20-mm cannon and four MG 17s in the nose

Wright's main engine was the R-1820 Cyclone, which in the Boeing B-17 had at last emerged with an operational turbo-supercharger which harnessed much of the otherwise wasted energy of the hot exhaust gases, and added 60 per cent to the power transmitted to the propellers above 25,000 ft altitude. In addition to this dramatic development, Wright had run the first R-2,600 (2,600 cu in) Double Cyclone in 1937. Rated at 1,500 hp, this was the most powerful conventional engine available and it was destined to partner the Double Wasp in powering many Allied bombers, starting with the Martin Baltimore and B-25 Mitchell.

Behind all the fierce competition with piston engines, the dogged proponents of the gas turbine had at last begun to achieve success. In Britain Frank Whittle, on special duty while detached from the RAF, had run his first experimental jet-propulsion engine in April 1937. It ran well, causing a sudden astonished re-think among the vast majority of 'experts' who had previously refused to show any interest in his work. In Germany, Ernst Heinkel ran a somewhat similar experimental unit a month later, designed by one of his young engineers, Pabst von Ohain. While Whittle at last received an Air Ministry contract for a flight engine, and Gloster began to build the aircraft for it, Heinkel moved much faster. By the end of August 1939 his jet, the He 178, had already flown.

Although the emergence of the potent Focke Wulf Fw 190 fighter, with its air-cooled radial engine, gave the RAF an unpleasant shock in 1941, the official view continued to be that liquid-cooled engines made aircraft go faster. So, in spite of the fallacy of this statement – and the fact that such engines were complicated, costly and vulnerable in battle – the Rolls-Royce Merlin continued throughout the war as the main engine type of the Lancaster and Mosquito, the two bombers that did the biggest and best job of all for RAF Bomber Command.

In contrast, the excellent Bristol Hercules was used in only a few Lancasters, in the Halifax and in such indifferent machines as the Stirling and Albemarle, while the splendid new 2,500–3,000 hp Centaurus was fitted to the Buckingham – which was never used as a bomber at all. Altogether, British bomber engines made little progress during the war apart from a steady refinement of existing designs.

In Germany the situation was exactly

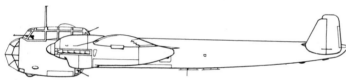

## DORNIER Do 215 B-1

**Gross weight:** 19,600 lb **Span:** 59 ft $\frac{2}{3}$ in **Length:** 51 ft 10 in **Engine:** 2×1,100 hp Daimler Benz DB 601Aa **Armament:** 6×7·92-mm machine-guns **Crew:** 4 **Speed:** 292 mph at 16,400 ft **Ceiling:** 31,170 ft **Range:** 965 miles **Bomb load:** 2,200 lb

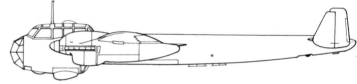

## DORNIER Do 215 B-4

Similar to the B-1 except in the type of equipment carried

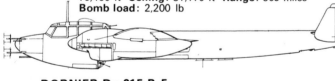

## DORNIER Do 215 B-5

A night-fighter version, whose solid nose contained one 20-mm MG FF cannon and three 7·9-mm machine-guns

## DORNIER Do 215 B-1

A reconnaissance-bomber which carried cameras as well as twenty 110-lb bombs

the reverse. There were by 1942 so many radical new designs and concepts that there had to be a ruthless pruning of more than forty engines and projects. Work was concentrated on the three major German engine types. The DB 601 had yielded to the faster-running (3,200 rpm) Jumo 213, rated at 1,750 hp. The excellent new BMW 801, the 14-cylinder air-cooled radial fitted to the Fw 190, had been put into production in various forms for several important bombers, beginning with the Do 217 and Ju 88. By 1942, German engines were marching forward almost too boldly, and the interminable problems that were encountered delayed not only the 1939 Bomber-B specification so much that it never got into production, but also the He 177.

The He 177 was planned in 1938 as an outstanding long-range heavy bomber, using two large propellers each driven by a pair of coupled engines. On top of this radical arrangement it was planned to use evaporative steam cooling, such as the RAF had sought for fighters in 1930, but this was soon abandoned in favour of ordinary radiators. The first engine used for the He 177 was the Daimler Benz DB 606, comprising a pair of DB 601 engines geared to a single propeller shaft, which had an unfortunate habit of overheating to the point of catching fire. By early 1943 the DB 606 had been replaced by the more

powerful DB 610, made up of a pair of DB 605s, but this behaved no better. Heinkel's later bombers, the He 274 and He 277, both used four individually mounted DB 603 engines, these being an enlarged and improved version of the 605 rated at 1,850 hp.

No gas-turbine engines were used in Allied bombers before the end of the war, but in Germany there was one true jet reconnaissance bomber and one unique jet-propelled pilotless bomber or guided missile. The former was the Arado 234, which went into service in September 1944 powered by two Jumo 004B turbojets. These engines had axial compressors and, bearing in mind the primitive state of the technology and the need to manufacture in vast quantities, they were outstanding pieces of engineering. Take-off thrust was 1,980 lb and reliability was good, though if the engine suffered a flame-out (extinction of combustion) it was essential to shut off the fuel before trying to relight.

The missile was, of course, the Fieseler Fi 103, commonly called the V-1 or Doodlebug. Its Argus pulse-jet was a simple assembly mainly welded from mild-steel sheet, giving a thrust of about 660/750 lb at low altitude. Although very noisy, inefficient and short-lived, it was ideal for its purpose, taking only about 50 man-hours to build and having remarkably high reliability in very arduous circumstances.

### HEINKEL He 177 A-5/R2

**Gross weight**: 68,343 lb **Span**: 103 ft 1¾ in **Length**: 66 ft 11 in **Engine**: 2×2,950 hp Daimler-Benz DB 610A-1/B-1 **Armament**: 3×7·9-mm; 3×13-mm machine-guns; 1×20-mm cannon **Crew**: 5 **Speed**: 303 mph at 21,500 ft **Ceiling**: 26,250 ft **Range**: 3,400 miles **Bomb load**: 4,964 lb + 2 mines, torpedoes or missiles

This variant of the He 177 was used extensively on the Russian Front for bombing, transport and ground-attack duties, fitted for the latter role with 50-mm or 75-mm cannon

### VERGELTUNGSWAFFE I FZG-76

**Gross weight**: 4,800 lb **Span**: 17 ft 6 in **Length**: 26 ft **Engine**: Pulse jet 600-lb thrust **Armament**: 1,870 lb warhead **Speed**: 390 mph **Range**: 150 miles

One of Hitler's secret weapons, the threat of the V1 was contained first by intercepting and bringing down the rockets and later by destroying their launching sites. Nevertheless, the V1 played an important part in the development of unguided and guided missiles

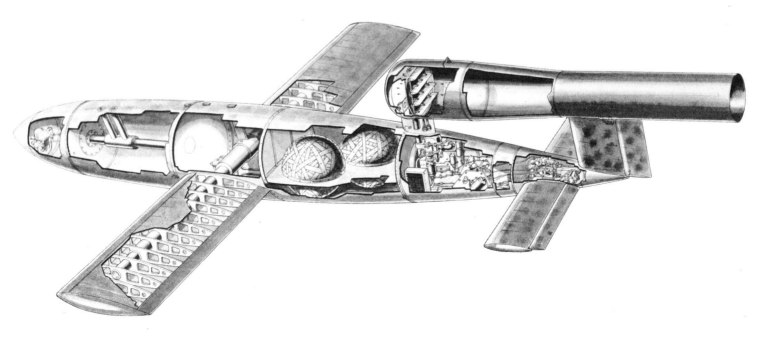

# The secret war

Air power was the dominant factor in the Second World War. Although psychological strategic bombing did not always achieve the results promised by its proponents on both the Allied and Axis sides, the use of bombers – against military and industrial targets, in tactical support of ground forces, and for anti-shipping strikes at sea – was fundamental to the final victory. This was largely due to the tremendous advances in technology – greater even than those made during the First World War.

These, in many ways, were simply improvements on First World War performances. Bombing was carried out on a far greater scale but it was primarily for the same purposes and against the same kind of targets. There was little new in bomber tactics; in fact, many of the lessons of the First World War, such as the necessity for fighter escorts during daylight raids, had to be re-learned. The US Eighth Air Force bombers raiding Occupied Europe from 1942 onwards could be escorted by fighters, but when, in 1943, they began to attack targets deep in Germany, beyond the range of any Allied fighter of that time, the losses became so great that the offensive had to be called off. It was only renewed in 1944 with the arrival of the Mustang long-range fighter which could escort the B-17s on their missions over Germany.

The only really new development, apart from the manufacture of the atomic bomb at the end of the war, was in radio and radar techniques. These made it much easier for bombers to reach and locate their targets, especially at night or in bad weather. But they also enabled fighter defences to locate enemy bombers. Thus, throughout the war, a separate battle of technology was fought as each side strove to develop its own navigational aids and at the same time to deny the enemy use of his.

During the 1930s scientists both in Britain and Germany were working on the newly discovered electronic aids to navigation, but because of the different objectives of the two countries, the one defensive and the other offensive, development followed two opposite courses. Thus, while the Germans gave priority to the use of radio aids for bombing, the British concentrated on building up a defensive system of radar, a field in which they were more advanced thanks to the pioneering work of R. A. Watson-Watt of the National Physics Laboratory.

The two German systems which emerged were based on the 'Lorenz' beam. This was basically two slightly over-lapping radio beams, one transmitting Morse dots and the other Morse dashes. Where the signals interlocked, an aircraft received a steady note; any deviation either way resulted in a changed signal. Thus the pilot could follow a predetermined course.

The simpler of the two methods was the 'bent leg' (*Knickebein*) which employed two Lorenz beams, one to hold the bomber on course right up to the target, while the second crossed the first at the point where the bombs should be released. At ranges of about 180 miles from the beam transmitters, this system gave the crew a 50 per cent chance of placing bombs inside a one-mile diameter circle, and its accuracy was even less at greater ranges.

Another method, the X-device, was developed for more precise bombing. This employed four Lorenz beams, each on a different frequency. One held the bomber on course to the target while the other three crossed it at precise intervals. The first intersection gave warning of approach while at the second and third, the navigator started the two hands of a special clock which rotated independently. When the second hand caught up with the first, an electrical contact was made and the bombs were released automatically. The combination of the clock and the Lorenz beams gave the X-device great accuracy since it provided data on the aircraft's speed, an essential requirement for precision bombing. A practised crew could place bombs within a 400-yard diameter circle at 180 miles from the transmitters. Because the X-device was so complicated to use however, it was confined to the specially trained crews of *Kampfgruppe* (wing) 100.

The disadvantage of the Lorenz beam was that it was vulnerable to radio countermeasures. By day of course, during the Battle of Britain, British fighters could be vectored on to incoming enemy bombers located by radar (reflected radio waves able to be picked up and seen on a cathode-ray tube). The chain of radar stations built along the southern and eastern coasts of England was a major contribution in defeating the *Luftwaffe's* attempt to bomb Britain into submission, providing a scientific means of interception which greatly increased the efficiency of the RAF's woefully low fighter strength. Friendly aircraft were distinguished by the transmission of a special radar pulse called IFF (Identification Friend or Foe). But it was during the German night raids which followed, once the Spitfires and Hurricanes of Fighter Command had overcome the threat by day, that the Lorenz systems came into use.

One counter-measure was to jam the beams, but a more successful method was

to deflect them so that the Germans unwittingly dropped their bombs at sea or in open country. For use against the bombers themselves, night-fighters were developed, carrying a new and – at that time – highly secret form of interception radar which enabled aircraft to be located in darkness. The first British aircraft to be equipped with AI (aircraft interception) was the Bristol Blenheim fast bomber but it was not really suited for the purpose. The Bristol Beaufighter and the American Douglas DB-7 Havoc, both conversions of light bombers, proved to be more effective night-fighters in late 1940. But although considerable success was achieved, it was not sufficient to provide a deterrent. The night-blitz on British towns was almost over before enough night-fighters became available in the summer of 1941.

Because efforts had necessarily concentrated on using radar for defensive operations, little was done in the early months of the war to provide electronic aids for bombers attacking targets in Germany and Occupied Europe. These had to rely on dead reckoning combined with visual fixes on known landmarks – an extremely unreliable method – and not until late 1941 did the RAF's first radar navigational aid became available.

### Getting closer

The Gee system was based on three ground transmitters, about 100 miles apart, which transmitted a complex train of radar pulses across the continent of Europe. By means of a special radar receiver in the aircraft which enabled the time difference between the various signals to be measured, and referring these differences to a special Gee map, the navigator could determine his position to within six miles when 400 miles from the most distant transmitter. This was an improvement on the Lorenz system but it had the same vulnerability to jamming by the Germans and by the end of 1942 it had reached the end of its usefulness.

A much more successful radar device was the H2S system which could be carried in aircraft and could therefore operate beyond the range of beacons. This system gave a representation of the terrain below the aircraft on a cathode-ray tube, by means of a radar beam tilted downwards. Its only drawback was that it required a skilled crew to make the best use of the results. As the *Luftwaffe* had learned in 1940 and 1941, it was advisable to have the equipment used by picked crews to start marker fires at the target to direct the main bomber force – the only problem being that if the markers went wide, so would the majority of the bombs. The H2S device brought about a marked improvement in the accuracy of attacks when it was introduced in 1942, but still too many bombs were missing their targets during night raids. This was one of the main reasons why the US Eighth Air Force, starting its European operations in mid-1942, concentrated on daylight precision bombing in spite of the high casualty rate. An area-bombing night sortie was regarded as successful if the bombs were released within three miles of the aiming point.

In a concerted effort to improve bombing accuracy, RAF Bomber Command in 1942 created the Pathfinder Force, under the command of D. C. T. Bennett. This coincided with the introduction of the new 'Oboe' radar system which, while limited in range, provided a remarkably fine degree of accuracy. Two radar transmitters were set up at Dover and Cromer, sending out a stream of pulses which could be picked up and returned by an airborne transmitter in the aircraft. Ground operators could direct the pilot by means of radio instructions to fly along any determined path to the target and compute the bombing release point for him. Bombing accuracies of 200 yards or so were achieved by this method. The maximum range for an aircraft flying at 28,000 ft was 270 miles, which covered most of the Ruhr but not many other targets in Germany. However, after the Normandy invasion, beacons were set up in France and during the latter stages of the war, the Oboe system could be used to reach almost every target of importance in Germany.

Another limitation was that each pair of ground transmitters could control only one aircraft at a time on its bombing run of about ten minutes. Consequently, Oboe was used only in the high-altitude Mosquito bombers of the Pathfinder Force.

With the development by the British and American air forces in 1943 of a round-the-clock bombing offensive against Germany, the *Luftwaffe* found itself in the same position as RAF Fighter Command in 1940. Its fighters could cope with the American daylight raids, but there were no specialised night-fighters to combat the British night bombers. A number of twin-engined fighter-bombers and medium bombers were converted for night fighting duties, beginning with the Messerschmitt Bf-110G, but the most widely used was the G6 variant of the versatile Junkers Ju 88, some of which carried an upward-firing cannon mounted in the central fuselage. What gave these night-fighters such a devastating offensive power was the introduction of a German form of airborne interception radar, known as the Lichtenstein system.

This in turn led the British to devise counter-measures to confuse the German radar screens and radio signals, both in the air and on the ground. Lengths of tinfoil – 'Window' – corresponding to the wavelengths of German radar were dropped in large quantities to produce a vast number of echoes which obscured the echoes made by the aircraft. 'Mandrel' was a device carried by pairs of circling British aircraft which radiated signals to jam the German radar before it could detect incoming bombers. 'Piperack' was a rearward-facing device which shielded bombers in a cone behind the jamming aircraft. And 'Jostle' was a large radio transmitter carried by an aircraft to emit a raucous note on the same wavelength as the German fighter control so that instructions could not be heard. These counter-measures were used in many different combinations to keep the Germans guessing. They were the crude forerunners of today's sophisticated ECM (electronic counter-measure), the bomber's main defence against guided surface-to-air missiles.

### BRISTOL BLENHEIM Mk IV

**Gross weight:** 12,500 lb  **Span:** 56 ft 4 in  **Length:** 42 ft 9 in
**Engine:** 2×920 hp Mercury XV  **Armament:** 5×·303 machine-guns
**Crew:** 3  **Speed:** 266 mph at 11,800 ft  **Ceiling:** 22,000 ft  **Range:** 1,450 miles  **Bomb load:** 1,320 lb

Primarily intended as a light bomber, the Blenheim served throughout the war in many roles for the RAF and was the first night-fighter to use airborne interception radar

# Incendiaries to atom bombs

British bombs changed very little between 1918 and 1937. Apart from the 20-lb Cooper used for practice, the standard high explosive bombs were 112-lb, 230 or 250-lb, and 520 or 550-lb. But in 1937 a completely new range of bombs was introduced with a better streamlined shape, improved ballistic properties (almost all were fitted with tail fins surrounded by an open cylindrical drum), and redesigned fuses and arming devices operated by a combination of pistol and a separate detonator. The pistol was merely a mechanical device, armed after the bomb's release, and containing a striker which was actuated on impact. The striker in turn fired an initiator cap in the detonator which, after a required period of delay by the burning of a pyrotechnic fuse, set off the main detonator and thus the explosive.

The new range included bombs of 2,000 lb nominal weight, one of which was specially designed for armour-piercing. In addition to the demolition bombs, used for the majority of bombing operations to produce either blast, fragmentation or mining effect (the explosive charge averaging some 50 per cent of the total weight), the RAF also introduced a comprehensive range of improved flares, flame-floats, smoke-floats and other pyrotechnic devices. A standard aerial mine was also produced, together with further versions of the long-lived 18-in torpedo weighing around 1,650 lb. By the outbreak of the Second World War, an incendiary bomb was also in production with a body of combustible magnesium – it had a hexagonal section so that clusters could be stacked in a minimum of space.

In 1939 the HE bombs produced in the greatest numbers were the 250 and 500-lb general purpose types, regarded as the most useful since they could be delivered by every RAF bomber then in service. Except for leaflets and, increasingly, incendiaries, they far outnumbered all other 'stores' dropped by the RAF during the first two years of the war. They were even the preferred weapons for attacks on the German Fleet which began soon after the outbreak of war, although it was obvious that they could cause little more than superficial damage to a capital ship. In fact, the results achieved were even less because a high proportion of the bombs in those early attacks failed to detonate. Not only was there a need for an improvement in navigation and in bombing tactics generally, there was also an urgent need for better bombs.

The range of bombs available to the *Luftwaffe* was slightly smaller, the standard HE weights being 50, 250 and 500 kg (110,

551 and 1,102 lb) and all having a straight-sided shape with braced tail fins. All were very simple to make, yet there is a well-authenticated story that in 1937 the *Luftwaffe* had not received a single modern bomb; to rectify the deficiency, Hitler himself suggested filling surplus gas cylinders with explosives. Only gradually did bomb production get into its stride, as the inevitability of a major war became recognised.

## Electrical fuses

At first the Germans used methods similar to the British for fusing their bombs, but then they turned to a new and more reliable form of electrically-fired fuse. This contained a condenser which was charged from an electrical supply in the aircraft when the bomb was released. A second condenser was charged from the first by means of a resistor whose capabilities could be altered to provide the length of delay required before the ignitor was fired. In Germany in 1939, production began on a cylindrical magnesium incendiary bomb, and also a high explosive bomb of 1,000 kg (2,205 lb), both with annular fins. The Heinkel He 111 could carry two of the latter weapons on side-by-side external racks, but the standard weapon used against British cities from May 1940 onwards was the parachute mine.

The use of mines against 'soft' land targets, although unconventional at first, was actually only common sense. The mine was a large container of high explosive which, when detonated, released far more energy than a traditional GP-type bomb, even if bombs were available in such a size (2,200 and 3,300 lb). For marine use, special triggering and fuse systems were necessary. During 1940, both the British and Germans began to perfect mines of much greater sophistication than the old horn contact type. Ultimately, mines were provided with fuses sensitive to the various effects made by a passing ship, such as distortion of the Earth's magnetic field, sound wave vibrations, or slight reduction of the hydrostatic pressure. (These patterns were called magnetic, acoustic and influence.)

For use against cities, the ingenious marine fuses were merely replaced by impact fuses, sometimes triggered by a long pistol probe before the case struck the ground or a building. The need for a parachute was a hang-over from the original role when each mine had to be laid in a pre-designated area and yet protected against the shock of hitting the water.

Later in the war, a new form of radio

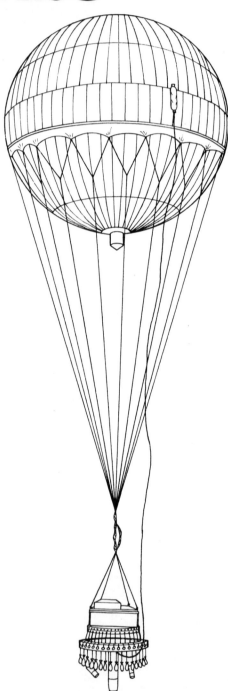

## JAPANESE BOMBING BALLOON

**Balloon volume:** 19,000 cu ft **Envelope diameter:** 32·81 ft **Fuse:** 64 ft; approx. 1 hr 22 min burning time **Bombs:** 2 incendiary bombs; 1×15 kg anti-personnel bomb

One of the more bizarre bombing ideas of the war, these balloons were released by the Japanese and intended to be carried by the wind to the United States. Some of them travelled as far as Alaska

proximity fuse was developed, making it possible for a bomb to explode at a given height above the ground. This operated by means of a radio-frequency oscillator which received a reflection of its own waves from the ground as the bomb fell.

As a result of experience in the first year of the war, the RAF realised a very important fact. Previously, the bomb had been regarded as an aerial artillery shell, made in much the same way and given a streamlined form to minimise drag when carried in the slipstream and to prevent random tumbling after release. In fact, all that was needed in most cases was to deliver the maximum amount of chemical energy to the target. The bomber was seen in this new light as no more than a trucking system, with high explosive as the payload. The metal part of a bomb was merely a necessary evil, to be reduced in weight as much as possible. As all the payloads of the planned heavy bombers were carried inside an enclosed bomb-bay, there was no need for streamlining. The logical result was a completely new type of bomb called the Light Case (LC), in principle no more than a glorified oil drum.

The first LC bomb had a nominal weight of 4,000 lb. It was a welded drum of thin mild steel, painted dark olive and provided with three windmill-armed fuses on its bluff front face. Empty cases were filled by pouring in molten RDX (sometimes called Exogen), an explosive which combined ease of manufacture and high safety qualities with a very powerful blast effect.

It had been discovered by a German as far back as the end of the 19th century, but it was only during the Second World War that a means was found of producing it in quantity. In use, the 4,000-pounder was soon dubbed the 'blockbuster' and it was the ideal partner to the incendiary in the area-bombing attacks on cities which were the staple diet of Bomber Command for the rest of the war. All the four-engined heavies' could carry the big bomb, as could the Wellington, and by 1944, even the Mosquito could take one as far as Berlin, in a specially enlarged bomb-bay.

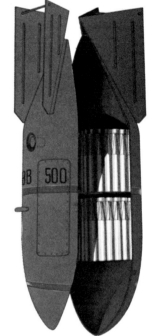

## GERMAN INCENDIARY
## CONTAINER BOMB

**Weight**: 500 kg **Length**: 5 ft 6 in **Diameter**: 18·5 in **Contents**: 120×1-kg incendiary bombs

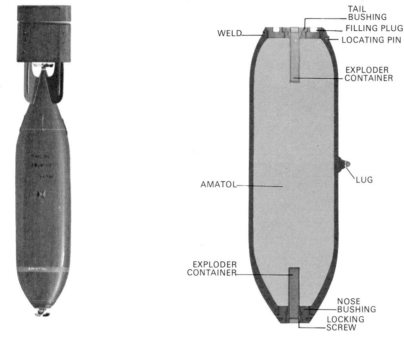

## BRITISH MC 500-lb BOMB

Drawing and section of one of the most common types of bomb dropped by the RAF in the early years of the Second World War. The explosive used was Amatol

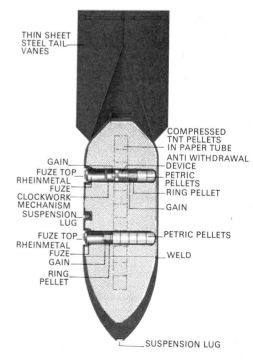

## GERMAN 250-kg HE BOMB

**Overall length**: 64·5 in **Body length**: 42 in **Overall width**: 18 in **Body diameter**: 14·5 in **Charge/weight ratio**: 52·2%

## JAPANESE BOMBS

Clockwise from top:
**Type 2 No 25 Mk 111**
**Weight**: 250 kg **Length**: 69 in
**Type 99 No 80**
**Weight**: 750 kg **Length**: 92·1 in
**Type 1 No 25**
**Weight**: 262 kg **Length**: 71·75 in
**Type 97 No 6**
**Weight**: 59 kg **Length**: 40 in

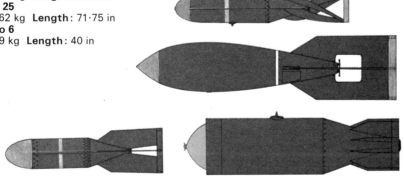

But the supreme exponent of blockbusting was the Lancaster, whose 33 ft long bomb-bay was completely unobstructed by any longitudinal or transverse structure. By 1942 the Lancaster was carrying 8,000-lb bombs, essentially two 4,000-pounders bolted together, and for special targets it also carried a triple-unit 12,000-pounder, the Tallboy. The ballistic properties of the 4,000-lb bomb rested mainly on keeping the centre of aerodynamic pressure aft of the centre of gravity, but when three were joined end-to-end it was deemed advisable to put on a stabilising tail.

Although the RAF dropped many other weapons during the war, the two most outstanding were both designed by a team at Vickers led by Barnes Wallis, who had earlier been responsible for the Wellington. The first was the famous device used by 617 Squadron, the Dambusters. To deliver high-explosive to the face of a dam protected by torpedo netting, Wallis invented a bomb in the form of a huge drum, carried semi-externally by a modified Lancaster. The drum was hung on crutches at left and right with its axis arranged transversely, and shortly before the attack was rotated at high speed by a separate auxiliary engine in the bomber's fuselage. Released at a height of sixty feet, the spinning bomb then skipped across the lake surface behind the dam, slowing when it had passed the final net and sinking down in contact with the face of the dam to be detonated by a hydrostatic fuse at the correct depth.

Wallis's other great contribution was a supreme form of conventional HE bomb, for use against the most hardened targets (notably U-boat pens, with reinforced-concrete roofs up to thirty-five feet thick). This was a free-falling (supersonic) bomb of beautiful streamline form, spun by four canted tail-fins, with a heavy-case body and pointed nose. Its weight was no less than 22,000 lb, and its name of 'Earthquake' was singularly appropriate. Dropped against large solid structures, such as the great railway viaduct at Bielefeld, an Earthquake detonated deep underground caused a tremor so severe that the structure was brought down by the shock waves.

An alternative weapon used against U-boat pens was the Disney bomb devised by Capt Terrell RN. This streamlined 4,500-lb hard-case bomb had a rocket in its tail ignited by a barometric fuse after a free fall from above 20,000 ft down to 5,000 ft. This boosted terminal velocity to about 2,400 feet a second, a speed greater than that of any weapon of such a size other than the V-2 rocket. Owing to stowage difficulty, it was finally agreed to deliver Disney bombs in pairs carried externally by the Boeing B-17s of the US Eighth Air Force. Later, the B-29 successfully carried external pairs of 22,000-lb Earthquakes. Apart from this, the bombs of the US services were generally conventional, though it was chiefly American research that in 1944 led to the use of napalm, a jellied mixture of petroleum fuels (the name stemmed from naphtha and palm oil) carried in a simple container – often the drop tank of a fighter.

Mention should be made, however, of the awe-inspiring technique of filling a complete B-17 with high explosive and directing it at a major target under radio control. This campaign, Project Aphrodite, was aimed at first against the huge 'V-weapon' structures in the Pas de Calais, and war-weary Fortresses, redesignated BQ-7, were turned into huge guided missiles containing 20,000 lb of Torpex triggered by an impact fuse system. A courageous pilot took off manually, in an open cockpit, and then bailed out by parachute near the British coast after setting the fuses. A director aircraft took over the steering of the BQ-7 by radio.

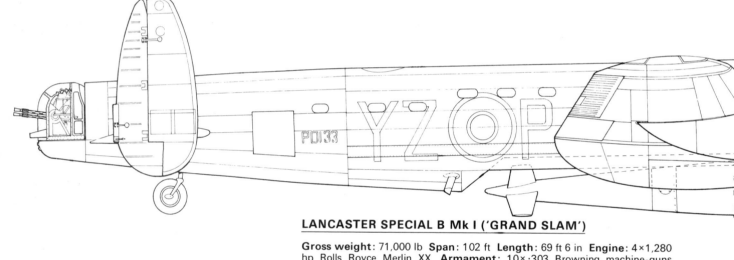

## LANCASTER SPECIAL B Mk I ('GRAND SLAM')

**Gross weight:** 71,000 lb **Span:** 102 ft **Length:** 69 ft 6 in **Engine:** 4×1,280 hp Rolls Royce Merlin XX **Armament:** 10×·303 Browning machine-guns **Crew:** 7 **Speed:** 280 mph at 18,500 ft **Ceiling:** 23,500 ft **Range:** 2,700 miles **Bomb load:** 22,000 lb

A special modification of the Lancaster, with the dorsal turret removed, to enable it to carry the 22,000-lb 'Earthquake' or Grand Slam bomb

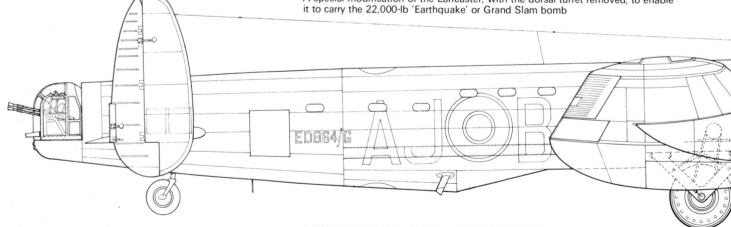

## LANCASTER SPECIAL B Mk III ('DAM BUSTER')

**Gross weight:** 70,000 lb **Span:** 102 ft **Length:** 69 ft 6 in **Engine:** 4× Packard-built Rolls Royce Merlin 224 **Armament:** 10×·303 Browning machine-guns **Crew:** 7 **Speed:** 287 mph at 11,500 ft **Ceiling:** 24,500 ft **Range:** 1,660 miles **Bomb load:** 14,000 lb

A Lancaster specially adapted to carry the spinning-drum bomb developed by Dr Barnes Wallis for the raid on the Mohne and Eder dams on 17 May 1943

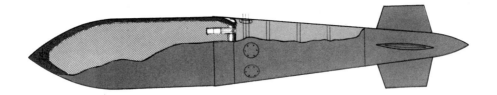

## 12,000-lb 'TALLBOY'

**Total weight:** 11,885 lb **Overall length:** 21 ft **Body length:** 10 ft 4 in **Body diameter:** 3 ft 2 in **Charge/weight ratio:** 45%

The Tallboy bomb was basically three 4,000-lb light-case bombs bolted together, with tail fins added for stability

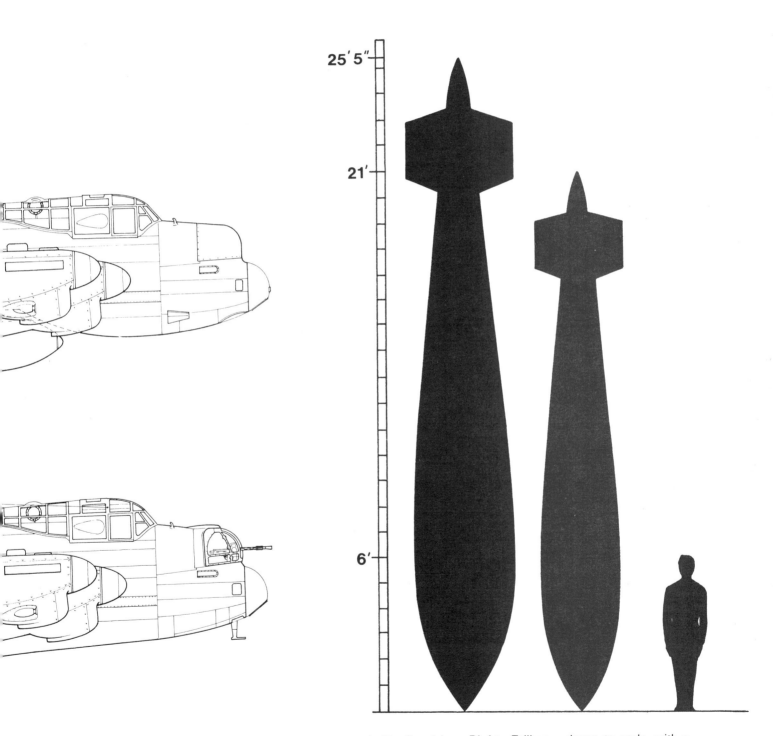

**Left:** Grandslam **Right:** Tallboy — drawn to scale, with a human figure added for comparison

Some indication of the explosive power of an Aphrodite attack is afforded by the fact that, when a converted US Navy Liberator commanded by Joseph P. Kennedy Jr (brother of the future US President) exploded prematurely with the crew on board at 15,000 ft over Blyth, Northumberland, severe blast damage on the ground extended over a radius of six miles.

These radio-controlled bombers were as far as the Allies got in creating the concept of the guided missile in the Second World War. The Germans, however, were more imaginative. Radio-controlled bombers existed in abundance under the Mistel programme, together with an even greater number which were merely set on course by a manned fighter carried on top of the bomber and then, after the fighter had uncoupled, left to fly to the target in a straight shallow dive on auto-pilot. The commonest Mistel missile was a rebuilt Ju 88 with a circular section front fuselage (warhead) triggered by a long nose probe.

Much more advanced in conception were the true guided missiles. Fritz X (FX) or 'PC 1400', an armour-piercing bomb with cruciform fixed wings, had a complicated tail fitted with guidance spoilers controlled by radio from the launch aircraft. The Hs 293 was a stand-off missile with less punch but longer range, and it took the form of a miniature aircraft, with rocket boost, steered by radio links similar to those of FX. Both weapons became operational in August 1943 with units of *Kampfgruppe* 100 equipped with special sub-types of Do 217, and the following month FX began sinking battleships with apparent ease (though HMS *Warspite* limped into Malta after being struck by three). There were many other *Luftwaffe* missiles, such as the Bv 246 which flew on wings of reinforced concrete and the L 10 equipment which converted a torpedo into an air-launched stand-off missile.

Perhaps the most unexpected development was the Japanese piloted missile, the Yokosuka MXY-7 Ohka (Cherry Blossom), flown in suicide attacks against US forces in the Pacific. Carried to within about fifty miles of its target by a conventional bomber (usually a G4M2 'Betty'), the Ohka was then released and flown by its pilot seated in a cramped cockpit between the stubby 5-metre (16 ft 5 in) wing and the twin-finned tail. Most of the distance to the target was covered in a fast glide, and though the device was incapable of much evasive action, it was small and difficult to intercept. Three miles from the target, the pilot ignited three solid-fuel rockets in the tail, arriving in a steep dive at something like 620 mph. The whole front half of the fuselage was a warhead containing 2,645 lb of tri-nitrol aminol. In its day, the Ohka was practically unstoppable once it had neared its target, but US fighters usually managed to destroy the mother-planes before the missiles were launched. Even so, many direct hits were scored in these suicide attacks, and they were taken more seriously than implied by the name given to them by the US Navy – *Baka*, the Japanese word for fool. Towards the end of the war, Japanese suicide attacks were carried out mostly by conventional aircraft filled with explosive.

The ultimate weapon of the Second World War was, of course, the atomic bomb, the outcome of combined work by British, American and French scientists. Based on the fission principle, the tremendous de-structive power of the two bombs dropped on Japan resulted from the release of energy which raised the air to a very high temperature and caused radiation on various wavelengths. The bombs were dropped from Boeing B-29s flying at about 30,000 ft, on Hiroshima on 6 August 1945 and on Nagasaki three days later. The first destroyed more than four square miles of the city and killed some 80,000 people, while the second destroyed one-and-a-half square miles, killing 40,000. In fact, the physical damage was not as great as that caused by the big incendiary raids on Germany when whole cities were set ablaze and swept by firestorms created by the upward rush of hot air.

But those first atomic bombs were only a foretaste of the much more powerful nuclear weapons which were to be developed later. So great is the energy released by an atomic bomb that it embraces all the known destructive principles, including blast and penetration, and some, resulting from radioactivity, whose delayed effect still cannot be measured precisely.

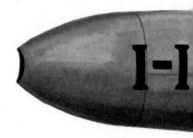

## YOKOSUKA MXY-7 OHKA

**Gross weight:** 4,718 lb **Span:** 16 ft 5 in **Length:** 19 ft 10 in **Engine:** 3×solid propellant rockets, 1,764 lb total thrust **Crew:** 1 **Speed:** 403 mph; terminal velocity 576 mph **Ceiling:** NA **Range:** 20 miles **Bomb load:** 2,645 lb

Japanese piloted suicide bomb, air-launched from specially adapted G4M2 motherplane

## HS 293 ROCKET BOOSTED GLIDE BOMB

**Total weight:** 1,730 lb **Span:** 10 ft 2·857 in **Length:** 10 ft 5·25 in **Fuse:** impact **Warhead:** 550 lb Trialen

A glide bomb designed to be released from a Dornier 217 bomber. It was radio controlled, and the rocket boost motor burned for 10 seconds after launch

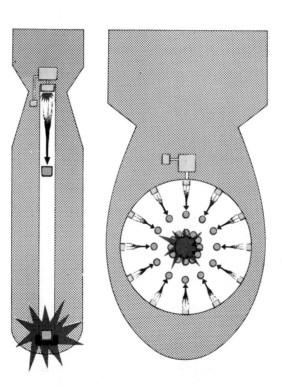

## 'LITTLE BOY' (left)

**Weight:** 9,000 lb **Diameter:** 28 in **Length:** 10 ft

The atom bomb dropped on Hiroshima, a 'gun-type' weapon in which one piece of Uranium 235 was fired into another, cup-shaped piece to produce the nuclear explosion

## 'FAT MAN' (right)

**Weight:** 10,000 lb **Length:** 10 ft 8 in

'Fat man', the bomb dropped on Nagasaki, used the implosion method, with a ring of 64 detonators shooting pieces of Plutonium together to create the explosion

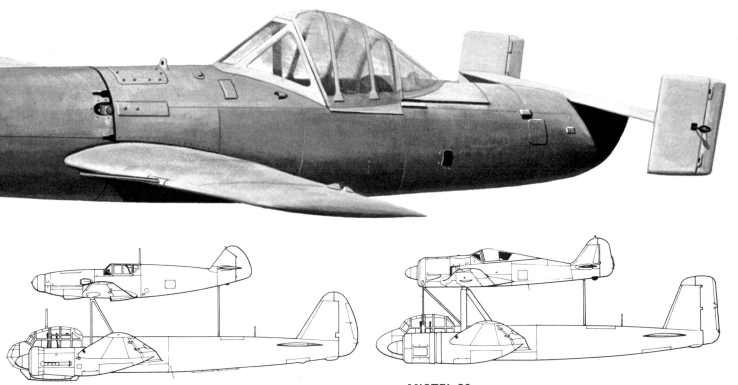

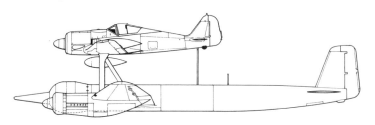

## MISTEL S1

**Bomber:** Ju 88 A-4 **Fighter:** Messerschmitt Bf 109F

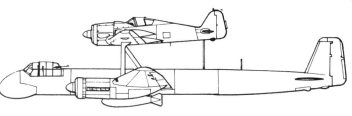

## MISTEL S2

**Bomber:** Ju 88 G-1 **Fighter:** Fw 190A-8

Under the Mistel programme various types of plane, most commonly a rebuilt Ju 88, had their forward fuselage filled with explosives were guided by an attached fighter towards their target, then released and either radio-controlled or left to dive onto the target on auto-pilot. They never made a significant impact on the war

## MISTEL 3C

**Bomber:** Ju 88 G-10 **Fighter:** Fw 190A-8

## 'FÜHRUNGSMACHINE'

**Bomber:** Ju 88 H-4 **Fighter:** Fw 190A-8

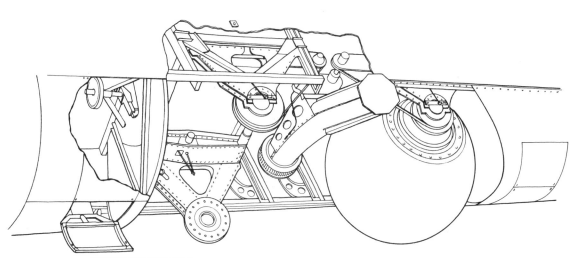

## TWIN HIGHBALL

experimental installation in Mosquito B-IV

The Highball bomb was designed by Barnes Wallis after the success of the Dambusters raid. It was an experimental installation of two bombs for use against Japanese shipping in the Far East, to be delivered by Mosquito bombers

## Bomb-sights

In 1939 the standard bomb-sights in use by all the major air forces were merely refined versions of the primitive sights used in the First World War. The sight was a purely mechanical device, usually fixed in the nose or belly of the bomber and operated by a bomb-aimer (bombardier in the USA) lying prone on his stomach. He would first set up the sight rather in the way that a rifleman of 1914 would set up his backsight, the difference being that instead of feeding in the range he would insert the height of the bomber and the true airspeed, and usually the wind velocity and direction as well. This would align a sighting scale with the apparent path of the target, seen moving towards the bomber far ahead and below. Almost always the bomb-aimer would have to aim by passing corrective instructions to the pilot, either by hand signals or intercom voice.

Markers or cross-hairs arranged transversely would indicate the correct moment for release, when the bomb-aimer would either press a button, pull a trigger, start a clockwork sequence release, or even, on some of the older types of aircraft, release the bombs by manual linkage. Most bombers offered alternatives for release in singles, spaced groups, clusters, or in a long stick covering perhaps a mile of ground.

A major drawback of these course-setting sights was that they were accurate only under ideal conditions, and were hopelessly upset if the bomber departed significantly from straight and level flight. The corrections to a bombing run had to be made quickly and deftly, often with flak bursting all around the bomber or while under attack from enemy fighters. Any evasive action by bank or sideslip might mean that either the bombing run had to be made again or that the bombs would fall wide of target. In fact, many of the early high-level attacks by both the *Luftwaffe* and the RAF were extremely inaccurate.

To rectify these shortcomings, improved types of sights were devised. In August 1942, the RAF Pathfinder Force introduced the Mk XIV sight in which the sighting head in the nose of the aircraft was supplemented by a computer unit further aft. The bomb-aimer sighted in the usual way, looking through an inclined optically-flat glass panel, with the sight set with true airspeed, ground speed, wind velocity and altitude; an electro-mechanical calculator box then kept the sight on target even while the bomber took evasive action. By 1943 the Mk XIV was in all Bomber Command heavies and, as the T-1, was also used by the USAAF.

### The Norden sight

The Americans, however, had an even more advanced sight, the Norden. In 1928 Carl L. Norden had begun to develop a sight that could be linked with an autopilot in such a way that the bombardier would aim the aircraft directly, with corrective actions made automatically. The SBAE (Stabilised Bomb Approach Equipment) was a special autopilot comprising gyros, servo-motors and feedback follow-up systems to impart control-surface movements proportional to the deviation of the bomber from the desired flight path.

The sight was mounted in the nose behind a flat glass panel in the usual way, but instead of mere crossed wires, the sighting head consisted of an optical telescope which could be trained on the target by the bombardier and kept in alignment by a drive motor and two-axis linkage. The act of keeping the sight pointing at the target automatically commanded the aircraft to steer for an accurate bombing run, making correctly banked turns. When the target reached the correct angular direction the bombs were released automatically. The bombardier would then cut out the sight and restore control to the pilot. Though the widely publicised claim that with this sight a B-17 or B-24 at 25,000 ft could 'drop a bomb in a pickle barrel' was obviously an exaggeration, the Norden was the most advanced high-altitude sight used in the war.

### Radar aids

German, Russian and Japanese sights were without exception conventional, apart from the special equipment used on the versions of the Do 217, Fw 200C and other aircraft equipped to launch FX or Hs293 guided bombs. As mentioned earlier, various radar and radio navigational aid systems were often used directly as an aid to bombing or accurate target-marking; indeed, the accuracy of Pathfinder Force, and consequently that of the entire might of Bomber Command, depended on Oboe. The British H2S radar, publicly revealed in 1944 by widespread stories in the American press, gave a picture of the terrain beneath on a PPI (plan-position indicator) cathode-ray tube face and was especially useful over regions of combined land (which looked bright) and water (which looked dark) so that river bridges and docks could be bombed with fair accuracy using this aid alone, even in conditions of total cloud cover.

At the other extreme of the technological scale was the sight used by 617 Squadron to breach the Mohne and Eder dams. The Lancasters were brought down to 60 ft above the water and held there by fixed spotlights at nose and tail, angled so that the spots coincided on the water when the planes were at exactly the right height. The bomb-aimer sighted on the twin towers of the dams, just 600 ft apart, by using a device like a boy's catapult with two upright prongs. According to Guy Gibson himself, the sight cost "a little less than the price of a postage stamp".

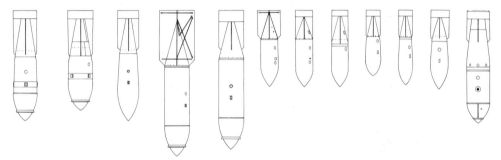

**GERMAN BOMBS** (to scale)

**Left to right:** SC 1,000 kg; SC 1,000 kg L; SD 1,400 kg (piercing); SC 1,800 kg; SC 2,000 kg; SC 500 kg; SC 500 kg; SD 500 kg A (splinter); SD 500 kg E (piercing); SD 500 kg II (piercing); SD 1,000 kg (piercing); BM 1,000 kg ('G' mine)

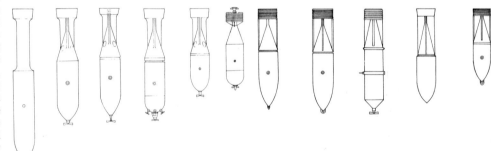

**ITALIAN BOMBS** (to scale)

**Left to right:** 800 kg; 500 kg; 500 kg C; 500 kg RO; 250 kg; 160 kg CS; 100 kg M; 100 kg T; 100 kg SP & SP1; 70 kg IP; 50 kg T

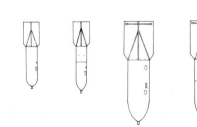

**GERMAN BOMBS** (to scale)

**Left to right:** SC 10 kg; SC 50 kg; SD 50 kg (splinter); SC 250 kg; SD 250 kg (splinter)

# BOMBER OPERATIONS

# Was it all worthwhile?

## BOEING B-29A SUPERFORTRESS

**Gross weight:** 141,100 lb **Span:** 141 ft 3 in **Length:** 99 ft **Engine:** 4×2,200 hp Wright Cyclone R-3350 **Armament:** 8×·50 machine-guns in remote-control turrets; 1×20-mm *or* 3×·50 machine-guns in tail **Crew:** 10 **Speed:** 358 mph at 25,000 ft **Ceiling:** 31,850 ft **Range:** 4,100 miles **Bomb load:** 20,000 lb

Designed to bomb Germany from bases in the USA should Britain have fallen, the B-29 came into service in 1944 and was used primarily against Japan. A B-29 carried the two atomic bombs dropped on Japan. After the war, it was the first bomber to equip the new USAF Strategic Air Command

The use of land-based bombers during the war was divided into distinct areas of operation – tactical co-operation with ground forces (including *Blitzkrieg* and airborne campaigns); strategic offensives against military, industrial and civilian targets in enemy homelands; and attacks at sea against shipping and submarines. In many instances these objectives overlapped, as in the bombing of U-boats at their coastal bases or assembly plants. And the bombers themselves were not confined to any single type of operation. The four-engined heavy strategic bombers sometimes took part in tactical missions, while such versatile aircraft as the de Havilland Mosquito and the Junkers Ju 88 played a part in all kinds of offensive, as well as serving in fighter roles. And while air power was the dominant factor of the war, it was most effective when used in combination with other forces.

This was well understood by the Germans in the early stages of the war. Their *Blitz-kriegs* on Poland in September 1939, and on Norway, Holland, Belgium and France in May 1940, depended on the initial destruction of opposing air forces, particularly by the bombing of aircraft on the ground, so that the *Luftwaffe* could concentrate on supporting German ground forces. It was with such aims in mind that the Germans devoted so much effort in the pre-war period to the production of effective dive-bombers.

The first German mistake was the attempt to use air power to defeat the French and British armies trapped in Dunkirk. In order to give the *Luftwaffe* free reign, the German tanks which could have accomplished the task were halted on the outskirts of the city. There was good reason for the Germans to be confident of their air supremacy. They

had easily overcome the largely obsolete aircraft which the French and British had previously put into the air, except for the Hurricane fighter which was available only in very small quantities. The Fairey Battle light bomber, for instance, suffered a 70 per cent loss rate in daylight sorties during the Battle of France.

But at Dunkirk the *Luftwaffe* for the first time came up against British air superiority in the shape of Spitfires and Hurricanes which were within range of Dunkirk from bases in England. The air cover they provided was an important factor in the escape of a major proportion of the British army.

### Fatal mistake

Had the Germans continued with their *Blitzkrieg* technique during the Battle of Britain, the later course of the war might have been very different. In spite of the effective combination of radar and well-organised fighter defences, the RAF was in a desperate situation by the end of August 1940 following German bomber attacks on airfields and the communications centres of Fighter Command. But at that point the Germans made another vital mistake – they changed from a tactical to a strategic offensive. After an RAF raid on Berlin on the night of 25 August, in retaliation for the bombing of London two nights previously, Hitler ordered the German air attacks to be concentrated against London.

Not only did this fail to shatter British morale, it also relieved the pressure on Fighter Command and gave time for the damaged airfields and radar bases to be repaired. Meanwhile, British bombers pounded the Channel ports where the German invasion forces had been gathered. By 13 October, it was clear that the plan to invade Britain was no longer possible. The

*Luftwaffe* had failed in its attempt to destroy the RAF, the necessary prelude to an invasion. Operation Sea Lion was 'postponed indefinitely', then cancelled altogether.

German bombers then switched from day to night operations, which the hastily improvised British night-fighters could do little to prevent at first. Indeed, if the Germans had continued their attacks on the docks of London and Liverpool in the spring of 1941, the disruption of supplies they caused might have forced Britain to surrender. But once again the opportunity was lost by a change of strategy – this time Hitler's decision to invade the Soviet Union. In June, most of the German bombers were transferred to the Eastern Front and the Battle of Britain petered out, won by the British fighter and radar defence system, lost by a curious failure of German intelligence to assess correctly the results of their air offensive, and to concentrate on the most vulnerable targets.

The German *Blitzkriegs* in the Balkans were a necessary preliminary to the invasion of Russia, firstly to secure and maintain the vital Rumanian oilfields, and secondly to protect the German flank from British bombers based on Crete (occupied by the British, following Italy's unsuccessful attempt to conquer Greece). Yugoslavia was attacked and overrun in April 1941, and by the end of that month Greece had also capitulated after a *Blitzkrieg* on the French pattern had destroyed key bridges and aircraft on the ground.

Two months later the British forces on Crete were overcome by a combination of landings by airborne troops and German air superiority. The German attack, though successful, was very costly. Many transport aircraft and a considerable proportion of the élite paratroop regiments were lost, and

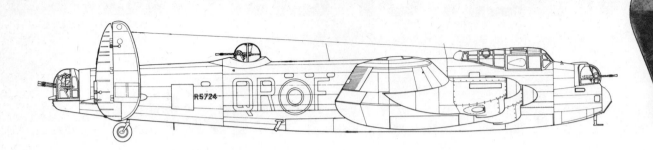

## AVRO LANCASTER Mk I

**Gross weight:** 63,000 lb **Span:** 102 ft **Length:** 69 ft 6 in **Engine:** 4×1,280 hp Rolls Royce Merlin XX **Armament:** 10×·303 Browning machine-guns **Crew:** 7 **Speed:** 280 mph at 18,500 ft **Ceiling:** 23,500 ft **Range:** 2,700 miles **Bomb load:** 14,000 lb

The most famous and successful heavy night bomber of the Second World War, the Lancaster first entered service with RAF Bomber Command in 1942

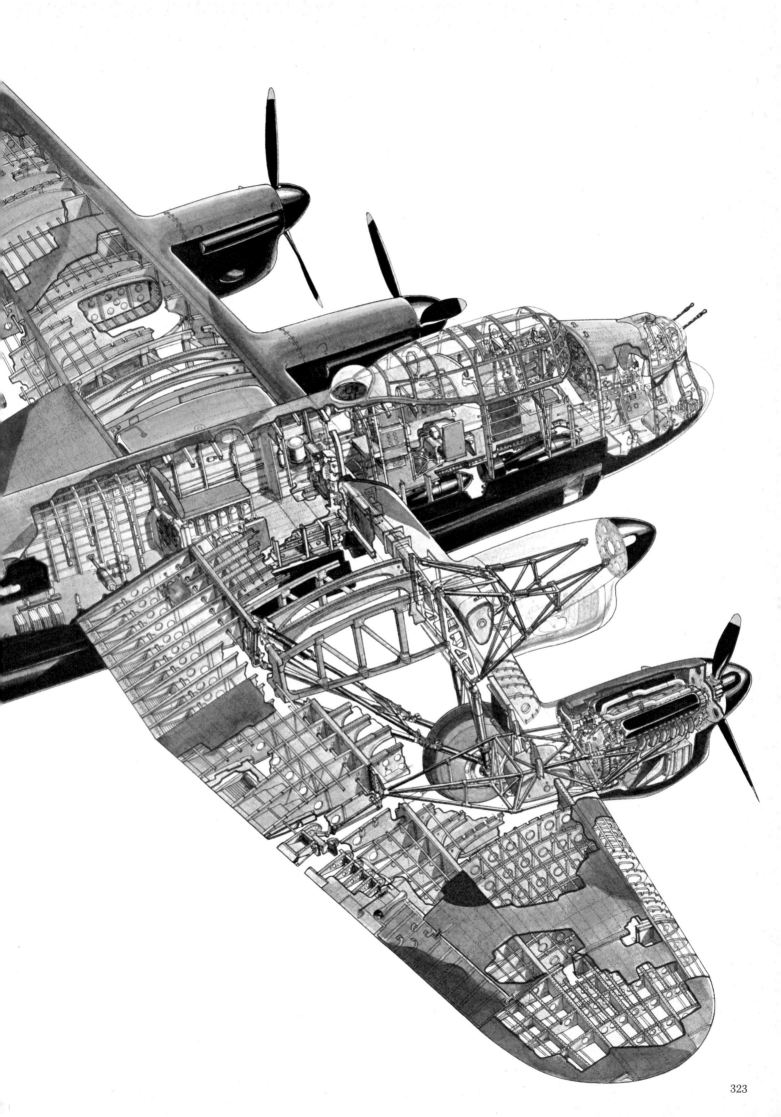

## COMMONWEALTH CA-11 WOOMERA

**Gross weight**: 22,287 lb  **Span**: 59 ft 2½ in  **Length**: 39 ft 7 in  **Engine**: 2×1,200 hp Pratt & Whitney Wasp  **Armament**: 2×20-mm cannon: 5×·303 machine-guns  **Crew**: 3  **Speed**: 282 mph  **Ceiling**: 23,500 ft  **Range**: 2,225 miles  **Bomb load**: 3,000 lb

A dive/torpedo/attack bomber developed in Australia towards the end of the war, but not brought into service. Unusual for its remote control barbettes behind the engine nacelles

## YERMOLAYEV Yer-2

**Gross weight**: 32,730 lb  **Span**: 75 ft 5½ in  **Length**: 54 ft 1½ in  **Engine**: 2×1,250 hp M-105  **Armament**: 1×20-mm; 2×12·7-mm machine-guns  **Crew**: 5  **Speed**: 311 mph at 19,000 ft  **Ceiling**: 25,260 ft  **Range**: 3,107 miles  **Bomb load**: 2,200 lb

This Russian long-range bomber was intended as a replacement for the Ilyushin DB-3, and carried out raids on Germany from 1941 onwards

## KYUSHU Q1W TOKAI

**Gross weight**: 11,755 lb  **Span**: 52 ft 6 in  **Length**: 39 ft 8 in  **Engine**: 2×610 hp Hitachi Tempu 31  **Armament**: 1×20-mm cannon; 1×7·7-mm machine-gun  **Crew**: 3–4  **Speed**: 200 mph at 4,400 ft  **Ceiling**: 14,730 ft  **Range**: 814 miles  **Bomb load**: 1,100 lb

A coastal patrol bomber introduced by the Japanese Navy in the latter part of the war in an effort to combat Allied submarines off the Japanese mainland

the Germans never really recovered. Crete was the last great German airborne campaign. Ironically, its success convinced the Allies of the value of airborne operations and they began to build up similar forces, with disastrous consequences at Arnhem in 1944.

After the Balkans, the Germans extended their *Blitzkrieg* technique to the invasion of Russia but this time, although initial successes were achieved, the *Luftwaffe* was unable to repeat its former brilliance. For one thing, the attacks were made on a broad front, instead of the short, sharp thrusts of the previous campaigns. And as the war continued, German bomber units were called upon to operate on too many widely-dispersed battlefronts, sapping their ability to achieve further decisive victories. From 1942 onwards, although it fought grimly on until the end, the *Luftwaffe* was increasingly fighting a defensive war for which it had not planned, and for which it lacked much of the necessary equipment.

The Japanese *Blitzkriegs* in the Pacific and south-east Asia in December 1941 relied even more heavily on the use of air power, although in their case it was primarily seaborne air power – a subject which needs a book to itself. The attack on Pearl Harbour which destroyed most of the US Pacific Fleet was carried out entirely by carrier-based aircraft. But in the simultaneous attacks on RAF bases in Malaya, as a prelude to amphibious landings, land-based Japanese bombers also played a part. The most dramatic success of the light bombers and torpedo-bombers operating from bases in southern Indo-China was the sinking of the British battleships *Prince of Wales* and *Repulse*, the first time that capital ships at sea had been destroyed by air attack.

Air strikes spearheaded all the Japanese invasions during the following months as their army and navy forces overran the Philippines, the Dutch East Indies, south-east Asia, and many of the islands of the south-west Pacific. From their newly-established bases in these areas, the Japanese could extend their bombing attacks against the Allied forces, including some bombing raids on Australia.

But the Allies were also learning the value of tactical air operations, especially with the arrival of suitable new aircraft. This was first evident in the Middle East where, in 1942, Baltimores and Mitchells replaced the outdated Blenheims and Marylands, and four-engined British and American bombers supplemented the faithful Wellingtons in attacks on enemy supply ports.

A mobile tactical air force was formed, and provided continuous air cover in the wide-ranging battlefields of the Western Desert. The Allied invasions of enemy

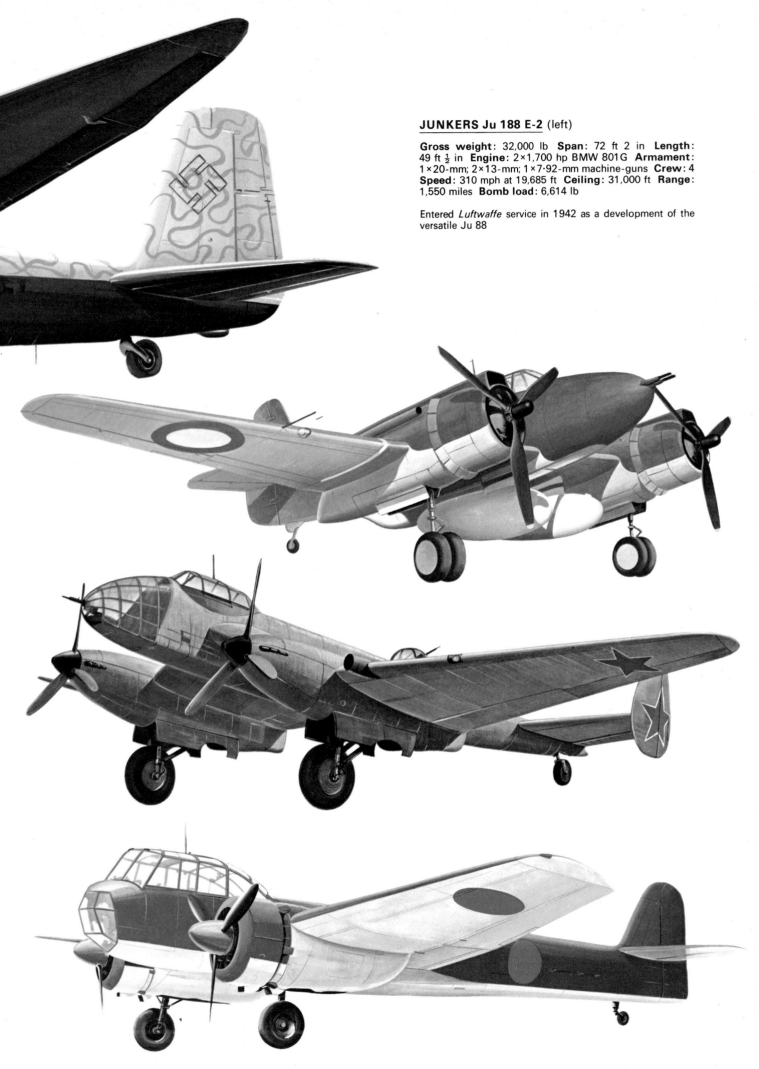

**JUNKERS Ju 188 E-2** (left)

**Gross weight:** 32,000 lb  **Span:** 72 ft 2 in  **Length:** 49 ft ½ in  **Engine:** 2×1,700 hp BMW 801G  **Armament:** 1×20-mm; 2×13-mm; 1×7·92-mm machine-guns  **Crew:** 4  **Speed:** 310 mph at 19,685 ft  **Ceiling:** 31,000 ft  **Range:** 1,550 miles  **Bomb load:** 6,614 lb

Entered *Luftwaffe* service in 1942 as a development of the versatile Ju 88

territory, beginning in North Africa and moving on to Sicily and finally Normandy in 1944, the greatest military assault in history, were spearheaded by air strikes which owed much to an appreciation of earlier German methods.

The turning point on the Eastern Front came in 1942 with the build-up of Russian tactical air forces to assist their ground troops. The Soviet aircraft industry had been moved back, behind the Urals, out of reach of German bombers – giving the Germans reason to regret the lack of a long-range strategic bomber. In Asia, tactical air power was a vital element in the Allied offensives from late 1943 onwards. The re-conquest of Burma was to a large extent made possible by the continuous operations of a tremendous Anglo/American air supply force, working under an air cover in which bombers assisted by bombing Japanese air bases.

Although it was not at first widely recognised, land-based bombers played a very important role in the war at sea: mines dropped by aircraft in European waters, for instance, sank more ships than those laid by surface vessels. From the beginning of the war the Germans, Italians and British all employed land-based bombers against enemy shipping with varying degrees of success, depending on the equipment available. Level bombing was generally found to be too inaccurate while the medium bombers used for that purpose were vulnerable to attacks by shore and ship-based fighters and to flak put up by convoys. Manoeuvrable fighter-bombers such as the Bristol Beaufighter and the Junkers Ju 88, adapted to carry torpedoes and attacking at low levels, were the most successful.

From the Allied point of view, however, U-boats posed the greatest threat to shipping. RAF Coastal Command suffered a serious lack of long-range patrol bombers during the first two years of the war, and large areas of the Atlantic were beyond their reach. But in 1942, long-range bombers began to become available, such as VLR (very long-range) Liberators, B-17s, and eventually de Havilland Mosquitos specially armed with a 57-mm cannon. These aircraft were able to cover the gap in the middle of the Atlantic and were assisted in their task of locating and attacking submarines by such aids as radar and Leigh lights; the latter was especially valuable in locating U-boats which had surfaced at night to re-charge their batteries or make a quicker passage.

The Germans had not developed the concept of air cover for their navy, nor did they have the long-range aircraft to provide it. Consequently the Allies won the Battle of the Atlantic, largely by superior air power. Whereas from 1939 to mid-1943, surface vessels sank 61 German and Italian submarines and assisted in 7 cases – while aircraft were responsible for only 9 – during the last two years of the war, surface vessels accounted for 224, land-based aircraft 239, ship-based aircraft 41, and a combination of surface and air forces 46. A further 64 U-boats were destroyed at their bases by bombing.

Ultimately, the war was won by Allied superiority in quantity and quality of equipment, based largely on the industrial might of the USA. Manpower played an important part of course, especially in the air where crews had to become increasingly skilled to use the ever more sophisticated instruments. Neither Germany nor Japan had the resources to match the Allied training programmes, under which, by late 1944, an RAF fighter pilot for instance had received nearly 300 hours of flying experience before being awarded his wings.

In terms of technology, however, the balance was more even. In some areas, such as jet aircraft and rocket-propelled bombs, the Germans were more advanced, and even made some progress towards the production of atomic weapons. Even the Japanese, at the end of 1944, were able to bring out a new bomber, the Mitsubishi Ki-67 Hiryu, which was such an advance on any of their previous aircraft that it came as an unpleasant shock to the Americans in the Pacific.

The build-up of Allied forces was a slow process, however: 83 per cent of the two million tons of bombs dropped on Germany throughout the war were dispatched in the last two years, when the RAF's early night raids were supplemented by daylight precision raids by the US Eighth Air Force. But the success of such raids was somewhat exaggerated by the Allied proponents of strategic bombing. Although nearly half of Cologne was devastated by the RAF's first thousand-bomber raid on the night of 30 May 1942, the city made a remarkable recovery.

The most damaging of the American daylight raids, on the ball-bearing works at Schweinfurt on 14 October 1943, caused only a temporary setback in production, while 60 of the 291 bombers involved were destroyed and 138 damaged, mainly by German fighters. The Americans were compelled to curtail their daylight bombing offensive until the introduction in the following spring of the long-range Mustang fighter which could provide an escort all the way to targets deep in Germany.

By 1944, the British and Americans had built up large bomber forces, and had the long-range fighters to protect them on the way to their targets. But having attacked most of the military targets available during and after the Normandy invasion, the Allies were reluctant to keep their bomber forces idle. In the last year of the war, therefore, they initiated vast strategic air offensives against Germany and Japan, aimed at bringing about a psychological victory by bombing cities and towns.

The moral validity of the raid on Dresden, the most destructive of the European war, is a matter of controversy to the present day. But there is evidence that such raids increased the determination of the German people to fight, no matter how hopeless the odds. Similarly, in spite of the terrible destruction of Japanese cities by the B-29 raids of the US Twentieth Air Force during the spring and summer of 1945, against practically no fighter opposition, when millions were made homeless and over 100 square miles of Tokyo and four other cities were destroyed, it was only with the dropping of the two atomic bombs on Hiroshima and Nagasaki that the Japanese felt compelled to surrender. Not only was the strategic value of these offensives questionable, they also compelled the defeated nations to concentrate their energies after the war on building up their destroyed cities and economies with aid provided by the United States. The resulting programme of reconstruction has become an increasing economic challenge to the victors in the post-war years.

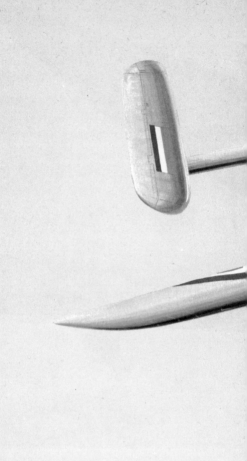

## TUPOLEV Tu-2

**Gross weight:** 24,232 lb **Span:** 62 ft **Length:** 43 ft 3⅔ in **Engine:** 2×1,400 hp AM 47 Mikulin **Armament:** 4×7·6-mm machine-guns; 2×20-mm cannon **Crew:** 4 **Speed:** 394 mph at 26,000 ft **Ceiling:** 34,775 ft **Range:** 880 miles **Bomb load:** 5,000 lb

Developed too late to see service during the Second World War but used widely afterwards by Russia and her satellite air forces

## NORTH AMERICAN B-25J MITCHELL

**Gross weight:** 33,500 lb **Span:** 67 ft 7 in **Length:** 52 ft 11 in
**Engine:** 2×1,700 hp Wright R-2600 **Armament:** 6×·50 Browning
machine-guns **Crew:** 6 **Speed:** 272 mph at 13,000 ft **Ceiling:**
24,200 ft **Range:** 1,350 miles **Bomb load:** 3,000 lb

The J version of the B-25 Mitchell had a dorsal turret close behind
the cockpit and three 0·50-in guns in the nose; some in fact had no
less than eight guns in that position, adding to ten carried elsewhere

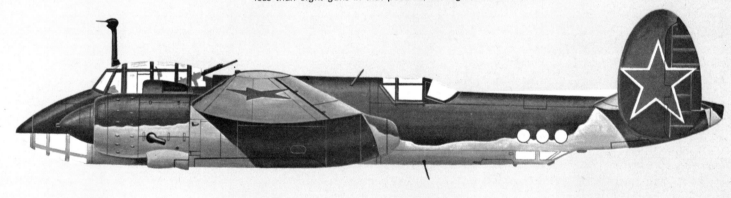

## JUNKERS Ju 288 C-1

**Gross weight:** 47,120 lb **Span:** 74 ft 4 in **Length:** 59 ft 5¾ in
**Engine:** 2×2,950 hp Daimler Benz B 610 **Armament:** 6×13-mm;
1×15-mm machine-guns **Crew:** 4 **Speed:** 407 mph at 22,300 ft
**Ceiling:** 34,000 ft **Range:** 1,615 miles **Bomb load:** 6,614 lb

Developed to make use of an engine of new design, but abandoned
before development work had been completed

*First operational jet bomber with the USAF, the North American XB-45 first flew in 1947*

559479

# JET BOMBERS
## *David Anderton*

From the Arado Blitz of the Second World War to the swing-wing bombers of the future, this chapter covers 35 years of jet bomber development. The experimental aircraft of the postwar years, the strategic bombers that carried the nuclear deterrent of the Cold War, and the supersonic generation of the sixties and seventies are all included, along with a discussion of bomber armament and a selection of attack bombers.

# CONTENTS

*A Hawker Siddeley Vulcan B Mk 2 flies over the Fylingdales early warning establishment*

This narrative traces the development of the manned jet bomber from its early beginnings in the last days of the Third Reich. The order is chronological, with the reference date being either the first flight of the bomber prototype, or of the development aircraft that immediately preceded it.

With few exceptions, the aircraft here described were designed from the start as bombers. And with two exceptions, all of them were originally designed around gas turbines for jet propulsion.

Serious thought had been given to rocket-propelled manned bombers by German scientists during the Second World War. But none of these ideas ever got beyond feasibility studies. Even the most optimistic assumptions could not produce anything but a gigantic and vastly complex machine with a miniscule payload.

The rocket-propelled bomber finally was developed: we know it today as the ballistic missile. It is unmanned and – since this history deals only with piloted aircraft – it is not considered further.

The gas turbine for jet propulsion is an old idea that has been rediscovered several times. It was first patented in 1791 by John Barber, an Englishman. Two more patents, issued in 1917, foresaw the possibility of aircraft powerplants being built around this principle.

But the pieces began to come together in the years just before the Second World War. Then a young German scientist and a young Royal Air Force cadet began studying the concept seriously and independently. The German was Pabst von Ohain, who developed a workable jet engine that ran in September 1937 and powered an aircraft in flight on 27 August 1939. The RAF cadet was Frank Whittle, and his ideas had been made available to the public in published patents as early as 1931. Whittle's first engine ran before the von Ohain unit, on 12 April 1937. But it did not power an aircraft until 15 May 1941. The delay made a great deal of difference, because Germany had operational jet fighters in quantity long before the British did.

Powering a fighter is one thing: thrusting a bomber through the air with jet engines is quite another. The useful load for a fighter is small, or it was then: some guns and ammunition, and fuel enough to enable it to defend a small area near its home base. A bomber, on the other hand, had to carry a load of bombs, equipment to drop them, large quantities of fuel to reach enemy targets behind their front lines and guns to defend itself against fighter attack. This added up to thousands of pounds, instead of the few-hundred-pound load of the fighter.

To carry several thousand pounds of load required an aeroplane at least three times as heavy when empty. The Avro Lancaster, for example, normally carried an 8000-lb bombload on long missions. Empty, it weighed about 37,000 lb. The Boeing B-29 carried 10,000 lb on its long-range raids, and weighed 70,000 lb empty.

These were piston engined bombers, the best of their day, and as efficient a pair of load carriers as could have been built in their time. They were powered by four of the most powerful piston engines available. So a jet bomber would need multiple engines and – since jet engines burned at least twice as much fuel per thrust horsepower as did a piston engine – they would also need at least double the fuel load. The sums worked against early development of jet bombers with load or range comparable to those of piston-engined equivalents.

That's one reason the jet bomber was slow in arriving on the scene, and why its development had to wait for several new concepts in technology to be discovered.

**WARTIME DEVELOPMENTS**

# THE STARTING PLACE

The advanced German aeronautical technology that spawned the jet engine and the fighter also gave birth to the jet bomber. But in contrast to the relatively major success enjoyed by the fighter programme, the jet bombers made little or no contribution to the long-range development of jet aircraft, or to the advancement of aeronautical technology.

This probably was due to Germany's wartime aeronautical policy, primarily one of whims, favouritism and offhand decisions. It is remarkable how much was accomplished, given the atmosphere of the times, the dreams and eccentricities of Hitler and the sycophantic officials around him.

The jet bomber programme did not produce a major striking force capable of inflicting telling damage on Allied targets. As an operational achievement, then, it was a complete failure. But as a technical achievement, it had some bright spots of interest to the historian of things aeronautical. One of these was the Arado 234, the world's first jet bomber.

*16 June 1943:*
### Arado Ar 234 V-1
The surprise German offensive through the Ardennes in December 1944 – the Battle of the Bulge – was the last major strike by combined forces of the Third Reich. Backing up the infantry and armoured units was every available aircraft the Germans could get into the air. Fighters and fighter-bombers flew ground-support missions as often as possible in the foul weather that dogged that offensive. They tangled with Mustangs and Tempests, Thunderbolts and Spitfires, in swirling dogfights or high-speed passes at low levels.

But there was a new sound in the skies over the Ardennes, and it came from jet engines. The Germans had committed their new jet fighter-bombers to that action, and among them were the aircraft of *Kampfgeschwader* 76.

KG 76 recently had completed training on the Arado 234B, a sleek twin-jet bomber, and they were taking the aircraft into operation for the first time. Streaking down in a gentle dive from the leaden skies, the Arados dropped their bombs against targets in the Allied lines.

Who dropped the first bomb from that first jet bomber may never be known. But that action, around the Ardennes, was the first in which jet bombers, designed as such, were committed to action against an enemy force.

The results were inconclusive. The maximum bomb load that could be carried by the Arado was about 2000 lb, essentially the same as the bomb load of the USAAF's Thunderbolts. The Arado's asset was its speed, which made for more difficulties on the bomb drop but which gave definite advantages in getting away from encounters with Allied fighters.

The first interceptions ended when the Arados accelerated away from their pursuers while taking an abnormal amount of hits from Allied machine-guns and cannon. This may have been due to the slightly heavier construction of jet aircraft, but it may have been because the bullets were striking at very shallow angles in the tail chase, and glancing off without doing much damage.

The war ended before the Arados really could show their abilities. More than 200 were built and there were plans to build them by the hundreds. But those plans were ended by the final destruction of Germany, and the bomber version must be judged as unsuccessful in terms of what it accomplished for the expended effort.

It was a fine design, using conventional aerodynamics and a pair of jet engines in clean underwing nacelles to achieve its performance. The original intention was to produce a high-altitude reconnaissance aircraft, and it was in that role that the Arado 234 Blitz (Lightning) was best employed. It flew an experimental series of high-altitude photographic missions over the beaches during and after the invasion of Normandy, and made other runs over Allied installations at its operating altitudes above 30,000 ft. But the reconnaissance role is outside the scope of this history.

The designers tried to strip all useless weight from the airframe to give the Blitz maximum performance. One of the useless things, they decided, was the landing gear. They planned to have the aircraft mounted on a takeoff trolley, jettisoned after the takeoff. Landing would be done on a skid gear mounted to the belly. Weight and drag savings had been calculated to be considerable. But using the belly in this way meant that no bomb could be carried underneath, and Hitler wanted every new aircraft to be a bomber or a fighter-bomber. So the prototype 234 was redesigned as a high-level bomber, with retractable landing gear and a bomb snuggled up against its belly in a recess that partially submerged the weapon.

It was a single-place bomber, with a unique control system for the bomb run. A standard level-flight bombsight was mounted in the nose of the aircraft behind an optically flat glazed panel. Further back was the cockpit. Before the bomb run the pilot trimmed the aircraft for steady flight, swung his control stick to one side and moved forward into the bombardier position, to lie face down and fly the airplane through the automatic pilot circuits of the bombsight.

A variety of bombs could be carried under the belly, from a single 2200-pounder to a cluster of incendiaries or fragmentation bombs. Auxiliary fuel tanks could be mounted under the engines; but if they were not used, 1100-lb bombs could be carried, one under each nacelle. The alternative method of bomb-dropping was in a shallow dive, typical of the standard fighter-bomber attack. Rocket packs were developed to assist the takeoff of the heavily loaded Blitz, and were used whenever the maximum bomb and fuel load were carried.

Developments of the Arado as long-range bomber, reconnaissance aircraft and night-fighter, with two or four engines and various special equipments, were planned, and were in various stages of development when Germany's surrender brought them to a halt.

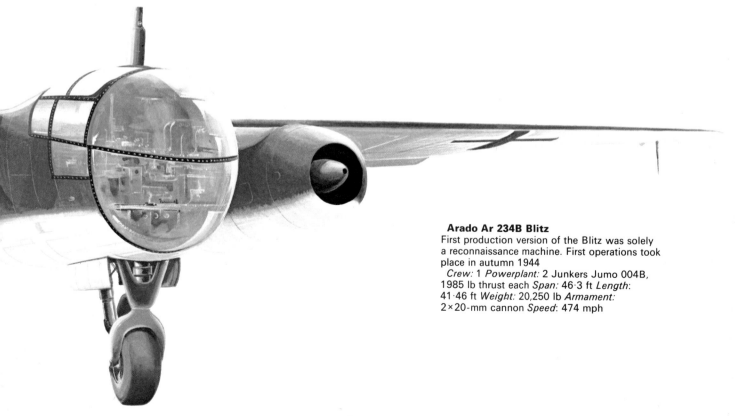

**Arado Ar 234B Blitz**
First production version of the Blitz was solely a reconnaissance machine. First operations took place in autumn 1944
*Crew:* 1 *Powerplant:* 2 Junkers Jumo 004B, 1985 lb thrust each *Span:* 46·3 ft *Length:* 41·46 ft *Weight:* 20,250 lb *Armament:* 2 × 20-mm cannon *Speed:* 474 mph

**Junkers Ju 287V-1**
*Crew:* 2 *Powerplant:* 4 Junkers Jumo 004B,
1985 lb thrust each *Span:* 66·9 ft *Length:*
60·02 ft *Weight:* 44,100 lb *Armament:* 2×13-mm
mg *Speed:* 346 mph

**Horten Ho IXV-2**
*Crew:* 1 *Powerplant:* 2 Junkers Jumo 004B,
1985 lb thrust each *Span:* 54·97 ft *Length:*
24·48 ft *Weight:* 18,742 lb *Estimated speed:*
540 mph (Data for Gotha Go 229 are similar)

*16 August 1944:*
## Junkers Ju 287 V-1
This flying test bed was the result of a 1943 study project by Junkers, intended to produce a heavy bomber with performance that would enable it to outrun any known or expected Allied fighter.

Swept wings were then coming into fashion in Germany as a result of theoretical studies and wind-tunnel tests at research institutes, and the first Junkers idea was to sweep the wings back in what would become the standard fashion. But the Junkers team finally elected to sweep the wings forward. They had their reasons.

Sweptback wings have poor low-speed characteristics, which is why they now are adorned with so many aerodynamic tricks. A plain swept wing will have very poor aileron control at low speed because the airflow tends to slide spanwise along the wing. By the time it gets to the tip, the combination of lateral flow and tip stall has just about wiped out that portion of the wing as a lifting and control surface.

By sweeping the wing forward, the Junkers team felt they could overcome this

dead weight would resist the aerodynamic twisting, and would – it was hoped – prevent disaster.

Understandably, nobody wanted to start an aircraft programme without testing this idea. So the Junkers team received an order to build a test vehicle employing the principle and to get it flying as soon as possible.

To do so, they put together a fuselage from the Heinkel He 177 heavy bomber, and the tail assembly from a Junkers Ju 388. The main landing gear was built from Ju 352 components and the nosewheels were salvaged from captured USAAF B-24 bombers. The wing of course was completely new, fabricated by Junkers to their own design.

Since the wing was to be typical of the final bomber proposal, it only carried two engines. But the huge test aircraft needed more power to get off the ground, and another pair of engines was mounted, one on each side of the forward fuselage. Still more thrust for takeoff came from podded rocket powerplants attached beneath each wing nacelle and jettisoned after takeoff.

abbreviation for *Versuchsmaschine*, or research machine. V-1 therefore would be the first research aircraft, or prototype, V-5 the fifth, and so on.) This aircraft was to be built and tested as a glider, which was the primary field of Horten expertise.

It flew during 1944, and work continued on a second prototype to be powered by a pair of Junkers 004B turbojets. It was built of wood except for a welded steel-tubing structure at the centre section and the engine bay area. It was covered with plywood, except for the surfaces around the engine, which were protected with the standard firewall steel panels used in German aircraft.

It was decided that the Hortens needed industrial assistance with their project, and so development of the Ho IX was handed over to Gothaer Waggonfabrik, designers and producers of the famous Gotha bombers during the First World War. Gotha had been active in aviation since then as a developer of troop-carrying gliders and as a sub-contractor for the German aircraft industry.

The Gotha team made some changes,

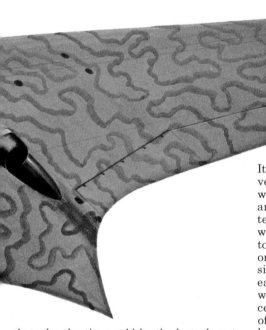

obstacle; the tip would be the last place to stall, and the lateral control should stay available. True – but there was a disadvantage.

Consider what happens when the wing starts to bend upward under gust loads. A swept-back wing tends to flex so that its trailing edge, near the tip, bends up more than the leading edge. In other words, the wing twists when it bends. In swept-back wings, that twist-when-bent is kept from being disastrous because the airflow tries to blow the trailing edge down again.

In a swept-forward wing, things are reversed. The leading edge of the wing tends to bend up higher than the trailing edge, and the airflow wants to get underneath the leading edge and increase the wing twist. When that happens, the wing twists more and more and is literally ripped off the fuselage.

But there are ways to prevent this. One of the most useful is to hang weights on the wings in key locations to resist the twist. Jet engine nacelles are such weights, and so the Ju 287 was designed with single nacelles, well out along the span. Their

Surprisingly, the test vehicle flew well. It was intended only for low-speed investigations, and whatever other flights were necessary to measure wing performance within the limited speed range of the test aircraft. Meanwhile, a second prototype was being developed for high-speed tests, to be powered by six jets instead of the four or less envisioned for the final design. The six were clustered in two nacelles of three each, with a triangular cross-section, and were mounted with their weight concentrated well ahead of the torsional axis of the wing. The Junkers designers had learned where to put the weight to counter the wing twist.

Construction had begun on a third prototype, and plans were at hand for the start of a pre-production run when advancing Russian troops arrived in the area and seized the aircraft and the designers. Both prototypes were taken to Russia and – so the story goes – were flown there in a test programme. But there is no further evidence of any Russian interest in the Ju 287's peculiar swept-forward wing geometry.

*Late January 1945:*
## Horten Ho IX V-2
The most daring of all the German jet bomber designs was a flying wing, developed by the Horten brothers who had achieved fame as designers of elegant flying-wing sailplanes.

Like most devotees of drag reduction, the Hortens thought in terms of speed, and fighters. So the first approach to their ninth design was to consider it as a jet-propelled fighter. They drew plans for a first prototype, the Ho IX V-1. (German identification of prototypes used the V-number designation; V in this case was the

converting the Ho IX to a fighter-bomber. On that basis, the new aircraft – redesignated Gotha 229 – is included in this history.

Gotha began a prototype line, which included an all-weather fighter and a trainer version as well as the fighter-bomber, and the Horten brothers continued to work to complete their powered prototype. That aircraft, the Horten IX V-2, flew a very conservative flight research programme, gradually working up from the low-speed end of the spectrum to higher speeds. By the spring of 1945 it was ready for high-speed tests, and did achieve one run at close to 500 mph. But in the approach to the field after that test, one engine flamed out, and the aircraft slammed into the ground in a ball of flame.

The programme never progressed further. The Gotha prototypes were not completed in time to fly before the end of the war, although one of them was almost ready when the factory was reached by Allied troops advancing into Germany.

Had the war lasted longer, there might have been other designs to describe. The combination of jet propulsion, new radar, sweepback and other technological advances had spurred German designers to a wide variety of proposed aircraft. And given Hitler's desire to see bombs strapped under the wings of everything that could fly, one must assume that bomber and fighter-bomber designs would have proliferated.

But they didn't, and that is perhaps the fortunate aspect of the German jet bomber programme. It only produced a limited number of operational bombers of one model, plus two flying prototypes of two others. In no sense did it make a major contribution to aeronautical progress.

# THE CLASS OF '47

U.S. AIR FORCE
461509

Streams of piston engined bombers being attacked by jet-powered fighters was a prospect too unpleasant to contemplate, and in mid-1944 the USAAF gave the go-ahead for a development programme that was to produce the first crop of postwar jet bombers – the 'Class of '47'

There was no doubt that the jet age had arrived. The ingenious, last-ditch German designs for bombers, night-fighters and interceptors had clinched the argument. The day of the reciprocating engine was well and truly over.

But the war wasn't, and there was an obvious need for the performance of those last products of Third Reich science to be matched, not only by counter-air fighters but also by their primary targets, the bombers.

The thought of a slow-moving bomber stream, harassed by darting fighters, was most unpleasant to strategic planners. Not all of them believed that the war would be brought to an abrupt and chilling conclusion by nuclear weapons, and few suspected the extent of the internal disintegration of Germany. So the word went out to get going fast on jet engine development, and on fighter and bomber designs using the new type of engines. Fighters were the first to benefit, with early engine design, development and production earmarked almost exclusively for them.

A few months after the initial successes of the jet fighter programme in the United States, its military and industry turned their attention to jet bomber development.

Not much could be done to modify existing types; replacing piston engines with jets on a Boeing B-29 just would not pay off in performance.

There was one possible design conversion from piston power to jet thrust: the Douglas XB-42 pusher bomber. But what really gave jet bomber design its initial impetus was an Air Materiel Command (USAAF) competition in mid-1944. It called for a jet bomber to carry a mixed bomb load, up to a single 22,000-lb 'county buster' bomb, and with the ability to drop it from a height of 40,000 ft.

Four competitors were chosen to design, build, and fly experimental prototypes: Boeing, Convair, Martin and North American. All their aircraft first flew during 1947, consequently becoming known to aviation history as the 'Class of 1947'.

In any typical American graduating class, one graduate was most likely to succeed – the star in this case being the radical Boeing B-47, last of the four to fly and really in a class by itself. (It will be described in the next chapter.) Then, late in the war, the Air Materiel Command gave the nod to Northrop for conversion of two piston-engined YB-35 flying wing bombers to jet power. That programme produced the YB-49 and the YRB-49A, both technical achievements, both operational duds.

This is the story of the Class of 1947.

## 17 May 1946:
### Douglas XB-43

First jet bomber to fly in the United States, the Douglas XB-43 was a development of a piston-engined pusher predecessor, the Douglas XB-42. That aircraft, nicknamed the 'Mixmaster' because of its rear-mounted counter-rotating propellers, was the culmination of a major engineering effort. The goal was to develop a smaller, lighter and cheaper bomber capable of doing the job of

heavyweights like the Boeing B-29, which was just entering service.

Early studies convinced Douglas that only a radical approach to aerodynamic layout would meet this goal. They settled on the pusher propeller scheme to leave the laminar-flow wing clean and free of any flow distributors such as portly engine nacelles. They picked a fairly high aspect ratio to reduce one component of drag, and they worked on the wing to make it a high-lift section and to perform at the low-speed end of the scale as well as at the high.

The first figures were exciting. Compared to the B-29 the Douglas design would carry a 2000-lb bomb load for more than 5000 miles, and burn only about 30% of the fuel needed by the bigger bomber. Further, the little bomber would be handled by a much smaller flight and ground crew and would cost less to buy and maintain.

A pair of prototypes were built and flown. They proved most of the basic points of the design, but failed to meet all requirements: the bomb load was much less because the airframe was overweight.

But time was running out for the reciprocating engine and the Douglas XB-42 was transformed into a hybrid aircraft with both piston and jet engines. The jets were little Westinghouse engines slung outboard, one under each wing, and – as is obvious with hindsight – the airplane was a dog. The performance was not much better with the jets. But this is getting ahead of the story, because the hybrid XB-42A did not fly until much later than the XB-43. It is included here to show the continuity of the programme.

In October 1943 the Air Materiel Command met with Douglas and worked out an agreement for an all-jet version of the XB-42. Douglas first had to choose between two engines, both originally developed by

**Douglas XB-43**
*Crew:* 3 *Powerplant:* 2 General Electric J35-GE-3, 4000 lb thrust each *Span:* 71·17 ft
*Length:* 51·42 ft *Weight:* 40,000 lb
*Bombload:* 8000 lb *Speed:* 515 mph

General Electric: the I-40, a 4000-lb thrust centrifugal-flow unit that owed many of its features to the early British Whittle engines, and the TG-180, an axial-flow engine also rated at 4000 lb of thrust.

Douglas preferred the slimmer TG-180 and revised the XB-42 airframe to take the jet engine. The new aircraft was designated XB-43 and the resemblance between it and its predecessor was remarkable. To all intents and purposes, the piston engines of the XB-42 had been replaced by the jets of the XB-43.

On 17 May 1946, after two taxi runs with short hops on each, the XB-43 first flew. By the time it had logged just under 13 flight hours it had been flown faster than 500 mph, the 'magic' speed for those days.

But engine delays plagued the programme, slowing the delivery of the second prototype, and the XB-43 never reached its planned production status. The second prototype flew almost one year after the first and it served for a variety of test purposes at Muroc Army Air Base (later Edwards AFB).
*17 March 1947:*

### North American XB-45 'Tornado'

Overshadowed by the faster and more numerous Boeing B-47, the North American B-45 has never received the full credit for its contributions to the development of jet bombardment in the USAF. But during its relatively short service life, it established a number of significant firsts that pioneered the strategic and tactical bombardment techniques of the jet age.

It was the first operational jet bomber in the United States Air Force. In March 1949 the first B-45A aircraft were accepted by the 84th and 85th Bombardment Squadrons of the 47th Bombardment Wing, and they served with those units for at least eight years.

It was the first jet aircraft to be refuelled in the air. During a test flight, the second RB-45C joined with a Boeing KB-29B 'flying boom' tanker to take on fuel in the first demonstration of this particular type of air-to-air refuelling.

It was the first jet bomber to drop a nuclear device.

And it served in a limited role as a jet reconnaissance aircraft in combat. During 1951, a pair of RB-45C aircraft were sent on detached duty to Korea to gain combat experience for the type. MiGs were faster and more manoeuvrable, and they shot up the slower and clumsier RB-45Cs. That first combat experience led to a restriction to flights with fighter escort only. The prohibition was later extended to forbid any daylight reconnaissance runs into northwestern Korea's infamous 'MiG Alley'.

North American's 130th design was entered in the mid-1944 competition, and was planned around the axial-flow TG-180 jet. This engine had a smaller diameter than the I-40 of the same thrust rating. Obviously, a multiple engine installation would be necessary to get the required performance.

North American stayed with the kind of design it knew well: unswept, tapered laminar-flow wings, and structure built in the conventional manner. Its simple appearance belied its performance requirements. That smallish airframe was to carry a 22,000-lb 'county-buster' bomb, or the new nuclear weapons in unspecified number and weight, or large numbers of 500-lb bombs. It was required to drop these weapons accurately, by radar or visual means, from a bombing altitude of 40,000 ft. Its limit speed was expressed in terms of Mach number, the ratio of that speed to the speed of sound. The B-45 was one of the first aircraft to have its limit speed so defined, and it was set at Mach 0·76, or 76% of the speed of sound.

The first XB-45 – three had been ordered, two for flight and the third for static tests on the airframe – first flew on St Patrick's Day 1947. It was the first of the Class of '47 to take to the air.

The Air Force liked the results of that first flight and subsequent testing, even though the XB-45 and its crew were lost in a tragic accident early in the test programme. A production run was ordered, and the first model B-45A was built in two slightly different versions. The first production block of B-45A-1 aircraft was powered by Allison-built J35-A-9 and -11 turbojets. The second block of B-45A-5 aircraft was powered by General Electric J47-GE-7 and -9 engines. The total order for the A model was 74, and all these early aircraft went to Air Training Command and Tactical Air Command.

The C models were equipped with inflight refuelling plus other changes, but their basic performance was unchanged. After the first ten C models had been produced, the Air Force told North American to finish the production run by modifying the remaining C models to a reconnaissance configuration. As RB-45C-1, the last 33 to be built were extensively changed internally. They were almost totally rebuilt for camera installations, including some that used new and different techniques for low-level high-speed photography.

The B-45s served well and long, and generally in a minor role. But for several years they were the only jet strategic deterrent force on station in Europe with the NATO forces, and they played that vital part well. But by 1958 they were being replaced in service, and the first were finding their way to the scrap heap.
*2 April 1947:*

### Convair XB-46

This was a lovely aircraft, with sleek, extended lines in its fuselage, and a long wing with a high aspect ratio. Inside its cylindrical belly was space for a 22,000-lb county-buster, or a mix of nuclear and non-nuclear weapons. A tail turret, mounting a pair of ·50-cal machine-guns remotely fired by radar control, was the only defensive armament to protect its three-man crew.

But its beauty was only skin deep. The Convair design was slow, or slower than other aircraft in that 1944 competition, and so it never reached production. Only a single prototype was built, and the money for the second and third prototypes was shunted into another unbuilt Convair design, the XB-53, a proposed jet bomber with swept-forward wings like those of the wartime German Ju 287.

With the same powerplants as the North American XB-45, the Convair entry was some five tons heavier. That reduced its performance below that of the XB-45, and the latter aircraft was chosen in preference.
*22 June 1947:*

### Martin XB-48

The third straight-winged member of the Class of '47 was the Martin XB-48. Its high-mounted, thin wing featured a high aspect ratio, although not quite so high as that of the Convair XB-46. Six jet engines were tucked under the wing, three to a side, in what looked like individual nacelles but which were actually integrated installations with contoured cooling passages between the engines.

This thin wing left no room for big-

*Prototype of the USAF's first operational jet bomber, the North American XB-45*

wheeled landing gear, and so the ingenious Martin engineers designed a tandem, dual-wheeled landing gear and installed it, bicycle fashion, in the fuselage. Wing outriggers were used to handle the low-speed and turning situations. The gear was tested earlier by Martin on a converted B-26H piston-engined bomber, nicknamed the 'Middle River Stump Jumper'. (Middle River was the location of the Martin plant near Baltimore, Maryland.)

Whether it was the tandem gear or the bomb load of 22,000 lb, or both, the fuselage of the XB-48 was bulky and ungraceful.

Like the others in its class, the XB-48 was to be flown by a three-man crew, and defended by a pair of remote-controlled ·50-cal machine-guns in the tail. Two prototypes were funded, but there is no record whether or not both were flown; only the first is documented to any extent.

The XB-48 eventually was chalked up as another learning experience. It scored a modest first by being the first six-jet aircraft to be built and flown. It is probably fair to credit its tandem landing gear layout with setting a fashion that the Boeing XB-47 was to use also. That was an ingenious engineer-ing solution to a tough problem, and was probably worth the cost of the entire programme.

*21 October 1947:*
**Northrop YB-49**
Although chronologically the Northrop flying wing YB-49 fits into the Class of '47, it was a very different aircraft, and was not an entry in the 1944 competition. It was basically a jet version of two of the 15 YB-35 piston-engined flying wing bombers developed by Northrop under contract to the Army. Further, the YB-49 was about twice the weight of its smaller contemporaries and was designed for much greater load capacity, range and crew requitements.

The fundamental premise was that a flying wing was all load-carrying ability with minimum drag. In the centre was the accommodation for the crew, with full stations for pilot, co-pilot, flight engineer, navigator, radio-operator, bombardier and gunner. Off-duty rest space was provided in the 'fuselage', a fairing behind the crew area. On either side of the crew compartment were three bomb bays, separated two and one by a bay for the landing gear. Fuel tanks filled the wings.

The obvious difference between the YB-35 and YB-49 designs was the replacement of the counter-rotating propellers and piston engines of the former with the eight Allison jets of the latter. But the pusher propellers had also functioned aerodynamically as

*The XB-45 is given a thorough polish in front of Dakotas, a Beech transport and a row of light planes*

Tornado in flight. B-45s served as tactical and strategic bombers and as reconnaissance aircraft with the USAF

vertical stabilising area. On the YB-49 the loss of their effect was compensated for by the addition of four vertical flow separators – anyone else would have called them fins – to increase the directional stability.

But apparently these fins were not successful in curing the permanent affliction of the YB-49: dynamic instability. Unofficial reports on the flight tests of the two prototypes indicated that they were continually hunting around one or more axes, seeking but never finding a single stabilised attitude. This ruined the chances of dropping a bomb with any accuracy.

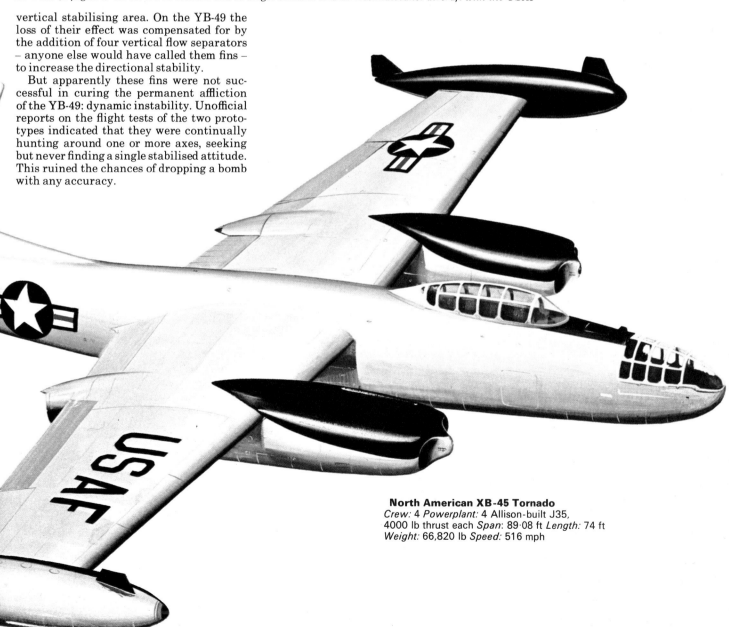

**North American XB-45 Tornado**
*Crew:* 4 *Powerplant:* 4 Allison-built J35, 4000 lb thrust each *Span*: 89·08 ft *Length:* 74 ft *Weight:* 66,820 lb *Speed:* 516 mph

*Although sleek and graceful, the performance of the Convair XB-46 was not up to standard, and only a single prototype was built*

So Northrop proposed to make the YB-49 a reconnaissance bomber, taking advantage of its long range – because of its low drag and, not incidentally, its huge fuel capacity. The Army approved the development of the YRB-49A. That version used four Allison jets mounted internally in the wing, and two hung below the wing in nacelles.

The sole YRB-49A took to the air on 4 May 1950, and it was the last hurrah for

**Convair XB-46**
*Crew:* 3 *Powerplant:* 4 Chevrolet-built J35-C-3, 4000 lb thrust each *Span:* 113 ft *Length:* 105·75 ft *Weight:* 91,000 lb *Bombload:* 22,000 lb *Armament:* 2 × ·5-in mg *Speed:* 491 mph

the flying wing. Development was halted after both YB-49 prototypes were lost in crashes, with their crews.

One major point about the flying wing designs was their low wing loading, of the order of 20 lb per square foot. By contrast, other members of the Class of '47 had wing loadings of between 62 and 72 lb per square foot. This made a considerable difference, in take-off and climb performance for example. The Northrop wings were able to fly from the mile-long runway of Northrop Field, at Inglewood, California, but the other bombers had to operate from the dry lake bed at Muroc Army Air Base because of their 7000–8000 ft take-off distance.

The wings had an interesting and unusual control system, featuring elevons which combined the functions of elevators and ailerons. There were no conventional rudders; split wingtip horizontal surfaces opened to cause drag at one wingtip or the other, producing the yawing motion typical of rudder application.

They were interesting aircraft, technically speaking, with some unusual design and engineering features. So history records the Northrop flying wing programme as exciting and advanced in the technological sense, but as an operational failure.

*The Martin XB-48 is hauled out on its tandem landing gear and wing outriggers at the Martin plant at Middle River, Baltimore*

**Martin XB-48**
*Crew:* 3 *Powerplant:* 6 Allison-built J35-A-7,
4000 lb thrust each *Span:* 108·33 ft *Length:*
85·75 ft *Weight:* 102,600 lb *Bombload:* 22,000 lb
*Armament:* 2×·5-in mg *Speed:* 516 mph

*The first six-jet aircraft to fly, the XB-48 gave the Martin engineers useful experience of jet bomber design*

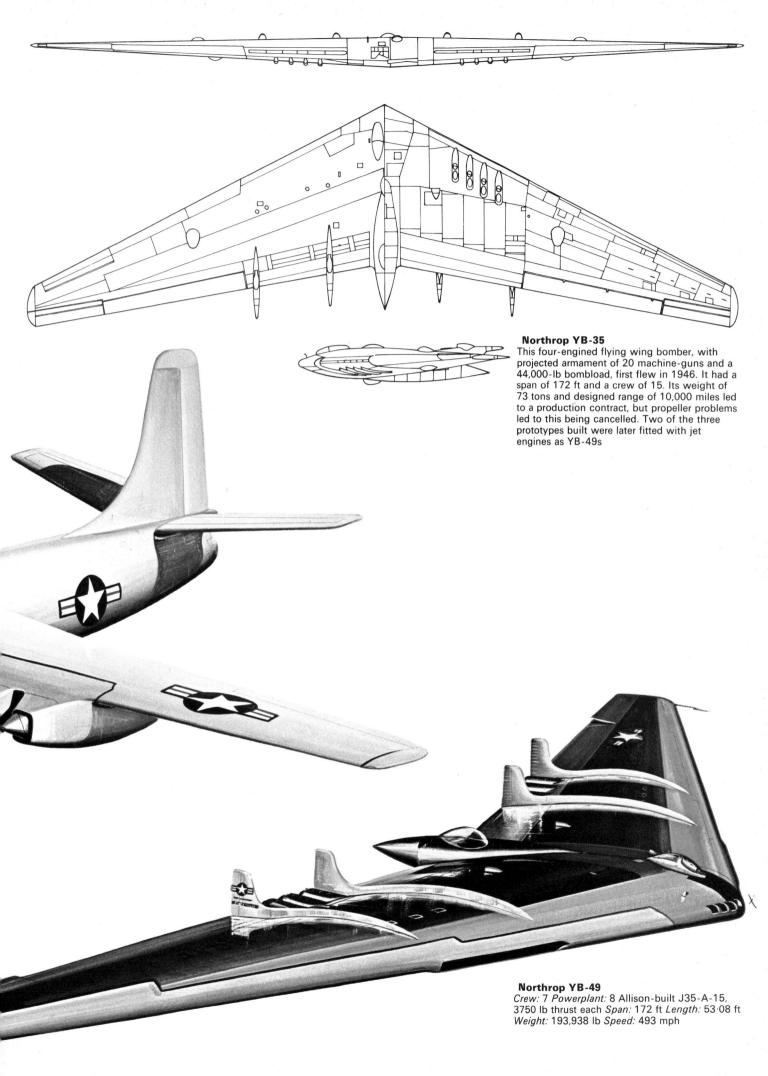

**Northrop YB-35**
This four-engined flying wing bomber, with projected armament of 20 machine-guns and a 44,000-lb bombload, first flew in 1946. It had a span of 172 ft and a crew of 15. Its weight of 73 tons and designed range of 10,000 miles led to a production contract, but propeller problems led to this being cancelled. Two of the three prototypes built were later fitted with jet engines as YB-49s

**Northrop YB-49**
*Crew:* 7 *Powerplant:* 8 Allison-built J35-A-15, 3750 lb thrust each *Span:* 172 ft *Length:* 53·08 ft *Weight:* 193,938 lb *Speed:* 493 mph

# IN A CLASS BY ITSELF

Although the Boeing XB-47 Stratojet was, chronologically, a member of the 'Class of '47', it was such a daring exploitation of technology that it deserves to be classed by itself as a milestone in aircraft design

It is probably true to say that the Boeing XB-47 was the first American aircraft to benefit from the wartime studies done in German research institutes on wing sweepback. Although it was beaten into the air by the first flight of the North American XP-

86 Sabre, which was both the first US sweptwing fighter and sweptwing aircraft, the difference in time required for the development of a fighter and a bomber almost guaranteed that the smaller Sabre would fly first.

The Sabre lifted off in October 1947; the Boeing XB-47 had been rolled out of the factory the previous month and readied for its first takeoff.

*17 December 1947:*
## Boeing XB-47 Stratojet
The choice of the day was appropriate, though there are those at Boeing who will argue to this day that it was entirely accidental that the XB-47 first flew on the 44th anniversary of the Wright brothers' first flight at Kitty Hawk.

But few would argue the importance of the event, not only to Boeing, but to the

USAF and the industry that was then trying to recover from its postwar doldrums. The sleek, sweptback shape – a revolutionary look in aircraft design – was as different from its predecessors as any aircraft had been since that December day in 1903.

It had started as a perfectly conventional design, following the practices that had built the Boeing company The mid-1944 competition had been preceded by a request from the Army Air Corps during 1943, asking industry to look at designs of jet-propelled aircraft for reconnaissance or bombing. Boeing's first step was the obvious one; they tried to modify the B-29 layout by scaling it down and replacing its piston engines with jets. But this was no good, as they discovered almost immediately.

Subsequent studies resulted in a design with the engines buried in the body. This was as good an arrangement as any, and the Army awarded a first-phase contract to Boeing to work on refining the design. The same kind of contract went to Convair, Martin and North American. Their results are described in the previous chapter.

In the midst of the development programme on the newer Boeing straight-winged configurations, the war in Europe

**Boeing B-47E Stratojet**
*Crew:* 3 *Powerplant:* 6 General Electric J47-GE-25, 5970 lb thrust each *Span:* 116 ft *Length:* 107·08 ft *Weight:* 175,000 lb *Bombload:* 18,875 lb *Armament:* 2×20-mm cannon *Speed:* 650 mph

began to wind down. The US military then organised systematic searches of German research, development, and production facilities, and took on civilian scientists and engineers as well as military personnel to make these searches and report on what was found.

One of the teams scanning the ruins of Germany's aeronautical research effort included George Schairer, a Boeing engineer whose grasp of new ideas was to be the key to the XB-47 development and to whole generations of aircraft that followed. Schairer saw some of the test data on wing sweepback, studied some of the reports, looked at the wind-tunnel models and put it all together in his mind. The result was a cablegram to Boeing, in effect telling them to drop everything and switch the XB-47 design to a swept-back wing layout.

Sweepback was not a new idea then – it had been openly discussed at technical conferences as far back as 1935. But German scientists had recognised earlier than others that there were advantages in sweepback and that its use could produce a wing design which delayed the onslaught of compressibility. By sweeping the wings back, oncoming air was eased over the wing surfaces much more gently than across a straight wing, and compressibility bubbles did not form until a much higher speed had been reached.

In effect, sweepback fooled the air into thinking it was passing over a much thinner wing. Thick wings are the enemy of air; they cause drag and as the speed increases they force a flow breakdown that causes the phenomenon described as compressibility. Shock waves form, the drag suddenly rises, and there is no way to go any faster.

Thin airfoils have much less drag and disturb the air so much less that they can be used to reach much higher speeds. But thin wings then were difficult to build, and difficult to make strong enough to respond to high flight loads and gusts without failure of the structure.

Boeing knew all this intuitively, and yet they had hesitated to make the big step until prodded by Schairer and the results of the German research. Then their engineers laid out a new design around a swept, graceful wing angled back at 35°, measured at the quarter-chord line, the correct way to define the amount of sweep. Wind-tunnel models were built, using the new wing but retaining the buried engines of the earlier design.

But the military had objections to the buried engines, and there was only one other logical place to put them. So Boeing laid out wing nacelles, paired engines inboard and single engines outboard in the six-jet installation.

Understand the magnitude of the problem. The XB-47 swept wing was as thin as Boeing's knowledgeable manufacturing personnel could guarantee it, but as thick as the stress engineers wanted it for reasons of strength. Nicely balanced between too thick and too thin, the wing was suddenly called upon to handle the installation of six hot, thrusting jet engines with their own and some new aerodynamic loads. It might have defeated the whole concept.

But some unsung hero, or heroes, at Boeing solved the problem by thinking of the engines as damping weights also. If they were properly positioned, they could reduce, rather than increase, the tendency of the thin wing to flutter and fail. By locating the engines properly, their dead weight could be used to relieve the bending movements in the wing due to upward air loads. Further, they could increase the torsional stiffness of the wing and minimise the chances of torsional flutter. And that is what happened. Instead of being a liability, the engines served as stress relievers on the wings.

The different wing demanded a different control system. (Spoilers were the primary lateral control surfaces on the B-47 series.) And because the wing was thin, there was no room for the main landing gear. Tandem bicycle gear was chosen, and outrigger wheels were installed for retracting into the inboard engine nacelles.

The early generation of jet engines had really bad characteristics at low airspeeds. They did not accelerate well, and it took a long time to get full power after the throttle was opened all the way. So the takeoff was comparatively long and hazardous. Consequently, additional take-off thrust was built into the XB-47 in the form of 18,000 lb of thrust from 18 auxiliary solid-propellant rocket units.

But that only solved half the slow-speed problem. The other half was that the early jets had no thrust reversers, as they do today. An aborted takeoff almost certainly meant an accident as the aircraft rapidly ran out of runway. Drag on demand was the need, and it was met by another wartime German innovation, the brake parachute. Streamed from the tail of the B-47, the high-speed ribbon chute was the equivalent of lots of thrust reversal and braking power, and it dragged the B-47s to a halt. The addition reduced the accident potential considerably.

The fuel tanks were in the fuselage – if there was no room for landing gear in the wings, there was less for fuel. Late models of the B-47 did have some wing fuel in self-sealing tanks, but the bulk of the fuel was carried in the fuselage, augmented by drop tanks, also carefully located on the wings to minimise their effects on drag, and on spanwise bending and torsional stiffness.

These were the basic design concepts that went into the XB-47: thin, swept wings; nacelles used productively to relieve normal and gusting wing loads; extra installed thrust from rockets, and installed drag from a brake parachute; tandem landing gear and fuel in the fuselage.

The weapons, meanwhile, had not been ignored. There was room in the fuselage for the early hulking nuclear weapons, their steel casings swollen by the moderating materials used to control the rate of the chain reaction.

The two experimental XB-47s went through their test programme; the Air Force liked what it saw and ordered the airplane into production. The first batch was of a service test version, ten B-47A models to be built at Boeing's Wichita, Kansas, plant, and to include a few changes, the most important being more powerful engines.

The first flight of the first B-47A took place almost unnoticed on 25 June 1950. Screaming headlines announced other news: that day the North Koreans had slashed across the border into South Korea and that unhappy conflict was launched.

A combination of events during the

Korean war moved B-47 production to a level of highest national priority. SAC's war-weary B-29 bombers were doing yeoman work in the Korean theatre, but they were far outclassed by the Russian-built MiGs. Only continuing development of tactics and the use of escorts and electronic warfare allowed the B-29s to survive in the hostile skies over Korea.

Jet speed and altitude were essential, and the B-47 programme swung into high gear. Following the pattern of the Boeing B-17 programme, Douglas and Lockheed helped build the newer B-47. Douglas built their versions at Tulsa, Oklahoma and Lockheed operated the huge wartime B-29 bomber plant at Marietta, Georgia. Boeing continued to turn out the B-47 at its Wichita plant, and sent components and personnel out to Douglas and Lockheed in the initial transition stages of their production of the new bomber.

With the A models delivered, the B-47B entered production, and nearly 400 were built before the next major model change was made. Again there was an increase in engine thrust, and in the aircraft weight and performance. The first few models produced by Douglas and Lockheed were B-47B types; the real effort was to come with the next production bomber model.

The B-47E was the high point of the entire series, and more than 1570 were built. Still more powerful engines were featured, and in-flight refuelling ability was included, so that the range of the B-47 was now limited only by the endurance of its three-man crew. That was dramatically demonstrated in November 1954, about three years after the bombers had made their debut with USAF's Strategic Air Command.

On 17 November that year, Col David A. Burchinal, then commanding the 43rd Bombardment Wing, took off from the long runway at Sidi Slimane, French Morocco, and headed back to Fairford RAF Station in England, where the 43rd was in the midst of a 90-day rotational tour outside the United States. By the time he was near the RAF base, the local weather had become so bad that Burchinal decided to head back to Sidi Slimane. That also had soaked in while he was airborne.

So Burchinal called for a tanker rendezvous, and decided to wait out the weather. He refuelled nine times in the air before the weather had cleared enough at Fairford to permit final approach and landing. The wheels of the B-47 touched down 47 hours and 35 minutes after they had lifted off the runway in Morocco. During the two-day flight, the B-47 had logged a distance of

**Boeing B-47E Stratojet**
John Batchelor and the Editor would like to thank the Boeing Company for their help in the preparation of this cutaway illustration

21,163 miles, nearly once around the world at the equator.

Few B-47 flights demanded as much as that one. Most of the time the training missions were accomplished with one re-fuelling on the outbound leg and one on the inbound.

Just a few months before the first B-47s had entered service with SAC, another Boeing product, the KC-97 tanker, had joined the Command. This combination made SAC a truly global and jet-propelled force. Earlier bombers had the range, but not the speed; the Boeing B-47 had the speed, but not the range. But combined with the KC-97s, stationed around the world at key locations for aerial refuelling, the range of the SAC bomber force was no longer limited.

The Korean truce had been established, but other crises were popping up. In June 1953, East Berlin erupted in a general strike by the workers. It was put down with Russian tanks and troops.

During those eventful weeks, SAC moved its first B-47 wing out of the United States to England on a standard 90-day overseas rotational mission. At the end of the tour, the wing was replaced by a second wing of B-47s, and a third followed that one, establishing a practice that continued until 1958. For five years there was always a combat-ready B-47 wing based in England.

The B-47E was the last 'pure' bomber version of the series to serve with SAC. Later models were primarily reconnaissance aircraft, built for intelligence gathering by photographic and electronic means. They retained a bomber capability, however, and

could have been used in that role. But the story of the reconnaissance B-47s, although equally fascinating, is not part of a history of jet bombers.

The original mission of the B-47 was to carry thermonuclear weapons to distant targets and drop them from high altitudes. When tactical considerations and enemy strengths in air defence made a second look at this approach necessary, a new tactic was evolved. The idea was a novel one. The bomber would come streaking in at probably less than 500 ft altitude, heading straight for the target. It would pull up in the start of a loop, and release its nuclear weapon on the upward zoom, in effect tossing it high into the air ahead of the bomber. The bomber would then continue through half of the loop and roll out at the top, breaking for low level again and a speedy departure from the scene.

The nuclear bomb, meanwhile, would have been flung thousands of feet above the flight path of the bomber, would slow to a halt at the top of its trajectory, and start down again, accelerating under gravity. By the time it was at the target, the bomber would be far away, safe from the searing blast of radiation that accompanied the bomb's detonation.

High-altitude and low-altitude flying demand different kinds of design, and the high-altitude B-47 had to be extensively modified to strengthen key structure to take the aerobatic manoeuvre.

The final total of B-47s of all types built exceeded 2000. They served for many years as the United States' credible strategic deterrent. By moving rapidly from the US to foreign bases on redeployment, they could emphasise the presence of SAC and the United States. It was battleship diplomacy brought into the jet age.

The first B-47s were delivered to SAC in October 1951. Only ten were delivered to the Command that year, but by 1958 SAC had equipped 29 full bombardment wings with a total of 1367 B-47 aircraft. They were replaced gradually by the newer and bigger Boeing B-52. But they soldiered on in the reconnaissance role, still able to double as a bomber, until the very last one was flown away to storage on 29 December 1967.

# LIGHT TWINS AND LOSERS

**Ilyushin Il-28 (Beagle)**
*Crew:* 3 *Powerplant:* 2 Klimov VK-1, 5954 lb
thrust each *Span:* 70·36 ft *Length:* 57·89 ft
*Weight:* 40,572 lb *Bombload:* 2205 lb *Armament:*
4×23-mm cannon *Speed:* 559 mph

Where the United States was first, Britain, France and Russia were soon to follow: the late 1940s saw the first jet bomber designs in each of those countries – all of them twin-engined, two of them successful and the remaining three doomed to be classed as unfortunate failures

In the early years after the Second World War, Britain, France and Russia also began development of jet bombers for their specific strategic and tactical needs.

In the United States the pace was faster, following a speedier start. But Britain, France and Russia had been hard hit by the war, in complete contrast to the United States, and there were tasks of higher priority than the development of new jet aircraft; people had to be housed and fed, and countrysides and cities had to be rebuilt or repaired. But even in the tightest times there has always been money, some-

where, for armament, and these years were no exceptions.

Five light bombers are described here. One of them, the English Electric Canberra, is deservedly a classic. A second, the Ilyushin 28, is an example of how mediocrity can be utilised, and how a lot of second-class bombers are a greater threat than a few first-class ones. The other three aircraft were, unfortunately perhaps, losers. This chapter tells their stories.

*8 August 1948:*
## Ilyushin Il-28
This long-lived light tactical bomber was Russia's first. It served in both the Soviet air force and navy, and with the military arms of satellite and Russian-influenced countries such as Poland, Czechoslovakia, the German Democratic Republic and China. More recently, it has turned up in the air arms of some African countries.

It has operated in every kind of environment from tropical jungle to Arctic wastes, and has been flown and maintained by pilots and ground crews of every level from intellectual to functional illiterate. It is a basic, simple design built around the conventional structural and aerodynamic practices of the late 1940s.

Design work on the aircraft that was to lead to the 28th Ilyushin design probably

began late in the Second World War, or soon after, when the Russians acquired a large number of German turbojets, technicians, reports and factories as a result of their systematic exploitation of their zones of occupation.

The Russians did not make much progress on the German foundation. But their salvation was at hand in the form of British permission to import a batch of Rolls-Royce Derwent and Nene turbojets. Although these were not the very latest British gas turbine designs, they were current enough

to show a major advance over any of the German-produced engines in the Russians' possession. The deal was welcomed, and it is cursed to this day by some knowledgeable observers for having given the Russians a healthy boost upward in their aeronautical development.

Russian designers welcomed the new powerplants and immediately set to sketching layouts around them. Sergei Ilyushin's design bureau competed with the offices of Andrei Tupolev in trying to meet the requirements of the Soviet air arm for a light tactical bomber powered by a pair of jets. Both bureaus developed their designs and produced prototypes which were evaluated by experienced flight crews who flew both types. The nod went to the Ilyushin aircraft, and it was placed in production.

But first the factory had to meet Stalin's order for 25 of the bombers to parade in the May Day fly-past in 1950 – just about a year away from the date Ilyushin was told his bomber had won the competition. Hammering out 25 new and untried jet bombers, a first in the country, getting them airworthy and training enough flight and ground crews to fly and maintain them for a huge public spectacle was a major accomplishment. Ilyushin's people delivered on time. The bombers might have been nearly empty shells, lacking operational equipment, and they may have been totally unfit for operational duties. But they were in the parade, and they were noticed by Western observers, which is what Stalin might have had in mind.

But it was to be 12 more years before they were really noticed in the West, except by interested military intelligence and technical personnel. The later event occurred during the 1962 Cuban confrontation. The Russians had begun to move missiles into Cuba and to site them in positions where they were obviously a threat to the United States. As if this wasn't bad enough, they

also began shipping in crated Ilyushin bombers, which were uncrated and started through an assembly process at a Cuban airfield. There, photographed by high-flying reconnaissance aircraft from the United States, rows of Il-28 fuselages, crated wings, tails and powerplants told their story. It was a shocking one: a tactical nuclear bombing force was being based 90 miles off the shores of the United States.

The dénouement of that crisis is well-known. The Russians backed down, the Ilyushins were put back in their crates, and left Cuba as deck cargo, photographed again in low-level passes by USN and USAF aircraft checking the contents of freighters leaving Cuban waters for Russian ports.

The Il-28 was a Russian technical success, enabling the rapid build-up and deployment of a strong tactical bombardment force that could have delivered nuclear weapons against any target in the NATO alliance. They were a formidable threat, and they forced some uncomfortable defensive postures and programmes on the West.
*1948:*

### Tupolev Tu-14

The Russians being the way they are, there is little hard data on this contemporary of the Ilyushin 28. Aleksandr Yakovlev's authorised – if not necessarily authoritative – book *Fifty Years of Soviet Aircraft Construction*, makes only a single reference to the '. . . three-engined bomber Tu-14 . . .' Yet pictures of the Tupolev design, which did see limited service with the Russian naval air arm, show it clearly to be similar to the Il-28, with an unswept wing, swept

**Tupolev Tu-14 (Bosun)**
*Crew:* 3 *Powerplant:* 2 Klimov VK-1, 5954 lb thrust each *Span:* 70 ft approx *Length:* 65 ft approx *Weight:* 46,297 lb *Bombload:* 6614 lb max *Armament:* 4×23-mm cannon *Speed:* 525 mph

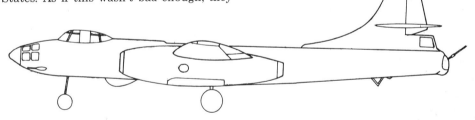

*A flight of Ilyushin Il-28 light tactical bombers, still in service with several air forces*

Novosti

349

tail, and definitely only two powerplants, housed in individual nacelles which are very similar to those of the Il-28.

It is known that the Tu-14, if that is even its correct designation, lost the competition to the Il-28. But a limited quantity was produced anyway, and they went to the Russian navy for torpedo dropping and similar activity.

The torpedo bomber was a casualty of the Second World War. When the war ended, so did the useful life of the type. They were just too vulnerable. So relegating the Tu-14 to this assignment, postwar and in the jet age, was some kind of an admission that it wasn't much of an airplane.

It must have been heavier than the Il-28, and it certainly had inferior performance, otherwise it might have been produced instead of the Il-28. With the same powerplants, and basically the same aerodynamic layout, one expects comparable performance.

Apparently the first deliveries were made to squadrons in 1949. There are a few pictures of the type that have been published, some heavily retouched, others pristine. What the true story is may never be known; for the purposes of this history, let us record that there was once a light bomber design from the Tupolev design bureau. It served with Russian naval units and it faded into history unhonoured and almost unsung.

*13 May 1949:*

**English Electric Canberra**

This elegant light jet bomber is the closest approximation to an immortal in its field. Conceived before the end of the Second World War, the Canberra was not to be blooded in combat until the Suez crisis of 1956. Today, it still serves actively in military air units, including those of Great Britain and the United States.

In the years that have passed since its first flight, the Canberra has been used in a number of roles other than the one for which it was originally designed. In them, as in its light bomber task, it has shown outstanding performance.

It started as a replacement for the piston-engined Avro Lincoln heavy bomber, itself a development of the famed Lancaster which was in turn an extrapolation of the ill-fated Manchester bomber. The original requirement was for a high-altitude, high-speed bomber, and it was apparent to the design staff at English Electric, working under the brilliant W E W Petter, that only the then-new jet engine could produce the performance.

But Petter's ideas moved at a faster pace than did the development of the jet engine,

and a continuing iteration process was necessary during the design. The Air Ministry had written Specification B 3/45 to spell out the desired characteristics. It seemed as if a single huge turbojet would do the job. But Rolls-Royce design teams were working at top speed to develop a really new and different jet engine, and it was around the first development of the line that Petter finally froze the basic Canberra.

The Rolls-Royce engines, then designated AJ 65 (referring, no doubt, to an axial-flow jet with a design thrust of 6500 lb), were slim and light, and two of them could be used in partially submerged nacelles out along the wingspan. Petter had selected a large wing area for altitude performance and manoeuvrability, coupled with a low aspect ratio to keep the rolling performance high. The Canberra is one of the most manoeuvrable of aircraft as a result of those basic design choices.

The design was submitted to the Ministry in mid-1945; the Ministry looked, discussed and finally got around to awarding a contract for four prototypes in January 1946. There was still some scepticism in high quarters about the ability of Rolls-Royce to deliver their advanced engines on time, so one of the prototypes was earmarked for the installation of Rolls-Royce Nenes as an alternate powerplant.

The original intention had been to use radar bombing only in the Canberra. But the development of the radar lagged behind that of the aircraft. Consequently, when the Air Ministry awarded a contract for the first bomber version in March 1949, it was for the Canberra B 2, a visual-systems bomber to Specification B 5/47.

The Korean war moved the Canberra high on the list of British priority programmes, along with its Rolls-Royce engine, now named the Avon. Both powerplant and airframe were sub-contracted extensively throughout the British industry.

Early in 1951, the Canberra was chosen to be the standard tactical bomber of the USAF, and negotiations were begun for its manufacture in the United States.

The Royal Air Force received its first jet bombers on 25 May 1951, with deliveries to 101 Squadron to replace its Avro Lincolns. At the peak of its deployment with the RAF, Canberras equipped at least 34 squadrons, including nine that were specifically established to use the type.

Canberras were called to the Middle East in 1956. The seizure of the Suez Canal rang the alarm, and Canberras were deployed from their British bases to stations on Cyprus and Malta. On 31 October the Canberras of No 10 Squadron dropped their first bombs in combat on Egyptian targets in the brief but intense action that marked the Suez war.

Canberras saw combat with the Australian forces in Southeast Asia, flying

**Martin XB-51**
*Crew:* 2 *Powerplant:* 3 General Electric J47-GE-13, 5200 lb thrust each *Span:* 53·08 ft *Length:* 85·08 ft *Weight:* 55,923 lb *Bombload:* 10,400 lb *Projected armament:* 8×20-mm cannon *Speed:* 645 mph

sorties against North Vietnamese targets in that air war. They also fought on both sides during the combat between India and Pakistan in 1965, because both countries operated the type.

*28 October 1949:*

**Martin XB-51**

The Martin XB-51 was most unconventional. It was loaded with advanced technical features, and that might have been the major reason it never got beyond the prototype stage. Only two were built and both were lost in crashes. Its planned place in production was taken by the Canberra.

The design had begun under an attack designation as one of a group of heavy attack aircraft planned toward the end of the war years. Its intended mission was ground support, the destruction of surface targets by bombs, gunfire, or both. Consequently, the XB-51 was designed with a battery of eight 20-mm cannon in its nose, and a bomb load of 10,400 lb.

Its swept wings had variable incidence, so that the wing could be set for the optimum angle of attack for takeoff, flight and landing. Because of the high wing loading, the layout included large-span flaps. Top-wing spoilers were used for lateral control instead of ailerons, but small ailerons were fitted to give some force feedback into the control system.

The empennage featured one of the first of the high Tee-tail layouts; Martin engineers located the horizontal stabiliser there to keep it out of the wake of the swept wing.

Two XB-51 prototypes were built, flown and in time crashed. That was the end of the programme. Viewed in retrospect, it may have been a case of engineers amusing themselves with an elegant design of little practical value; but that seems to be given the lie by the knowledge that spoiler controls, variable-incidence wings and Tee-tails have since been used on successful aircraft and have proved their value.

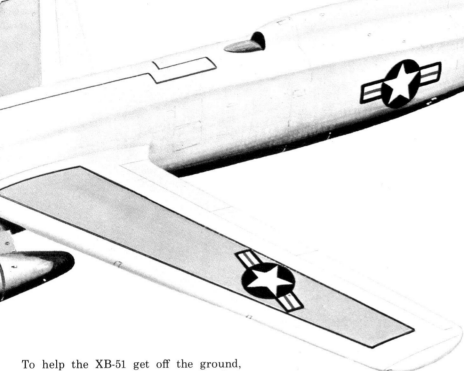

To help the XB-51 get off the ground, four solid-propellant rockets were used to augment the thrust of the three jet engines. And a brake parachute was fitted for reducing speed on landing.

The interesting touch was the bomb bay design. High speeds and flows around open bomb bays had been giving trouble in other aircraft. The bombs literally would not drop out of the bay, but would float there, to the great distraction of the crew. Two Martin engineers invented and patented the rotary bomb-bay door, a structure mounted on a fore-and-aft axis. Before the drop, it was rotated so that its bombs, which were mounted on the inner surface of the structure, were exposed to the air. No hinged doors and no open cavity were there to produce ejection or drop problems. It was a unique approach to a difficult problem, and one which was later to see service on the redesigned Canberras built in the United States as the Martin B-57.

*15 March 1951:*
### SNCASO SO 4000
The postwar French aircraft industry reminded some observers of the excited cavalryman who leaped on his horse and galloped off in all directions. The pent-up dreams of French designers showed in a stream of aircraft, most of which never got beyond the prototype stage.

The SO 4000 was one of a pair of jet bombers that had been chosen, out of the welter of designs available, to be developed for the French Armée de l'Air. Its only competitor was dropped before it reached the prototype stage.

At this point the French industry was making a comeback from the devastation of the war; experimental facilities were lacking, and neither the British nor the Americans were about to help France

become a strong competitor for future aircraft markets. So the French had to go it alone, and they went sometimes cautiously and sometimes recklessly.

Caution was the watchword on the SO 4000 programme. It was intended to start with gliding tests of a half-scale unpowered model, to be followed by powered flights with a second half-scale model. This was the only way the French then had of getting aerodynamic data without a huge wind tunnel, which they did not have.

The glider was beaten into the air by the powered model, which first flew in April 1949. In September the glider was launched on its test programme.

The following March the SO 4000 was rolled out and began its ground runs and taxi tests. It was a slick looking aircraft, with its two turbojets mounted side-by-side in its wide-oval fuselage. There didn't seem to be straight lines anywhere on the fuselage, from its forward tandem cockpit for the crew to the flattened elliptical shape that faired around the two engine tailpipes.

The wing was huge, by comparison, and with moderate sweep. Its strange-looking landing gear consisted of four independent main wheels and struts retracting into the wing root, and a high nosewheel gear. During taxi tests, the gear failed; the bomber settled ignominiously to the runway in a shower of sparks and with great grinding sounds. It took the better part of a year to repair the plane.

Daniel Rastel, who had flown the very first French jet aircraft, was the pilot for the first flight, and it turned out to be the only flight of the plane. The flight was successful, and Rastel landed safely. But irreparable harm had been done to the French aircraft development programme by budget cuts as the government found it was living beyond its means. Besides, better and newer ideas were coming along for bomber design.

So the SO 4000 programme was stopped cold. The prototype was wheeled off the ramp and into a corner of the hangar, eventually to be scrapped.

**English Electric Canberra B Mk 2**
*Crew:* 3 *Powerplant:* 2 Rolls-Royce Avon 101, 6500 lb thrust each *Span:* 63·96 ft *Length:* 65·5 ft *Weight:* 46,000 lb *Bombload:* 6000 lb *Speed:* 570 mph

# TWO NEW
# TWO FOR INSURANCE

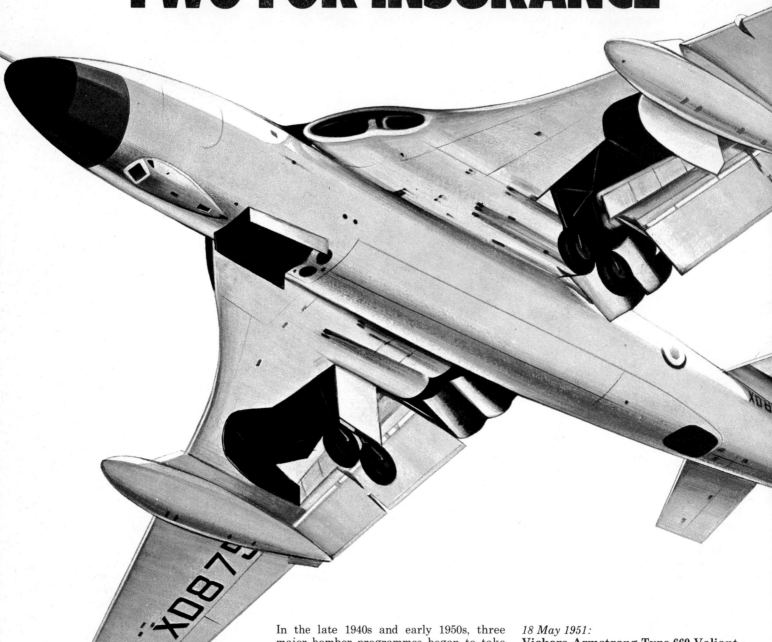

As the builders of jet bombers gained experience in the years around 1950, significant advances were made in the art of their design. But governments were still wary. Back-up programmes were called for: the Short Sperrin flew in the same summer as the Vickers Valiant, and the Convair YB-60 took to the air within days of the first XB-52

In the late 1940s and early 1950s, three major bomber programmes began to take shape. In Britain, the first specifications leading to the V-bomber programme were drawn up and issued. In the United States, the heavy follow-up to the Boeing B-47 was in design and development. In Russia, the first-generation medium bomber was being designed around native powerplants that were to startle observers in the West.

Two of the bombers to be described in this chapter were new designs, built to specific needs that advanced the art of large jet aircraft: the Vickers-Armstrong Valiant, and the Tupolev 16. And two of them were back-up programmes, insurance against delays or failures in other advanced designs: the Short Sperrin and the Convair YB-60.

The same period saw the high point of heavy jet bomber design, the Boeing XB-52. Because that bomber, like its predecessor, the Boeing XB-47, was – and remains – in a class by itself, it is considered separately.

*18 May 1951:*
**Vickers-Armstrong Type 660 Valiant**
The Valiant was developed as an interim bomber to fill the time gap between the RAF's piston-engined bombers and a new generation of jet-propelled, high-altitude V-bombers. The interim stretched to a service life of about ten years, during which Valiants dropped Britain's first nuclear and thermonuclear devices and saw combat in the Suez crisis. They might still be in service if the right model had been produced, or if they had continued to be used in the manner for which they were designed.

The Valiant grew out of a 1946 specification for a high-altitude, high-speed bomber capable of carrying nuclear weapons at speeds approaching that of sound. The major British companies responded with a large collection of designs. Two were chosen for development. A third had been placed under contract to an earlier specification in case the fancier

**Vickers Valiant B Mk 1**
*Crew:* 5 *Powerplant:* 4 Rolls-Royce Avon, 10,000 lb thrust each *Span:* 114·33 ft *Length:* 108·25 ft *Weight:* 138,000 lb *Bombload:* 21,000 lb *Speed:* 567 mph

designs did not pan out. The Vickers design was rejected, but management would not accept the refusal, and continued to argue the merits of the Type 660 design.

Yes, it was fairly conventional. But it was powered by the same four Rolls-Royce Avons that were being considered for the advanced bombers. And it had essentially the same speed over the target, the same bomb load, the same available volume for advanced electronic systems for bombing and navigation. All it lacked was the desired unrefuelled range. The arguments were persuasive, finally. A new specification was written around the Type 660 and Vickers went into production.

Two prototypes were built. The first was lost about eight months into its flight test programme, but by then the performance had impressed officialdom and the second prototype was nearing its flight date. The programme was continued, aiming at early service of the B Mk 1 version.

A second type was ordered later, which was intended as a low-level pathfinder bomber. It was the subject of a major re-design to add necessary structure to take the high air loads that result from going fast at low altitudes. It was painted black all over and had a distinctive shape because its wheels rotated to retract into trailing-edge pods. It was a fine performer at low levels, and the B Mk 2, as events proved, was the bomber that should have been ordered into production.

Valiants began to equip RAF units late in 1955, and the force began transition to its first jet bomber capable of serving as a strategic deterrent force. Hardly a year had passed before the Valiants were in action, striking with conventional weapons against Egyptian targets during the Suez campaign. They were the first British bombers to see combat in that brief action.

Meanwhile, a Valiant squadron (No 49) had been involved in the dropping and monitoring of Britain's first atomic bomb at Maralinga, Australia. That drop and the first strike by Valiants at Suez took place on the same day, 11 October 1956. On 15 May of the following year a Valiant from No 49 Squadron dropped the first British thermonuclear device against a target array on Malden Island. Two later drops of the British H-bomb were made, also by Valiants of 49 Squadron.

One measure of the Valiant's capabilities was its placing near the top of the list in the USAF Strategic Air Command's bombing, navigation and reconnaissance competitions. These events, staged on an annual basis by SAC to improve its own capabilities, were opened to British crews by invitation. The Valiant teams did well in their first competition and better in their second.

Early in 1964, the Royal Air Force switched its bombing attack from high- to low-altitude. The change signified that the effective ground-to-air missile had arrived. The change in tactics was evolved to avoid long-range anti-aircraft missiles, and to come in under enemy radar coverage so that there would be no warning time in which to ready the short-range AA defences.

The whole V-bomber force (by then including Vulcans and Victors as well as Valiants) began to operate near the ground at high speed. Any aircraft designed for

*Only prototype of the Valiant B Mk 2, designed for low-level pathfinding missions, which first flew on 4 September 1953. Structural strengthening needed for the low-level role involved using underslung pods for the backward-retracting bogie undercarriage*

Vickers Ltd

**Short SA 4 Sperrin**
*Crew:* 5 *Powerplant:* 4 Rolls-Royce RA 3 Avon, 6500 lb thrust each *Span:* 114·33 ft *Length:* 102·2 ft *Weight:* 115,000 lb *Speed:* 564 mph

high-altitude flight will suffer if it is flown low and fast, and the Valiants were no exception. They had not been designed for this kind of attack; only the single B Mk 2 prototype had been developed for low-level bombing runs – and it had not been ordered.

Predictably, the wing spars began to crack. There was a choice: fix the damage through extensive rebuilding, or scrap the force. The latter decision was made. The entire Valiant force was first grounded and then, in official parlance, 'reduced to produce', which is a nice way of saying that they were sold for junk. They deserved better than that.

*10 August 1951:*
### Short SA 4 Sperrin
This conventional straight-winged jet bomber with an unconventional engine installation was Britain's insurance policy against the failure of its advanced V-bomber programme. Like many insurance policies, it was never needed and the premiums paid were wasted, though fortunately small.

In these days of multi-billion dollar (or pound or franc or rouble) costs for aircraft development, it is interesting to look back at the Sperrin and see that the whole programme cost 3·5 million pounds sterling. That included the design, development, tooling, construction and flight testing of two prototypes. Today, that kind of money will hardly buy a major piece of the Rockwell International B-1.

The Sperrin was designed around a specification dated 1946, but issued in 1948. By that time, the Boeing XB-47 had flown and was in volume production. But the British, showing their first signs of jet lag, opted for low-risk projects at a time when just one daring leap forward might have made all the difference to the state of their aircraft industry today.

Hindsight, however, is a commodity very useful to commentators on history, but unavailable to the planner whose neck is out with every decision.

The Sperrin could have been a piston-engined bomber just as well, with its shoulder-height straight wing, its huge fuselage, its conventional tail and high ground clearance. But it used four of the newly available Rolls-Royce Avon jets, stacked in an over-and-under installation in two nacelles which were split by the wing.

By the time the Sperrin flew, almost three months after the Valiant, it was already a loser: a loser in terms of production contracts, but definitely not in terms of how smoothly the flight test programme went. The Sperrin had no real troubles; it handled nicely at altitude, showed a fair turn of speed for its size, power and layout and had a substantial range with the equivalent of its intended nuclear weapon.

The second prototype was flown a year later, and both then went into general

flight research, testing new bombing systems and dropping models of proposed bomber missiles. The second prototype was cannibalised in 1956 to repair the first, and a year later the first prototype was grounded, never to fly again, when the experimental engine it was flying on tests was dropped from the list of approved projects.

*18 April 1952:*
### Convair YB-60
Like any other aircraft company, Convair tried to stretch the life of its product line as far as possible. They proposed the B-36G as a sweptwing, jet-propelled version of the gigantic B-36 piston-engined, jet-boosted bomber. It would use almost three-quarters of the parts of a standard production B-36. That would make it a relatively economical project, and the jet engines would give it a major increment of performance. The Air Force approved the idea, and ordered a pair of prototypes in March 1951, giving them the new designation of YB-60.

Convair redesigned the big wing of the B-36, maintaining most of its lines but sweeping it back to the standard 35° mark. They designed a new vertical and horizontal tail, slung eight Pratt & Whitney J57 engines in four nacelles under the wings, and they had a new bomber. Unfortunately,

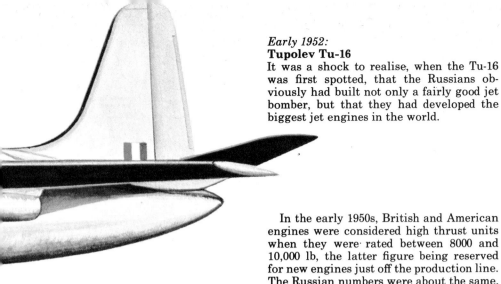

they had the fuselage of the old one, and that was a big fuselage. And they may not have spent enough time thinking through the wing design, because they ended up with an aircraft that was almost 100 mph slower than the Boeing XB-52. Both aircraft had about the same empty weight, both had the same engines. But the best performance for the YB-60 was just a shade over 500 mph, while the XB-52 could go close to 600 mph at the same altitude. That was too big a difference to be ignored in favour of economy in production.

Convair engineers had put together, creditably, a multi-jet bomber that would carry up to the 72,000-lb bomb load of its piston-engined parent. It would have been flown by a crew of ten. Its defensive armament was a quintet of remote-controlled turrets, each with a pair of 20-mm cannon and 360 rounds per gun.

Axiom: it's very difficult to adapt an airplane to do something that it was not originally intended to do. High-altitude bombers do not work well in low-level dashes; interceptors do not make good ground-support aircraft; and piston-engined planes create whole new sets of problems when they are converted to jet power.

The Convair YB-60 proved that axiom all over again.

*Early 1952:*
## Tupolev Tu-16

It was a shock to realise, when the Tu-16 was first spotted, that the Russians obviously had built not only a fairly good jet bomber, but that they had developed the biggest jet engines in the world.

In the early 1950s, British and American engines were considered high thrust units when they were rated between 8000 and 10,000 lb, the latter figure being reserved for new engines just off the production line. The Russian numbers were about the same, but their thrusts were measured in kilograms, and the big new Mikulin powerplants that thrust the Tu-16 through the air were rated somewhere between 19,000 and 21,000 lb thrust.

That was the real significance of the Tu-16. It had been a foregone conclusion that the Russians would develop a jet bomber. But most observers expected that it would have four or more powerplants.

Today, the Tu-16 looks back on a service life of more than 20 years, during which time it has served with Russian air force and naval units. Some were exported to Iraq, Indonesia and Egypt, and the Egyptians used them, with air-to-surface missiles, in the 1973 war between Israel and the allied Arabian states.

Hard data on the Tu-16 is difficult to come by. Aleksandr Yakovlev, in *Fifty Years of Aircraft Construction*, has this to say:

'A N Tupolev's Tu-16 was powered by two AM-3 engines, each with a thrust of 8750 kg, mounted laterally where the wings were joined to the fuselage. With a weight of 72 tons, the Tu-16 could carry a bomb load of 3 tons, having a range of 5760 km. Its maximum speed was almost 1000 kmph. The six-man crew had powerful defensive armament, seven 23-mm cannon. Later the Tu-16 became a terrifying rocket carrier able to destroy ground targets without entering the enemy's air defence zone.'

Several versions of the Tu-16 have been noted by Western observers and, like other Russian aircraft, have been given NATO code names in a sort of spoken or written shorthand. The Tu-16 basic name is Badger. Badger-A was the first bomber version observed, with a standard bombardier's compartment in the nose. Badger-E, -F and -G are similar but contain special installations: E has bomb-bay cameras, F carries electronic intelligence gathering pods, and G has rocket-powered air-to-surface missiles. Badger-C is an anti-shipping version which carries another kind of air-to-surface missile beneath its wings, and Badger-D is similar, but is used for maritime reconnaissance. Some of these versions may be really only one basic model with hard points

**Convair B-36D**
Basis for the giant Convair YB-60 eight-engined jet bomber *Crew:* 15 *Powerplant:* 6 Pratt & Whitney R-4360, 3500 lb thrust each; 4 General Electric J47, 5600 lb thrust each *Weight:* 370,000 lb *Bombload:* 72,000 lb *Armament:* 16×20-mm cannon *Speed:* 406 mph

*Top: A Badger-D electronic surveillance version of the Tu-16 flies over HMS* Royal Oak. *Above left: Badger-F, a maritime reconnaissance type with under-wing electronic pods. Above right: Soviet airmen parade in front of their Badger-B anti-shipping missile carriers*

under the wings for alternate mission loadings of missiles and electronic pods.

These aircraft have been well documented photographically by British and American pilots who have flown formation with them as the Russians scouted NATO fleet manoeuvres or patrols. Badgers have been photographed from the decks of US carriers, escorted by close formations of F-4 or F-8 fighters. So it is likely that there is much detailed knowledge about the weapons and electronics installations on board these bombers, and considerable other knowledge that can be gleaned from the external appearance.

The Badger is becoming obsolete, but it continues to form part of Russia's medium-bomber strength. One current estimate suggests that 500 are so employed, out of about 2000 believed to have been built.

### Tupolev Tu-16 (Badger A)
Initial production version of the Soviet medium-range strategic bomber, shown here in Egyptian markings *Crew:* 6/7 *Powerplant:* 2 Mikulin AM-3M, 19, 290 lb thrust each *Span:* 109·88 ft *Length:* 120·93 ft *Weight:* 170,000 lb *Bombload:* 19,800 lb max *Armament:* 7×23-mm cannon *Speed:* 586 mph

*Badger-E maritime reconnaissance aircraft, with camera equipment in their weapon bays, demonstrate their wingtip-to-wingtip refuelling*

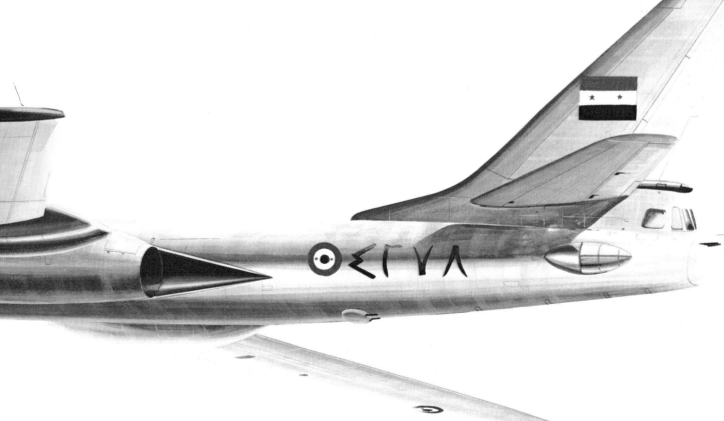

# HEAVYWEIGHT WARRIOR

All-time classic among heavy jet bombers, the Boeing B-52 Stratofortress has served the US Air Force's Strategic Air Command for more than two decades – and its days are not yet numbered

It is *the* heavy bomber. There has been no other aircraft like it in the world and there is not likely to be.

Its preliminary design was done in a Dayton, Ohio, hotel over a long weekend in October 1948. Since then, it has been continually refined so that more than 25 years after its first flight and more than 20 years after its introduction into opera-

tional service, the Boeing B-52 remains a modern aircraft.

It has set world records, officially and unofficially, for speed and distance in its class. It has carried rocket-powered research aircraft aloft to probe the near reaches of space. And recent modification programmes have reworked and rebuilt some of the B-52 fleet to extend their useful operational life into the 1980s.

This is part of the story of the USAF's heavyweight warrior.

*15 April 1952*
## Boeing YB-52 Stratofortress
For more than 20 years this huge bomber has been the mainstay of Strategic Air Command, USAF. Its swept wings and long fuselage have become the symbol of strategic deterrence. Designed to carry nuclear weapons, particularly the big thermonuclear weapons just entering develop-

ment in the early 1950s, the B-52 had to be modified to carry enough 'iron' bombs to make the game worth the candle in Southeast Asia.

Its record is a proud one. In their most effective aerial battle, against military targets in North Vietnam at the end of 1972, B-52s struck and destroyed with great precision. The number of civilian deaths was minimal according to official North Vietnamese casualty figures. In fact, so low were these figures that no attempt was made to propagandise against the United States by using them.

In the mid-1970s there were B-52s of the Strategic Air Command flying practice missions, honing the skills of the highly trained crews in continuing exercises and simulated runs on scattered targets. And modification programmes will guarantee the life of the B-52 well into the 1980s, giving the bomber a useful operational life of 30 years or more.

**Boeing B-52D Stratofortress**
*Crew:* 6 *Powerplant:* 8 Pratt & Whitney
J57-P-29W, 10,900 lb thrust each *Span:* 185 ft
*Length:* 157·58 ft *Weight:* 450,000 lb
*Bombload:* 70,000 lb *Armament:* 2×20-mm
cannon *Speed:* 612 mph

To put that on another time scale, it is as if the British and Americans had entered the Second World War still using de Havilland DH 4s as their primary heavy bomber, and continued to use them until the end of the war and beyond.

The Boeing design started as an attempt to meet an Air Materiel Command requirement of 1945 for a second-generation long-range bomber. The first-generation contract had just been awarded to Convair for what was to become the B-36. Convair and Boeing competed to meet the new requirement, and Boeing won with a conservative approach. They substituted turboprop engines for the piston engines of the B-50, itself only a more powerful B-29. It was a case of trying to stretch the stretched, and it was destined to fail. But before it did, Boeing got a contract to develop the model, by then designated the XB-52.

Beginning in April 1946, Boeing designers laid out a new airframe around four giant Wright turboprop engines, and began to detail the major components. By then it was becoming apparent that the straight wing and the propeller were approaching obsolescence faster than they could be improved. Swept wings and jet engines were the answer, and the turboprop XB-52 was doomed before it ever got off the drawing board. Besides, Wright was beginning to slip into its slough of mediocrity, and the Boeing team, ever alert to the possibility of being let down by somebody outside of Seattle, began to look for alternatives.

They found a new Pratt & Whitney jet engine which combined the thrust they would need with a promise of fuel economies on a scale then unheard of.

Boeing began a parallel study of the bomber requirement with company funds, designing a sweptwing aircraft with the new P & W jets. They continued work on the turboprop version as well. When the Air Force experts began to consider that turboprop bomber in more detail, they grew less satisfied. The performance increment wasn't enough of a jump over a re-engined B-36.

So the Air Force wrote a new require-

ment during mid-1948, handed it to the Boeing team that was visiting Air Materiel Command in October that year and sat back to see what would happen.

In one of the classic stories of the aircraft industry, the Boeing team headed for their Dayton hotel to study the requirement. The next day – a Friday – they telephoned to tell the AMC they would submit their official proposal on the following Monday.

The first three-view drawings were completed on Saturday, and a balsa model was made with materials purchased from a Dayton hobby shop that day. The proposal was typed, the model was completed, and a red-eyed Boeing team confronted the USAF planners on Monday morning with their answer to the requirement.

Now this is not to imply that the XB-52 was 'designed' in four days, any more than one is expected to believe the stories about fighters that were designed on the backs of convenient envelopes. But the team that represented Boeing to the Air Force had been so immersed in heavy bomber requirements and possibilities that among them they knew all the basic answers. They could, and did, do a fairly complete preliminary design of the XB-52, choosing its basic layout, dimensions and powerplants, and estimating its weights and performance. And they could do that in a matter of days, because of the close professional relations that existed among the team. It was a remarkable technical and engineering feat.

Boeing already had contracts for two prototype bombers to the original requirement and design, and it did not take much to shuffle paperwork around various desks to make those same contracts cover the new XB-52.

Less than four years after that concentrated weekend in Dayton, the YB-52 was rolled out, taxied and flown for the first time. It preceded the XB-52 into the air in a reverse of the usual procedure. The XB-52 was known to need some improvements that had been planned for, and built into, the YB version from the start. Even so, the XB-52 made its first flight on 2 October 1952, still under four years from that famous weekend.

There were similarities between the new bomber and the B-47 series. They shared swept wings, podded powerplants, and bicycle landing gear in the fuselage. Everything was, of course, bigger and more developed than it had been on the B-47. The two prototypes featured a bubble canopy for the pilot and co-pilot, similar to the cockpit layout of the B-47. This was one of the first things to be changed in subsequent models.

Boeing had received a contract for a block of 13 service test aircraft, designated B-52A. The first three were actually used by Boeing as developmental aircraft, and they pioneered the new flight deck layout, with its blunted multi-paned nose and the pilot and co-pilot seated side-by-side instead of in tandem. The first of the B-52A models flew on 5 August 1954.

The remaining ten of the original B-52A contract were redesignated B-52B and incorporated further improvements. These were the first aircraft to go to the USAF. The 12th B-52B, by then including provisions for reconnaissance and' officially designated RN-52B (Serial No 52-8711) was the first Stratofortress delivered to the Strategic Air Command. It stands today in the museum that borders on Offutt AFB, near Omaha, Nebraska, headquarters of SAC.

On 25 June 1975 General Russell E Dougherty, SAC commander, was joined in front of that bomber by retired Major General William E Eubank Jr. Eubank had made the flyaway delivery of the plane exactly 20 years earlier while commander of the wing at Castle AFB, California.

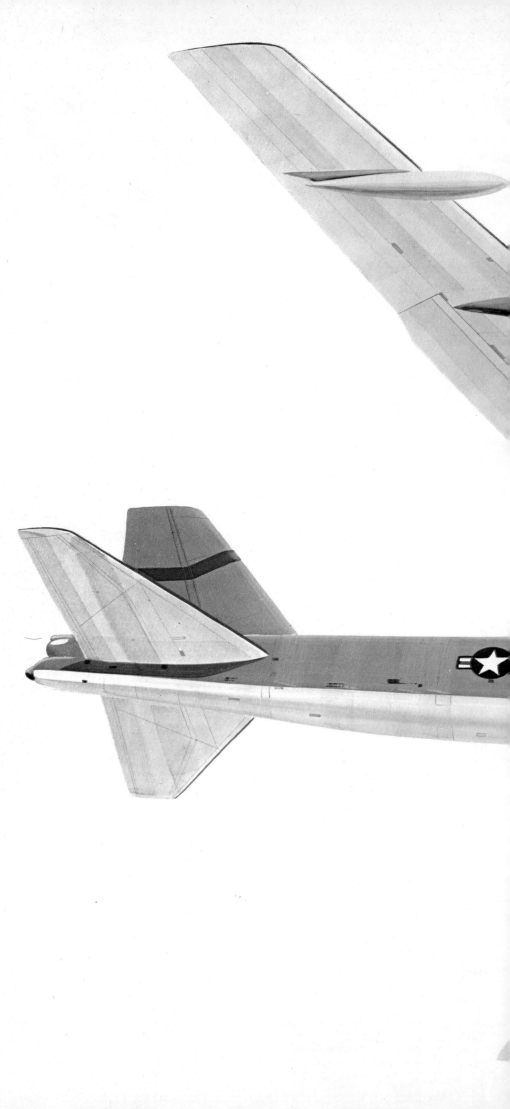

**Boeing B-52G Stratofortress**
*Crew:* 6 *Powerplant:* 8 Pratt & Whitney J57-P-43W, 11,200 lb (13,750 lb with water injection) thrust each *Span:* 185 ft *Length:* 157·67 ft *Weight:* 480,000 lb *Bombload:* 60,000 lb plus 2 Hound Dog stand-off ASMs *Armament:* 4 × ·5-in mg *Speed:* 665 mph

General Dougherty commented, 'The aircraft that General Bill Eubank delivered to the 93rd Bomb Wing 20 years ago is a far different aircraft from the one he arrived in today . . . The one Bill used to fly, like aircraft 711 [the USAF call sign for RB-52B 52-8711], was designed to operate in the stratosphere at altitudes around 40,000 ft, and deliver weapons from that high altitude.

'He came up here this morning in the later G-model, from Barksdale AFB, Louisiana. And a portion of that flight was flown at 400 ft, the stratum in which we plan to operate today if we must.'

There is no better simplified account of the change in the B-52 over the years than Dougherty's comments. Designed to fly and operate at 40,000 ft, the B-52 now is a ground-hugger flown in low-level penetrations of target areas, under the far-seeing eyes of radar. It still retains the ability to fly and fight at high altitudes, but to this it has added the capability to fly near the ground, guided by a wide variety of new electronic and infrared systems and by the skills and high competence of its crew.

As mentioned earlier in this history, flying at low level is a sure way to place high stresses on an airframe, and particularly one that was never designed to fly that kind of mission. The B-52 was planned for manoeuvres at high altitude, where the dynamic pressure and the air loads are comparatively low. There have been major modification programmes on the entire SAC fleet of B-52s to make them capable of coping with these new and higher stresses.

Currently, 80 of the B-52D models are being cycled through a 200-million-dollar effort to replace primary wing and fuselage structures, and to make other additions and changes in systems and structures to enable the D models to handle low-level missions.

One earlier modification programme arose out of the needs of the Southeast Asia theatre, where military requirements called for saturation bombing of fairly large areas. The B-52s, designed for a pair of thermonuclear weapons, were hastily adapted to the 'iron' bomb, but their capacity was limited. A major modification, called 'Big Belly', was carried out on the B-52D fleet. Before this rework, the standard internal iron-bomb load of a B-52 was 27 of the 750-lb bombs, for a nominal bomb load of 20,000 lb. 'Big Belly' increased that number to 66 of the 750-pounders, and an additional 24 slung on pylon racks under the wings. The total iron-bomb load then was 90, or a nominal weight of 67,500 lb, more than triple the original design bomb load.

'Big Belly' also provided for an alternative loading of 84 × 500-lb bombs on internal racks, plus 24 × 750-lb bombs externally, for a total of just under 61,000 lb.

SAC is now equipped with B-52D and B-52G models which are powered by versions of the original Pratt & Whitney J57 engines that sparked off the design, and the B-52H,

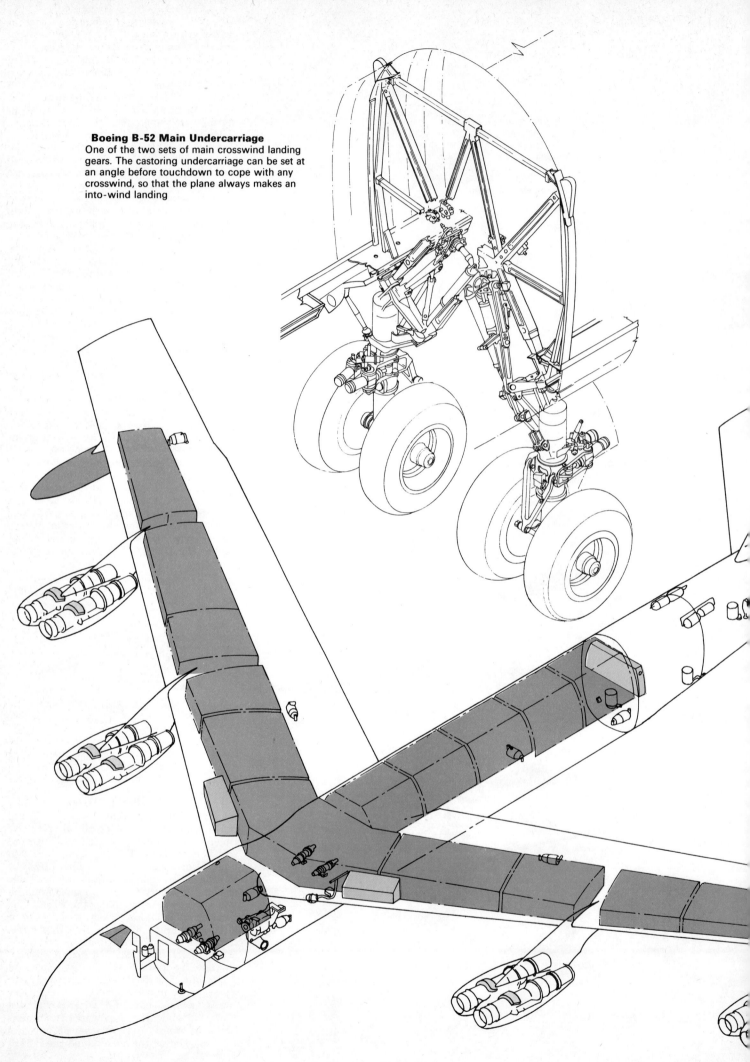

**Boeing B-52 Main Undercarriage**
One of the two sets of main crosswind landing
gears. The castoring undercarriage can be set at
an angle before touchdown to cope with any
crosswind, so that the plane always makes an
into-wind landing

*A B-52 Stratofortress refuels from a KC-135 Stratotanker*

**Boeing B-52 Fuel Distribution**
Designed as a long-range strategic bomber, the B-52's massive fuel load gives it an official unrefuelled range of over 9000 nautical miles — and one B-52H made a record unrefuelled flight of 12,532 nautical miles

- ▓ Fuel tanks
- ░ Water injection tanks
- ░ Engine oil tanks
- ▓ Air refuelling point

which is powered by a higher-thrust unit, the P & W TF33 turbofan engine. The new powerplant adds almost 50% to the installed thrust and improves the performance of the B-52 considerably.

The H model was designed to be the carrier aircraft for the Skybolt ballistic missile, a programme to be jointly developed for, and funded by, the United States and Great Britain. The US cancelled the programme unilaterally, amid much feeling of annoyance by the British and the RAF. They had spent much effort and many pounds as their share of the programme, and were at best vexed when Skybolt was dropped.

The B-52H now carries a pair of Hound Dog missiles which resemble small fighter aircraft, but are loaded with nuclear warheads. More recently, SRAMs (Short-Range Attack Missiles) are replacing Hound Dogs.

The strong point of the B-52H is high-altitude range. They are all equipped for in-flight refuelling, as is standard these days; but the official unrefuelled range figure is in excess of 9000 nautical miles. The validity of that figure can be understood in the frame of reference of a record-setting flight by an unrefuelled B-52H, which logged a point-to-point flight of 12,532 nautical miles.

To board a B-52H, you enter just behind the nose radome through a belly hatch, climb to the first level, turn right and climb to the flight deck. Racks of equipment are on your right as you move forward, a bunk for crew rest is at your left, then comes a check-pilot's seat and then the two ejection seats for pilot and co-pilot.

Between them, the eight throttles are mounted on a console, and they are ingeniously placed so that each engine can be handled individually, or as one of a group of up to all eight engines at once. The co-pilot and pilot have the usual primary display instruments; the co-pilot has more engine instrumentation because he does a lot of the tasks that would be done by a flight engineer on a transport. He also has the separate console for starting and launching the Hound Dog missiles.

Each flight panel mounts a 6-in terrain-avoidance scope in its centre which gives a presentation of the terrain at three different slant ranges ahead of the flight, and also shows the profile at right angles to the flight path. The B-52H is flown manually using indications from these scopes for terrain avoidance.

All B-52s have crosswind landing gear. The crab angle of the gear can be set before touchdown to cope with whatever crosswinds there are. If the nose is angled to the left the co-pilot makes the touchdown; to the right it is the pilot's job.

At the rear of the flight deck are side-by-side positions for the electronics warfare officer (starboard) and the tail gunner (port side). Earlier models of the B-52 had manned turrets; the H model has a remote-controlled one.

In a compartment one level below the flight deck sit the navigator and the radar-operator, who is actually the bombardier. It is a comment on modern technology and the complexity of the mechanics of the bomb that it is still not possible to drop weapons accurately without preceding the drop with a long, stabilised, constant-speed, constant-altitude run into the target area.

Behind the entrance hatch are the two bomb bays. Racks for the nuclear weapons are stowed flat against the roof of the bay itself. Iron bombs are slung on vertical racks at three or more stations in each of the bays. Other kinds of stores can be carried and dropped or launched from these cavernous bomb bays, such as decoy missiles like the Quail. SRAM missiles are mounted on a single rotary launcher in the bomb bay, holding eight missiles. Twelve more can be carried, six to a side, on underwing pylon racks.

The B-52Ds, with their 'Big Belly' modifications, are even more versatile. Their official mission includes ocean surveillance and other naval support, so it is likely they can carry a variety of mines and other anti-shipping arms.

Both the B-52G and H aircraft will get new detection and bombing systems for the low-level, all-weather penetration attack. An electro-optical system is now being delivered, using infrared and low-level-light television to present a picture of the area ahead of the bomber. A further advance in electronic countermeasures has been planned, and will be installed.

SAC's fleet now consists of about 300 of the late model G and H aircraft, plus the 80 D bombers. They will remain operational, according to plans current in the mid-1970s, until the 1980s.

363

# THE LAST OF THE SUBSONIC BOMBERS

**Hawker Siddeley Vulcan B Mk 2**
*Crew:* 5 *Powerplant:* 4 Rolls-Royce Bristol Olympus 301, 20,000 lb thrust each *Span:* 111 ft *Length:* 99·92 ft *Weight:* 190,000 lb (estimated max) *Bombload:* 21×1000-lb or 1 Blue Steel stand-off bomb (shown) *Speed:* 645 mph at 40,000 ft

The ultimate challenge for bomber designers was the other side of the sound barrier, but in the meantime the technology of the 1940s and 1950s was taken to full stretch by the last crop of subsonic bombers – each of them marking the end of its own particular line of development

There was to be one last fling for the subsonic bomber as designers in the United States, Britain and Russia took one more swing at the basic idea. There was good reason: it was still a workable concept, and the supersonic bomber – the logical next step – could not be achieved with the technology then available.

But there were improved jet engines, and some new aerodynamic tricks. So, during the 1950s a handful of heavy and light bombers exploited that technology and achieved a significant advance in performance in the subsonic regime.

Here are the stories of six such bombers, using widely different design approaches to refine their performance. But each one was the end of the line, each was the last of its type and each marked, in its own way, the end of the subsonic bomber era.

*30 August 1952:*

## Avro Type 698 Vulcan

The flying displays of the Society of British Aircraft Constructors were high points in the aeronautical years of the early 1950s. One of the most thrilling years was 1952. Off the black-topped runway at RAE Farnborough roared a new shape, a large, white triangular aircraft, looking sometimes like a manta ray, sometimes like a racing yacht as

it banked over the field in an amazing low-level demonstration of manoeuvrability.

It was the Avro Vulcan, a four-jet bomber, then one of two new and heavy V-bombers for the Royal Air Force.

The Vulcan grew out of the 1946 specification that also produced the Handley Page Victor, about which more later. The delta-winged geometry – so called because the basic wing shape is a triangle and so is the Greek letter *delta* – evolved from the classical sweptwing formula. It started as a sweptwing aircraft with long and tapered wings and a swept tail, looking remarkably like the Boeing B-47, at least in the general aerodynamic conception.

But structually it would have to be different. American factories had the machine tools necessary to build strength into that kind of aircraft, but the British didn't. Their tooling was almost completely

limited to that used during the Second World War. It was fine for making airplanes out of bits and pieces of thin-gauge alloys, but it could not handle the heavier stuff.

So the bomber designers at Avro began by reducing the span so that they might be able to build a lighter structure. But that didn't work out either, even when the span was reduced to a point where the range began to fall off dramatically. They compromised on a very short-span swept wing and then filled in the area behind it to complete the triangular surface. With some aerodynamic tricks of their own, and their existing tooling, they were able to produce a radical aircraft with conventional structure.

A Blue Steel-armed Vulcan strategic bomber. By 1974 most remaining Vulcans had been switched to the overland strike role

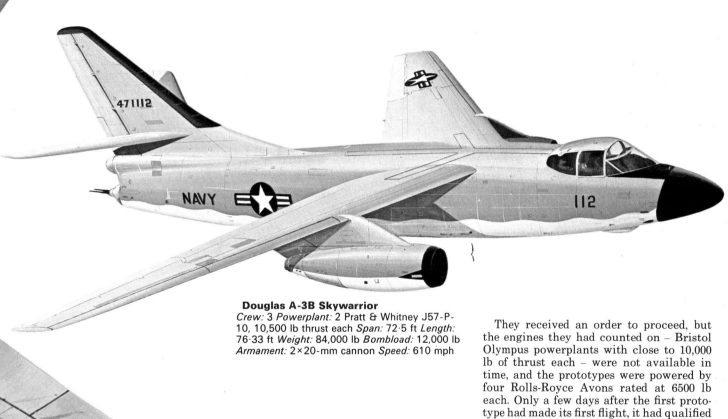

**Douglas A-3B Skywarrior**
*Crew:* 3 *Powerplant:* 2 Pratt & Whitney J57-P-10, 10,500 lb thrust each *Span:* 72·5 ft *Length:* 76·33 ft *Weight:* 84,000 lb *Bombload:* 12,000 lb *Armament:* 2×20-mm cannon *Speed:* 610 mph

They received an order to proceed, but the engines they had counted on – Bristol Olympus powerplants with close to 10,000 lb of thrust each – were not available in time, and the prototypes were powered by four Rolls-Royce Avons rated at 6500 lb each. Only a few days after the first prototype had made its first flight, it had qualified to fly at Farnborough in the display. (The rule was that a type needed ten hours of flight time to be allowed to take part in the programme.)

Parallel to the bomber programme, Avro had designed, built and flown a number of small delta-winged aircraft, with the intent of learning about this new shape, particularly its handling qualities. The value of those small deltas would seem now to have been small. Too much time was spent in learning things about them that were never going to be used on the full-scale aircraft, and they were not true scaled models anyway.

But the brightly coloured shapes, escorting the white prototype in flight at Farnborough – remember the 'delta of deltas'? – were a visual delight and may have paid their way in advertising alone.

**Handley Page Victor B Mk 2**
*Crew:* 5 *Powerplant:* 4 Bristol Siddeley Sapphire 202, 11,000 lb thrust each *Span:* 110 ft *Length:* 114·92 ft *Weight:* 180,000 lb *Bombload:* 35×1000-lb *Speed:* 630 mph

One of the things learned soon enough on the big aircraft was that the straight leading edge was not a very good answer to the problems of flight. So the leading edge was cranked part-way along the span, and then cranked again, by adding a new section that projected ahead of the straight leading edge. It added more area, it moved the aerodynamic centre of forces and it added camber to the airfoil section. The wing shape was further refined in the Mk 2 version of the Vulcan to show gracefully curved wingtips. By that time there was hardly a straight line on the wing for a distance greater than a few feet; the leading edge was kinked, curved and twisted and the tips were no longer the truncated triangle they had been. The Vulcan grew into an effective high-altitude bomber, and held its own nicely in competition with Strategic Air Command's bombers in the annual USAF bombing, navigation and reconnaissance competitions of those years.

Vulcans began to enter service early in 1957 with operational conversion units, and went to regular squadrons in the RAF, where they still serve today.

*28 October 1952:*
### Douglas XA3D-1 Skywarrior
This was the US Navy's first strategic bomber, masquerading under the attack designation to preserve the official stipulation that strategic bombing was the exclusive province of the United States Air Force. But nobody was fooled by a Navy requirement for a carrier-based bomber that could carry nuclear weapons for long distances. That requirement began to take form in 1947, when the Navy and Douglas began talking together about how to design such an aircraft.

Some preliminary thinking by the Navy had resulted in weights that were astro-

nomical for any operations from any carrier then in existence or projected. Successive rethinking of the problem had enabled the Navy to get estimated gross weights down near 100,000 lb, and if they got the big carrier they were working towards that weight would be acceptable.

But Douglas designers were realists. They figured it would be better to design an airplane that could be used on existing carriers than to go all out on a design for use on a carrier that might never be launched. They were right. The super-carrier never was launched, and Douglas were well ahead in their design for the smaller aircraft.

The goal was to save weight everywhere. Every pared pound actually saved more than its own weight by permitting a reduction in the weight of structure needed to support it, the weight of the fuel needed to carry it around and the weight of the wing needed to lift it. In fact, for the design that Douglas finally settled on, one pound saved was worth more than six pounds overall. This was the first real recognition of the 'growth factor', and it was a major contribution to the designer's art.

Dozens of 'paper' airplanes were developed to the stage where they could be either rejected or saved for a further study. Finally, the team reached its last evolutionary shape. It was a sweptwing aircraft, powered by a pair of turbojets slung underneath the wings. It was a very clean aircraft; much attention was devoted to keeping the drag low, and the overall aerodynamic efficiency high.

The pilot and co-pilot sat side-by-side in a large cockpit, and their navigator-gunner sat behind the pilot, facing aft. They did not have ejection seats; a slide led out of the fuselage belly behind an escape door, rammed open by cartridge-actuated systems.

The bomb bay was designed for the huge weapons then being turned out by the nuclear explosive people. Because nuclear weapon design was changing continually then, the general procedure was to design a big bomb bay that would be able to hold the Hiroshima or Nagasaki bombs – the theory being that those two were probably the biggest and heaviest that were ever going to be developed.

But Douglas added an idea to make the bomb bay more versatile. About halfway up its height they installed a removable platform. Above it they made a place for an auxiliary fuel tank; below it they hung bombs on ejector racks, one bomb per rack. Ahead of the bomb bay they attached a retractable, perforated flap; before the bomb bay was opened, the flap was extended so that the buffeting of the air into the bomb bay would be reduced. The ejector racks were for positive separation of the bomb from the rack and out of the open bomb bay.

The total bomb load was 12,000 lb. In later years, the bomb load was a versatile mix of depth charges, mines, bombs and any other kind of weapon that could be adapted to the cavernous bomb bay of the Skywarrior.

They served with heavy attack squadrons of the US Navy, and they were the basis for a conversion to a model for the USAF, which will be discussed later in this chapter.

US military aircraft were redesignated in 1962, following a system established by the Department of Defense. The A3D-1 Skywarrior became the A-3A; the standard bomber version, the A3D-2, became the A-3B. There were other models developed for photographic and electronic reconnaissance; they retained some attack capability, but every cubic inch of available

*A Victor K Mk 2 tanker refuels a Hawker Siddeley Buccaneer. Inset: The refuelling probe of a Phantom FGR 2 in a Victor tanker's drogue*

MOD

**Myasischev M-4 (Bison-C)**
*Crew:* 5/6 *Powerplant:* 4 Soloviev D-15,
28,660 lb thrust each *Span:* 170 ft *Length:* 162 ft
*Weight:* 363,760 lb max *Bombload:* NA
*Armament:* 6×23-mm cannon *Speed:* 625 mph

Inboard, the wing was highly swept; partway outboard, the sweep angle decreased; near the tips, the sweep angle was low. The critical Mach number – that high speed at which shock waves begin to form, with resulting sudden high drag – was kept constant over the entire wing. In contrast, earlier sweptwing shapes would have, typically, a small section of the span where the Mach number would go critical early and trigger turbulence over a much greater area. Once the critical Mach number is exceeded drag increases enormously, and no amount of power will produce much more speed.

reached its operational units. It served in two basic bomber versions, Mk 1 and Mk 2. The Mk 2 was somewhat larger of fuselage, and carried more things to confuse and annoy the enemy.

As a bomber, the Victor could carry the same nuclear weapons as the Vulcan, and the same Blue Steel stand-off missile. But its capacious bomb bay could carry more of the conventional weapons: 35 of the 1000-lb bombs compared to 21 × 1000-lb in the Vulcan.

Victors have gone through a number of modification programmes and now serve as strategic-reconnaissance bombers and as tankers with the RAF.

*1953:*
### Myasishchev M-4
This four-jet Russian strategic bomber, their first big jet bomber, was a contemporary of the Boeing B-52. They entered design at about the same time, must have flown at about the same time and have been in service with their countries' air forces since their first deliveries.

volume was crammed with special installations for the reconnaissance role.

The Douglas Skywarrior bombers served well with the Navy, and the Douglas team contributed greatly to the art of aircraft design when they recognised and applied the growth factor concept.

*24 December 1952:*
### Handley Page HP 80 Victor
The Handley Page Victor, an unconventional design, is one of the best examples of subsonic bomber design technology in the world, and one of the least appreciated.

In its heyday as a strategic bomber, official RAF figures stated that the Victor cruised at Mach 0·94 at 55,000 ft, an astounding performance for an aircraft designed to meet a 1946 specification. It also is capable of low-level flight in a terrain-following mode; can carry and launch the Blue Steel stand-off missile from either low or high altitude; can fly strategic reconnaissance missions; and can refuel Britain's bombers and fighters.

This versatility grew out of necessity, and out of a programme that was the classic example of 'penny-wise and pound-foolish' funding that characterised many of Britain's potentially great aircraft.

It began, as did the other British heavies, with the 1946 specification for a strategic bomber. Handley Page's designers had discovered some exciting aerodynamic ideas during their trips to Germany immediately after hostilities ceased, and among them was the concept of what was to become known as the crescent wing. It was shaped, one fancied, rather like a crescent, and that was an easier term of definition than to call it correctly a constant critical Mach number wing.

So the Handley Page designers adapted this wing, calculating the shape to meet their specific requirements. Remember that this was before the automatic computer, and that all calculations were done with slide rules, logarithm tables and desk calculators of the mechanical type.

The British aerodynamicists of that era were superb at this kind of work. Handley Page and Royal Aircraft Establishment engineers worked out the wing shape, the fuselage contours to go with such an optimised wing, and the characteristic tail that topped off the shape. Even by today's standards the Victor continues to look strange. Yet its basic geometry has not changed since the design was frozen for production, and every line has a specific reason for being that way.

Confronted with two strange shapes, the Victor's 'crescent' wing and the Avro

Vulcan's delta, British officialdom hesitated. Unable to choose between the two designs, they bought both. But like Orwell's equal animals, the Vulcan turned out to be 'more equal' than the Victor. Handley Page had to struggle along under a series of annoying small contracts, buying a few of these and a few of those, instead of being able to place large orders well in advance of the needs of the ongoing production line.

Finally, later than the Vulcan, the Victor

The M-4 was not built in the quantities of the B-52, however, and it definitely is not today the first-line bomber with the Russian forces. It has been relegated to the maritime reconnaissance role, and is generally seen being escorted by British or US interceptors as they fly over NATO fleet units on manoeuvres.

The M-4, or Bison, to give it the NATO code name, was built around four of the big Mikulin engines that powered the earlier

Bison-C, above, is the definitive production version of the M-4, with a redesigned nose and Soloviev D-15 turbofans. Bison-B, below, is the maritime reconnaissance version, with nose radome and a bulged weapons bay to accommodate in-flight refuelling equipment

# NIGHT INTRUDER

**Martin B-57B Night Intruder**
*Crew:* 2 *Powerplant:* 2 Wright J65-W-5,
7220 lb thrust each *Span:* 63·96 ft *Length:*
65·5 ft *Weight:* 55,000 lb *Bombload:* 6000 lb
*Armament:* 8×·5-in mg *Speed:* 582 mph

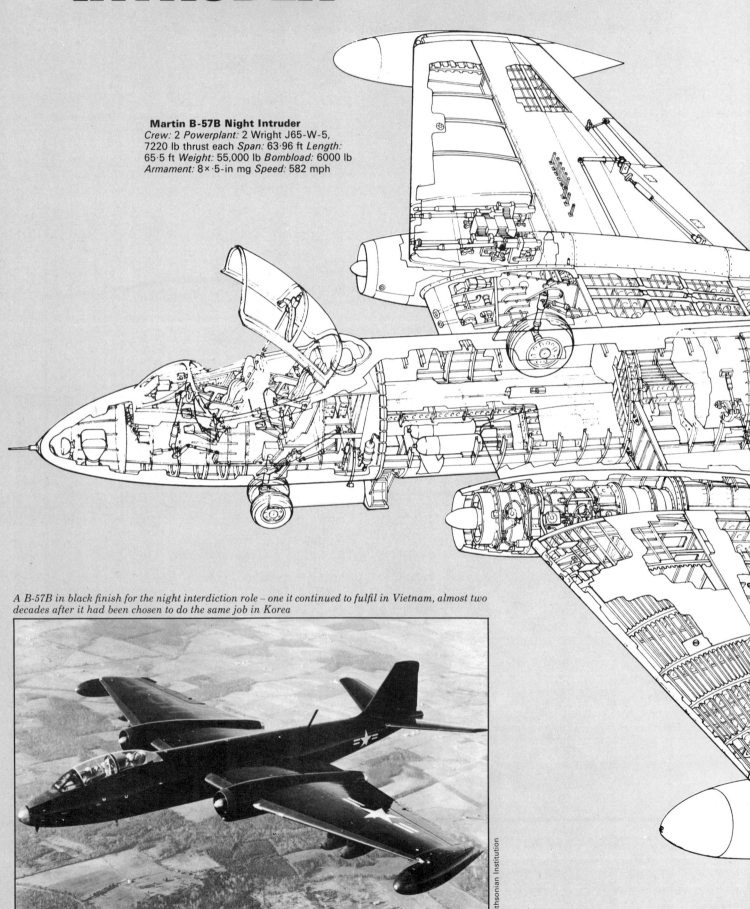

*A B-57B in black finish for the night interdiction role – one it continued to fulfil in Vietnam, almost two decades after it had been chosen to do the same job in Korea*

Smithsonian Institution

*Air brakes out and undercarriage down, a B-57E comes in to land. Target-towing apparatus was the main distinguishing feature of this mark*

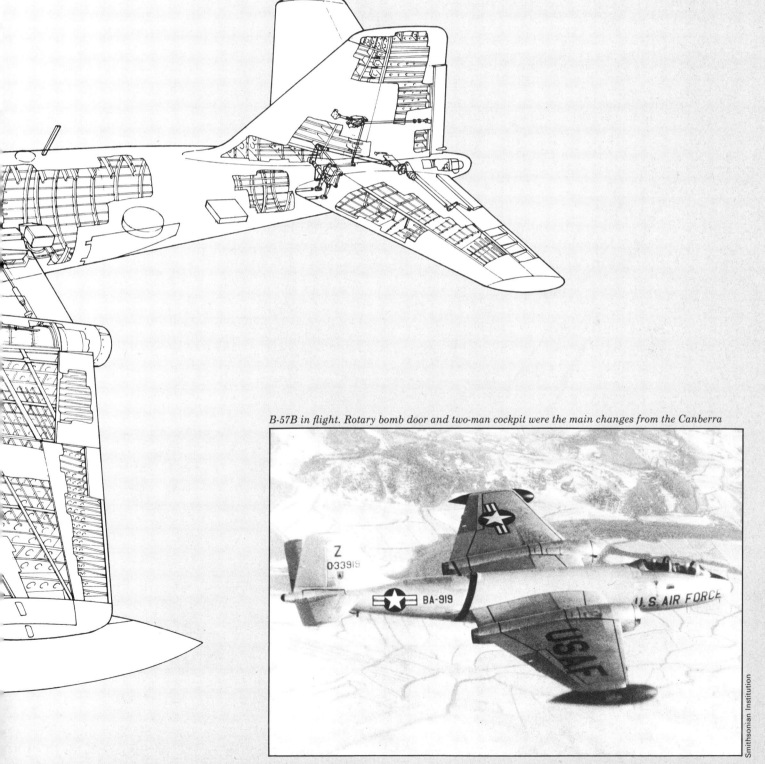

*B-57B in flight. Rotary bomb door and two-man cockpit were the main changes from the Canberra*

Badger. The layout is rather conventional, and really should not have caused the furore it did when it was first seen publicly over Moscow in the May Day parade of 1954. The real worry was, of course, that the Russians had a strategic jet bomber and that it probably had range enough to carry nuclear weapons to targets in the United States.

Three different versions of the M-4 have been observed. Bison-A is the standard strategic bomber, armed with nuclear or conventional bombs. Bison-B is a maritime reconnaissance version, in which the bombardier's nose position, multi-paned and equipped with an optically flat visual bomb-aiming window, has been replaced with a solid nose. An in-flight refuelling probe has been added above the nose. New bulges on the fuselage are fairings over electronic reconnaissance equipment, mounted inside and in the bomb bay.

Bison-C has a larger nose radar than Bison-B, for increased range in its maritime search mission. Other modifications include a standard in-flight refuelling installation which can be carried as an alternative bomb bay load. The probe-and-drogue system is used to refuel other strategic bombers of the Russian air force.

There is a wide variation in the estimates for sizes, weights and performances of the Bison, as there is for other Russian military aircraft. Those cited in this book have been taken from (one hopes) reliable sources.

One interesting point: Yakovlev's book makes no mention at all of Myasishchev, the M-4, or the designer's later work. Non-mention in official Soviet histories often indicates that the person not mentioned has become a non-person.

*20 July 1953:*
## Martin B-57A Night Intruder
Does it look familiar? It should, because its outlines mark it as a near-twin to the English Electric Canberra. It started as the Canberra, bought from Great Britain by the United States to serve as a tactical bomber. It was redesigned to suit American tooling and production methods, and was operated primarily by USAF's Tactical Air Command. It served in combat in Southeast Asia as a day tactical bomber and later as a night intruder.

The requirement originated during the Korean war, when light tactical bombing was being done by the ageing Douglas B-26 (redesigned from A-26, and not to be confused with the Martin B-26 Marauder of the Second World War). There was only one trained squadron operating B-26s on night interdiction missions; daylight operations were suicidal for one thing, because of the intense anti-aircraft fire, and futile for another, because so much of the logistics supply for the North Korean armies moved at night by road and trail.

A number of aircraft were evaluated, and only the British Canberra met the primary requirement: lots of internal volume for the

special equipment needed for the night and bad-weather interdiction missions. Those systems didn't exist then in much more than paper form, but the decision was made to get an airplane big enough to handle them when they did become available, and not to have to try to compromise the electronics by fitting them into inadequate space later.

Despite the geometric similarity, the missions and details of design were quite different. The B-57 was designed for low-level missions against moving targets at night, which called for low-altitude manoeuvrability, ability to loiter in the target areas and range to reach out to the enemy's supply lines, well behind his front-line structure.

The USAF flew the airplane with two men, instead of the RAF's three-man crew, and added machine-guns to the unarmed Canberra for forward fire. The cockpit was completely redone, powerplants were changed – to another British-designed engine, the licence-built Armstrong-Sideley Sapphire – and a rotary bomb door was installed.

It was to be some years before the B-57 actually was used in the night-interdiction role for which it was designed, although some were used in daylight tactical bombing in Vietnam. A small group – 16 identified aircraft – were modified to include the latest technology for night attack, including forward-looking infrared sensors, laser rangefinding, low-light-level television, and

**Douglas B-66B Destroyer**
*Crew:* 3 *Powerplant:* 2 Allison J71-A-13, 10,200 lb thrust each *Span:* 72·5 ft *Length:* 75·17 ft *Weight:* 83,000 lb *Bombload:* 10,000 lb *Armament:* 2×20-mm cannon *Speed:* 631 mph

a rapid-firing belly gun target. Designated B-57G, 11 of these aircraft were deployed to Vietnam in September 1970 to work out against targets on the Ho Chi Minh trail at night. But time had caught up with the B-57; USAF reports said it was underpowered for the mission, and its last combat record was not impressive.

*28 June 1954:*

### Douglas RB-66A Destroyer

This was one of those cases where the Air Force decided to buy a Naval aircraft because only minimal changes would be required to meet USAF specifications.

But the changes were not minimal. New engines, new wings, reworked fuselage and dozens of detail changes added up to an aircraft that cost more to buy and modify than it might have cost to design, develop and produce from a clean sheet of paper.

It was to be the USAF's version of the A3D Skywarrior, but most of the 294 Destroyers bought were assigned to the mixed mission of reconnaissance and bombing, and they were issued primarily to Tactical Air Command Reconnaissance squadrons. Only 72 of the type were completed for the bomber mission alone, and most of these were either converted or eased out of service after a short life.

The reconnaissance version stayed in service, and did yeoman work during the Vietnam war, loaded with advanced electronic systems to navigate flights of fighters and to raise an electronic shield of confusion around them to improve their chances of surviving the trip. That is another story.

One example should serve to illustrate the point about the Air Force's minor changes. The empty weight of the Navy's A3D-2 bomber was about 37,000 lb. The empty weight of the Air Force version was about 42,500 lb, nearly three tons heavier.

The B-66 was designed around a variety of internally carried weapons from nuclear to conventional, and could carry up to 15,000 lb of them in the bomb bay. Official Air Force data claims a better performance for the B-66 than Navy figures claim for the A3D, and that is strange. The Air Force version was not only heavier; it had engines that delivered considerably less thrust and at a lower altitude. Official figures have been used for performance of the A3D/B-66 series in the specifications, there being no other choice without an aerodynamic staff and accurate thrust and drag data.

*An RB-66 reconnaissance version of the Destroyer leads four Republic F-105D Thunderchiefs on a radar bombing mission over Vietnam*

US Air Force

373

# THE SUPERSONIC GENERATION

Tactical and strategic considerations – not to mention the improving efficiency of anti-aircraft missiles – have forced bombers from the stratosphere down to tree-top level, while still demanding ever higher performance. So far, the only solution has been the variable-sweep wing configuration

Over the last 20 years, new aerodynamic tricks have been learned, new structural approaches and new materials have been discovered, and new powerplants have been developed.

Concurrently, new airborne electronic systems have taken over many of the multi-faceted tasks of flying an aircraft.

Automatic flight is a reality in some modes, and automation is proving a great aid in others where the pilot still retains control of the overall flight system.

During these years, bomber design philosophy has changed from requiring supersonic speeds at high altitude to demanding high subsonic or low supersonic speeds just above the contours of the terrain. This change in basic requirements has forced a major change in the design of aircraft because, increasingly, the variable-geometry design appears to offer the only feasible answer. It was pioneered with the General Dynamics F-III series, and currently graces the lithe forms of the two major advanced bombers in the United States and Russia.

Will the next generation of bombers keep that kind of polymorphous shape, or will some entirely new kind of aerodynamic layout appear? That's a difficult question

to answer. But there is an even more difficult question: will there be a next generation of bombers?

*11 November 1956:*

**General Dynamics/Convair B-58A Hustler**

This delta-winged jet was the world's first supersonic bomber to enter service. During a ten-year lifetime with the Strategic Air Command, the Hustler set 19 world records. It received five major aviation trophies for outstanding flights: the Bendix, Blériot, Harmon, Mackay and Thompson.

The Hustler was a unique airplane. It carried its weapons outside its fuselage, slung underneath in pods. Reconnaissance systems and fuel were the two most common loads after strategic weapons.

It was designed to perform a high-altitude, high-speed delivery, dashing at Mach 2 into enemy territory, launching or dropping the pod and banking away in a

### General Dynamics/Fort Worth B-58A Hustler

*Crew:* 3 *Powerplant:* 4 General Electric J79-GE-5B, 16,000 lb thrust each with afterburning *Span:* 56·83 ft *Length:* 96·75 ft *Weight:* 163,000 lb *Bombload:* Multi-systems under-fuselage pod *Armament:* 1×20-mm Vulcan cannon *Speed:* Mach 2

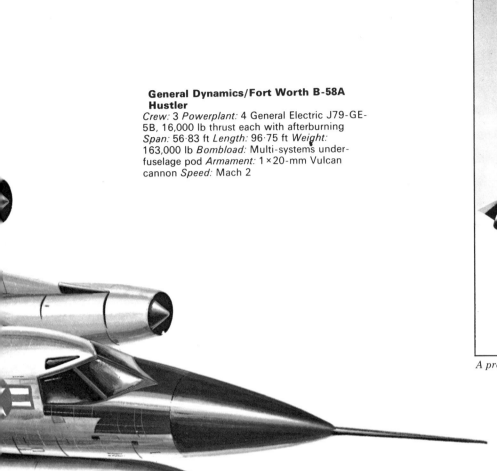

*A prototype B-58 Hustler in flight*

blaze of power and speed to avoid the radiation blast damage.

But when the time came to go in on the deck below the watchful eyes of radar, the B-58 turned to that mission as if it had been designed for it. It could speed along just under the velocity of sound when it was making those low-altitude bombing runs, and no bomber since then has had such performance.

It could also toss-bomb, using the half-loop, half-roll technique described earlier. Its pods included both free-fall and rocket-powered types, so that various options were available to SAC crews for weapons delivery.

The external pods made for a small bomber; its wing span was just under 57 ft. But it stood on towering landing gear to clear the pods, and you always had to look up to look at the B-58.

Under its wings were snuggled a quartet of General Electric J79 powerplants. These engines were cleared for two hours of operation at maximum thrust, compared to the much briefer ratings of most earlier engines. This made it possible for the Hustler to get its dash performance over more than 1500 miles instead of the more normal few hundred.

After a long and noisy takeoff, it could head for altitude faster than many contemporary fighters, with an initial rate of climb, fully loaded, of 17,000 ft per minute. It could nearly triple this figure when lightly loaded.

The list of pioneering achievements of the B-58 programme is lengthy. For one, it was the first weapon system contract, that term designating the practice of naming a contractor and giving him complete responsibility for every part of the programme through design, development and early flight test. In this case, General Dynamics/Convair made sure that the aircraft, pods, ground support equipment and spare parts were completely integrated into a single weapon system.

The B-58 originated in 1949 as a parasite bomber, to be carried by another bomber to some location within target range, dropped, and then recovered after the attack if all went well. It was small, with a two-man crew, an external weapons pod and just enough fuel to complete the mission. But it grew after the concept changed to a self-launched bomber: the Hustler was built around a three-man crew, lots of fuel, and in-flight refuelling.

Considering that the original ideas began to take form in 1949, one can appreciate what an advanced design the B-58 was. Only 15 months or so earlier, the Bell X-I had first flown in supersonic flight, a shade above Mach 1; already Convair were proposing that a delta-winged bomber be built to fly at twice the speed of sound. Nobody had flown at that speed before; there was no experience in flight research, and there was not that much confidence in the findings of wind-tunnel tests.

But by 1952 General Dynamics had worked out a winner, and were awarded the development contract. Two years and two months after the day the first drawings were released to the shop, the first B-58 flew. It was a remarkable achievement.

The delta-wing design looked deceptively simple. It had a straight-edged wing, but close examination revealed that its leading edge curved downward, increasingly toward the wingtip, so that the cambered surface resembled a section of a long and finely tapered cone. This gave the Hustler the aerodynamic qualities it needed at both the low- and high-speed ends of the flight regime.

The Hustler had no other high-lift devices. The trailing-edge surfaces were elevons, combining the functions of ailerons and elevators, because the B-58 had no horizontal tail surface.

Aerodynamic refinement was everywhere. The fuselage was area-ruled to minimise drag in the critical trans-sonic flow region, and to reduce the power and time needed to accelerate from subsonic to supersonic speed.

One problem with the B-58 may have been that its dash range was considered inadequate compared with the penetration range of the B-52. Although the B-58 could fly a lot faster than the B-52 in low-level missions speed is not the primary criterion; range is. And, since effective mission range begins from the point at which the last pre-strike refuelling ends, the B-58 could not strike as deeply as could the B-52.

Still, their blistering speed and unique capabilities made the Hustlers a valuable part of the strategic deterrent force. Further, they had a unique reconnaissance role. The pod could carry complex, major electronic and photographic reconnaissance systems. The Hustler could slash through the upper atmosphere, at speeds of Mach 2, on a high-altitude reconnaissance mission in many ways comparable to that of the Lockheed SR-71 that arrived on the scene much later.

Remember the B-58, then, as a first-class weapon system that pioneered much advanced technology, served as the prototype for a new contracting approach, and was the world's first supersonic bomber. It was a pace-setter and a record-breaker, and much of its performance envelope still remains unchallenged by later and newer aircraft designs.

*31 August 1958:*

### North American A3J-1 Vigilante

The Navy's first Mach 2 strategic bomber – although never officially called that – originated as a company-funded proposal during 1954. North American suggested that the Navy consider an aircraft to deliver weapons in low-level attacks at high subsonic speeds.

Fine, said the Navy about a year later, but also let's go for Mach 2 at altitude, and let's be able to launch when there is no wind over the deck of our carriers.

Adding two requirements like that changed the design considerably, and it emerged later as the A3J-1, with a wing that was really too big for the high-speed role and too small for the high-lift requirement of the zero-wind launch. But the wing was a fine compromise, and only one of the many advanced technical features of the A3J design.

High speed at low level is best handled with a small and thin wing; the ride is more comfortable, the drag is lower. On the other hand, a zero-wind launch of a heavy airplane needs high lift, and that means a large and thick wing, speaking aerodynamically. The North American compromise design resulted in a thin wing, with a system of blown flaps at the trailing edge. Blown flaps are rather conventional flaps that

*A-5A Vigilantes on the flight deck of USS* Enterprise, *the biggest warship ever built*

have their effectiveness increased by blasting air at high pressure around their leading edges. The source of that air is the jet engines, which are bled of a small percentage of their high-pressure air which is then ducted to the flaps.

Additionally, drooped leading edges on the wings could be deflected downward about 30° for takeoff and landing, or only a few degrees for an improvement in the cruise flight regime.

Roll control was handled by spoilers. The horizontal tail was divided into a left and right slab surface; they were deflected together for pitch control, and differentially to trim any rolling tendency of the aircraft. The vertical tail was a one-piece surface.

All of these complex controls, plus navigation, weapons delivery and reconnaissance data, were tied together in a single system designed by North American's Autonetics division. The system featured inertial navigation for one of the first times in an airborne application, and a miniaturised digital computer to tie it all together.

The fuselage was different. The weapons

**North American Rockwell A-5A Vigilante**
*Crew:* 2 *Powerplant:* 2 General Electric J79-GE-4, 16,150 lb thrust each with afterburning *Span:* 53 ft *Length:* 73 ft *Weight:* 49,500 lb *Bombload:* Internal nuclear or conventional weapons *Speed:* Mach 2·1

are ejected through a tunnel that leads out the tail of the plane, between the engines. So the fuselage is simply an enclosure for three large, cylindrical spaces. The outboard ones are the engine bays, and the inner one is the weapons bay. The nuclear weapon was attached to two auxiliary fuel tanks; when they were emptied, they were retained to be dropped as part of the weapon.

The Vigilante was developed under the weapon system management concept, with North American responsible for everything except the government-furnished engines. It was redesignated as the A-5 series in 1962.

About then, because of continual arguments in high places about whether or not the Navy had a strategic bombing mission, the Vigilante was converted to a reconnaissance aircraft. Its capacious bomb bay was filled with electronics; its flat-topped fuselage was curved to cover additional fuel volume, and the latest and most advanced reconnaissance systems – photographic and electronic – were installed. That's where our story of the Vigilante ends, because that is where its deployment as a jet bomber ended.

*1958(?):*

### Tupolev Tu-22

This was Russia's first supersonic bomber, and its elegant shape told a story of sophisticated design techniques. It was area-ruled, long and sleek, and from every angle it looked like a beauty.

And Beauty it was called, for a short while, very unofficially; NATO later chose the code name of Blinder. It was first flown publicly at the 1961 Tushino air display.

At that show, two different models of Blinder appeared. One was a conventional bomber; the other was carrying an air-to-surface missile, partially submerged in its belly, and was equipped with a refuelling probe. Its nose radome was considerably larger than that of the other bomber, hinting at the search and guidance role for the airborne missile.

When the Blinder was first analysed from photos and other intelligence data, it appeared that it would have a top speed of about Mach 2. But later data show that the top speed is a more modest Mach 1·4, and that it is most likely a dash speed, not a sustained cruise at altitude.

Other notes about the bomber say that the range is deficient, and for that reason the Blinder has been relegated to the maritime reconnaissance role, where it can remain on station over those waters reachable by a combination of internal fuel and in-flight refuelling.

There are reports that some Blinders have been seen with extensive electronic intelligence installations, and that there is also a training version, with a second cockpit mounted in a stepped configuration aft of the standard tandem three-place cockpit.

The biggest batch of Blinders was seen at the 1967 air display at Domodedovo, an airport near Moscow, on the occasion of a

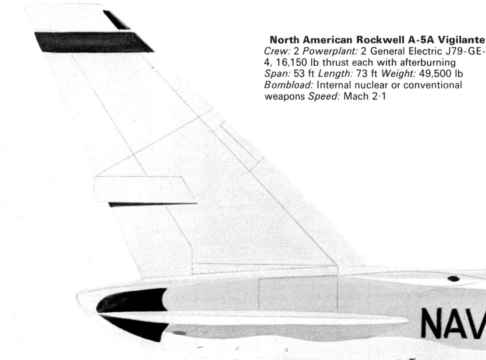

*An early model A3J-1 Vigilante comes in to land. All USAF and Navy aircraft were redesignated on 18 September 1962 – the A3J-1 becoming the A-5A*

Soviet show of strength. Formations of the bombers flew past, totalling 22 aircraft, and observers noted that most of them were the missile-carrier variety.

Less than 200 of the bombers are believed to have been built, and about one third of them went to the Russian naval forces for maritime reconnaissance.

Blinders may have suffered from some of the same ailments that limited the B-58: deficient unrefuelled range at low level, and limited payload under those conditions. Whatever the reason, Blinder should not now be taken too seriously as a threat, or as a major part of Russian strategic forces in being.

*17 June 1959:*

## Dassault Mirage IV-01

If there is a more efficient group of aircraft

designers anywhere than in Marcel Dassault's employ, they must be well hidden. Time after time, his teams of brilliant designers and engineers have come up with new answers to a wide variety of requirements.

One of their brightest efforts was the first French strategic jet bomber, the Mirage IVA. It started simply enough, with a 1956 requirement for an aircraft that could carry a bomb, weighing about 7000 lb and some 16 ft long, for a distance of about 1200 miles without refuelling. And, added the specification, it ought to do this at supersonic speed.

The last part was the tough one. But Dassault and two competitors produced preliminary designs, and Dassault's was chosen. They began detail design in April

1957; two years and two months later the first prototype flew.

Dassault had basic experience with the thin delta wing in the Mirage III fighter as a foundation. The company always has made its best progress by doing the most with the current state of the art in technology. They made some changes to the basic shape of the delta wing, enlarged the dimensions by about half, powered it with a pair of engines instead of one, and produced the first prototype.

But meantime, the programme itself had changed. French officials wanted a bigger and better bomber, and an enlarged and developed Mirage IVB was to be the answer. Dassault and SNECMA, the engine suppliers, went through a whole routine with the new design and arranged licence rights

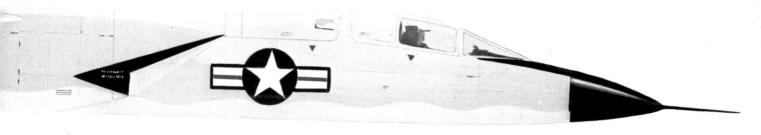

to a new Pratt & Whitney engine, and then the project was dropped by the government. It was back to the Mirage IVA as before, with some minor changes.

Three pre-production prototypes were built and tested, the first of these flying for the first time on 12 October 1961, two years and a month after the contract date for the three planes.

Now, the Mirage IVA operates as an air-refuelled bomber, drawing its sustenance from one of 14 Boeing KC-135 tankers the French bought from the United States. Additional fuel can be taken on board via a buddy system from another Mirage IVA.

Tactically, the bombers operate as a pair, one with a bomb and one with only auxiliary fuel tanks and the buddy system. They take off, and top up from a tanker. They then fly together to a pre-plotted mission point and the bomber refuels from the other Mirage IV, leaving it enough fuel to get home or, in the event of war, draining its tanks completely. That way the Mirages can guarantee at least a one-way trip deep into Russia, the theoretical enemy postulated in all their studies.

Like all jet bombers of 15 years or so ago, the Mirage was designed for high-speed, high-altitude strikes, and in the case of the Mirage IV, high speed meant Mach 2. But when an examination of the real effectiveness of anti-aircraft missiles forced the conclusion that attacks would have to be mounted at low level, the Mirages required only minor changes.

The production run was not a large one; Dassault built 62 production Mirage IVA bombers, plus the three pre-production

his design of the M-4 Bison, his name no longer figures in the current pantheon of aircraft designers.

Bounder achieved notoriety when it first showed over Tushino airport in the 1961 air show. It roared over the field, propelled by four powerful turbojets, and then it faded into history.

In retrospect, Myasishchev did almost everything wrong. We can't tell what requirement he was trying to meet, so it is possible that some anonymous Comrade Planner is the real person to blame. What it seems is that he set out to build a long-range jet bomber that could carry the Russian thermonuclear weapons and other stores, and that would have a supersonic dash performance over the target.

Assuming that requirement, the first thing he did wrong was to choose a poor wing shape. That truncated delta with high taper is no good for high-speed design. The next thing he did wrong was to design that hulk of a fuselage, and to ignore the whole concept of the area rule, which by then was known to everybody in the world who studied aeronautics.

**Dassault Mirage IV-A**
*Crew:* 2 *Powerplant:* 2 SNECMA Atar 9K, 15,435 lb thrust each with afterburning *Span:* 28·88 ft *Length:* 77·08 ft *Weight:* 73,800 lb *Bombload:* 1×70-kiloton nuclear device or 16×1000-lb *Speed:* Mach 2

aircraft and the single Mirage IV-01 prototype, a smaller but geometrically similar aircraft. They now form a major component of France's strategic deterrent force.

*1960(?):*

### Myasishchev M-50

If there is a single reason why Vladimir Myasishchev is not currently listed among the heroes of Soviet aviation, this bomber is it. NATO-named Bounder, it should have been called Blunder, for that is certainly what it was.

As an attempt to build a big, fast jet bomber it was an outstanding failure which revealed an embarrassing lack of understanding of the problems of high-speed flight. Although Myasishchev and his design bureau had won a Lenin Prize in 1957 for

Realise that by that time, Convair had been flying the B-58 for a couple of years, breaking records. The French were well along with their supersonic Mirage IV bomber, and were flying it when Myasishchev probably was watching his prototype take form.

It might not have been too late, even then, to have learned something from the Convair and Dassault designs, but Myasishchev obviously could not read the handwriting on the wall, even though it was in large Cyrillic letters.

Bounder flew, probably in 1960, and the Russians decided to show it at the 1961 Tushino display. The aircraft that performed that day used afterburners on its inboard engines, but not on the outboard ones. That might have been one last attempt to get the M-50 to accelerate a little faster through the speed of sound.

So little is known, and so much has been conjectured, about the Bounder that there is no more to be said about it at this point. It was never much of a threat, although it was made to seem real enough and threatening enough at the time.

But Bounder has not been heard of since, and neither had Vladimir Myasishchev.

*21 September 1964:*

### North American XB-70A Valkyrie

The gleaming white XB-70A, first of two prototypes built and flown, stands today in the Air Force Museum at Wright-Patterson AFB, Dayton, Ohio. In the mid-1950s, it was conceived at Wright-Patterson in a requirement for a new weapon system. And if the original schedule had reached reality, the B-70 would, about now, be leaving service with Strategic Air Command, and most likely one of them would be earmarked for display at the Air Force Museum.

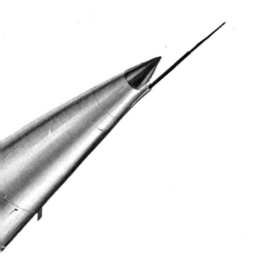

But that is not the way the story went. Instead, it encompassed policy changes, technical problems, financial cutbacks and restorations, and a spectacular fatal accident that helped speed the end of the programme.

In late 1954, Strategic Air Command was looking for a bomber to follow the B-52, and to use the same bomber environment that SAC had developed – runway requirements, ground facilities and crew composition. By mid-1955, SAC's needs had been translated into a requirement for Weapons System 110A, a bomber with subsonic cruise and maximum possible penetration speed, able to carry thermonuclear bombs over long ranges.

Boeing and North American responded, submitting their designs in October 1956. But the designs were stretched to the outer limits of technology. There seemed to be no way to meet the WS-110A requirement without repealing the laws of aerodynamics and thermodynamics.

And this was attempted. Both companies knew about boron hydride programmes in the rocket field. These new fuels had a heating content, per pound, about twice that of the hydrocarbon fuels. They promised rocket combustion performance of a new order, and they looked adaptable to turbojet designs. There was enough push behind them to set up pilot production plants for fuels, and to start developing turbojets to burn what would become known as 'zip' fuels.

The laws of aerodynamics were a little harder to repeal. But North American found an obscure technical note published by the National Advisory Committee for Aeronautics (NACA, the predecessor of today's NASA, the National Aeronautics and Space Administration). It described a phenomenon which would become called 'compression lift' as one way of improving the lifting efficiency and reducing the drag at supersonic speeds. Armed with that knowledge, North American laid out their re-studied

**Tupolev Tu-22 (Blinder B)**
*Crew:* 3 *Powerplant:* 2 afterburning turbojets, 26,000 lb thrust each (est) *Span:* 90·88 ft *Length:* 132·96 ft *Weight:* 185,000 lb *Bombload:* 1×AS-4 (Kitchen) stand-off bomb *Armament:* 1×23-mm cannon *Speed:* Mach 1·4

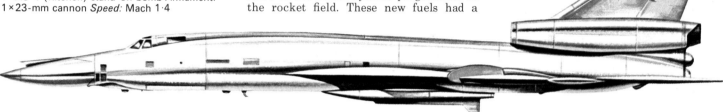

*A Tu-22 leads a formation of Su-11 interceptors in a fly-past at Domodedovo, July 1967*

concept of WS-110A, and in December 1957 were awarded a contract for development of their design.

But the 'zip' fuels had been oversold and under-investigated, and that programme was cancelled. A funded programme at North American for the F-108 long-range interceptor, which was sharing some of the high costs of supersonic research and development, was also cancelled; the whole developmental cost burden then shifted to the B-70 programme. In December 1959, because of escalating costs, the B-70 programme was cut to the design and flight tests of a single prototype. In October 1960 it was re-instated as a complete weapon system, after a lot of forceful arguments by the Air Force. In April 1961 it was cut back to three prototypes. In March 1964 it was cut to two prototypes, its final form.

That is no way to run a development programme. The original USAF schedule had projected that the bomber would be entering service with SAC in 1965. One year from that date, the airplane still had not flown, and the programme was going to produce only two prototypes.

But what prototypes they were! Their unique features were many, from the downward-folding wingtips to the brazed structure of honeycomb panels.

To get compression lift, the shape of the fuselage was critical. So all six of the big General Electric J93 turbojets were laid out side by side in a single underwing body, contoured to trigger a shock wave at a key location. That shock wave was confined by swinging the wingtips downward through 65° in flight, boxing in the flow so that the B-70 would ride on the compressed mass of air. The top of the wing was flat; paired vertical surfaces were located just outboard of the engine body, but above the wing. The fuselage projected ahead of the wing, and

was shaped in a curving set of contours that gave the plane the nickname of 'Cecil, the Seasick Sea Serpent', after a TV puppet character of the 1950s.

The purpose of the flight test programme was to achieve the predicted performance, which included a cruising speed of Mach 3 at altitudes between 70,000 and 80,000 ft. The prototypes were heavily instrumented, the second much more intensively than the

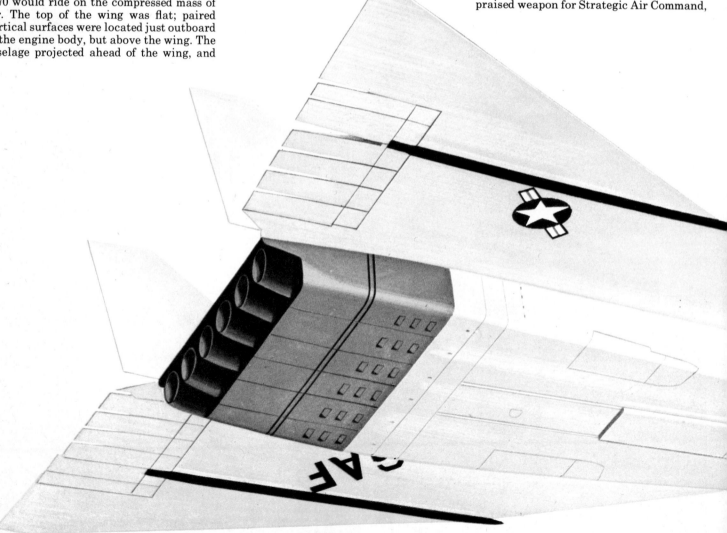

first. Almost every aspect of the proposed flight envelope was a new one, with altitudes and speeds reached previously only by special research aircraft. The two prototypes were flight research vehicles, never intended to be bomber prototypes. Crew stations were different, and there was none of the many features that transform an experimental aircraft into a service bomber. That had been reserved for the third prototype, cancelled in March 1964.

Both prototypes flew. Towards the end of their first phase of flight tests, the second prototype led a formation of aircraft for an air-to-air photograph. There was a mid-air collision; the second XB-70A and its co-pilot, Major Carl Cross, were lost, as was a Lockheed F-104 flown by NASA pilot Joe Walker, also killed in the collision.

The fiery spiral of the stricken Valkyrie was its funeral pyre. The programme went downhill from there. The first prototype had

to be extensively instrumented to replace the test equipment lost with the second, and it required much modification. By then, the plane had been handed over to NASA for research on the supersonic transport programme. By the time they retired the test airplane, it had logged only a few hours of flying time at Mach 3, hardly enough to have made any difference to the knowledge required for a supersonic transport design.

Now the XB-70A first prototype is a museum piece, the bomber that never was. And neither its compression lift concept, nor its general layout, nor 'zip' fuels have been used again.

*25 September 1969:*

### General Dynamics FB-111A

The FB-111A variable-sweep medium-range bomber originated as a tactical fighter-bomber which was the centre of a swirling political controversy for more than seven years. But out of all this came a highly praised weapon for Strategic Air Command,

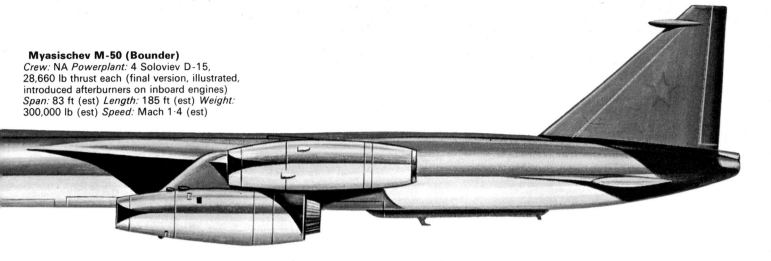

**Myasischev M-50 (Bounder)**
*Crew:* NA *Powerplant:* 4 Soloviev D-15,
28,660 lb thrust each (final version, illustrated,
introduced afterburners on inboard engines)
*Span:* 83 ft (est) *Length:* 185 ft (est) *Weight:*
300,000 lb (est) *Speed:* Mach 1·4 (est)

and, in the words of one SAC pilot, 'The only people who knock this airplane are the ones who haven't flown it!'

One reason for choosing the FB-111A as an SAC bomber was its ability to get right down to the ground in a high-speed terrain-following ride that was automatic and that left the crew free to concentrate on other problems. Armed with up to 50 Mk 117 bombs, which weigh nominally 750 lb each, or with up to six nuclear weapons including SRAM missiles, an FB-111A is a formidable force by itself.

It currently equips two of SAC's bomber wings and it is not likely to see any wider service. Only 76 of the bomber version were built, and the Air Force has no further plans or funds for more.

The variable-sweep wing is the key to the performance of the FB-111A. Wings spread, it can take off and climb; add a little sweep for high-speed cruise, more sweep to go supersonic, and nearly all the sweepback for those on-the-deck dashes.

They began to enter service with an SAC training wing in the autumn of 1969. In May 1970, two of them went to the RAF's bombing and navigation competition, to which SAC had been invited. An FB-111A

was leading the field in the navigation event after three days, but finally lost it to one of the B-52s. But in November that year, in SAC's own competition, an FB-111A crew carried off the individual bombing trophy.

It is equipped for in-flight refuelling, a standard SAC requirement. But it is classed as only a medium-range bomber because of its unrefuelled range, and it has been assigned to targets near the enemy borders, rather than far into the interior. On its low-level penetrations, it can fly at about 660 mph just 200 ft above the ground.

Aside from the terrain-following radar which gives the FB-111A this capability, the plane has an inertial navigation system

which is so accurate that it can be used to taxi and position planes on the ramps. They start all missions from a surveyed mark on the ramp where the aircraft has been pre-positioned, and from which the inertial system measures all future displacements.

For defence, there are on-board electronic countermeasures and infrared detectors mounted at the tail to warn of fighters or missiles to the rear. Automatic jammers, chaff and flares will further confuse the enemy radars and weapons.

On a typical bombing exercise mission, three FB-111A aircraft will take off and climb to 30,000 ft, rendezvous with a tanker and top their tanks. Then just before they

*The massive propelling nozzles of the Valkyrie's six J93 turbojets, designed to propel it at Mach 3 on a compressed mass of air – now an expensive exhibit at the US Air Force Museum*

John Batchelor

**North American XB-70A Valkyrie**
*Crew:* 2 *Powerplant:* 6 General Electric YJ93-GE-3, 31,000 lb thrust each with afterburning
*Span:* 105 ft *Length:* 185 ft *Weight:* 525,000 lb
*Speed:* Mach 3

**General Dynamics/Convair FB-111A**
*Crew:* 2 *Powerplant:* 2 Pratt & Whitney TF30-
P-7, 20,350 lb thrust each with afterburning
*Span:* 70 ft (wings spread)/33·9 ft (wings fully
swept) *Length:* 73·5 ft *Weight:* 81,537 lb
*Bombload:* 37,500 lb or 6 Boeing AGM-69A
SRAMs *Speed:* Mach 2·5

*An F-111A folds back its wings from fully extended to fully swept in preparation for an on-the-deck dash.
The FB-111's wings are larger, and are those originally designed for the cancelled US Navy F-111B*

General Dynamics

reach the target area – low-level runs
surveyed in barren and remote country in
the western United States – they refuel
again. They drop down to the 200-ft level and,
with wings swept to the 55° or 60° mark, they
blast along the run at about Mach 0·65. For
the actual simulated bomb run or weapon
launch, they accelerate to Mach 0·75.

Four targets are 'hit' during a typical
mission by each bomber, and then they
climb back to altitude and the cruise home.
*1970(?):*

## Tupolev variable-geometry bomber 'Backfire'

Proving the principle of parallel invention,
the USSR has a large variable-geometry
bomber in development, production and
perhaps in early operational deployment.
In size, shape and performance it compares
to the Rockwell International B-1. But there
is a major difference: the timetable.

The B-1 had only flown a handful of times
by mid-1975. Backfire, as NATO calls the
new Russian bomber, was observed via
satellite in mid-1970, and was expected to
become operational with the Soviet air
force during 1974.

Given the general knowledge of aircraft
design and the state of the art, plus a few –
even one good example – satellite or SR-71
photographs, some fairly accurate estimates
can be made of Backfire's weights and
performance. Pentagon data credit it with
dash speeds between Mach 2·2 and 2·5 at
altitude, low supersonic speed in low-level
dashes, and a combat radius close to 4000
miles.

Tupolev's design bureau has long been a
leader in the design and development of big
Russian jet aircraft, both civilian and
military. By Russian standards, they have
been very successful aircraft. But on the
military side of the ledger, Backfire is an
outstanding technical achievement, by any
standard. Assuming that there are no
troubles during early deployment, the
Russians should have their bomber in
operational units by now.

The Russian concept of variable-geometry
designs differs from that of the US. The
Soviet teams, typically, have started their

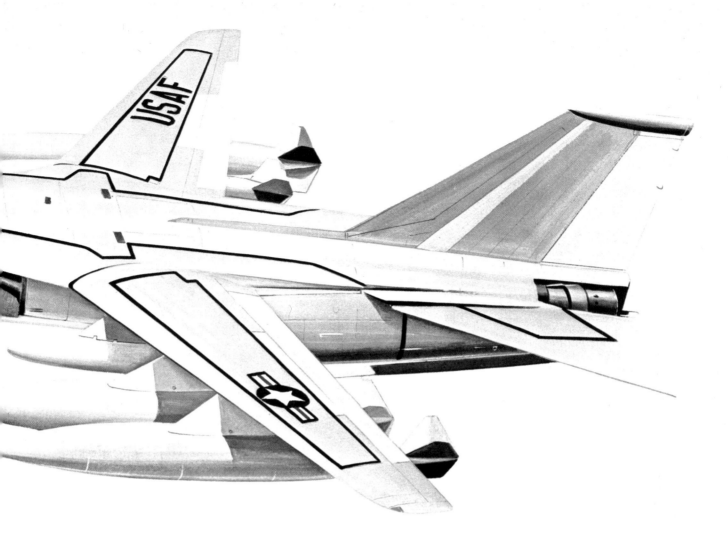

variable-geometry portion of the wing well outboard, much farther along the span than in Western designs. This feature of Backfire's design may result from Tupolev's commitment to the landing gear, retracted aft into a wing pod, and almost a Tupolev trademark. Part of it may be the desire, or the need, to keep structural loads on the wing pivot lower than the loads that Rockwell International find acceptable.

Backfire's weapons load is the usual Russian range of iron and thermonuclear bombs, plus a pair of new air-to-surface missiles which are believed to have been developed as part of the Backfire weapon system and to have a range of more than 450 miles.

*21 December 1974:*
## Rockwell International B-1A
Currently America's most controversial weapon system development programme, the Rockwell International B-1A jet bomber is intended to replace late-model Boeing B-52G and B-52H aircraft in the SAC fleet.

The genesis of the B-1A lies in the basic argument for a bomber fleet as part of a balanced deterrent force. The opponents argue that missiles can and should do the job. But a missile is an all-out weapon, not subject to recall, and targeted best against enemy cities and strategic complexes. Hardened point targets are another matter indeed, and it seems that the only way to get them is to go in with a bomber and drop the thermonuclear weapon accurately right on top of the target.

That is what the B-1 was designed to do.

And more. The B-52 fleet is geared to a quick reaction time measured in terms of only a few minutes. But even that short a time can be too long, in the event of a real

crisis, and an airborne alert – maintaining a portion of the bomber fleet constantly aloft – is a quick way to national poverty, if not bankruptcy.

What is needed is a bomber that can take off in less than a minute after receipt of the early warning, that can operate from dispersed fields rather than the known and targeted SAC bases, that can carry a huge load of nuclear weapons and that has a penetration range of many thousands of miles, unrefuelled.

That is the B-1A.

Currently, three prototype aircraft are being built to fly, and a fourth will be completed to the extent necessary to perform static tests. Originally, there were to have been five flying prototypes and two static test articles. That reduction is one that has hit the B-1A programme hard as the real costs of research and development begin to mount.

The only layout that would meet all the requirements for performance was a variable-sweep geometry. With its wings spread to the fullest, the B-1A can take off, fully loaded from a runway of the order of 6000 ft long, and there are hundreds of runways that length in the US. Further, its tyres of relatively low pressure mean that the footprint of the big bomber will be light enough to use the average 6000-ft runway, which was not designed to bear SAC heavyweights in the first place. The B-1A will cruise at high altitudes with its wings partially swept, for maximum efficiency in cruise. The extreme sweepback will be reserved for the low-level penetration, when the B-1A will be slashing along a few hundred feet above the terrain just under the speed of sound.

The B-1A will carry up to 24 SRAM

missiles internally, and another eight externally. Additionally, it will be equipped with electronic countermeasures and penetration aids, designed to fool the enemy, decoy his attacks, and raise havoc with his defences.

The crew will include pilot, co-pilot, defensive systems officer and offensive systems officer.

As of mid-1975, the B-1A programme was moving slowly but efficiently. There is much test data to be gathered and analysed before the next flight can be undertaken, to avoid repetition of data or to repeat a portion of the mission if some data were missed on the previous flight.

After seven months on flight status, the first B-1A has flown ten test flights for a total of a little over 41 hours in the air. It has reached a top speed of Mach 1·25, and a maximum altitude of 29,500 ft, both well below the desired maximum speed above Mach 2 and flight altitudes of about 60,000 ft.

If the programme proceeds as planned, there will soon be a decision on whether or not to order production. The cost of doing so has been estimated widely, but in current dollars it seems like a $20,000 million programme just for procurement of the B-1A itself. Other figures for supporting and flying the B-1A over its projected 30-year life span exceed $90,000 million.

The deciding factor will finally be whether or not such a bomber is judged as a necessary programme for the defence of the US. If it is, then it will cost what it will cost.

The story of the B-1A is still to be written. It will be interesting to look back from, say, 1980, and review the programme again from that vantage point.

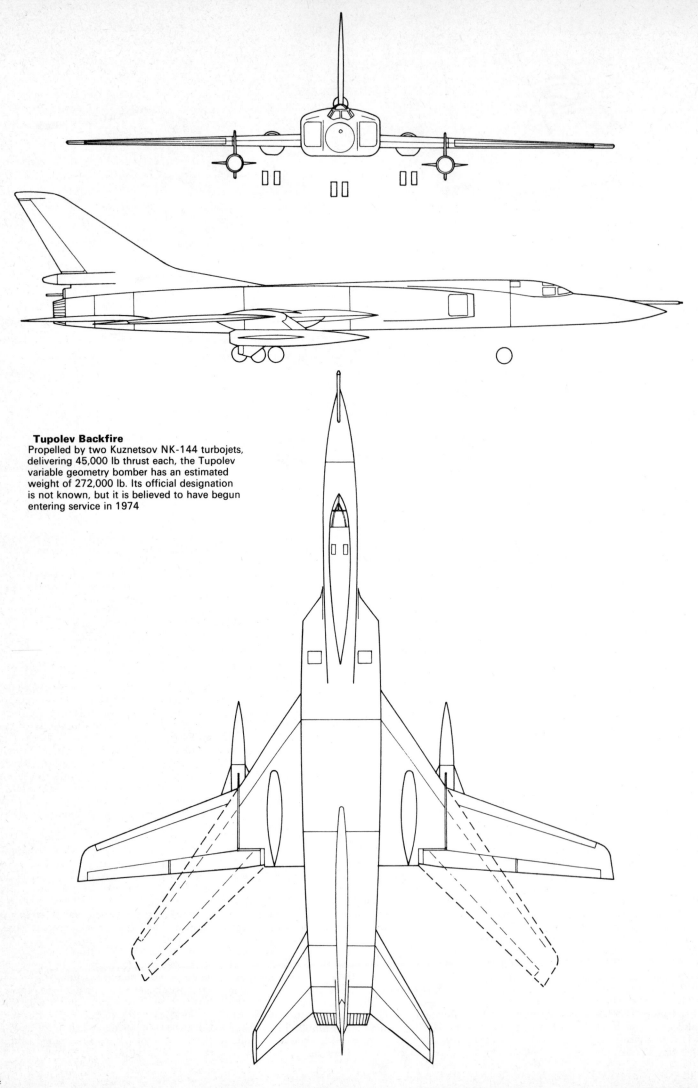

**Tupolev Backfire**
Propelled by two Kuznetsov NK-144 turbojets, delivering 45,000 lb thrust each, the Tupolev variable geometry bomber has an estimated weight of 272,000 lb. Its official designation is not known, but it is believed to have begun entering service in 1974

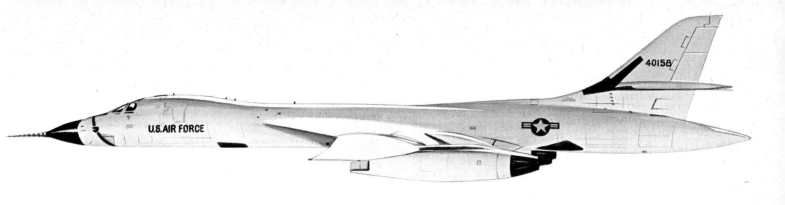

**Rockwell International B-1A**
*Crew:* 4 *Powerplant:* 4 General Electric YF101-GE-100, 30,000 lb thrust each with afterburning *Span:* 136·7 ft (wings spread)/78·2 ft (wings fully swept) *Length:* 143·3 ft *Weight:* 395,000 lb *Bombload:* 32 SRAMs *Speed:* Mach 2+

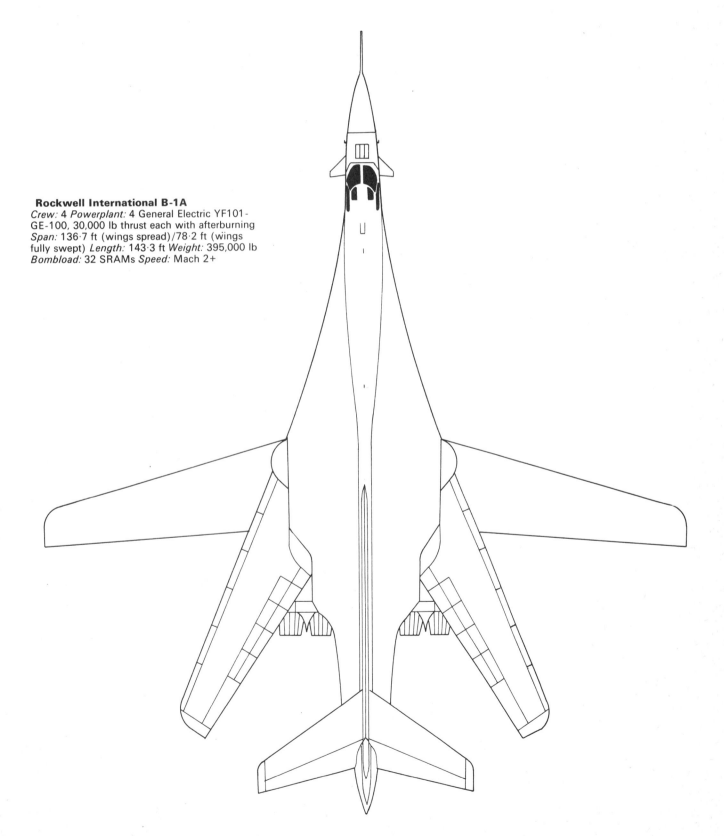

# ATTACK BOMBERS

Wars since 1945 have not seen much demand for heavy strategic bombers to perform their designed role, and over the years the fighter-bombers of the Second World War have evolved into specialised ground-attack and close support aircraft

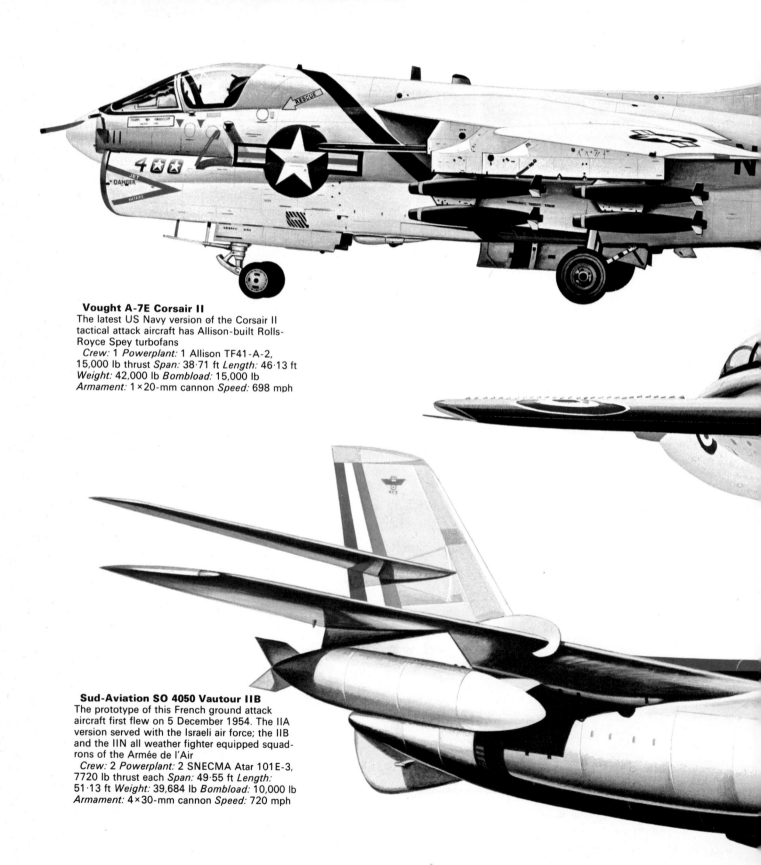

**Vought A-7E Corsair II**
The latest US Navy version of the Corsair II tactical attack aircraft has Allison-built Rolls-Royce Spey turbofans
 *Crew:* 1 *Powerplant:* 1 Allison TF41-A-2, 15,000 lb thrust *Span:* 38·71 ft *Length:* 46·13 ft *Weight:* 42,000 lb *Bombload:* 15,000 lb *Armament:* 1×20-mm cannon *Speed:* 698 mph

**Sud-Aviation SO 4050 Vautour IIB**
The prototype of this French ground attack aircraft first flew on 5 December 1954. The IIA version served with the Israeli air force; the IIB and the IIN all weather fighter equipped squadrons of the Armée de l'Air
 *Crew:* 2 *Powerplant:* 2 SNECMA Atar 101E-3, 7720 lb thrust each *Span:* 49·55 ft *Length:* 51·13 ft *Weight:* 39,684 lb *Bombload:* 10,000 lb *Armament:* 4×30-mm cannon *Speed:* 720 mph

**Blackburn Buccaneer S Mk 1**
XK 486, first of 20 development aircraft, flew
on 30 April 1958. Originally built for the Royal
Navy, Buccaneers now serve as low-level high-
speed strike aircraft with the RAF
 *Crew:* 2 *Powerplant:* 2 Bristol Siddeley Gyron
Junior 101, 7100 lb thrust each *Span:* 44 ft
*Length:* 63·42 ft *Weight:* 62,000 lb *Bombload:*
16,000 lb *Speed:* 720 mph

x *air fix model if possible*

XK 486

VA-195

USS KITTY HAWK

+ *air fix also.*

313

### Fairchild A-10A
First prototype of this close-support attack
aircraft flew on 10 May 1972. The two prototypes flew in a USAF competition for a specialised
ground attack aircraft against two Northrop
A-9As in late 1962. The Fairchild was declared
the winner, and the first 22 of an envisaged total
of more than 700 A-10s were ordered in
December 1974
  *Crew:* 1 *Powerplant:* 2 General Electric
TF34-GE-100, 8985 lb thrust each *Span:* 55 ft
*Length:* 52·65 ft *Weight:* 44,547 lb *Bombload:*
18,500 lb *Armament:* 1×20-mm Vulcan cannon
(production aircraft to have 1×30-mm GAU-8/A
cannon) *Speed:* 500 mph (est at sea level)

### Northrop A-9A
The loser in the USAF's A-X close-support
aircraft competition first flew on 30 May 1974
  *Crew:* 1 *Powerplant:* 2 Lycoming ALF502,
6000 lb thrust each *Span:* 57 ft *Length:* 53·5 ft
*Weight:* 39,570 lb *Bombload:* 18,500 lb
*Armament:* 1×30-mm GAU-8/A cannon

11369

A-9A

11367

# PLUTONIUM TO IRON

US Air Force

Bombing strategy divides itself neatly into two problems: one for the attacker, one for the defender. The attacker must get the bombs on the target; the defender must prevent that from happening. On these two commandments hang all the offensive and defensive systems discussed below.

The practical jet bomber arrived on the scene soon after the atomic bomb. The general concept then was that nuclear

**Low Approach Bombing System**
To evade anti-aircraft missiles, the bomber approaches the target in a low-level dash, pulls up into a loop and releases the bomb on the upward zoom. The bomb is thrown up and towards the target, while the bomber rolls out of the top of the loop and breaks for safety

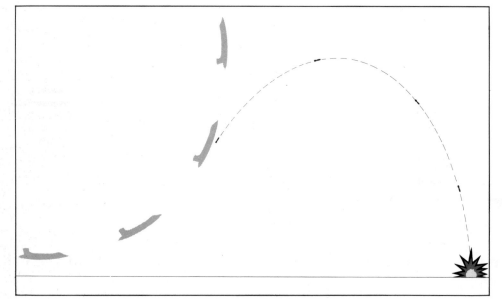

weapons were the be-all and end-all of bombing, and all jet bombers were designed to carry atomic bombs. That was easier said than done for two reasons: advances in weapons design, and security.

The former kept the shapes and weights changing, as well as the requirements for bomb-bay heating and for electrical power. The latter kept anybody from knowing about the former. Consequently, designers actually were planning bombers and missiles around weapons and warheads whose exact dimensions and weights they did not know. Nor did they know where such weapons might be grabbed, or fastened, or supported. They did not know where the fuses would be, where the centre of gravity would be located, how the bombs were to be

*Boeing B-52 with Hound Dog stand-off cruise missiles. Powered by a J52 turbojet and weighing 1100 lb, Hound Dog has a range of 600 miles and a speed of Mach 1.6 carrying a thermonuclear warhead*

armed or what kind of clearances would be needed in the bomb bay.

So early bombers tended to have big bomb bays. For one thing, the bombs were big. For another, the bomb design was so fluid that designers assumed the worst and chose bigger than usual dimensions for the bomb bays.

This choice provided later bonuses, although at first carrying all that extra weight and space didn't do anything for performance. Later, that big bomb bay was used to carry much smaller thermonuclear weapons, bombs of a more reasonable size. The space behind or ahead of them could be used for an extra fuel tank, or for something else to make the bomber a bit more of an adversary.

One of the ideas for doing this developed a series of decoy missiles, of which one – the Quail – saw service with USAF's Strategic Air Command. The idea is simple: you build a tiny airplane with the speed and other performance of the bomber that carries it. You build in some special electronic reflectors so that it looks just like the big bomber to enemy radars. And then you take it along on your ride to the target. As you approach you release the decoy, which soon appears as a second bomber to the enemy radar operator. Now it starts to turn away from the mother ship and the enemy has two tracks to follow. Are both bombers? He has no way of knowing, and by the time he finds out it may be too late.

Suppose that by luck the enemy picks out the bomber instead of the decoy, and he calls the warning to his anti-aircraft section. They are ready for action when the bomber gets within range. The bomber knows this, and he goes upstairs, heading for as much altitude as he can get. For one thing, he can get above the range of artillery-type anti-aircraft fire. For another, he can get some time warning of anti-aircraft missiles fired at him, and maybe take evasive action or decoy or destroy the missile.

But missiles kept getting better, and faster, and they had a greater range, and it was going to be suicidal to try to sail in undisturbed above the level of the effective fire. That level was invariably going to be several miles above the best bomber altitude.

That leaves two choices. The bomber can try to put up an elaborate electronic screen, jamming and confusing the enemy search and guidance radars to the point where their chances of success are very low. The risks are going to be acceptable in such a case, and the bomber goes in high.

Alternatively, the bomber can get out of the whole mess by heading down instead of up, heading for the protective cover of low altitude. There, he is below any effective anti-aircraft missile fire, and if he is really low he only exposes himself to the fire of automatic heavy machine-guns or light cannon for the briefest instant before he is out of range again.

He may have other missiles to deal with, fired by fighters or other aircraft flying above him and able to look down with their radars and see him against the clutter of the ground. The bomber still has some options remaining; he may be carrying, in addition to the electronic jammers, some window or chaff, tiny aluminium foil ribbons which, when dropped, make a huge bright spot on any radar system and tend to make the missile home on the cloud of foil rather than on the plane. If the on-board detectors tell the bomber that an infrared guided missile is coming his way, he may fire a flare, whose hot burning will attract the missile more strongly than the infrared signature of the bomber, and the enemy is fooled again.

This cat-and-mouse game describes the defensive counters that the bomber can use in a passive sense. Active defences include the conventional gun turrets, which in the early days of jet bombers were manned, and which now are almost universally remotely controlled by semi-automatic radar and a gunner. A further development is the bomber defence missile, proposed but never adopted, which would give the bomber a complement of air-to-air missiles that could be fired from turrets or launchers against enemy fighters or missiles.

But there is more and more reliance on using electronic means to counter the enemy defences, and the B-52 provides one outstanding example. One electronics warfare officer sits at a console surrounded, as far as the hand can reach and the eye can scan, by black boxes, their faces covered with knobs, dials and switches. Each of these can do something very well; each is a specialised bit of defence, able to jam an enemy radar, or overload a locked-on guidance system or fool a homing device. All over the bomber are antennae, reaching out with their electronic eyes to detect something other than empty air around and the ground below. Once those antennae spot an incoming missile or fighter, or detect and pinpoint an operating radar, the

**Concrete Dibber Bomb**
The Israeli bomb that was reported to have wrecked Egyptian airfields in 1967 and 1973. Three-tenths of a second after its release (at about 330 ft) form Vautour or Mirage aircraft, the four solid-propellant outward-angled retro rockets fire, and 0·6 seconds later a cruciform drogue parachute opens, stabilising the bomb at an attack angle of 60°–80°. Finally, 4·7 seconds after launch, four booster rockets ignite, accelerating the bomb to 525 ft per second, ploughing its 365-lb explosive charge into the runway

electronic countermeasures go into action.

But that is not the end of the defensive system. The electronics warfare officer is a walking tape deck and reel, with memorised sounds of radars catalogued in his memory. An early-warning radar of a certain type buzzes. A height-finder radar produces a beeping sound. Search radars for anti-aircraft missiles play an interrupted tone while on the lookout; when they find something, the tone shifts to a higher note. The missile radar is higher still. Radar fire-control systems also screech. And a data link sounds like a touch-telephone tone.

All these chirps and buzzes and squeaks have to be heard, identified and acted upon in split seconds. Some of this is done automatically by electronic systems, but an amazingly large part of it can only be done by the electronics warfare officer.

Meanwhile, bombs have been augmented and in some cases replaced by newer families of weapons. Most bombers are equipped with one or more types of stand-off bombs, so-called because they are fired while the bomber is still well out of enemy reach, outside his borders and far from his defences. They can be used to saturate a defensive zone, or to strike at targets en route to the main objective.

They generally started as rocket-powered winged bombs, looking much like miniature aircraft. In the early days of their development, they were too big for the bomb bays, which could hardly squeeze in an unadorned bomb, let alone one to which wings, tail surfaces and a powerplant had been added. Later versions were submerged in the fuselage, their smaller size reflecting later technology. The latest are concealed completely within the bomb bay, or slung underneath the wings on pylon racks.

These air-launched missiles have been of two basic types: the boosted ballistic, and the cruise. The former, like the ill-fated Skybolt, was an attempt to combine ballistic missile performance with the advantages of launching from a position of speed and altitude. This would reduce the initial requirements for launch thrust, and reduce the size of the weapon. The cruise missiles, like the Hound Dogs carried under the wings of the B-52s, are miniature aircraft, jet powered like their carriers, and generally able to perform over a much wider range of

parameters because they are pilotless, and small, and can be designed to be much stronger than the bomber.

The Hound Dog, for example, can streak along the ground with its belly practically scraping the earth, at speeds that would demolish the average bomber. And it is a safe bet that the Hound Dog carries some electronic trickery so that it could double as a decoy for its B-52.

One of the newest of the current crop of bomber weapons is the Boeing/USAF SRAM, which stands for Short-Range Attack Missile, and is pronounced as if it were spelled SHRAM. These missiles are rocket-propelled, carry nuclear warheads, and are endowed with amazing performance and manoeuvrability. Once launched from the B-52 or the FB-111A, the SRAM is inertially guided to its target. It can turn to hit targets abeam or even somewhat behind the position of the carrier aircraft, lending an element of suspense to the defence problem. It can operate along a variety of mission flight profiles, from low to high altitude. Because of its inertial guidance system, which does not need any external signals from electronic systems once launched, no electronic countermeasures can be applied to it, to jam its guidance, or to decoy it. Further, its radar and infrared signatures are so small, because of its size, that it is unlikely to be detected until it is much too late. SRAM is as sure of getting through to its target as you are of reading to the end of this sentence.

SRAM can be mounted in a rotary launcher, much like the chamber of a revolver. Installed in the bomb bay of a B-52, the launcher will index one of its eight SRAMs every five seconds, release it and start the launch sequence. After a brief free-fall to clear the aircraft, the solid-propellant motor fires: SRAM is on its way.

Development work continues on such weapons, as adjuncts to the gravity-propelled bombs that bombers now carry, and the other forms of weapons they employ.

But SRAM and its generation may be the last of the bomber missiles, because if there are no more bombers, there will be no more need for bombs and such weapons.

But this is the real world, isn't it? Shall we look at jet bomber history again, say, in ten years?

*Vought A-7D Corsair II attack bombers in formation over a Pacific island during exercises*